THE PHYSIOLOGY *of* HUMAN SURVIVAL

Edited by

Dr. O. G. EDHOLM
National Institute for Medical Research, Medical Research Council Laboratories London

AND

A. L. BACHARACH
Formerly of Glaxo Laboratories Limited, Greenford, Middlesex, England

1965

ACADEMIC PRESS · LONDON · NEW YORK

ACADEMIC PRESS INC. (LONDON) LTD
Berkeley Square House
Berkeley Square
London, W.1

U.S. Edition published by
ACADEMIC PRESS INC.
111 Fifth Avenue
New York, New York 10003

Library of Congress Catalog Card Number: 65-27088

Printed in Great Britain by
W. & J. Mackay & Co Ltd, Chatham, Kent

List of Contributors

A. L. Bacharach, M.A.(Cantab.), F.R.I.C., 26 Willow Road, London, formerly of Glaxo Laboratories Limited, Greenford, Middlesex, England.

Peter B. Bennett, B.Sc.(Lond.), Ph.D.(S'ton.), Senior Scientific Officer, Royal Naval Physiological Laboratory, Fort Road, Alverstoke, Hampshire, England.

R. W. Brauer, B.A.(Col.), M.Sc., Ph.D.(Roch.), Head, Pharmacology Branch, Biomedical Division, U.S. Naval Radiological Defense Laboratory, San Francisco, California, U.S.A.

Loren D. Carlson, Ph.D.(Iowa), Professor and Chairman, Department of Physiology and Biophysics, Medical Center, University of Kentucky, Lexington, Kentucky, U.S.A.

J. V. G. A. Durnin, M.A., M.B., Ch.B.(Aberd.), D.Sc.(Glas.), M.R.C.P.(Glas.), Senior Lecturer, Institute of Physiology, University of Glasgow, Glasgow, Scotland.

O. G. Edholm, B.Sc., M.B., B.S. (Lond.), Head of Division of Human Physiology, National Institute for Medical Research, Medical Research Council Laboratories, London.

R. H. Fox, M.B., B.S., Ph.D.(Lond.), M.R.C.S., L.R.C.P., Senior Scientific Officer, Division of Human Physiology, National Institute for Medical Research, Medical Research Council Laboratories, London.

J. Donald Hatcher, M.D. (West.Ont.), Ph.D.(West.Ont.), Professor and Head, Department of Physiology, Queen's University, Kingston, Ontario, Canada.

S. M. Hilton, M.B., B.Chir., Ph.D.(Cantab.), Professor, Department of Physiology, The Medical School, University of Birmingham, Birmingham, England.

Peter Howard, Ph.D.(Lond.), R.A.F. Consultant in Aviation Physiology (Wing Commander, R.A.F. Medical Branch), R.A.F. Institute of Aviation Medicine, Royal Air Force, Farnborough, Hampshire, England.

Arnold C. L. Hsieh, M.D. (St.John's), Senior Lecturer in Physiology, Department of Physiology, University of Hong Kong, Hong Kong.

† F. E. Hytten, M.D. (Sydney), Ph.D.(Aberd.), Scientific Staff, Obstetric Medicine Research Unit (Medical Research Council), University of Aberdeen, Aberdeen, Scotland.

Ladell, W. S. S., M.A., M.B., B.Ch.(Cantab.), M.R.C.S., L.R.C.P., Assistant Director (Medical), Medical Division, Chemical Defence Experimental Establishment, Porton Down, Salisbury, Wiltshire, England.

Mary C. Lobban, M.R.C.V.S.(Lond.), M.A., Ph.D.(Cantab.), Scientific Staff, Division of Human Physiology, National Institute for Medical Research, Medical Research Council Laboratories, London.

Robert A. McCance, C.B.E., F.R.S., M.A., Ph.D.(Cantab.), M.D., F.R.C.P., Professor of Experimental Medicine, Medical Research Council Department of Experimental Medicine, University of Cambridge, Tennis Court Road, Cambridge, England.

J. P. Masterton, M.B., Ch.B.(Glas.), F.R.C.S., Senior Lecturer, Department of Surgery, Monash University, The Alfred Hospital, Melbourne, Victoria, Australia.

† Present address: Medical Research Council, Reproduction and Growth Research Unit, Princess Mary Maternity Hospital, Newcastle-upon-Tyne, Northumberland, England.

L. G. C. E. PUGH, M.A., B.M., B.Ch.(Oxon.), Division of Human Physiology, National Institute for Medical Research, Medical Research Council Laboratories, London.

H. JOHN TAYLOR, B.Sc., Ph.D.(Lond.), Superintendent, Royal Naval Physiological Laboratory, Fort Road, Alverstoke, Hampshire, England.

† A. M. THOMSON, B.Sc.(Glas.), M.B.(Glas.), D.P.H.(Lond.), Honorary Deputy Director, Obstetric Medical Research Unit (Medical Research Council), University of Aberdeen, Aberdeen, Scotland.

A. T. WELFORD, M.A. (Princeton), Sc.D.(Cantab.), Fellow of St. John's College and University Lecturer in Experimental Psychology, University of Cambridge, St. John's College, Cambridge, England.

ELSIE M. WIDDOWSON, D.Sc.(Lond.), Assistant Director of Department, Medical Research Council Department of Experimental Medicine, University of Cambridge, Tennis Court Road, Cambridge, England.

‡ ROBERT T. WILKINSON, Ph.D.(Cantab.), Department of Clinical and Social Psychology, Division of Neuropsychiatry, Walter Reed Army Institute of Research, Washington, D.C., U.S.A.

† Present address: Medical Research Council, Reproduction and Growth Research Unit, Princess Mary Maternity Hospital, Newcastle-upon-Tyne, Northumberland, England.

‡ Correspondence to Dr. R. T. Wilkinson, M.R.C. Applied Psychology Research Unit, 15 Chaucer Road, Cambridge, England.

Preface

In the first chapter of this book, called "Introductory", we have tried to indicate its target – that is to say, for whom it has been written and why. There we have broadly indicated the scope of its contents and have also discussed the reason for our choice of title. In this Preface, therefore, we shall confine ourselves mainly to dealing with certain matters of convention or logistics, so as, if possible, to answer in advance some questions that might otherwise spring to the minds of our readers (including our reviewers). What we regard as explanations, they may regard as apologetics, but this can't be helped.

Weights and Measures

In a border-line subject such as human physiology, where several disciplines may overlap as they undoubtedly do in this book, the likelihood of all authors using the same system of units is less than in any single branch of basic science, such as biology, chemistry or physics. Here authors have had occasion to refer to, and often to quote from publications originating all over the scientific world and in several languages. Some of these sources, more especially those dealing with applied sciences such as engineering and meteorology, use the c.g.s. and the "Imperial" systems alternatively, even though no one investigator may have used both indiscriminately. At best this makes comparisons of results by different workers, if not impossible, at least a nuisance to subsequent inquiriers. We have tried to help reduce this nuisance. Hoping, as we do, that the book may be of interest outside the countries where the yard and the pound flourish, we have thought it essential to avoid using the two systems as unconnected alternatives. Although the practice of giving in parentheses the equivalent in feet of a number of metres (and vice versa) and similar information in the text does unpleasantly hold up the flow of reading, we see no other equally helpful procedure for a book of this kind; the question will no longer arise when the much heralded conversion of all British units (and presumably all American ones also) to metric has been completed and people have become accustomed to the change. On such matters, however, it is necessary to be pragmatic rather than dogmatic. At least, that is our view; it should explain why in Chapter 6, Part II, where the author has cited from the publications he quotes values in p.s.i. (pounds per square inch), we have thought it best not to interrupt the text continually with interpolated values in grams (or dynes) per square centimetre, but have provided a conversion table at the end (p. 181). For similar reasons we have supplied the same chapter with an appendix giving the values in metres of those numbers of feet specifically mentioned in the text of the chapter. For general usage, however, we have reproduced in the book (p. xiii), with the kind permission of the publishers, Messrs John Wright and Sons, Stonebridge Press, Bath Road, Bristol 4, the more complete table for converting feet to metres and metres to feet, compiled by us for "Exploration Medicine" (Chapter XI, "High altitudes"), recently published by the house named and jointly edited by us.

Units, Symbols and Abbreviations

On p. xiii we have collected together all the abbreviations and symbols used in this book, except some highly specialised ones confined to, and explained in individual chapters. For the sake of completeness, we have included standard abbreviations known to everyone, such as those for inch, centimetre, gram, pound; in doing so we have derived some mild amusement from the fact that the so-called c.g.s. system misuses in its own description the accepted abbreviations for centimetre (cm) and second (sec)!

Temperatures are all given in degrees Celsius (centigrade) unless otherwise stated; the Fahrenheit equivalent has been given when we thought it would be useful, but we have not thought it necessary to supply a table of degree equivalents.

The abbreviations of the titles of journals are, as far as we have judged practicable, based on those of the "World List", though we have naturally been unable to anticipate any additions or other changes to be made in the edition known to be under preparation.

On this matter of symbols and abbreviations – and particularly the latter – we have found ourselves in somewhat of a quandary. Even for scientific terms in the English language, let alone those in international usage, there appears to be no single body who can speak with such authority on the matter as to leave no author (or editor) any room for "deviation". We should have thought that here lay an opportunity for joint action by ISO (the International Organization for Standardization), presumably acting through its British and United States affiliates, with the Royal Society and the National Academy of Sciences. It would seem essential for all English-speaking scientists, both pure and applied, collectively and jointly to put their houses in order on this matter and to devise an officially acceptable and unambiguous set of symbols and abbreviations.

In the apparent absence of any such agreed list, individual editors have perforce sometimes to decide for themselves. We have tried, when obliged to do this, to base our choice on "the greatest comprehension of the greatest number", with the scales deliberately weighted on the side of the physiologist when we have found any difference in practice. (The same principle has guided us in some matters of spelling; see below.) Take, for example, the abbreviation "c", with or without modifications; it has been variously used to stand for *centi-*, *cubic*, *circa*, *cum*, *concentration*, *heat capacity per cubic mass*, *speed of light* and, more recently, *curie*; we had to decide how to abbreviate any of these terms likely to occur in the text. Fortunately, several of them do not turn up at all here; fortunately also the fact that m stands both for *milli-* and *metre* is unlikely to cause trouble, since its meaning is always clear from its position. In the event, we believe that our usage, as summarised in the table on p. xiii, represents a reasonable compromise as to what is the "best practice" and will commend itself to scientific "men of good will"!

Spelling

We refer now to a few matters of spelling, if only in the hope that by doing so we shall prevent others from wasting their time by commenting on them. First, we have deliberately cut the Gordian knot involved in the decision whether a particular verb should end in "-ise" or "-ize". We are well aware of the etymological basis of the difference and that the correct (if that is the right word to use on this matter) usage can be established for those who have forgotten any Greek they knew by consulting relevant books of reference. But what a bother it all is. Our American colleagues have cut the knot in their own way, and use the letter "z" for everything. We confess that our dislike of this is largely instinctive, but it is much reinforced when we find that the practice leads inevitably to such atrocities as "analyze" (though still "analysis") and "hydrolyze" (though still "hydrolysis"). The French, in our view, have made a better choice; they have abandoned the "z" and use "-iser" throughout. We have followed their example for all verbs, and their parts, except for such obvious exceptions as "size" which is clearly "another story".

Secondly, after much heart-searching, we have decided to use the term "dilatation", rather than "dilation", to meet the obvious preferences of physiologists on both sides of the Atlantic. But we can offer no logical defence for the use of this hypertrophied abstract noun while we are all still writing "dilate", "dilating', "dilatory" (not, be it noted, "dilatatory") and so on. It is indeed puzzling to know where the extra syllable comes from and who put it there. But the O.E.D. does indeed give "dilatation" and illuminatingly adds "(& irreg. dilation)"!

Thirdly – and on this matter we have perhaps "stuck our necks out" even more riskily – we have refused to make use of a second "r" in arhythmia and arhythmic. Some competitors have seemed to prefer this to the more ordinarily occurring spelling with two "r's"; indeed, only one author questioned the matter at all. To use the simpler form seems more logical and just as etymologically valid, but we admit to using it without "benefit of dictionary". The two we have consulted do not give either spelling, for they omit the word altogether.

These decisions which we have had to make without any guiding light except our native intelligence (or pigheadedness) have again, as on the matter of abbreviations and symbols, made us wonder who should be accepted as in any way an authority on these topics and when such an authority is likely to come into being.

Finally, about transatlantic differences in spelling, we have on purely pragmatic grounds used the British alternative throughout, since the book has been printed in England and we thought it might somewhat ease the compositor's task in setting a book already containing a large number of, to him, exotic words and esoteric matter. So we have plumped for programme, colour and so on and, again in the compositor's interest, as well as those of ourselves and other proof readers, decided to use English spelling throughout the book, whether the author is British or American. With this explanation of what might to some of our contributors appear to be yet another example of British orthographical chauvinism, we trust that our United States and Canadian colleagues will decide not to take the matter amiss.

Terminology and a *Façon de Parler*

The biologically active steroids of the adrenal cortex are mentioned several times in the book, whether because they had been used in some experimental work cited or because they were thought to take part (or not to take part) in some relevant physiological process under discussion. They (and the semi-synthetic "analogues" produced industrially for therapeutic use) have been variously called simply "steroids" or "corticosteroids" or "corticoids", with or without a preceding "adreno-"; there is even lack of uniformity in the naming of individual compounds (e.g., hydrocortisone and cortisol, although the latter term is not recognised officially in Great Britain, in spite of its shortness and simplicity). The problem, if there is one, of standardising this nomenclature is primarily one for endocrinologists and biochemists, not for physiologists; our enquiries have, however, not let us to find any generally accepted body of authority on these matters who have laid down, or whose instructions would necessarily be followed if they did lay down rules for the guidance of those who wish to refer to the substances in question. We have therefore thought it wisest, or at least safest, at the risk of being charged with pusillanimity, to let the various contributors in this book use their own form of nomenclature, always provided that it was clear to just what substances allusion was being made in any particular chapter.

Finally, we want to admit to having no patience with the still widely held view that it is a sign of objectivity in a scientific writer to refer to himself as "the author" rather than as "I". We believe that, in fact, this, like the epitaph on the Negro's tombstone, "don't deceive no one but hisself". Objectivity will be judged by contents and manner of expression, not by the avoidance of the first person pronoun, whether in the singular or the plural. Moreover, circumstances can arise in which this avoidance can involve ambiguity, as when the works of another author have been cited or in a book review. Therefore, on the few occasions when it was called for, we have got rid of the clumsy and pointless circumlocution in question; we believe that in so doing we have allowed every author not only to stand on his own feet but also to be seen to do so!

July 1965

O. G. EDHOLM
A. L. BACHARACH

Symbols and Abbreviations

acceleration due to gravity	*g*	kilo	k
atmosphere	atm	litre(s)	l
basal metabolic rate	BMR	median lethal dose	LD_{50}
British thermal unit	B.T.U.	metre(s)	m
calorie(s)	cal	micro (or micron)	μ
Celsius (centigrade)	C	microcurie	mc
central nervous system	CNS	milli	m
centi	c	minute(s)	min
Earth's gravitational force	*G*	ounce(s)	oz
electrocardiogram	ECG	partial pressure of x	*P*x
electroencephalogram	EEG	per	/
equivalent	eq	pound(s)	lb
Fahrenheit	F	relative humidity	r.h.
foot (feet)	ft	roentgen	r
gram(me)(s)	g	second(s)	sec
haemoconcentration	Hct	specific gravity (density)	s.g.
haemoglobin	Hb	temperature	temp
hour	hr	time	t
inch(es)	in	unit(s) (international)	u
		week(s)	wk

HEIGHT CONVERSION TABLE

(approximate values)

Feet	*Metres*	*Metres*	*Feet*
100	*c.* 30	100	*c.* 350
200	60	200	700
500	155	500	1650
1000	300	1000	3300
2000	600	1500	4950
3000	900	2000	6600
4000	1200	2500	8200
5000	1550	3000	9850
7500	2300	3500	11,500
10,000	3050	4000	13,100
15,000	4500	4500	14,750
20,000	6100	5000	16,400
25,000	7600	6000	19,700
28,000	8550	7000	22,950
		8000	26,250

From "Exploration Medicine" (1964) (O. G. Edholm and A. L. Bacharach, eds.), John Wright & Sons, Bristol, England.

Acknowledgements

Authors, editors and publishers wish to record here their thanks to the many colleagues who have courteously given permission for the reproduction, with or without modification, of illustrations, diagrams and tabular matter appearing in this book. References to the individual sources will be found in the legends and in the lists of References at the ends of the chapters. If it happens that any matter has been reproduced here without the consent of the copyright holder because this has not been sought, we assure him that the omission is not to be laid at the doors of author or publishers, but is solely due to editorial oversight, for which we have no excuse, but do indeed offer our apologies.

We should be failing in our duty as well as denying ourselves a pleasure if we did not here also express our gratitude to Mrs. M. Strong, secretary to one of us (O.G.E.), for her persistence and ingenuity in tracking down obscure references and correcting inaccurate ones, for reconciling bibliographical inconsistencies and, in general, for keeping us on the straight and narrow way of editorial rectitude.

O. G. EDHOLM
A. L. BACHARACH

Contents

Chapter 4

Chapter 5

Chapter 6

Chapter 7

Chapter 8

Chapter 9

Chapter 10

Chapter 15

Chapter 16

Chapter 17

CHAPTER 1

Introductory

O. G. EDHOLM and A. L. BACHARACH

A. Origins and Titles

This book arose from some discussions between the editors about the factors affecting the survival of the individual and his ability to adapt to changing conditions. Subsequently, following a suggestion by Professor J. S. Weiner, a symposium with the title "The Biology of Survival" (1964) was organised. It was held in 1963 in London at the offices of The Zoological Society, and organised by The Physiological Society, The Society for the Study of Human Biology and The Zoological Society of London. This symposium was planned to cover a wide range of subjects, including survival of wild animals in a variety of environmental conditions, among them captivity, as well as domesticated animals. As it was necessary to keep the symposium within reasonable limits, only four papers dealing with man were included. The Proceedings of this symposium have been published by The Zoological Society of London, but the papers in it about man were restricted to abstracts. This book was planned in conjunction to use the relevant material submitted to the symposium, but also to provide a wider and more detailed account of work on man.

The contents of the book and the selection of contributors were under continuous consideration, and even after these had been settled, its title came in for plenty of discussion. We had to avoid some false implications, not least the evolutionary ones of the word "survival", to say nothing of the theological ones of the phrase "human survival". Although we fully realise that the human species cannot survive unless its individuals do, we have tried to ensure that any bearing of this book on species survival is incidental and implicit, rather than direct and explicit. This was because the pursuit of human physiology, both in the laboratory and, sometimes literally, in the field, has been extremely prolific of results during the last two decades, and we believed it to be time that someone collected together the results of what has been achieved and tried to forecast to some extent where it was leading.

Moreover, in this first year of the International Biological Programme (I.B.P.) there is a strong tendency to move away from anything that looks like a purely anthropocentric attitude towards natural phenomena as a whole. We do not, therefore, feel apologetic for a book whose general tendency may be to some extent in the opposite direction, dealing solely with man's position in the world, and indeed with that of the individual man and woman, in an environment exhibiting various degrees of hostility, often exacerbated by its "natural" or contrived departure from the norm. In the final analysis, the work of the I.B.P. is going to be of practical interest only in so far as men can apply it to Man.

The human adaptability section of the I.B.P. is concerned with a considerably wider area than that covered by this book. Here we have not dealt with genetics or anthropology, and we have paid most attention to the more strictly physiological aspects. However, two chapters have been contributed by authors working in the field of experimental psychology; it seems clear that experimental psychology and physiology are closely related subjects. Indeed, in the *Annual Review of Physiology*, articles have of recent times frequently appeared with titles such as: "Physiological bases of memory"; "Physiological psychology"; "Higher functions of the nervous system".

In the interest of brevity, and to avoid what might be to some readers irksome repetition, we have omitted any reference to man in the titles of individual chapters, holding that the title of the book itself makes clear that it is not directly concerned with other species. By that same token we have not, in spite of the suggestion of some of our contributors, referred in the chapter titles either to stresses or to responses to them, since that is the subject of the whole book as well as of each chapter.

One of the most valuable aspects of the human adaptability section of the I.B.P. is the stimulus provided to breaking down the artificial barriers that have tended to separate various aspects of human biology. Although we could, and perhaps should have gone further in this direction, we are glad it was possible to include here the chapters on "Fatigue" by A. T. Welford and "Sleep Deprivation" by R. T. Wilkinson.

In a final search for a title, we rejected the one used for the symposium for the simple and obvious reason that it did not imply restriction of the book's subject-matter to the effects of environment on Man, nor could we call it "The Biology of Human Survival", since this might have led to expectations of a discussion on the evolution of the species. In our view, the title we finally chose is likely to be the least misunderstood, because only the individual, and not any group of individuals—family, clan, species or race—can be said literally to have a "physiology".

Whether we have succeeded in bringing together between two covers all matters of importance currently available on the subject will be for others to decide. Moreover, we hope not to be accused of shirking our editorial responsibilities if we say that this was in any event only one, and perhaps the less important, of our two objectives, and that we shall be more than half satisfied if we have been able to succeed in the second alone. This was to call attention to existing gaps in our knowledge of how the human organism meets various external and internal stresses and so to suggest fruitful lines of further research. We hope that the book, if not a complete guide to existing knowledge, will at least prove a useful one to gaps in it.

B. Semantic Problems

Perhaps one of the reasons for some of the gaps is (as so often) the purely semantic one that even physiologists may be confused by inconsistencies in terminology. Like almost everyone, we suppose, who has worked or read in this field, we have found ourselves pondering much, even if to little purpose, on this matter. We would have liked, however unpopular it might have made us with some contributors, to have urged on them a standard use of such terms as *response*, *reaction*, *adjustment*, *adaptation* (not *adaption*), *acclimatisation*, *acclimation*, *modification* and others, used to mean changes, and to indicate their nature and extent, occurring in response to changes in environment. It seemed, moreover, that such changes could be given some kind of hierarchical classification, from the most rapid of reflexes through shorter- and longer-term reversible changes to those more extreme alterations that are to be regarded as permanent, and in some instances even pathological. Unfortunately there is no authority anywhere, national let alone international, for relating any particular term with any particular place in this hierarchy, and we are sure that it is not for us to rush in where such authorities apparently have feared to tread.

So we have left it to our individual contributors to determine individually how to use the several terms at their disposal, in the hope, and indeed in the belief, that definition or context or both will make their intentions clear to the reader. All the same, we do enter a strong plea for some kind of collaborative action over this matter of terminology – at any rate among English-speaking workers in the field.

Another problem in semantics arises from the ever-increasing use of the word "stress". Physiologists have legitimately taken it over from their engineering colleagues and apply it to some change in environment, external or internal, that produces a counterbalancing change in the individual organism. Unfortunately its meaning has been extended by many authors to mean the state of the organism while it is responding (or before it responds) to some particular stress—that is, environmental change. A few physiologists have properly referred to environmental "stress" and the resulting physiological "strain". We did, in fact, ask the authors of this book as far as possible to avoid using the word stress altogether and at any rate only to use it in the first and (as we think) legitimate sense, when its meaning would be unequivocal. However, there is no doubt that it is sometimes extremely difficult without considerable circumlocution to do this, and we have, although reluctantly, let the word stand unaltered on various occasions.

C. Human Diversity

There cannot be much doubt that the ability to adapt enhances individual survival, but there is usually a price to pay, and the cost of adaptation needs to be assessed. Much is known about the immediate effects of exposure to hostile conditions, and a fair amount about the subsequent physiological changes that develop more slowly, but still reduce the effects of exposure. The effects of heat and cold, high altitudes and high barometric pressure have been widely studied, but many of the associated problems remain, and long-term effects have not been determined with the same degree of thoroughness. The number of studies of indigenous or immigrant populations in hot or cold climates, or at high altitudes, is still rather small. Indeed, this is one of the fields in which study in detail is planned by the I.B.P. There are many apparently distinct groups of mankind, and it might be expected that there would be considerable physiological differences between them. From the present, admittedly scanty, information the rather surprising conclusion reached is that there is a high degree of similarity rather than of distinction existing between widely different peoples. One of the major difficulties of establishing with confidence dissimilarity of physiological response is the marked heterogeneity of any one group of human beings. The range of variation of response to any particular conditions is large, and this makes it necessary to study many subjects from different groups if significant changes are to be demonstrated. It also makes it all the more important to link genetic and anthropological studies with physiological measurements, again a declared aim of the I.B.P.

We have conceived as one function of this book the need at the present time to bring together accounts of our knowledge of the physiology of man in relation to his environment. Such an attempt could easily have resulted in an unduly large book or even in a series of books. It was decided at an early stage, whatever the gaps, to resist the temptation to be encyclopaedic and to try to produce a book that could be read without too much difficulty and in a reasonable time. We were greatly helped in this resolve by knowing the plans, at that time already in an advanced stage, for a volume of the "Handbook of Physiology" of the American Physiological Society to be entitled *Adaptation to the Environment* (1964). This will undoubtedly remain for many years a standard work of reference. The scope of the book is great, covering as it does all forms of animal life. It is indeed a magnificent volume. It is hoped that the present book may be regarded, to some extent, as complementary, with its concentration on man. Inevitably there is some overlap, but this has turned out to be less than might have been feared. One reason for this was an early decision to invite contributions from workers who were personally known to one of us. Although this may be considered a limitation, it has certainly made the task of editing a pleasant and rewarding one. Many of the authors have worked professionally together, and all have, at one time or another, visited the Division of Human Physiology of the National Institute for Medical Research, London.

In the notes prepared for the guidance of authors it was pointed out that man was the essential animal to be considered, but that it was often clearly impossible to exclude discussion of work on other animals. It was also emphasised that the objective was not to produce review articles but rather to concentrate on the author's own field of work, though it was fully realised that the interpretation of such advice was bound to vary with the topics.

D. Physical Factors

The first chapters deal with the effects of physical factors in the environment. L. D. Carlson and A. C. L. Hsieh discuss those of cold on man, giving the equations for heat exchange and describing the physiological mechanisms that modify heat exchange in the cold. A critical account is given of the studies on adaptation to cold, and the authors consider that "no conclusive evidence can be presented of an ethnic or racial difference in response to cold exposure".

The next chapter, by R. H. Fox, deals with the effects of heat and includes an account of the historical development of the subject. He criticises the oft-quoted statement that man is a tropical animal. Further sections deal with the indices of heat and the many attempts to integrate all the factors contributing to the heat load or the physiological effect of a hot climate. His final section is about acclimatisation to heat; in it he describes the technique of controlled hyperthermia that he and his colleagues have used so successfully.

The effects on life at high altitudes are mainly due to the reduced partial pressure of atmospheric oxygen, and the chapter on man at high altitudes is preceded by one on acute anoxic anoxia by J. D. Hatcher, who includes an account of animal work, mainly on dogs. It was thought that such a description was essential, particularly for assessing some of the necessarily less complete work on man. The mechanisms responsible for the increased cardiac output in anoxia are discussed, and evidence is produced that a humoral agent or agents may be responsible. It seems that the cardiotonic agent has both adrenal and renal components. A cardiotonic agent appears to be produced in severe anaemia and to be responsible for the increased cardiac output, although there are some results to suggest that the two are not quite similar. Hatcher and his colleagues have shown that a combination of angiotensin, adrenaline and aldosterone can produce a considerable increase in cardiac output, although in the quantities used each hormone given separately had only slight effects on the circulation. Anoxia only produces a small increase in man at rest, but there is a greater sensitivity to carbon dioxide during anoxia, although there is also a marked increase in the work of breathing, due to the rise in airway resistance.

L. G. C. E. Pugh, in his chapter on survival at high altitudes, points out that the number of people who live at 3600 m (12,600 ft) or above is estimated to be 10 million. Apart from the effects of a reduced partial pressure of oxygen the main difficulties encountered are low temperatures and infertility. It is difficult to mimic in a decompression chamber all the characteristics of high altitudes, and virtually all studies have been carried out in the field, apart from the classic experiment of Houston and Riley. Until recently, physiologists have concentrated their work on the process of acclimatisation of lowlanders to high altitudes. Most of the features have been known for some time, and increased blood haemoglobin levels were first described in 1891. However, the extensive and detailed studies carried out at an altitude of 5800 m (19,000 ft) by the Himalayan Scientific Expedition of 1960 to 1961, led by Dr Pugh, have shown that this change is due not only to increased erythropoiesis but also to an initial fall in plasma volume, which only slowly returns towards sea-level values.

Although observations have been made on permanent inhabitants, we know more about the physiological changes in newcomers. Recent work in the Andes has shown the considerable problems that await elucidation. Pulmonary hypertension, for example, seems to be invariable, but its mechanism is still unknown. There are apparently also considerable differences between the acclimatised lowlander and the permanent inhabitant. It is still an open question whether these differences are essentially genetic.

At the other end of the scale come the effects of high pressures. So far, man has only stayed for a few hours or merely minutes at a time under such conditions, although recently longer term underwater residence has been practised. Certainly there is no group comparable to high-altitude inhabitants; however, diving instructors may be exposed regularly to high pressures. H. Taylor describes the hazards of diving to great depths and in particular the problem of "bends" due to bubble formation during decompression when the diver ascends. As a result of much experimental work, dives to depths of 150 m (500 ft) can now be safely accomplished.

Another hazard of diving is the narcotic effect of inert gases at high pressures, which is discussed by P. B. Bennett, a colleague of Dr Taylor at the Royal Naval Physiological Laboratory. The narcosis does not appear to be due to a raised central carbon dioxide tension, but is related to the partial pressure and density of the gas. It seems possible that work in this field, particularly on the use of drugs to counteract the narcosis, may provide not only methods for protecting divers but also information about the mechanisms of anaesthesia.

Although there is some evidence of adaptation to high pressures, as shown by the increased tolerance for divers to raised partial pressures of carbon dioxide, adaptive responses to high pressures are generally slight or absent. The effect of increased gravitational force appears to be another environmental factor to which man does not adapt. P. Howard describes the physiological effects of positive and negative acceleration: the most striking are the circulatory changes, although there are also important effects on respiration. Weightlessness appears to be well tolerated, but performance of tasks may be impaired in the absence of many of the usual sensory clues. Although there is as yet no evidence for adaptation to weightlessness or acceleration, Dr Howard points out that man, by adopting and adapting to an upright posture, has already some advantage over other animals. It is curious that the circulatory adaptation involved is fairly easily lost during prolonged bed rest and after short periods of weightlessness. How this happens is still unknown.

These six chapters deal with the way in which man can adapt and survive in spite of large changes in his physical environment. The rest of the book is concerned with alterations in man's internal environment brought about by patterns of life, be they nutritional, metabolic, emotional or in time.

E. Nutrition and Work

Man's average nutritional needs can be clearly defined, but it is certainly true that most of mankind has, until recently, often or even regularly suffered from more or less severe deficiencies. We still do not know if there can be a true physiological adaptation to a chronic shortage of calories, but R. A. McCance and Elsie M. Widdowson point out that today there is for a large number of people the hazard of a surfeit of calories. In primitive man, the opportunity to eat excess food and to lay down fat "enabled him to provide himself in periods of plenty with an internal store of food that would stand him in good stead in a time of scarcity." But this adaptation "wholly beneficial to primitive man" can have disastrous consequences for modern man.

In many parts of the world nutritional deficiencies still persist, and the physiological consequences are complex. An exceedingly interesting aspect requiring more study is the effect of combined dietary defects and changes in environmental temperature. Young rats provided with a low-protein diet "eat little, do not grow and die in a few weeks", but they eat vigorously if the temperature is lowered to 5° and obtain enough protein to grow and survive. "These multiple responses to simultaneous changes in the environment, having an overall survival value, would all be reckoned deleterious individually and require further study."

The problems of salt and water requirement are described briefly in this chapter, but are discussed in greater detail by W. S. S. Ladell in the next one. It was considered important to deal with this topic at length, since there are many controversial aspects, especially the practical problem of whether man can or cannot adapt physiologically to chronic dehydration. As Dr Ladell points out, it is impossible to define normal requirements for water and salt; they are dependent on "social custom, personal habits and the environmental demand". Water and salt are lost in sweat, and variations in the sweat rate can be large, so that the effects of the environment can be prepotent. The contents of this chapter not only overlap with those of the preceding one, but should also be read in conjunction with R. H. Fox's account of the effects of heat on man.

Energy expenditure can vary within wide limits, but until relatively recently almost all of the world's population was engaged in agriculture, and physical work made an unavoidable demand. Today, in the highly developed countries muscular work is rapidly diminishing as a necessity, and in consequence it is necessary to enquire whether exercise is important or not. J. V. G. A. Durnin, with some certainty, states in his chapter "The element of doubt about the benefits of exercise for health is surely minimal." He draws on Passmore's division of mankind into "*homo laborans*, *homo sedentarius* and *homo sportivus*", emphasising that the second is growing and the first declining, and pleads for encouragement of the third. In his lively and provocative chapter, Dr Durnin examines some of the long-term effects of exercise on the circulation and respiration, as well as the differences in life expectancy of muscularly active and inactive individuals, and makes out an impressive case for the benefits of muscular excrcise. This is a subject that arouses considerable emotional response, making objective judgement correspondingly difficult.

Although physical work may be declining, and with it hours of work, shift-work is increasing, particularly in technologically advanced industries, where many processes are continuous and cannot be stopped except at great expense. More and more people will be engaged in shift-work, and this alone would make it important to study the effects of time and alterations in time on human physiology. Mary C. Lobban, in her chapter, describes some of the recent work in this field, with particular reference to studies on the renal diurnal rhythms. There is a related article by J. Aschoff in "The Biology of Survival" on "Survival value of diurnal rhythms", where he mentions observations by Pizzarello *et al.* on X-radiation of rats. Rats given a dose of 900 r at 9 a.m. survived for more than 130 days; the same dose administered at 9 p.m. resulted in death within 13 days. This is a particularly striking example of a diurnal change, but virtually all functions of the body, both in animals and man, show a marked periodicity. What is not yet known is the extent to which the various diurnal rhythms are related to performance, or what are the mechanisms responsible for adaptation of the rhythms to a new time pattern. Dr Lobban describes some of the experiments she carried out at Spitsbergen, both on groups subjected to unusual day lengths and on the miners who work a three-shift system. Spitsbergen has proved to be an important laboratory, since it has a relatively large permanent population; it is at a high latitude and therefore has continuous light in the summer and continuous night in the winter. It is relatively easy in Spitsbergen for isolated groups to adhere to bizarre time schedules without social difficulties, and Dr Lobban has well exploited these possibilities.

F. Sleep and Fatigue

Sleep can be regarded as one of the most obvious diurnal phenomena, and J. P. Masterton describes the observations he and his colleagues have made on sleep patterns. In conditions when individuals were free to sleep as often and as long as they liked, it was found that the average time amounted to the 8 hr traditionally assigned to sleep. Amongst hospital staff, the house-surgeons studied averaged only 6 hr sleep per 24 hr for periods of 6 months. But, as Mr Masterton points out, we do not know if prolonged periods of shortened sleep have any ill effects, and further studies in this field are urgently needed.

Deprivation of sleep for several days has been more thoroughly investigated, as described by R. T. Wilkinson. It is surprising how difficult it has been to demonstrate unequivocal physiological or biochemical changes occurring even with prolonged sleep deprivation. Measurements of performance in the past have also shown little impairment, but recent work has clearly demonstrated the reason for this. It is characteristic that the sleep-deprived individual can, when called upon, make a sufficient effort so that for a short time his performance will be normal; however, if the test administered is prolonged, then performance will deteriorate, particularly if the test is not an interesting one. Since complex tasks are frequently more satisfying, they may suffer less impairment of performance than simple but dull tasks. What is clear from Dr Wilkinson's account is the need to correlate physiological and psychological observations; the first alone can frequently be meaningless. It is of considerable interest that even after 100 hr of sleep deprivation, subjects seldom sleep for more than 9 to 10 hr. Similar observations are made by Mr Masterton on partial sleep deprivation: when the opportunity for indefinite sleep is given there is incomplete repayment of the sleep debt. Although it is remarkable how little is known about such an everyday (or every night) phenomenon, rapid strides are now being made in psychological studies of it.

Closely allied to the effects of sleep deprivation are those of fatigue, and A. Welford discusses both physiological and psychological aspects. The use of the term "fatigue" is almost as unsatisfactory as is that of "stress", to which we have already referred above. But Dr Welford has restated clearly the problems involved in a study of fatigue. It is, as he says, a blanket term covering a variety of processes in many different bodily mechanisms. The term is used, for example, to describe the sensations and effects evoked by continued heavy muscular work or by continued mental work. With mental work there is the problem of distinguishing between boredom and fatigue, although the two conditions appear to arise from underloading and overloading, respectively, of the central nervous system. The effects of chronic fatigue are still virtually unknown, and the various phenomena associated with the term require to be studied by a combination of physiological and psychological techniques. It is frequently not realised by physiologists that there have been very impressive developments in the experimental techniques used by psychologists and that rigorous investigation of this subject can now be undertaken.

G. Emotional Strains

Although physiologists have been, in general, reluctant to work on psychological problems, Cannon's classic work (1929), described in his book "Bodily Changes in Pain, Hunger, Fear and Rage", provides an important exception. We were fortunate in persuading S. M. Hilton to write a chapter on the present state of the physiology of the emotions, and we agreed that it was essential to include in this chapter an account of animal work, as studies on man are still relatively scanty, although there has been a revival of interest in recent years. Professor Hilton and his colleagues have mapped out the areas of the hypothalamus that on stimulation result in the production of the defence reaction and the active vasodilatation in skeletal muscle with which it is associated. As he says, "by use of some of the simplest techniques known to physiology, we can begin to explore the mechanisms responsible for phenomena that have often been thought the preserve of the psychologist or even the philosopher."

The physiological consequences of the defence reaction are profound, and represent "a radical departure from the status quo". The cardiovascular response includes the muscle vasodilatation already mentioned and an inhibition of the baroreceptor reflexes. "Thus, the hypothalamus imposes its own pattern of reaction and overrides the homoeostatic reflex organised at a lower (medullary) level of the neuraxis."

The rather limited observations on man have yielded results consistent with those obtained from experimental animals. Although it is not yet possible to extrapolate from physiology to pathology, it is reasonable to suppose that the many and diverse changes observed in the defence reaction can lead, if repeated, to more permanent and potentially pathological states.

H. Pregnancy; Ageing

The physiological changes associated with pregnancy include such dramatic alterations that it was felt essential to include an account of them in a book dealing with the physiology of survival; it is so obvious as sometimes to be overlooked that they represent an essential feature in the survival of the species. A. M. Thomson and F. E. Hytten point out that in pregnancy the changes are in the internal environment and are often anticipatory rather than adaptive. The authors have, amongst their many studies, followed the changes in body composition during pregnancy and have shown clearly that most of the maternal weight gain, excluding the weight of foetus and products of conception, is due to the deposition of body fat. This occurs early in pregnancy, before the mother is impeded by the pregnancy, and may be regarded as a valuable adaptive change, providing an energy reserve to be drawn upon in later months when capacity for work and food gathering in primitive societies diminishes. The authors conclude that "pregnancy is the most common physiological stress to which women are exposed. The study of the adaptations that occur during pregnancy is not only of interest in itself but also may help to increase understanding of the nature of responses to other forms of stress."

The physiological changes associated with ageing are of fundamental importance to any consideration of survival. However, after many discussions about the need for a chapter specifically devoted to age effects, it was agreed that it would be of greater interest to consider the evidence for persistence of change after the removal of some particular environmental stimulus. Acclimatisation to heat, for example, can be readily demonstrated and has been studied extensively. But what happens when the acclimatised individual returns to a temperate climate has received less attention. An important question is whether previous acclimatisation affects the development of subsequent acclimatisation. In general, the evidence is that the effect of subsequent exposure to heat is similar to the first, that there is, in fact, no persistent effect of the first acclimatisation. But this may be too facile a conclusion and dependent on inadequate or insufficiently detailed investigation. R. W. Brauer describes a number of examples of adaptation that appear to leave permanent or at any rate long-lasting effects after the original stimulus has been removed. And inevitably this brings us to the problem of ageing, for the subject re-exposed to hot climates or high altitudes differs in one important respect on the second occasion; he is older.

I. Knowledge to Come

In his chapter Dr Brauer in effect reviews many of the problems discussed in previous sections of the book. He points out that there are "a number of mechanisms by which environmentally imposed changes can become fixed." He emphasises that growth development and ageing should be regarded together as continuous processes and that the manner in which environmental effects are manifested and produce more or less irreversible effects can depend critically upon the state of development of the individual. There is at present a lack of information about human populations of similar genetic constitution living in different environments. One of the major objectives of the I.B.P. is to obtain such information.

Some may be surprised – or even disappointed – to find no chapter in this book specifically about the reaction of *Homo sapiens* in an increasingly noisy age to a level of sound rising so rapidly as to make legitimate doubts about the validity of the specific attached to the generic name. We decided also to omit any special discussion of light and glare; these omissions are due in part to the existence of good accounts of these subjects. Moreover there are various references to the effects of sound during short-term experiments in the concluding chapters of the book. However, there can be little doubt that the next few years should, with their promise of much supersonic flying, give us both more practically based information – though long-term effects can by their nature hardly be assessed on the basis of what happens in a decade or two – and also an increased devotion to the fundamental research as badly needed in this field as in any.

Throughout this book, and particularly in the concluding chapter, there are many specific examples of the crying need for detailed studies of the various human populations who live at present in widely different conditions. There are many situations in which opportunities exist for observing the effects of population movements in which circumstances – for instance, as a result of war, political changes or disaster – often resemble the conditions of a controlled laboratory experiment. As human biologists, we must be prepared to seize such occasions and to study the phenomena intensively, so helping to increase our understanding of mankind's variability. This, as is clearly brought out in Professor K. Mather's recent "Human Diversity" (1964), is by no means determined exclusively by genetic differences: in unravelling the combined effects of environmental and inherited influences, the physiologist has one of the most important parts to play.

REFERENCES

"The Biology of Survival" (1964). *Symp. zool. Soc.* No. 13.

Cannon, W. B. (1929). "Bodily Changes in Pain, Hunger, Fear and Rage." 2nd Ed., Appleton, New York.

Handbook of Physiology (1964). Section 4. "Adaptation to the environment." American Physiological Society, Washington, D.C.

Mather, K. H. (1964). "Human Diversity." Oliver and Boyd, Edinburgh.

CHAPTER 2

Cold

LOREN D. CARLSON and ARNOLD C. L. HSIEH

A. Introductory

In an overall survey of the animal kingdom, man may be generally placed as a tropical animal. Man has, however, explored the coldest regions of the earth and can live in arctic and subarctic regions. The Eskimo, the Arctic Indian and the Lapp have long been symbols of man's ability to live at low temperatures. Living in areas where winter temperatures drop to $-40°$ for long periods, with extremes as low as $-68°$, these ethnic groups were a marvel to early explorers. That the Eskimo is not unique in his ability to survive at low temperatures is evidenced by the reports from many expeditions. C. F. Hall (1879) lived with the Eskimos along the Northwest Passage route for four years. He lived in Eskimo style and went on many sledge journeys lasting as long as three months. The temperature at the beginning of one two-month journey made with low provisions was $-62°$. Greely (1886) reported sledge trips for ten days with a mean temperature of $-41°$ and for sixty days with temperatures frequently between $-35°$ and $-45°$. In Scott's first expedition, a fifty-nine day trip was made by twelve men manhauling four sledges at temperatures as low as $-40°$ (Scott, 1905). Cherry-Gerrard (1930) reports a five-week midwinter journey for penguin eggs, manhauling two sledges and camping in a tent, with temperatures mainly at $-51°$ and as low as $-61°$. It is certain that clothing and shelter play a major part in making habitation possible in a cold environment, and the critical question relates to the microclimate within clothing or the degree of cold exposure. However, it is difficult to imagine that the men in the adventures described were not cold. Even the best clothing fails to protect the face or the hands and feet or the man himself when at rest.

Man also lives poorly clad in environments that are cool for long periods and cold in the winter. The Australian aborigine in Central Australia sleeps nearly nude at temperatures of 0° to 5°. Charles Darwin (1839) observed the people of Tierra del Fuego and marvelled "in this wretched climate subject to such extreme cold, is it not most wonderful that human beings should be able to exist unclothed and without shelter". Darwin described the snowflakes melting on the bosoms of the native women.

There are examples of extraordinary survival under conditions in which many would surely perish. From exploits of explorers and accounts of flyers downed in the Arctic (Howard, 1953) and survivors of shipwrecks at sea (Molnar, 1946; McCance, Ungley, Crosfill and Widdowson, 1956) to the recent account of the pilgrim seen by the 1962 Himalayan expedition (Pugh, 1963), these are accomplishments that call for the human attribute, motivation.

Failure to meet the environmental stress results in cold injury, frostbite, hypothermia and death. Meryman (1957) has reviewed the subject of cold injury and frostbite. Hypothermia is a relative term, but failure of temperature regulation occurs at 33° and death at some point below 25° (DuBois, 1948). We are limiting this chapter to considerations of successful competition with the environment.

In attempts to describe the range of human adaptation, man (more precisely we should say the human male adult) has been extensively studied under a variety of environmental conditions. In general there are three categories of studies. (*a*) Those of inhabitants of cold environments (Australian aborigines, Kalahari bushmen, Eskimos, Indians, Bantu and Laplander); (*b*) those of Caucasians under field conditions; and (*c*) those in cold chambers. The conditions of these experiments are not standardised. Man's physiological limits remain poorly documented, and the extent to which he adapts is ill-defined owing to difficulties in terminology and the variety of conditions and criteria used to demonstrate his ability to do so.

Man's reactions in various environments have been the subject of classical books (Newburgh, 1949; Burton and Edholm, 1955; Hensel, 1955) and have been discussed in recent symposia (Parkes, 1961; Smith, 1963; Hardy, 1963). In a bibliographical summary Carlson and Thursh (1960) gave nineteen categories of responses to cold and at least sixty-five different tests. This chapter will cover only the basic group of responses that are primary in the reaction to cold temperature or a negative heat load.

B. Classical Concepts of Heat Exchange

The classical concepts of heat exchanges at temperatures from 5° to 40° are illustrated in Fig. 1. As a negative heat load (one hour exposure at temperature) is applied to the semi-nude human, the surface of the body cools differentially, the skin of the hands and feet following ambient temperature more closely than the skin of head or trunk. The inserted manikins in Fig. 1 illustrate that a decrease in skin temperature is accompanied by an increase in depth of cooling. An average skin temperature is calculated by weighting for surface area. The core temperature has traditionally been rectal temperature, but recent work emphasises tympanum and oral temperatures as indices of the temperature that is regulated (Benzinger, 1959). Rectal temperature changes slightly over the temperature range and for the time of exposure given in Fig 1. By weighting the average skin temperature and rectal temperature in proportion to the mass they represent, the tissue cooling (storage) may be estimated. In the range plotted in Fig. 1 this heat derived from or accepted in storage may be a positive 180 kcal/m^2 (increase in body temperature) at 45° or a negative 180 kcal/m^2 at $-3°$ for a 63 kg man. From the Fourier equation for heat flow, conductivity may be calculated for the heat loss from the body surface. The inverse of this value is a measure of the insulation. The maximal value is indicative of the tissue insulation and is related to body fat content (Canon and Keatinge, 1960). Lowered insulation is a measure of a convective (blood flow) change. Calculations based on the values given in Fig. 1 indicate a tenfold change in insulation with a maximum at 15°. Below 15° insulation decreases, indicating an increase in convective loss.

Blood flow to the periphery (convective loss), as illustrated by studying the hand, approaches zero at 20°, with marked increases as temperature rises. Blood flow in the arm or leg shows less marked reduction with temperature change. Thus blood flow to skin (hand) is minimal when tissue insulation is maximal. Evaporative losses in a cool environment are from skin and lungs. Since a considerable heat exchange occurs in the nasal passages, there is a conservation of heat in the respiratory exchange. Insensible loss changes little

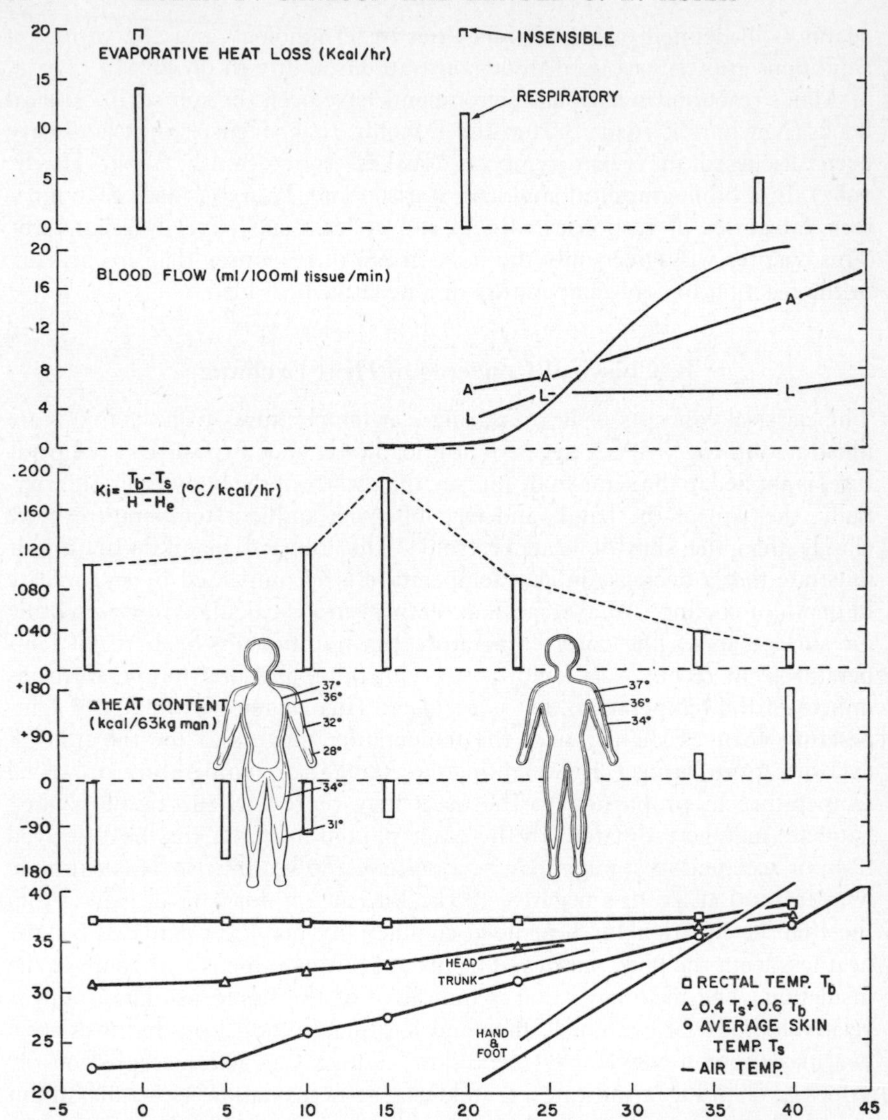

FIGURE 1. Abscissa, standard temperature in °C, which combines air temperature and air movement; ordinate, variables that vary with temperature. From bottom upward; rectal and skin temperatures and calculated average body temperature (from Hardy and DuBois, 1938; Wezler, 1950); calculated change in heat content, the inserted manikins indicating the increase in depth of cooling (from Aschoff and Wever, 1958); insulation calculated from skin and rectal temperatures given and heat production for a 63 kg man, H less 24% for evaporative loss, H_e. Values below 10° increased as indicated by Spurr, Hutt and Horvath (1957); blood flow to hand, smooth curve (Forster, Ferris and Day, 1946), and to arm (A) and leg (L) (Carlson, 1961); evaporative loss with a calculated respiratory loss based in the results of Webb (1951).

below 25°. It is apparent that cooling involves multiple factors, the important aspects being the extent to which the skin cools, the increase in insulation that may be produced and the heat production.

Another concept basic to the discussion of thermal balance is the site of production of heat. Whereas muscle is 50% of the body mass, it contributes only 20% of the heat production at rest (Fig. 2). Muscle may increase heat production tenfold, but this brings the heat production near the body surface and increases convective heat loss.

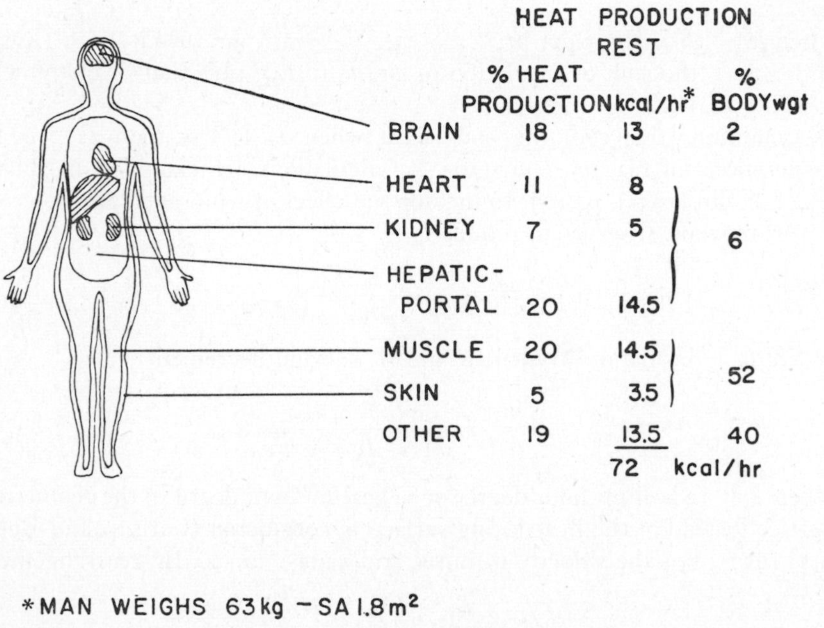

HEAT PRODUCTION REST

	% HEAT PRODUCTION	kcal/hr*	% BODYwgt
BRAIN	18	13	2
HEART	11	8	
KIDNEY	7	5	6
HEPATIC-PORTAL	20	14.5	
MUSCLE	20	14.5	52
SKIN	5	3.5	
OTHER	19	13.5	40
		72 kcal/hr	

*MAN WEIGHS 63 kg – SA 1.8 m²

FIGURE 2. Heat production of various tissues and organ systems at rest. Relative heat production, kcal/hr for a 63 kg man, and relative weight of systems are shown in columns from left to right. (From Aschoff and Wever, 1958; Bard, 1961.)

1. *Skin to environment system*

Exposure to cold is characterised by air temperature, air velocity and the radiant environment in the cold and will be modified by clothing or shelter: it is a negative heat load. The exposure causes a change in the homoeotherm brought about by the stress—the alteration in physiological function to maintain temperature or the degree to which this change is effective, or both. If the approximations of Gagge, Burton and Bazett (1941) for the human body are used, heat loss to the environment, H, is

$$H = K_o (T_s - T_a), \qquad (1)$$

where T_s = the average skin temperature, T_a = air temperature, qualified according to air movement, $K_o = h_r + h_a$, which are the heat-transfer coefficients for radiation and convection, respectively, and $H = \bar{H} - H_e$ in steady state (H_e in non-sweating state, i.e. below skin temperature of 33° is assumed to be $0{\cdot}24\bar{H}$; $\bar{H}$ is total body heat loss).

If H at a steady state with a fixed heat production and standard clothing is considered,

$$H = \frac{1}{0{\cdot}32}(T_s - T_a), \tag{2}$$

when H = 38 kcal/m² per hr, T_s = 33°, T_a = 21°, air velocity is 0·1m/sec, and 0·32 = the sum of insulation of air, I_a (0·14), plus that of clothing I_{cl} (0·18).

Conditions that cool the body exist when T_a is less than 21°, on the assumptions of Gagge *et al.* (1941). Calculations are extended in Burton and Edholm (1955), p. 107, to include the effect of wind and radiation.

Air movement serves to reduce I_a

$$H = \frac{T_s - T_a}{I_{cl} + I_a - W}, \tag{3}$$

where I_a = insulation by still air and W = wind decrement.

$$W = I_{sa}\frac{I_a\,\beta\,\sqrt{v/v_o}}{1 + I_a\,\beta\,\sqrt{v/v_o}}, \tag{4}$$

where β = 12 kcal/m² hour degree or 24 kcal/m² hour degree if the geometrical surface instead of the heat-losing surface is considered (Carlson and Buettner, 1957); v is the velocity in m/sec and v_o is 1 m/sec. By rearrangement,

$$H = \frac{T_s - (T_a - HW)}{I_{cl} + I_a}. \tag{5}$$

As I_{cl} becomes large, the significance of the wind decrement decreases. Thus, with the standard clothing value, the total insulation could change by 45% as the wind increased; with four times the standard clothing, the change would not exceed 16%.

To continue the analysis of environmental stress, the final consideration at air temperatures below 21° involves the heat gain from solar radiation (R). This heat will be added at the clothing surface:

$$H = \frac{T_s - T_{cl}}{I_{cl}} \qquad H + R = \frac{T_{cl} - T_a}{I_a}$$

$$H = \frac{T_s - (T_a + RI_a)}{I_{cl} + I_a}. \tag{6}$$

As I_{cl} increases, the addition from radiation will also decrease in significance.

Thus, a statement of the environmental conditions at temperatures below 21° should include air temperature, wind velocity, significant radiant temperatures and body position and an evaluation of the clothing worn and heat production.

Numerous methods of combining air temperature, wind decrement and radiation into such terms as temperature resultant, operative temperature, cooling power, standard operative temperature, cooling temperature and wind chill index have been proposed; none are adequate (Yaglou, 1949).

2. *Internal body to skin system*

Non-evaporative heat loss from the body depends on skin (surface) temperature. This is the critical surface factor, whether the individual is clothed or not. Skin temperature reflects blood flow to the skin, but is also dependent on the blood temperature, the insulation between the blood and the skin and the heat flow from the surface.

The heat lost from the surface per unit time, H_s, is

$$H_s = \frac{1}{I_a + I_{cl}} (T_s - T_a) + E_{ins}, \tag{7}$$

where I_{cl} is not uniform over the body; I_a is dependent on air and body motion; E_{ins} = insensible perspiration.

The heat transferred from the central body to the skin per unit time, H_{int}, is the amount of heat conducted through the tissues plus the amount of heat brought to the surface by the circulating blood:

$$H_{int} = \frac{1}{I_t} (T_b - T_s) + F\,S\,(T_b - T_s), \tag{8}$$

where I_t = insulation of tissues, F = blood flow per unit time, S = specific heat of blood. The conductance, I_t, changes in parallel with blood flow and as F becomes minimal, I_t has maximal effect.

When $H_{int} = H_s$

$$F\,.\,S + \frac{1}{I_t}(T_b - T_a) = \frac{1}{I_a + I_{cl}} (T_s - T_a). \tag{9}$$

Neglecting $\frac{1}{I_t}$, which is a constant conductance in the parallel circuit,

$$F = K\frac{T_s - T_a}{T_b - T_s}. \tag{10}$$

Similar approximations have been suggested by Burton (1934) and Hardy and Soderstrom (1938).

In studies of skin blood flow in the extremities, equation (10) has been found to fit approximately the values found. The exponential equation

$$T_s = K_1 - K_{2e} - \alpha F \tag{11}$$

is applicable over a wider range of blood flows more characteristic of the parallel assumption. For an appendage like the rabbit ear the constants α, K_1 and K_2 are temperature dependent (Honda, Carlson and Judy, 1963).

Although the tissue insulation was neglected for convenience in illustrating the role of circulation as a variable conductance in parallel, there is a large individual variation in tissue insulation, which is important when blood flow to the skin is reduced. The calculated insulation has been found to be directly proportional to the percentage of fat in the individual (Carlson, Hsieh, Fullington and Elsner, 1958). Subcutaneous fat is a means of protection in swimmers (Pugh and Edholm, 1955). Pugh, Edholm, Fox, Wolff, Hervey, Hammond, Tanner and Whitehouse (1960) have stressed the importance of subcutaneous fat (skinfold thickness) in preventing rapid cooling of the body during swimming in cold water. They estimate that an extra 1 mm of subcutaneous fat may be equivalent to raising the temperature of the water by as much as 1·5°. The amount of subcutaneous fat is also an important factor in explaining individual differences in skin temperature (LeBlanc, 1954) and should be considered when correlating the shivering and metabolic response to lowered skin temperature (Daniels and Baker, 1961) with the environmental temperature (Buskirk, Thompson and Whedon, 1963).

In the non-steady state, H_s is greater than H_{int} by the amount of heat provided from the cooling of tissues. The amount of heat that can be supplied from storage is a significant part of the early response to cold exposure. Accurate estimation of storage or heat capacity depends upon a knowledge of the average temperature of the body. Burton (1935) calculated a theoretical weighting of 0·8 for interior and 0·2 for surface if the temperature distribution is parabolic to a 2 in depth. As the assumed depth increases, the contribution of the internal temperature decreases. By making approximations for legs and arms, a more suitable equation was found to be $T_A = 0{\cdot}65\ T_b + 0{\cdot}35\ T_s$. With calorimeter results from experiments at 23·3 air temperature, the minimum error of estimation fell between 0·65 and 0·80 as a factor for internal temperature, and Burton (1935) proposed the factors 0·65 for rectal and 0·35 for skin temperature. Burton and Bazett (1936) confirmed these factors in water-bath experiments, and Bazett (1949a) has suggested the use of $0{\cdot}6\ T_b + 0{\cdot}4\ T_s$. In calorimeter experiments at 22° to 35° Hardy and DuBois (1938) found their results best represented by the equation $0{\cdot}8\ T_b + 0{\cdot}2\ T_s$.

None of the experimental verifications of these estimates of average body temperature are at temperatures of importance for the considerations in this chapter. It is apparent that thermal gradients along the surface of arm and leg and also gradients from surface to interior vary greatly. Pennes (1948) points out that the temperature of the legs and arms at rest is the result of warm blood flowing to these areas rather than of tissue metabolism. Further, the internal temperature is not uniform (Eichna, 1949; Bazett, Love, Newton, Eisenberg, Day and Forster, 1948). The participation of core and shell and the implication that cooled shell serves as insulation is a critical point in current discussions of heat transfer in acclimatisation. In spite of considerable evidence to the contrary, investigators continue to use weighting formulae under conditions to which they are not applicable.

C. Basic Physiological Mechanisms of Heat Exchange

Blood flow to the periphery is a primary factor in heat loss, convecting heat from the sites of production and conserving heat by local vasoconstriction and internal heat exchanging. When the limits of the circulatory adjustment are reached, the body surface has cooled and body heat content has changed. Beyond these limits an increase in heat production is required to prevent continued cooling. The circulatory and metabolic mechanisms are under neural and endocrine control. The definition of these as primary mechanisms does not exclude a number of secondary adjustments, such as change in hormone level, blood volume or food consumption. Here the discussion will be limited to the primary mechanisms for emphasis.

1. *Peripheral circulation*

Blood flow through the extremities has been studied by various modifications of the venous occlusion plethysmographic techniques proposed by Hewlett and Van Zwaluwenburg (1909) and Lewis and Grant (1925). These methods are limited to the extremities. A photoelectric method has been applied to other areas of the body (Hertzman, 1938). Generally the extremity is kept at heart level. Both water and air plethysmographs are used, and the temperature reported is often restricted to that of the plethysmograph, the rest of the body being at "room" temperature.

The results of Spealman (1945), Ferris, Forster, Pillion and Christensen (1947), Brown and Page (1952) and Krog, Folkow, Fox and Andersen (1960) indicate that over a range of 15° to 45° hand temperature is linearly related to the logarithm of rate of blood flow. However, this relationship is greatly modified by the thermal status of the subjects. The results shown in Fig. 3 have been divided into three groups according to the information about the subjective response to the environment given by the authors: uncomfortably warm, comfortable or cool, and cold. Hand-skin temperature is assumed to be bath temperature, a reasonable assumption above 15°. When thermal conditions of the body are held constant, an increase of 25° (from 15° to 40°) in hand temperature results in a fourfold to fourteenfold increase in blood

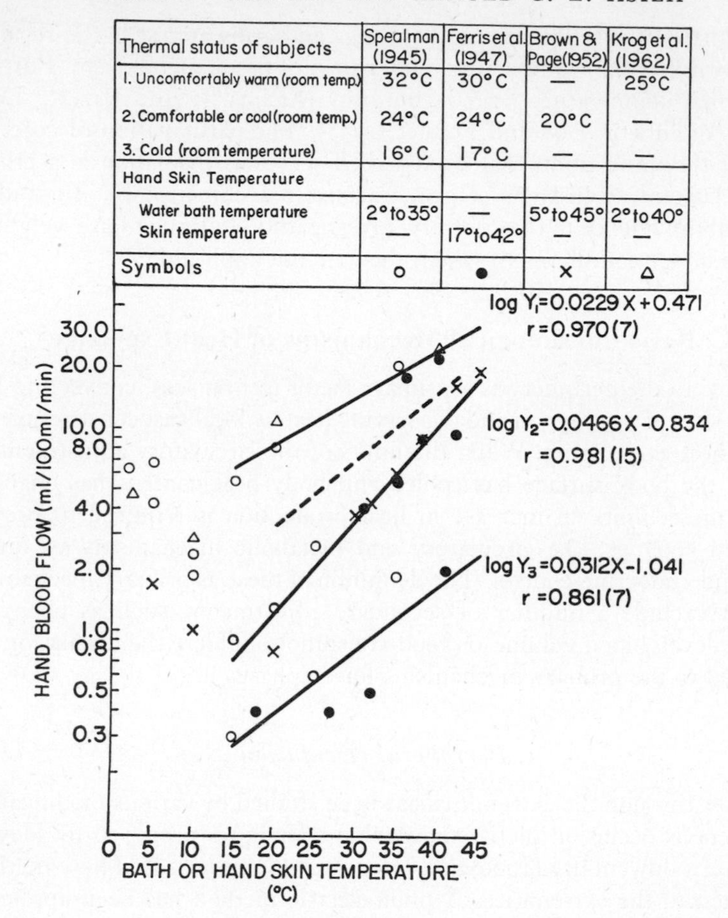

Thermal status of subjects	Spealman (1945)	Ferris et al. (1947)	Brown & Page (1952)	Krog et al. (1962)
1. Uncomfortably warm (room temp.)	32°C	30°C	—	25°C
2. Comfortable or cool (room temp.)	24°C	24°C	20°C	—
3. Cold (room temperature)	16°C	17°C	—	—
Hand Skin Temperature				
Water bath temperature	2° to 35°	—	5° to 45°	2° to 40°
Skin temperature	—	17° to 42°	—	—
Symbols	○	●	×	△

FIGURE 3. Blood flow in ml/100 ml hand per min plotted against bath or hand skin temperature. References and nature of specific experiments given in the inserted table. The values are grouped in relation to the subjective evaluation of the thermal state of the body. The broken line indicates results obtained from Eskimos by Brown and Page (1952).

flow. When hand temperatures are held constant a shift in thermal status of the subjects from cold to uncomfortably warm, which according to Spealman (1945) represents an increase of only 3° in trunk temperature (32·5° to 35·5°), leads to a fifteenfold to twentyfive-fold increase in blood flow. It is clear that blood flow through the hand is modified by reflex mechanisms that originate outside the hand and that these mechanisms can "override" local effects of temperature.

Allwood and Burry (1954), using plethysmograph temperatures of 26°, obtained values of 0·53 – 1·3 ml/100 ml foot per min for blood flow through the feet of comfortably warm subjects. Schnapper, Johnson, Touhy and Freis (1951) obtained similar values in cold subjects (wearing only shorts at an ambient temperature of 20°), but with plethysmograph temperatures of about 32°. Thus, it is possible that blood flow through the foot is also dependent upon the interaction between local and general reflex mechanisms.

A point of practical importance in this connection is the necessity for accurate assessment or control of the thermal status of the experimental subjects when seeking differences in blood-flow patterns between individuals.

2. *Counter-current heat exchanger*

The ability of the human circulatory system to act as a counter-current heat exchanger was suggested by Bazett (1949b) and documented for other animals by Scholander (1955). Aschoff (1957) and Aschoff and Wever (1958) have constructed a model to represent the exchanger system in the human hand (also see Schmidt-Nielsen, 1963). It is apparent from the model that the amounts of heat exchanged and heat lost are related to blood flow. The exchanger can only be dependent on blood flow if it is independent of room, core (body) or ambient temperature.

Honda and Carlson (see Carlson, 1963) speculated on the exchanger function in peripheral circulation. The heat, H_b, that arterial blood loses as it flows from the interior to the hand surface is

$$H_b = S \cdot (T_b - T_A) \cdot F, \tag{12}$$

where S = specific heat, T_b = rectal (body) temperature, T_A = arterial temperature entering the hand, F = flow.

Heat lost from the surface, H_a, is

$$H_a = S\,(T_A - T_V)\,F, \tag{13}$$

where T_V = venous outflow temperature.

If we assume that $T_V = T_s$, then

$$T_a = \frac{H_a}{S\,.\,F} + T_s \tag{14}$$

and

$$\frac{H_b}{H_a} = \frac{T_b - T_A}{T_A - T_V}. \tag{15}$$

This ratio increases greatly as environmental temperature falls below 20°. Further considerations of the exchanger are possible from the values shown in Fig. 3. At a hand temperature of 15° blood flow is about 0·27 ml/100 ml hand per min when the subject is cold and about 6·7 ml/100 ml hand per min when the subject is uncomfortably warm (Fig. 3). Assuming that heat lost from the hand comes from the circulating blood and $S = 1$, the following equations may be applied

$$H_{aw} = (T_{Aw} - T_{Vw}) \times 6{\cdot}7 \tag{16}$$

$$H_{ac} = (T_{Ac}) \times 0{\cdot}27, \tag{17}$$

subscripts indicating the thermal conditions of the subjects, $_w$ = uncomfortably warm, $_c$ = cold. Assuming $T_{Vw} = T_{Vc} = 15°$ and $H_{aw} = H_{ac}$,

$$T_{Ac} = 24{\cdot}8\, T_{Aw} - 357. \tag{18}$$

The range of T_{Aw} that gives reasonable levels for T_{Ac} is 15° to 15·9°. These values lead to T_{Ac} levels ranging from 15° to 37·3°. If equation (18) is valid, then the blood entering the hand held at 15° will become cooler when the subject's body is heated.

The validity of these calculations, which seem paradoxical, depends on the assumption that $T_V = T_s$. Support for the hypothesis requires experiments that measure at least three of the four variables in equations (16) and (17).

3. *Skin temperature and heat loss*

a. Body surface. Though there are various methods for measuring skin temperature, the usual means is by thermocouples attached to the skin with adhesive tape. The various methods are discussed by Prouty and Hardy (1950). An average skin temperature is calculated by weighting regional temperatures for the relative area represented. Weighting factors for seven areas (Hardy and DuBois, 1938) and even ten areas have been used (Iampietro, 1961). The average skin temperature sets the condition for heat loss, depending on the insulation of air and clothing, and appears in equation (1) and in Fig. 1.

b. Head surface. The temperature distribution over the human head has been studied in detail by Edwards and Burton (1960). Ears, nose, cheeks and chin appear to be regulated in a manner similar to the hands and feet. Froese and Burton (1957) concluded from calorimetric studies that insulation on the head did not change with environmental temperature and was independent of the thermal state of the body. Apparently, over the temperature range used, the contribution of the face and ears to heat loss is small. The importance of the head in heat loss is emphasised by their calculation that heat loss from the head at −4° may amount to half the total resting heat production of the man.

c. Extremity surface. Hand temperature and hand blood flow correlate extremely well in subjects under similar thermal conditions (Fig. 3). However, the effects of thermal status of the subjects on blood flow and the possibility of blood being pre-cooled at lower environmental temperature before entering the hand introduces problems of interpretation that must not be overlooked (see Figs. 1 and 2).

The skin of extremities exposed to low temperatures cools more rapidly if the subject feels cold (Freeman and Nickerson, 1938). It seems logical to expect the hands and feet of warm subjects to cool less rapidly or even not at all. This has been shown to be so by Rapaport, Fetcher, Shuab and Hall (1949), who were able to maintain at 21° the temperature of bare hands exposed to air temperatures of approximately −35° for one hour by supplying heat to the body. Wyndham and Wilson-Dickson (1951) have noted that the hands and feet of well-clothed subjects sitting quietly on the flight deck of an aircraft carrier with ambient temperatures about −12° and wind velocity 3 to 30 knots cooled to about 4° to 12° in about three hours. After exercise (step climbing) for one hour, rectal temperatures rose by about 1° and the temperature of the extremities to 28° to 34°. Though at least part of the increase in blood flow to the extremities during exercise may be due to circulatory adjustments to increased cardiac output, these experiments suggest that both exogenous application of heat to the body and endogenous heat production can increase blood flow to the extremities.

Heat loss from the hand has been determined by calorimetric methods, as described by Stewart (1911) and Greenfield and Scarborough (1949). In a water calorimeter at approximately 30°, heat loss from the hand varied from 17 cal/100 ml hand per min to 161 cal/100 ml hand per min, depending on the thermal state of the subject. Heat loss was correlated with blood flow and averaged 4 cal/ml blood (Cooper, Cross, Greenfield, Hamilton and Scarborough, 1949). Studies of the effects of cold on heat loss from the hand have been confined to exposures to water at about 5°. The procedures involve first placing the hand in warm water (30°) for about 30 min and then plunging it into a large thermos flask containing water at 4° to 5°. Heat loss is calculated from the resulting increase in water temperature during the succeeding 30 min by the equation

$$H = (W + E) \times T \tag{19}$$

where $H =$ total heat lost from the hand in cal/30 min, $W =$ amount of water in the calorimeter in ml, $E =$ water equivalent of the calorimeter obtained by adding a measured amount (approximately equal to the volume of the hand) of water of known temperature (usually about 35°) and noting the rise in temperature, $T =$ the increase in temperature of the water in °C/30 min.

By continuously recording the temperature of the water in the calorimeter the changes in rate of heat loss can be followed.

This method yields heat flow rates from the hands of comfortably warm control subjects (subjects not previously exposed to cold) that vary from

2840 cal/100 ml hand per 30 min (Elsner, Nelms and Irving, 1960b) to 4041 cal/100 ml hand per 30 min (Hellström and Andersen, 1960). Since the hand is first warmed, a correction for stored heat must be made or heat flow determined when stored heat is lost. When these corrections are made the figures of Elsner *et al.* (1960b) give values of 30 cal/100 ml hand per min and those of Hellström and Andersen (1960) 70 cal/100 ml hand per min. Thus rates of heat flow estimated by identical techniques give values that differ by a factor of two. A major source of variation in these experiments may be the closeness of the final calorimeter temperatures to the temperature at which cold-induced vasodilatation occurs. Blood flow through the hand is minimal at about 10°. Below this temperature vasodilatation occurs. Differences in the final calorimeter temperature, which are reported to be as much as 5°, must contribute to the experimental error. Further, the results of LeBlanc, Hildes and Héroux (1960) indicate that the regression of heat flow (in cal/30 min) on hand volume (in ml) does not pass through the origin. Thus expressing heat loss in terms of hand volume introduces an error.

An air-filled hand plethysmo-calorimeter was devised by Forster, Ferris and Day (1946). Over the ambient temperature range 15° to 38° heat lost from the hand varied from 6 cal/hand per min to 60 cal/hand per min. Blood flow increased from 0·15 ml/100 ml hand per min to 32 ml/100 ml hand per min. Loss of heat per ml of blood decreased as temperature increased.

4. *Cold-induced vasodilatation (CIVD)*

The response of the circulation in the finger to local cooling was first studied by Lewis (1930). After 5 to 10 min immersion in water at temperatures lower than 15° to 18°, there were repeated transient rises of skin temperature of a few degrees. Greenfield and Shepherd (1950) used a finger calorimeter to relate heat loss to blood flow. The rise in finger temperature has been ascribed to opening of arterio-venous anastamoses (Grant and Bland, 1930). Recent support for this hypothesis has come from Fox and Wyatt (1962), who found CIVD to occur in all areas in which arterio-venous anastomoses are known to exist. Thus what happens in the finger may well represent the response in the toes, ears, nose, cheeks and chin. The mechanisms underlying CIVD are not known. Although the response to cold does not depend upon the integrity of somatic sensory nerve fibres, the responses are much larger when these fibres are intact (Greenfield, Shepherd and Whelan, 1951).

The standard method now used to produce CIVD is to place the finger in a well-stirred bath containing melting crushed ice. Temperatures are usually measured by thermocouples attached to the pulp of the finger-tip. The experimental procedure is extremely simple, but complete analysis of the patterns of response has involved determining several variables during the first cycle of CIVD: rate of fall in finger temperature, minimum temperature attained, time until appearance of the rise in temperature, rate of increase, magnitude of the increase and final temperature of the finger.

Yoshimura and Iida (1950), using the temperature of the finger at the time of dilatation, the time of onset of dilatation and the mean temperature of the finger during the last 25 min of a 30 min immersion period, derived a "resistance index". With this index they showed that the pattern of response was influenced by the general thermal condition of the subjects (the index was increased after a hot bath or a hot meal and was reduced after a cold bath). Edwards and Burton (1960) found blood flow during CIVD to be 38 to 78 ml/100 ml finger per min in warm subjects, but only 17 to 20 ml/100 ml finger per min in cold subjects. Keatinge (1957) has shown that the appearance of CIVD is delayed in moderately cold subjects, but hastened in cold subjects (sitting in a bath with water at 6° up to the chest). Initial finger temperatures (or subject thermal state) also modify the time course of the response (Adams and Smith, 1962). The basal metabolic rate of the subject influences the mean temperature of the finger when immersed in ice water (Eagan, 1963). Moreover, emotional stress can modify the response profoundly (Meehan, 1957). Varying intensities of pain are always associated with the experiments and must be included in the list of modifying factors (Kunkle, 1949).

5. *Heat production*

Metabolic processes are the sole source of heat production. There are several mechanisms for increasing metabolism as well as several sites at which the increases may be produced. Chemical (metabolic) regulation was contrasted with physical regulation by Rubner (1902). Although it is often assumed that Rubner used chemical to designate a non-shivering mechanism, he, in fact, used the term to designate all heat production by metabolic processes. Rubner (1902), p. 221) states:

> Das Bestehen der chemischen Regulation und ihre Definition hat mit der Frage, woher die dadurch erreget Wärme stammt, gar nichts zu thun. Unter der chemischen Regulation verstehe ich nur jene biologischen Vorkommnisse, bei welchen die Erhaltung der Eigentemperatur durch die Vermehrung der Wärmeproduction beim ruhenden Thier erzielt wurde. Das Wie dieses Entstehens hat mit der Thatsache des Bestehens einer solchen keinen derartigen Zusammenhang, dass die Modificationen der ersteren die letztere in Frage stellen könnte.

The principal mechanism for increasing metabolism to combat a negative heat load is by activation of muscles in a characteristic phasic manner called shivering. In recent years a second mechanism, termed non-shivering thermogenesis, has been more clearly defined (Carlson, 1960). This non-shivering thermogenesis is related to the release of catecholamines. Chemical thermal regulation is therefore divided into shivering and non-shivering thermogenesis, the latter term being used for increases in metabolism resulting from cold exposure that are not accompanied by a visible or electromyographically detectable muscle activity.

Cannon, Querido, Britton and Bright (1927) inferred from the fact of adrenaline being liberated into the blood stream with cold exposure that the calorigenic effect of adrenaline was a part of chemical regulation. It is not yet clear to what extent this mechanism participates without prior chronic cold exposure. Adrenaline and noradrenaline are both calorigenic after chronic cold exposure in the rat (Hsieh and Carlson, 1957) and in man (Joy, 1963).

In general it may be said that in the human subject increased metabolism resulting from exposure is accompanied by shivering. The body must cool (acquire a heat deficit) to elicit shivering; the rate of cooling and extent of

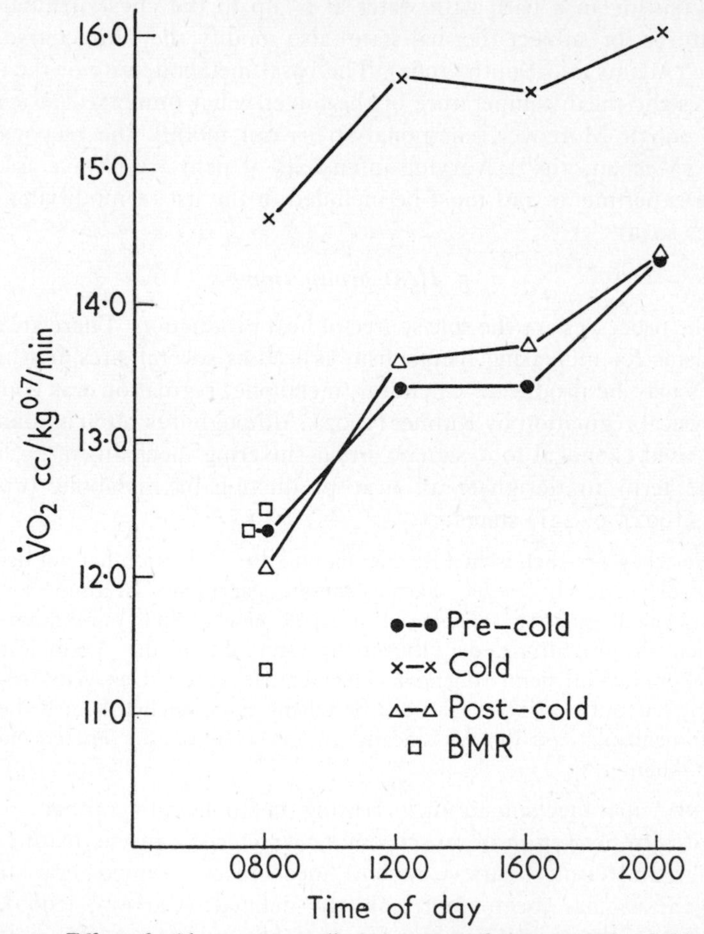

FIGURE 4. Effect of cold exposure on diurnal pattern of oxygen uptake. Cold exposure was at 15°. (From Iampietro, Goldman, Buskirk and Bass, 1959.)

change of internal temperature are both factors in the control (Carlson, 1954; Spurr, Hutt and Horvath, 1957).

Metabolism may be increased fourfold to fivefold by shivering, but the body temperature is not maintained. The efficiency of shivering in terms of maintenance of body heat is only 11% (Horvath, Spurr, Hutt and Hamilton, 1956). This is probably a consequence of the loss of insulation due to the increased circulation to the active muscles as well as to the increased convective loss due to body motion. Indeed, Keatinge (1961) has shown that physical exertion accelerates the rate of fall in rectal temperature of men immersed in cold water (5° to 15°). It is interesting to speculate that a shift in heat production to a non-shivering mechanism or to a more central region of the body would be more efficient in maintaining body temperature due to the lack of convection and the maximum use of tissue insulation. At rest, muscle metabolisn is 20% of total metabolism (Fig. 2). If the fivefold increase in metabolism in shivering is due to muscle alone, the muscle metabolism must increase twentyfold. A non-shivering efficient heat production will accomplish the same effect with only a 50% increase over the resting metabolism.

Metabolism increases with acute exposure in air or water, depending on time and temperature. Iampietro, Bass and Buskirk (1957) have shown that prolonged exposure to temperatures of 15° brings about an increase of metabolism superimposed on the diurnal pattern (Fig. 4). The extent of the stress of prolonged exposure at other temperatures is usually deduced from food-consumption studies. Values for army personnel (Fig. 5) have been studied most carefully and are summarised by Welch, Buskirk and Iampietro (1958). The efficacy of shelter and clothing and moderate activity is clearly indicated by the requirement of 47 to 49 kcal/kg body weight per day over a temperature range from −30° to +32°. It is surprising that no figures exist for food consumption of native populations in colder climates, although accounts of Eskimo diets by explorers sometimes present a fantastic picture (Lyon, 1824).†

† Lyon, on p. 143 of his Private Journal, states:

"Both sexes eat in the same manner, although not in equal proportions; the females very seldom, and the men very frequently stuffing until quite stupefied. . . . In this manner a meal continues a long time, as each eats, or rather bolts several pounds, and the pots are in consequence frequently replenished. In the intermediate time, the convives suck their fingers, or indulge in a few lumps of delicate raw blubber. . . . On all occasions the children are stuffed almost to suffocation."

On p. 396 Lyon invites an Eskimo to lunch and notes that the items consumed were:

Solids		*Fluids*	
Bread-dust and train oil	1 lb 10 oz	Rich walrus soup	2 quarts
Walrus flesh, boiled	7 lb 1 oz	Water above	4 quarts
Seal and bread	1 lb 0 oz		
Two candles	3 oz	Total	6 quarts
Bread and butter	1 oz		
Total	9 lb 15 oz		

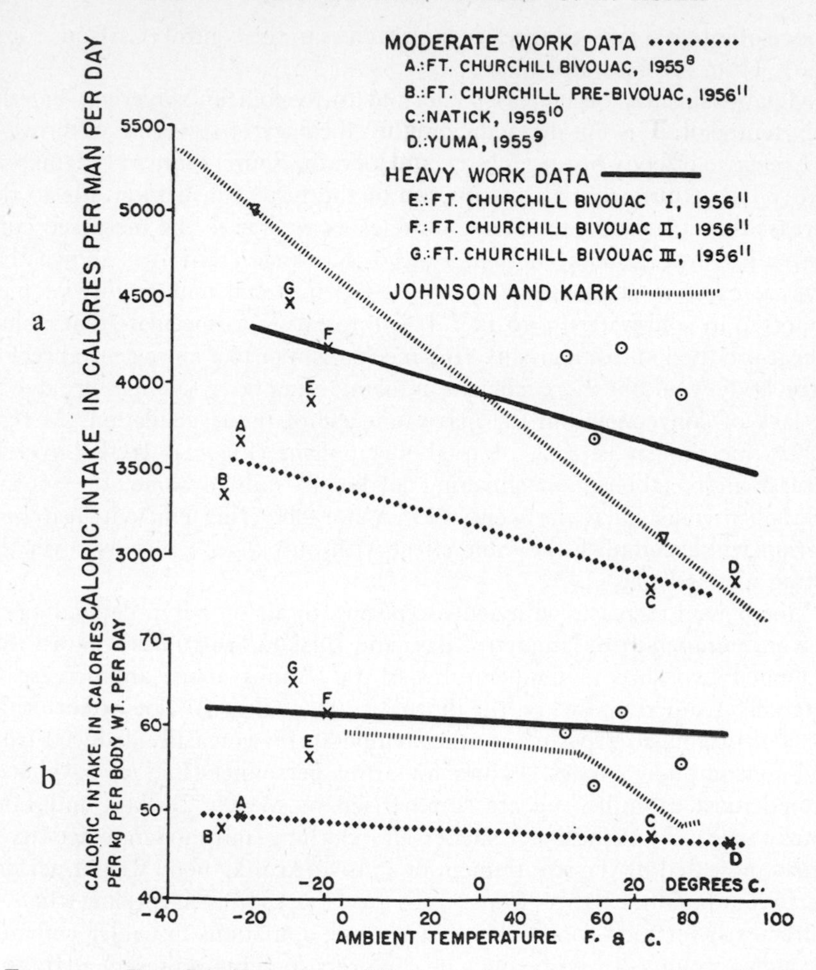

FIGURE 5. Top. Calorific intake at different outdoor temperatures. Heavy line represents experiments in which there was daily heavy work; the dotted line, daily moderate work. Earlier work of Johnson and Kark (1947) indicates a higher intake at low temperature (stippled line).

Bottom. Same results expressed in kcal/kg body weight per day. (From Welch *et al.*, 1958.)

6. *Temperature regulation*

Thermoregulation in the homoeotherm has been extensively discussed in several reviews (Hensel, 1959; Ström, 1960; Hardy, 1961; von Euler, 1961; Hannon and Viereck, 1961). It is most suitable here to present an operational definition of temperature regulation, which will be the basis of discussion of adaptive mechanisms. In response to cold, the primary system has a peripheral input as well as a central set or thermoceptive point. Whereas the peripheral and central inputs are clearly primary for temperature regulation, they certainly do not represent the only mechanisms. Their interaction is not completely understood. Schematically the interaction is shown in Fig. 6, which indicates that the internal temperature at which increased metabolism occurs is affected by the peripheral temperature. The recent work of Benzinger (1959) and Benzinger, Pratt and Kitzinger (1961) has directed attention to the critical question about what temperature is regulated. Benzinger correctly infers that hypothalamic (or internal cranial) temperature is more likely than rectal or oral to be regulated.

The physiology of the peripheral receptors has been reported in detail (Hensel, 1955). Cold receptor fibres may interact at various levels in the

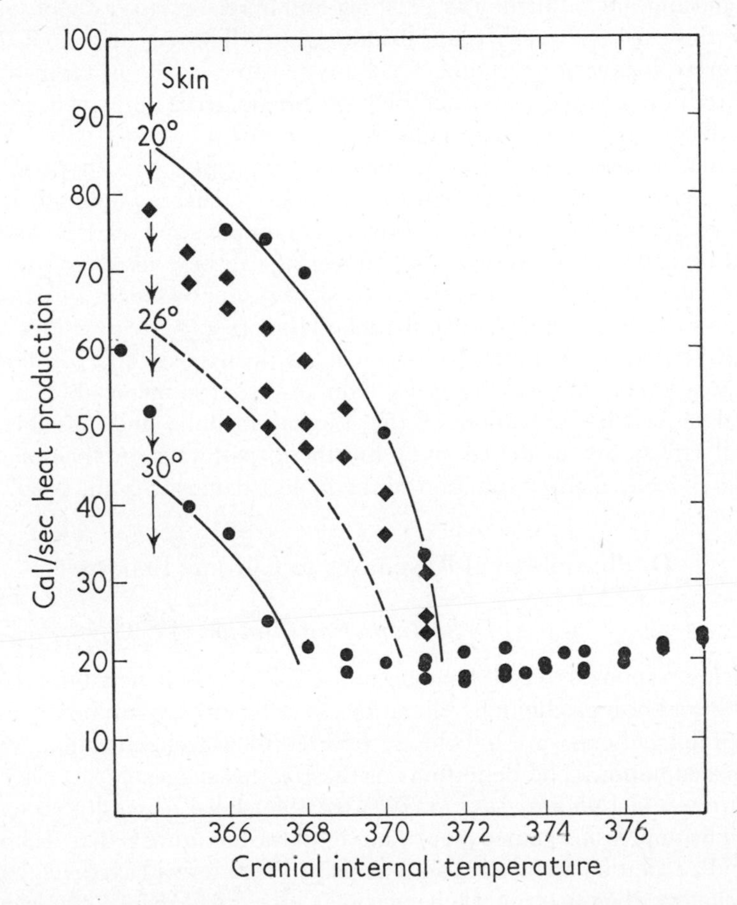

FIGURE 6. The relationship between heat production and cranial internal temperature as affected by average skin temperature of 20°, 26° and 30°. As peripheral temperature decreases, the metabolism at a given internal temperature is increased. (Redrawn from Benzinger, Pratt and Kitzinger, 1961.)

nervous system, ascend through the lateral spinothalamic tract and must eventually reach the posterior hypothalamus. The integrative and nervous effector systems are certainly involved in other homoeostatic and co-ordinating mechanisms, such as regulation of blood pressure, water balance, posture and muscular movement. A certain temperature receptive input may result in different effector outputs, depending on the interaction of the various homoeostatic systems (Ström, 1960).

Vasomotor control of blood vessels in temperature regulation is primarily in the skin and is most dominant in the skin of the hand and feet. The blood vessels of the extremities are normally subjected to a high degree of vasoconstrictor tone (i.e. nerve block leads to an increase in flow). The skin of the hand appears to function as a distinct unit in relation to vasomotor control below the wrist (Barcroft, 1960; Roddie, 1961). The vasculature of the skin contains α-constrictor receptors that are sensitive to both adrenaline and noradrenaline and are innervated by sympathetic nerve fibres. There are few if any dilator receptors (Green and Kepchar, 1959).

Neural pathways for evoking shivering originate in the hypothalamus; the descending pathway through the midbrain is lateral to the red nucleus, lateral to the reticular formation of the midbrain pons and medulla and in the lateral funiculus of the spinal cord. Shivering can be evoked by stimulation of the lateral hypothalamus, and it can be suppressed by warming the anterior hypothalamus or by stimulation in the preoptive region (Hemingway and Stuart, 1963). Voluntary movement, as do hypoxia and inspiration of 100% oxygen, suppresses shivering. Humoral control of metabolism is also neutral in origin. Activation of the adrenal medulla and possibly more general effects are mediated by sympathetic pathways. Activation of the thyroid by cold is also a function of the hypothalamus (Ström, 1960).

D. Physiological Responses to Chronic Exposure

1. *Definition of adaptation*

The acute responses to cold have been discussed in some detail in relation to heat loss and heat production. Alterations or differences in responses resulting from prolonged exposure to cold are referred to as acclimatisation, acclimation or adaptation. The definitions of these terms suggested by Hart (cited by Burton and Edholm, 1955 p. 162) have not been generally accepted or used. The impression gained from a survey of the literature is that acclimatisation and adaptation are used synonymously. These terms have been confined to physiological or anatomical changes that appear to enhance man's chances of survival on exposure to cold, but this seems to have been too restrictive. It is now tacitly accepted that any change in function or morphology resulting from exposure to cold should be called an adaptive change. Since this definition includes both long-term and short-term changes, the time course of these changes should be clearly described.

Exposure to cold can be brought about either by controlled laboratory experiments or by living in a cold climate. The type of cold exposure may lead to differences in both the quality and the quantity of the observed alterations. Whatever the procedures used the stimulus for adaptation or acclimatisation to cold is the increased heat loss and lowered skin temperature.

The signs and symptoms of adaptation to cold may be sought by looking for (i) deviations from mean values of variables whose normal distribution has been established in persons not previously exposed to cold (e.g. BMR, critical temperature, food intake, skin and body temperature, weight-surface area relationship, body composition, manual dexterity), and (ii) changes in the patterns of physiological responses to cold stress (e.g. patterns of heat loss, blood flow, heat production, endocrine activity). The first procedure seeks differences in general levels and may be regarded as a quantitative test, the second seeks changes in the quality of responses. Both methods require that determinations be made under comparable conditions.

2. *Ethnic differences*

Because man lives in many climates, there are marked differences in the social and cultural adjustments to his environment (Wulsin, 1949). There has been considerable effort to determine physiological differences in response to cold among various ethnic groups and to ascertain if these differences are genetic. The greatest problem in interpreting the physiological and morphological adjustments to environmental extremes is to decide whether they are plasticity responses or the result of selective processes (Newman, 1961). In a review on climate, culture and evolution, Paul Baker (1960) states that if Charles Darwin were to join a discussion on climate and human evolution, he would be perfectly at home with his 1870 concepts, indicating the difficulty in quantitatively describing the effect of climate on evolution. The reviews of Baker (1960), Newman (1961) and Barnicot (1959) are excellent discussions of the climatic factors in evolution. Although certain inferences have been drawn by these authors, in the discussion below no conclusive evidence can be presented of an ethnic or racial difference in response to cold exposure.

3. *Basal metabolic rate*

The basal metabolic rate is negatively correlated with average temperature of the climate (Roberts, 1952). The significance of this is not entirely ethnic, since the correlation exists within ethnic groups over a wide climatic range (Roberts, 1952) and a seasonal effect has been shown (Osiba, 1957). As a cause of the BMR differences the effect of diet has not been eliminated. For the Eskimo, in whom most of the evidence favours an elevated metabolic rate at rest, the question of diet has been raised by Rodahl (1952).

Body weight is negatively correlated with the average temperature of the climate (Roberts, 1953); paradoxically, skin folds indicate less subdermal fat in northern groups (Milan, Hannon and Evonuk, 1963). It seems that chronic exposure to cold leads to an increase in the ratio of "active tissue mass" to total body weight. Expressing basal metabolic rate in terms of lean body mass reduces but does not entirely obliterate the differences between the rates of Eskimos and Caucasians (Covino, 1961). Thus it is possible that a true increase in BMR does take place.

The significance of an elevated BMR as a response to continued cold exposure is not clear. However, the consequences of such an increase should not be overlooked. Theoretical considerations that follow from equation (1) lead one to expect an increase in average skin temperature if the metabolic rate is raised. An increase of 20% in the BMR without an accompanying rise in evaporative heat loss may lead to an increase in average skin temperature of about 1° and an increase in toe temperature of as much as 6° (Sheard and Williams, 1940). Thus increases in surface temperatures cannot be ascribed entirely to a direct effect of chronic cold exposure if the BMR is elevated. The level of the BMR should also be taken into consideration when discussing alterations in the responsiveness of heat production to cold exposure.

4. *Summary of studies*

Besides studies made under laboratory conditions, the summer-winter difference in temperate climates and the responses of inhabitants and migrants to cold regions serve as a source of informations for analysis of adaptive changes. The climatic differences are shown as zones on Fig. 7, which is an illustration taken from U.S. Army Quartermaster World Guide to Clothing Requirements (Sprague and Ross, 1959). Table I summarises studies that have covered metabolism, shivering and skin and rectal temperatures. Test subject, test condition, criteria and results are tabulated. Test conditions fall into two categories; studies of cooling in a controlled environment and overnight studies. In the light of the variety of test conditions and criteria used, it is not surprising that a consistent pattern of physiological adjustment has not been obtained. Though mean rectal and mean skin temperatures during a standard exposure to 14° showed no seasonal variation, the time to onset of shivering lengthened and heat production fell in winter compared with the end of summer (Davis, 1961; Davis and Johnston, 1961). Similar results were obtained after a daily exposure to cold (Davis, 1961). This result seems paradoxical unless one assumes that heat production becomes more effective in maintaining temperature. Other studies involving cooling experiments show increased peripheral temperatures, both with and without an increased metabolic response.

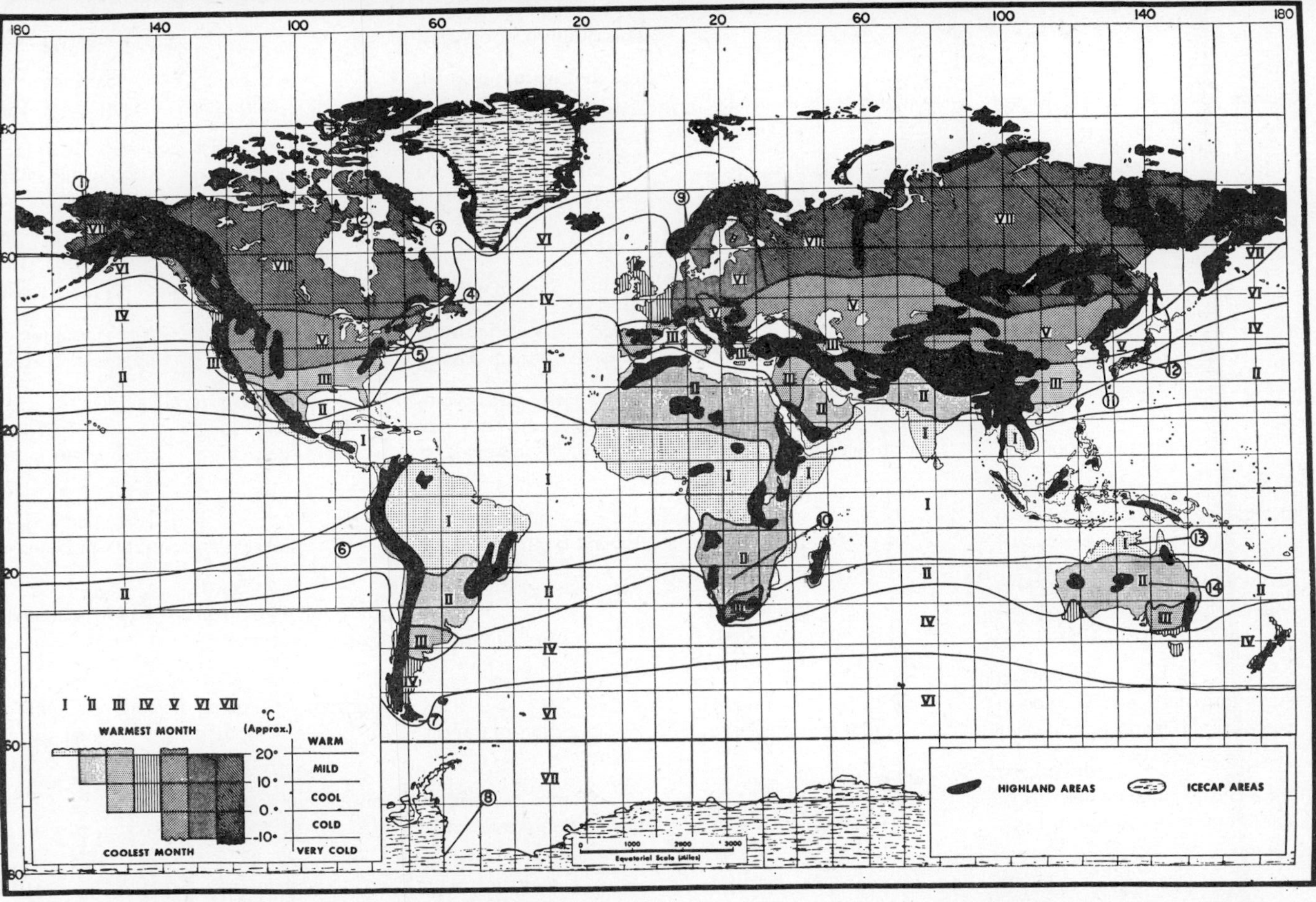

FIGURE 7. World map divided into seven zones (Roman numerals) of temperature, showing warmest and coolest month. Studies in various areas are designated by arabic numerals, which follow the reference in Tables 1 and 2. (From Sprague and Ross, 1959.)

Table I. *Summary of Studies on Adaptation to Cold – Whole Body*

Reference†	*Test Subjects*	*Test Condition*	*Criteria*	*Result*
Burton and Bazett (1936) 5	Caucasians summer and winter	Water bath – immersed to neck 30° and 35°	Index of conductivity	Shift in minimum conductivity to smaller gradient of rectal to bath temp
Goldby, Hicks, O'Connor and Sinclair (1938) 14	Caucasians – Australian aborigines	Tested in field	Skin and oral temp blood flow	Decrease in metabolic response to cold. Active control of skin circulation
Newburgh and Spealman (1943) 5	Caucasians – lightly clad at 15° for 2 and 4 wk	During exposure	Skin and rectal temp metabolism	No change in rectal temp no change in skin temp during exposure. Metabolism increased
Balke, Geiner, Kramer and Reichel (1944) 16	Caucasians	4 weeks at 1200 metres at 0° camping	Skin temp	Increase in peripheral temps (hand and foot) and decrease in body temp
Adolph and Molnar (1946) 5	Caucasians during Sept., Oct. and Nov. in Rochester, N.Y.	Exposure in sun, shade and wind at rest and with bicycle ergometer exercise	Rectal and skin temp, Hct, haemoglobin and plasma refractive index	Haemoconcentration
Horvath, Freidman and Goldern (1947) 5	Caucasians 8 days at −29°	Tested during exposure and control period	Oxygen consumption	No difference
Glaser (1949) 15	Caucasians — 72 hr exposure to −1° to +4°	1st and 3rd day	Rectal and skin temp	Rise in skin temp
Carlson, Burns, Holmes and Webb (1953) 1	Caucasians at beginning and end of 2 wk exposure	1 hr exposure clothed at environmental temp −5° to −17°	Skin and rectal temp heat production heat loss	Extremities warmer, decreased metabolic response
Brown, Bird, Boag, Boag, Delahaye, Green, Hatcher and Page (1954) 2	Eskimos Caucasians	Exposure in air at 20° hand and forearm in water bath 5° to 33° immersion of feet	Skin temp and blood flow	Eskimo has higher blood flow at given temp

Elsner (1955) 1	Caucasians, 3½ wk exposure −40° to 26°	Exposure at 15° with and without blanket	Skin temperature shivering	Increased extremity temp decreased shivering threshold
Meehan (1955a) 1	Eskimos Caucasians	Exposure on cot at 6° to 7°	Rectal and skin temp oxygen consumption	Eskimo showed greater increase in MR. No difference rectal and skin temp
LeBlanc (1956) 4	Caucasians test group: −1° to −35° 12 hr per day control: ½ hr per day	1 hr at 8° in shorts	Skin and rectal temp oxygen consumption	Decreased metabolic response to cold. Increased cooling
Iampietro *et al.* (1957) 5	Caucasians	Wearing shorts at 17° for 14 days	Oxygen consumption rectal temp	Higher morning temp no change in BMR
Scholander, Andersen, Krog, Lorent-Zen and Steen (1957) 9	Lapps Caucasians	Exposure at various temps 0° to 30°	"Critical temp" i.e. temp at which metabolism must increase	No difference
Adam (1958) 8	Antarctic expedition 15 months or more, new arrival	Sledging	Thermal comfort report related to skin temp	No difference
Adams and Covino (1958) 1	Negroes Caucasians Eskimos	2 hr in shorts on wire mesh cot at 17°	Rectal and skin temp metabolic rate sweating rate electromyograph	Eskimo maintained higher rectal and skin temp and higher metabolic rate. Negro has decreased metabolic response to cold
Adams and Heberling (1958) 1	Caucasian group before and after fitness regimen	1 hr on cot at 10°	Rectal and skin temp metabolic rate	Fitness occasioned higher heat production, lower rectal and higher skin temps
Scholander, Hammel, Andersen and Löyning (1958a) 9	Caucasians – 6 wk at 3° to 5°	All-night test – in sleeping bag at 3°; in room 20°	Skin and rectal temp oxygen consumption	Decreased fall in skin temp increase heat production

† The numbers refer to the general geographic region in Figure 7.

Table I. *Summary of Studies on Adaptation to Cold – Whole Body* (continued)

Reference†	*Test Subjects*	*Test Condition*	*Criteria*	*Result*
Scholander, Hammel, Hart, LeMessurier and Steen (1958b) 14	Caucasians Australian aborigines	All-night test – in +10° to 0° Radiant −15° to −20°	Rectal and skin temp oxygen consumption	Natives cool more, less increase in metabolism
Wyndham and Morrison (1958) 10	Caucasians Kalahari bushmen	Sitting at 16° to 12°	Skin and oral temp	Indefinite with respect to physiological responses
Rennie (1958) 1	Caucasians 4 wk field trip −35° to −12°	Exposure nude at 10° for 100 min	Skin and rectal temp metabolism and shivering	Increase in peripheral temp no change in metabolic response
Hammel, Elsner, LeMessurier and Andersen (1959) 13, 14	Australia aborigine central tropical	Overnight at 5°	Rectal and skin temp oxygen consumption	Central tropical Caucasian in magnitude of metabolic response and skin and rectal temp
Elsner, Andersen and Hermansen (1960a) 1	Arctic Indian – Seasonal	Overnight at 0° to 3° after 1 hr under blanket	Rectal and skin temp oxygen consumption	No winter difference
Irving (1960a) 1	Caucasians poorly clothed in Alaska	Standing at 0°	Skin temps shivering	Higher temp reduced shivering
Irving, Andersen, Bolstad, Elsner, Hildes, Löyning, Nelms, and Whaley (1960) 1	Arctic Athabaskan Indians Caucasians	Sleep at 0°	Oyygen consumption rectal temp	Higher initial oxygen consumption on SA basis
Andersen, Löyning, Nelms, Wilson, Fox and Bolstad (1960) 9	Lapps Caucasians	Overnight at 0°	Metabolic rate skin and rectal temp	Lapps had no metabolic response; higher skin temp and greater fall in rectal temp
Andersen and Hellström (1960) 9	Lapps Caucasians	Overnight at 0°	Oxygen consumption rectal and skin temp	Lowered rectal temp less increase in metabolism

Davis and Johnston (1961) 5	Caucasians – varied season	1 hr at 14·1°	Rectal and skin temp oxygen consumption	Winter group – less shivering less increase in metabolism no difference in temp
Davis (1961) 5	Caucasians, chamber at 11·8° and 13·5° daily 8 hrs	In chamber – 2 hr test	Rectal and skin temp oxygen consumption	Skin temp differences in summer; less shivering
Hart, Sabean, Hildes, Depocas, Hammel, Andersen, Irving and Foy, (1962) 3	Eskimos	Overnight on cot at 4° to 6° blankets removed	Rectal and skin temp oxygen consumption	Elevated metabolism higher peripheral temp
Budd (1962) 8	Antarctic research team	10° for 95 min	Rectal temp	Increased ability to maintain rectal temp
Rennie, Covino, Blair and Rodahl (1962) 1	Eskimos Non-Eskimos	Nude in air 23° to 35° for 3 hr	Tissue insulation rectal and skin temp	High oxygen consumption and skin temp at all temps below 28° greater fall in rectal temp
Palmai (1962a) 8	Caucasians	Antartic for one year	Thermal comfort	Felt warmer with increase in duration of stay
Palmai (1962b) 8	Caucasians	Antartic seasons	Oral temp	Difference in magnitude and character of circadian rhythm
Milan *et al.* (1963) 1	Eskimos Arctic Indians Caucasians	Water bath immersed to neck 30·5°, 33°, 35°, for 1 hr	Tissue insulation rectal temp	Eskimo Indian Caucasian no difference in rectal temp
Hong (1963) 11	Korean diving women. Non-diving women	Seasonal	Oxygen consumption onset of shivering	BMR increased in winter increased shivering threshold
Elsner (1963) 6	Andean Indians	Sleeping at 0° to 5° in bags	Skin and rectal temps	High skin and foot temp low rectal temp

† The numbers refer to the general geographic region in Figure 7.

Special studies have been made of finger and hand blood flow and temperature (Table II), but only at temperatures of 1° to 5°. With one exception (Hellström and Andersen, 1960) these studies show a reduction in cold-induced vasoconstriction in the hands of groups continuously exposed to cold.

5. *Criteria for differences*

The determinations of physiological responses at one temperature are insufficient either for adequate description of the response patterns or for use of the elegant statistical methods available for determining the precision of the experimental results. As an example of this, the results shown in Fig. 3 illustrate the difference in blood flow with temperature depending on the thermal state of the subjects. A response is easily imagined such that the temperature curves might have the same origin but differ in slope. Measurements of the variable at different temperatures provide more information, which may alter the conclusions.

Ideally, experiments should be performed over as wide a range of cold stress as possible, so that both upper and lower limits of response are determined. The slope of the linearised regression of a response metameter on cold stress will have units (change/unit of cold stress) that accurately describe the response. The ratio, slope of the test group: slope of the control group, gives the relative responsiveness of the test group referred to the control group. Proper experimental design will permit calculations of the confidence limits of this ratio. If either an increase or a decrease in the slope occurs after a period of cold exposure, then adaptation can be said to have taken place.

For those who wish to apply labels to the types of responses, an increase in the slope may be called "sensitisation" and a decrease "habituation". Sensitisation to cold has not been demonstrated, although its existence is theoretically possible. Habituation is generally used to describe a diminution in responsiveness of the CNS to sensory stimuli (Glaser, Hall and Whittow, 1959). A reduction in response may be due to reduced responsiveness of peripheral receptors, the CNS or effector end organs.

Four examples of the application of this method of analysis are given below.

(i) The values for heat production from Carlson, Young, Burns and Quinton (1951) and from Davis and Johnston (1961), when plotted against skin temperature, indicate a reduced responsiveness of the heat-production system to skin cooling after acclimatisation to cold (Fig. 8). Results from chamber cold acclimatisation show a similar trend (Davis, 1961). In these experiments, the responses to a cool environment in semi-nude individuals and to a cold environment in clothed individuals were determined. This method is to be compared or contrasted with experiments in which overnight cooling and heat production are measured.

Table II. *Summary of Studies on Adaptation to Cold – Hand*

Reference†	*Test Subjects*	*Test Condition*	*Criteria*	*Result*
Yoshimura and Iida (1951) 12	Japanese – seasonal	Finger in water at 0° for 30 min	"Resistance Index" – mean temp for last 20 min, time of onset of rise, maximum rise	Slight change
Brown and Page (1952) 2	Eskimos Caucasians	Hand in water at various temps, 5° to 45°	Hand blood flow	Eskimo Caucasian at all temps
Meehan (1955b) 1	Caucasians Negroes Native Alaskans	Hand in water at 0° for 30 min	Finger-cooling curves	Native Alaskans highest, Caucasians intermediate, Negroes lowest
Iampietro *et al.* (1959) 5	Caucasians Negroes	Hand in water at 0° for 30 min	Finger temp	Negro Caucasian
Krog *et al.* (1960) 9	Lapps Norwegian fishermen or expedition members	Hand in water at various temps 40°, 20°, 10° and 0°	Hand blood flow hand calorimetry	Earlier vasodilatation no difference in blood flow
Hellström and Andersen (1960) 9	Arctic fishermen non-fishermen	Hand in water at 4°	Heat loss by calorimetry	No difference
LeBlanc *et al.* (1960) 4	Gaspé fishermen non-fishermen	Immersion of one hand for 10 min in 2·5° water after 10 min in 30° water	Skin temp pressor response	Higher finger temp less cold pressor response
Elsner *et al.* (1960b) 1	Arctic Indians urban Caucasians	Hand in 4° to 5° water for 30 min	Heat loss by calorimetry skin temp	Indians transfer more heat, higher skin temps more rapid rewarming
LeBlanc (1962) 4	Gaspé fishermen winter and summer	Hand in water at 2·5°	Skin temp pressor response	No seasonal change
Nelms and Soper (1962) 15	British fish filleters Nov. through Mar. non-filleters	Hand in water at 0°	Skin temp	Temp higher earlier vasodilatation

† The numbers refer to the general geographic region in Figure 7.

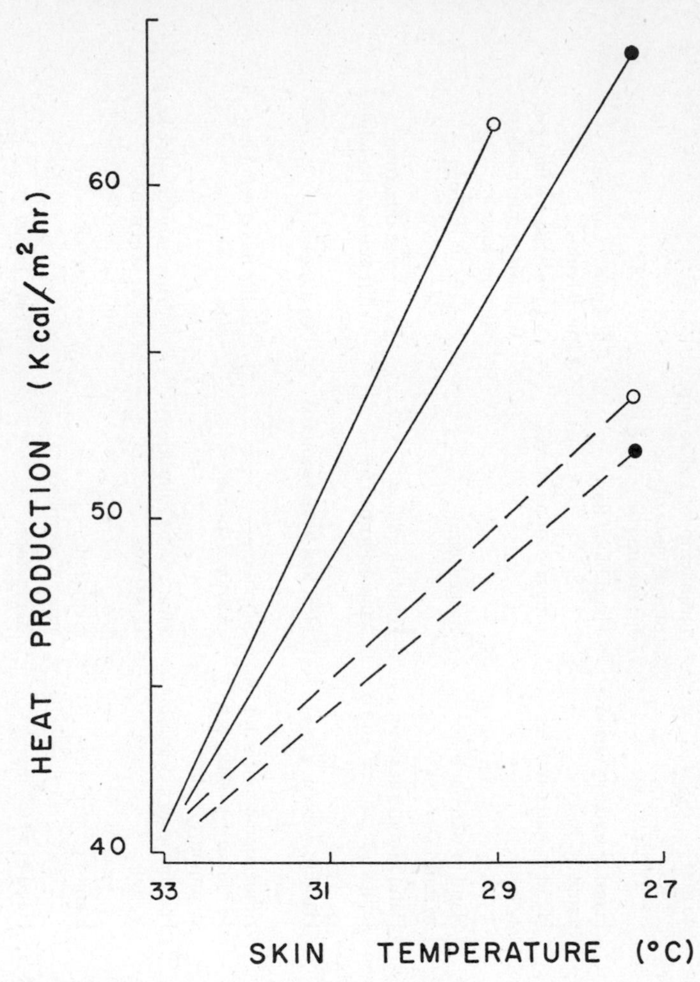

FIGURE 8. Open circles represent heat production after one hour cold exposure of clothed men, before (solid line) and after (broken line) two weeks of daily cold exposure. (From Carlson *et. al.*, 1951). Closed circles represent heat production after one hour cold exposure semi-nude, in summer (solid line) and winter (broken line). (Davis, 1961.)

(ii) The regression lines of heat production on average skin temperature from results obtained by Hart, Sabean, Hildes, Depocas, Hammel, Andersen, Irving and Foy (1962) are shown in Fig. 9. All these come from overnight experiments. The Alacaluf and aborigine were able to sleep, but the Eskimo and the white apparently did not. Over the range of 33° to 28° skin temperature, the rise in heat production in response to lowered skin temperature appears to be reduced in Eskimos and nil in Alacaluf Indians and Australian aborigines (Fig. 9a). A complete study of the responses would include within-

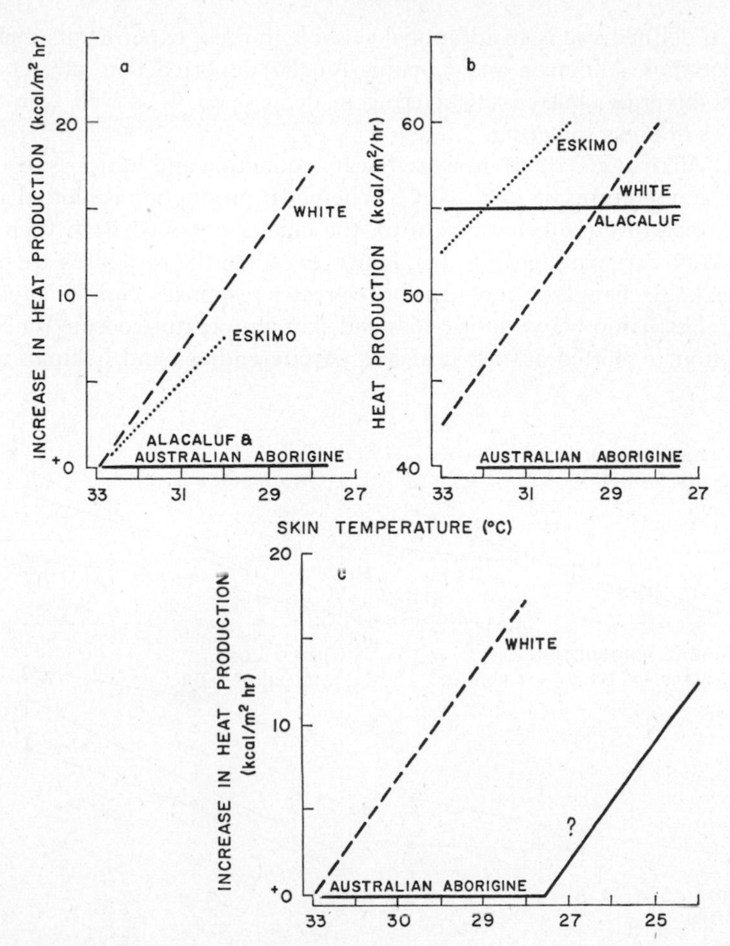

FIGURE 9. Results compiled by Hart *et al.* (1962). In (a) the increase in heat production with decreasing skin temperature is plotted. In (b) the total heat production is given. In (c) curve (a) is redrawn to indicate that a matter of interest is whether the skin temperature at which an increased heat production occurs is lowered in the aborigine or whether the magnitude of the response may also change.

group analyses of covariance, to determine any effects of such variables as basal metabolic rate, initial body heat content and body fat content on the responses. There is a considerable difference between the initial metabolic rates of the various groups (Fig. 9b). However, the response lines of the Eskimos and Alacalufs (who appear to have similar initial metabolic rates) have different slopes. The skin temperatures at which responses will occur in the Alacaluf and the Australian aborigine and the nature of the responses have not been determined (Fig. 9c). The effect of sleep on temperature has not been defined and is an additional variable in these experiments. Scholander, Hammel, Andersen and Løyning (1958a) reported that subjects sleep though shivering, but cease shivering in deep sleep. It is well known that sedatives depress shivering.

(iii) Milan *et al.* (1963) measured heat production and heat loss in a water bath at temperatures of 30·5° to 35°. When heat production is plotted against skin temperature (bath temperature), the curves obtained from Caucasians and Eskimos are parallel (Fig. 10). However, when the responses are plotted as percentage changes, Caucasians have greater responses. Similar alterations in the relationship between heat loss and skin temperature occur. When heat production is plotted against heat loss, Arctic Indians and Eskimos appear

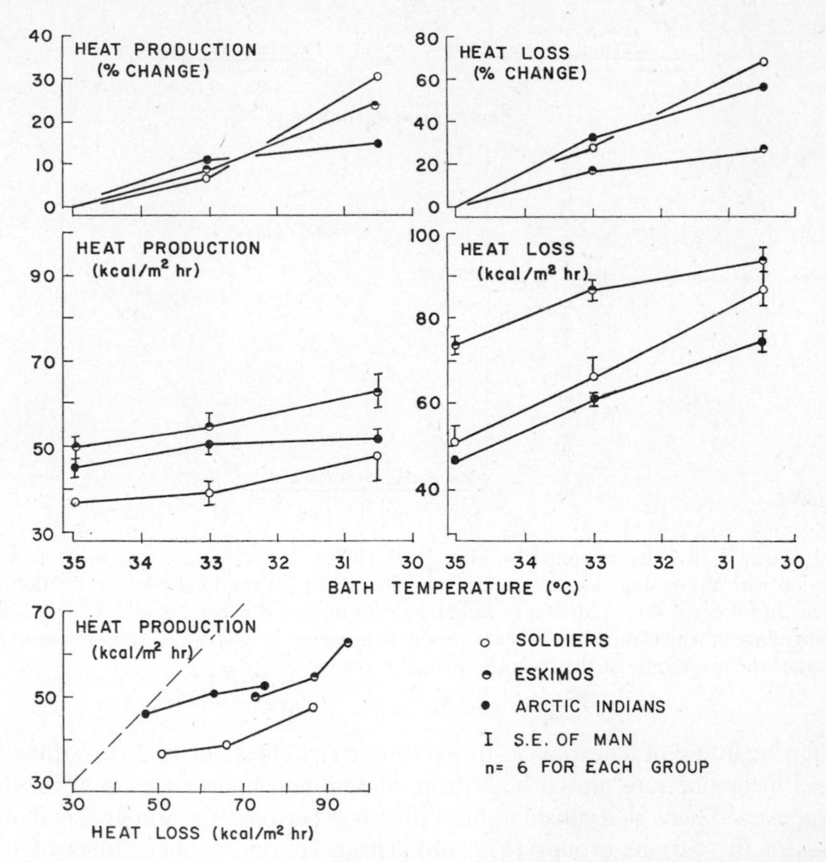

FIGURE 10. Heat production and heat loss at various skin temperatures obtained from soldiers, Eskimos and Arctic Indians in a regulated water bath. In upper section heat production and heat loss are shown as percentage changes, and the values are given in the centre section. In the lower left, heat production is plotted against heat loss. (From Milan *et al.*, 1963; additional values kindly supplied by the authors.)

to behave in a similar manner, Caucasians responding at a lower level. It should be noted that in all three groups heat production was less than heat loss. Thus the subjects were not in a steady state.

(iv) Wyndham, Strydom, Morrison, Peter, Williams, Bredell and Joffe (1963) suggest plotting the dependent variables against air temperature. A systematic comparison over a range of temperatures may elucidate or suggest the mechanisms affected.

E. Conclusions

It will be evident that we wish to avoid drawing any definitive conclusion about adaptation and acclimatisation, unless it is that tests must be made over a sufficient range of temperatures with controlled conditions to characterise a response. Thus it would seem reasonable to accept the fact that cold exposure alters the circulatory response in the hands (Table 2), but tests over a temperature range, as shown in Fig. 3 or as presented by Wyndam *et al.* (1963), more clearly characterise the response. The possibility exists that the primary effect is on metabolism rather than on peripheral circulation.

Whole-body responses show no distinctive pattern because of differing techniques and criteria (Table 1). At present one is forced to conclude that physiological responses to cold in man can be altered by cold exposure; that groups chronically exposed to cold are able to sleep in the cold, some of them shivering, with subsequent raised metabolism, and others allowing body and skin temperatures to fall. Though the rigid requirements of comparable control groups are not feasible to achieve, field studies should incorporate tests that indicate the thermal state, the physiological adjustment over a temperature range and perhaps the magnitude of exposure to cold that can be successfully met.

In conclusion we would favour experiments that include a standard exposure over a range of temperature with sufficient measurements to describe the body heat economy. The range of temperatures should be sufficiently great to elicit a strain. The variables of body heat economy include heat production, heat loss, skin temperatures, internal temperatures and blood flow to the periphery.

REFERENCES

Adam, J. M. (1958). *J. Physiol.* **145**, 26P.
Adams, T., and Covino, B. (1958). *J. appl. Physiol.* **12**, 9.
Adams, T., and Heberling, E. J. (1958). *J. appl. Physiol.* **13**, 226.
Adams, T., and Smith, R. E. (1962). *J. appl. Physiol.* **17**, 317.
Adolph, E. F., and Molnar, G. W. (1946). *Amer. J. Physiol.* **146**, 507.
Allwood, M. J., and Burry, H. S. (1954). *J. Physiol.* **124**, 345.
Andersen, K. L., and Hellström, B. (1960). *Acta. physiol. scand.* **50**, 88.

Andersen, K. L., Løyning, Y., Nelms, J. D., Wilson, O., Fox, R. H., and Bolstad, A. (1960). *J. appl. Physiol.* **15**, 649.
Aschoff, J. (1957). *Pflüg. Arch. ges. Physiol.* **264**, 260.
Aschoff, J., and Wever, R. (1958), *Naturwissenshaften*, **45**, 477.
Baker, P. T. (1960). *Hum. Biol.* **32**, 3.
Balke, B., Cremer, H. D., Kramer, K., and Reichel, H. (1944). *Klin. Wschr.* **23**, 204.
Barcroft, H. (1960). *Physiol. Rev.* **40** (Suppl. No. 4), p. 81.
Bard, P. (1961). *In* "Medical Physiology", 11th Edition, p. 240. Mosby, St. Louis.
Barnicot, N. A. (1959). *Cold Spring Harbor Symposia on Quantitative Biology*, **24**, 115.
Bazett, H. C. (1949a). *In* "Physiology of Heat Production and the Science of Clothing" (L. H. Newburgh, ed.), p. 128. Saunders, Philadelphia and London.
Bazett, H. C. (1949b). *In* "Physiology of Heat Production and the Science of Clothing" (L. H. Newburgh, ed.), p. 147. Saunders, Philadelphia and London.
Bazett, H. C., Love, L., Newton, M., Eisenberg, L., Day, R., and Forster, R., (1948). *J. appl. Physiol.* **1**, 3.
Benzinger, T. H. (1959). *Proc. nat. Acad. Sci. Wash.* **45**, 645.
Benzinger, T. H., Pratt, A. W., and Kitzinger, C. (1961). *Proc. nat. Acad. Sci. Wash.* **47**, 730.
Brown, G. M., Bird, G. S., Boag, T. J., Boag, L. M., Delahaye, J. D., Green, J. E., Hatcher, J. D., and Page J. (1954). *Circulation* **9**, 813.
Brown, G. M., and Page, J. (1952). *J. appl. Physiol.* **5**, 221.
Budd, G. M. (1962). *Nature, Lond.* **193**, 886.
Burton, A. C. (1934). *J. Nutr.* **7**, 497.
Burton, A. C. (1935). *J. Nutr.* **9**, 261.
Burton, A. C., and Bazett, H. C. (1936). *Amer. J. Physiol.* **117**, 36.
Burton, A. C., and Edholm, O. G. (1955). "Man in a Cold Environment", p. 107. Arnold, London.
Buskirk, E. R., Thompson, R. H., and Whedon, G. D. (1963). *J. appl. Physiol.* **18**, 603.
Cannon, P., and Keatinge, W. R. (1960). *J. Physiol.* **154**, 329.
Cannon, W. B., Querido, A., Britton, S. W., and Bright, E. M. (1927). *Amer. J. Physiol.* **79**, 466.
Carlson, L. D. (1954). *Proc. Soc. exp. Biol.* N.Y. **85**, 303.
Carlson, L. D. (1960). *Fed. Proc.* **19**, 25.
Carlson, L. D. (1961). ATC report 61–43. School of Aviation Med., Texas.
Carlson, L. D. (1963). *Fed. Proc.* **22**, 925.
Carlson, L. D., and Buettner, K. J. K. (1957). *Fed. Proc.* **16**, 609.
Carlson, L. D., Young, A. C., Burns, H. L., and Quinton, W. F. (1951). AF Technical Rep. No. 6247. Wright-Patterson Air Force Base, Dayton, Ohio.
Carlson, L. D., and Thursh, H. L. (1960). "Human Acclimatization to Cold", AAL-TR 59–18. Ladd Air Force Base, Alaska.
Carlson, L. D., Burns, H. L., Holmes, T. H., and Webb, P. P. (1953). *J. appl. Physiol.* **5**, 672.
Carlson, L. D., Hsieh, A. C. L., Fullington, F., and Elsner, R. W. (1958). *J. Aviat. Med.* **29**, 145.
Cherry-Gerrard, A. C. B. (1930). "The Worst Journey in the World". Dial Press, New York.
Cooper, K. E., Gross, K. W., Greenfield, A. D. M. Hamilton, D. McK., and Scarborough, H. (1949). *Clin. Sci.* **8**, 217.
Covino, B. G. (1961). *Fed. Proc.* **20**, 209.
Daniels, F., Jr., and Baker, P. T. (1961). *J. appl. Physiol.* **16**, 421.
Darwin, C. (1839). *In* "Journal of Researches into Geology and Natural History of the Various Countries visited by H.M.S. *Beagle*", p. 235. Colgurn, London.

Davis, T. R. A. (1961). *J. appl. Physiol.* **16**, 1011.
Davis, T. R. A., and Johnston, D. R. (1961). *J. appl. Physiol.* **16**, 231.
DuBois, E. F. (1948). *In* "Fever and the Regulation of Body Temperature", p. 9. Thomas, Springfield, Illinois.
Eagan, C. J. (1963). *Fed. Proc.* **22**, 947.
Edwards, M., and Burton, A. C. (1960). *J. appl. Physiol.* **15**, 201, 209.
Eichna, L. W. (1949). *Arch. physical Med.* **30**, 584.
Elsner, R. W. (1955). AAL Rep. No. 1. Arctic Aeromedical Lab., Fort Wainwright, Alaska.
Elsner, R. W. (1963). *Fed. Proc.* **22**, 840.
Elsner, R. W., Andersen, K. L., and Hermansen, L. (1960a). *J. appl. Physiol.* **15**, 659.
Elsner, R. W., Nelms, J. D., and Irving, L. (1960b). *J. appl. Physiol.* **15**, 662.
Ferris, B. G., Forster, R. E., Pillion, E. L., and Christensen, W. R. (1947). *Amer. J. Physiol.* **105**, 304.
Forster, R. E., Ferris, B. G., and Day, R. (1946). *Amer. J. Physiol.* **146**, 600.
Fox, R. H., and Wyatt, H. T. (1962). *J. Physiol.* **162**, 289.
Freeman, H., and Nickerson, R. F. (1938). *J. Nutr.* **15**, 597.
Froese, G., and Burton, A. C. (1957). *J. appl. Physiol.* **10**, 235.
Gagge, A. P., Burton, A. C., and Bazett, H. C. (1941). *Science* **94**, 428.
Glaser, E. M. (1949). *J. Physiol.* **110**, 330.
Glaser, E. M., Hall, M. S., and Whittow, G. C. (1959). *J. Physiol.* **146**, 152.
Goldby, F., Hicks, S., O'Connor, W. J. and Sinclair, D. A. (1938). *Aust. J. exp. Biol. Med. Sci.* **16**, 29.
Grant, R. T., and Bland, E. F. (1930). *Heart*, **15**, 385.
Greely, A. W. (1886). "Three Years of Arctic Service". Bentley, London.
Green, H. D., and Kepchar, J. H. (1959). *Physiol. Rev.* **39**, 617.
Greenfield, A. D. M., and Scarborough, H. (1949). *Clin. Sci.* **8**, 211.
Greenfield, A. D. M., and Shepherd, J. T. (1950). *Clin. Sci.* **9**, 323.
Greenfield, A. D. M., Shepherd, J. T., and Whelan, R. F. (1951). *Clin. Sci.* **10**, 347.
Hall, C. F. (1879). "Narrative of the second Arctic expedition." Government Printing Office, Washington, D.C.
Hammel, H. T., Elsner, R. W., LeMessurier, D. H., Andersen, K. L., and Milan, F. A. (1959). *J. appl. Physiol.* **14**, 605.
Hannon, J. P., and Viereck, E. (1961). "Neural Aspects of Temperature Regulation." Arctic Aeromedical Lab., Fort Wainwright, Alaska.
Hardy, J. D. (1961). *Physiol. Rev.* **41**, 521.
Hardy, J. D., ed. (1963). "Temperature, its Measurement and Control in Science and Industry, Part 3, Biology and Medicine." Reinhold, New York.
Hardy, J. D., and DuBois, E. F. (1938). *J. Nutr.* **15**, 477.
Hardy, J. D., and Soderström, G. F. (1938), *J. Nutr.* **16**, 493.
Hart, J. S., Sabean, H. B., Hildes, J. A., Depocas, F., Hammel, H. T., Andersen, K. L., Irving, L., and Foy, G. (1962). *J. appl Physiol.* **17**, 953.
Hellström, B., and Andersen, K. L. (1960). *J. appl. Physiol.* **15**, 771.
Hemingway, A. and Stuart, D. G. (1963). *In* "Temperature, its Measurement and Control in Science and Industry, Part 3, Biology and Medicine" (J. D. Hardy, ed.), p. 407. Reinhold, New York.
Hensel, H. (1955). *In* "Temperatur und Leben" (H. Precht, J. Christophersen and H. Hensel, eds.), p. 329. Springer-Verlag, Berlin, Göttingen and Heidelberg.
Hensel, H. (1959). *Annu. Rev. Physiol.* **21**, 91.
Hertzman, A. B. (1938). *Amer. J. Physiol.* **124**, 328.
Hewlett, A. W., and Van Zwaluwenburg, J. G. (1909). *Heart* **1**, 87.
Honda, N., Carlson, L. D., and Judy, W. V. (1963). *J. appl. Physiol.* **204**, 615.

Hong, S. K. (1963). *Fed. Proc.* **22**, 831.
Horvath, S. M., Freedman, A., and Golden, H. (1947). *Amer. J. Physiol.* **150**, 99.
Horvath, S. M., Spurr, B. G., Hutt, B. K., and Hamilton, L. H. (1956). *J. appl. Physiol.*, **8**, 595.
Howard, R. A. (1953). "Down in the North", ADTIC publication No. A-103. Maxwell Air Force Base, Alabama.
Hsieh, A. C. L., and Carlson, L. D. (1957). *Amer. J. Physiol.* **190**, 243.
Iampietro, P. F. (1961). *J. appl. Physiol.* **16**, 405.
Iampietro, P. F., Bass, D. E., and Buskirk, E. R. (1957). *J. appl. Physiol.* **10**, 398.
Iampietro, P. F., Goldman, R. F., Buskirk, E. R., and Bass, D. E. (1959). *J. appl. Physiol.* **14**, 798.
Irving, L. (1960). *Nature, Lond.* **185**, 572.
Irving, L., Andersen, K. L., Bolstad, A., Elsner, R., Hildes, J. A., Løyning, Y., Nelms, J. D., Peyton, L. J., and Whaley, R. D. (1960). *J. appl. Physiol.* **15**, 635.
Johnson, R. E., and Kark, R. M. (1947). *Science* **105**, 378.
Joy, R. J. T. (1963). *J. appl. Physiol.* (In press.)
Keatinge, W. R. (1957). *J. Physiol.* **139**, 497.
Keatinge, W. R. (1961). *Quart. J. exp. Physiol.* **46**, 69.
Krog, J., Folkow, B., Fox, R. H., and Andersen, K. L. (1960). *J. appl. Physiol.* **15**, 654.
Kunkle, E. C. (1949). *J. appl. Physiol.* **1**, 811.
LeBlanc, J. (1954). *Can. J. Biochem. Physiol.* **32**, 354.
LeBlanc, J. (1956). *J. appl. Physiol.* **9**, 395.
LeBlanc, J. (1962). *J. appl. Physiol.* **17**, 950.
LeBlanc, J., Hildes, J. A., and Héroux, O. (1960). *J. appl. Physiol.* **15**, 1031.
Lewis, T. (1930). *Heart* **15**, 177.
Lewis, T., and Grant, R. (1925). *Heart* **12**, 73.
Lyon, G. Y. (1824). "The Private Journal of Captain G. F. Lyon." Murray, London.
McCance, R. A., Ungley, C. C., Crosfill, J. W. L., and Widdowson, E. M. (1956). "The Hazards of Men in Ships Lost at Sea 1940–44", *M. R. C. Spec. Rep. Ser.* No. 291.
Meehan, J. P. (1955a). *J. appl. Physiol.* **7**, 537.
Meehan, J. P. (1955b). *Milit. Med.* **116**, 330.
Meehan, J. P. (1957). *In* "Protection and Functioning of the Hands in Cold Climates" (F. R. Fisher, ed.), p. 55. Nat. Acad. Sci.-Nat. Res. Coun., Washington, D.C.
Meryman, H. T. (1957). *Physiol Rev.* **37**, 233.
Milan, F. A., Hannon, J. P., and Evonuk, E. (1963). *J. appl. Physiol.* 18, 378.
Molnar, G. W. (1946). *J. amer. med. Ass.* **131**, 1046.
Nelms, J. D., and Soper, J. G. (1962). *J. appl. Physiol.* **17**, 444.
Newburgh, L. H., ed. (1949). "Physiology of Heat Regulation and the Science of Clothing." Saunders, Philadelphia and London.
Newburgh, L. H., and Spealman, C. R. (1943). Nat. Res. Coun. Rep. No. 241. Nat. Res. Coun., Washington, D.C.
Newman, M. T. (1961). *Ann. N.Y. Acad. Sci.* **91**, 617.
Osiba, S. (1957). *Jap. J. Physiol.* **7**, 1.
Palmai, G. (1962a). *Med. J. Aust.* i, 9.
Palmai, G. (1962b). *Med. J. Aust.* ii, 989.
Parkes, A. S., ed. (1961). "Hypothermia and the Effects of Cold", *Brit. med. Bull.* **17**, 1.
Pennes, H. H. (1948). *J. appl. Physiol.* **1**, 93.
Prouty, L. R., and Hardy, J. D. (1950). *In* "Biophysical Research Methods" (F. M. Uber, ed.), p. 152. Interscience, New York and London.
Pugh, L. G. C. (1963). *J. appl. Physiol.* **18**, 1234.

Pugh, L. G. C. E., and Edholm, O. G. (1955). *Lancet* ii, 761.
Pugh, L. G. C. E., Edholm, O. G., Fox, R. H., Wolff, H. S., Herver, G. R., Hammond, W. H., Tanner, J. M., and Whitehouse, R. H. (1960). *Clin. Sci.* **19**, 257.
Rapaport, S. I., Fetcher, E. S., Shuab, H. G., and Hall, J. F. (1949). *J. appl. Physiol.* **2**, 61.
Rennie, D. W. (1958). *In* "Cold injury" (M. I. Ferrer, ed.), p. 253. Josiah Macy, New York.
Rennie, D. W., Covino, B. G., Blair, M. R., and Rodahl, K. (1962). *J. appl. Physiol.* **17**, 326.
Roberts, D. F. (1952). *J. R. anthrop. Inst.* **82**, 169.
Roberts, D. F. (1953). *Amer. J. phys. anthrop.* **11**, 533.
Rodahl, K. (1952). *J. Nutr.* **48**, 359.
Roddie, I. C. (1961). *In* "Neural Aspects of Temperature Regulation" (J. P. Hannon and E. Viereck, eds.), p. 113. Arctic Aerometical Lab., Alaska.
Rubner, M. (1902). "Die Gesetze des Energieverbrauchs bei der Ernährung." Franz Deuticke, Leipzig and Wien.
Schmidt-Nielsen, K. (1963). *In* "Temperature, its Measurement and Control in Science and Industry, Part 3, Biology and Medicine" (J. D. Hardy, ed.), p. 143. Reinhold, New York.
Schnapper, H. W., Johnson, R. L., Tuohy, E. B., and Freis, E. D. (1951). *J. clin. Invest.* **30**, 786.
Scholander, P. F. (1955). *Evolution* **9**, 15.
Scholander, P. F., Andersen, K. L., Krog, J., Lorent-Zen, F. V., and Steen, J. (1957). *J. appl. Physiol.* **10**, 231.
Scholander, P. F., Hammel, H. T., Andersen, K. L., and Løyning, Y. (1958a). *J. appl. Physiol.* **12**, 1.
Scholander, P. F., Hammel, H. T., Hart, J. S., LeMessurier, D. H., and Steen, J. (1958b). *J. appl. Physiol.* **13**, 211.
Scott, R. F. (1905). "The Voyage of the *Discovery*." Scribners, New York.
Sheard, C., and Williams, M. M. D. (1940). *Proc. Mayo Clin.* **15**, 758.
Smith, R. E., ed. (1963). "Proceedings of the International Symposium on Temperature Acclimation", *Fed. Proc.* **22**, 687.
Spealman, C. R. (1945). *Amer. J. Physiol.* **145**, 218.
Sprague, M. E., and Ross, C. W. (1959). "World Guide to Field Clothing Requirements." Quartermaster Research and Engineering Center, Natick, Massachusetts.
Spurr, G. B., Hutt, B. K., and Horvath, S. M. (1957). *J. appl. Physiol.* **11**, 58.
Stewart, G. N. (1911). *Heart* **3**, 33.
Ström, G. (1960). *In* "Handbook of Physiology. Section 1" (J. Field, ed.), p. 1173. Amer. Physiol Soc., Washington, D.C.
von Euler, C. (1961). *Pharmacol. Rev.* **13**, 361.
Webb, P. (1951). *J. appl. Physiol.* **4**, 378.
Welch, B. E., Buskirk, E. R., and Iampietro, P. F. (1958). *Metabolism* **7**, 141.
Wezler, K. (1950). *In* "German Aviation Medicine, World War II", p. 792. Government Printing Office, Washington, D.C.
Wulsin, F. R. (1949). *In* "Physiology of Heat Regulation and the Science of Clothing" (L. H. Newburgh, ed.), p. 3. Saunders, Philadelphia and London.
Wyndham, C. H., and Morrison, J. F. (1958). *J. appl. Physiol.* **13**, 219.
Wyndham, C. H., Strydom, N. B., Morrison, J. F., Peter J., Williams, C. G., Bredell, G. A. G., and Joffe, A. (1963). *J. appl. Physiol.* (In press.)
Wyndham, C. H., and Wilson-Dickson, W. G. (1951). *J. appl. Physiol.* **4**, 199.
Yaglou, C. P. (1949). *In* "Physiology of Heat Regulation and the Science of Clothing" (L. H. Newburgh, ed.), p. 286. Saunders, Philadelphia and London.
Yoshimura, H., and Iida, T. (1950). *Jap. J. Physiol.* **1**, 147.
Yoshimura, H., and Iida, T. (1951). *Jap. J. Physiol.* **2**, 177.

CHAPTER 3

Heat

R. H. Fox

A. Introductory

Man's ability to survive and work in hostile climates, whether hot or cold, depends not only on his physiological mechanisms for temperature regulation but also on his behavioural response—in other words, his ingenuity in controlling the microclimate surrounding his body. The relative importance of these two factors in the maintenance of homoeothermy differs at the two climatic extremes, ingenuity playing the major role in cold climates and physiology in hot climates. Physical laws determine this difference. Heat is continuously generated inside the body; for homoeothermy there must always be a net flow of heat from the body to the environment in both hot and cold climates. In cold climates the rate of heat loss can be relatively easily modulated by interposing more or less insulation to impede heat flow, whereas in hot climates the heat must be extracted against the natural gradient by some form of heat pump; it is easier both to design and to wear an overcoat than a refrigerator.

It is therefore probably unwise to conclude that, because man possesses a thermoregulating system apparently well suited for life in warm climates, he originated as a tropical animal. It may simply indicate that the threat of hyperthermia represents a more serious problem than that of hypothermia, so that he has therefore evolved a greater reserve of heat-eliminating than of heat-conserving capacity (Hardy, 1961).

In this chapter an attempt is made to trace the development, from their inception to the present day, of some ideas about man's responses to exposure to heat. We can see how certain ideas and phenomena came into pre-eminence early and others, perhaps more important ones, at a much later stage. We can also see how much has depended on the ease with which certain of the phenomena can be measured and observed, how chance observations altered the course of events, how the attitudes and approaches of individual workers influenced them and, above all, how technical innovations redirected them. Perhaps, too, an understanding of what has gone before may help in reading more clearly the signposts toward future work.

B. Historical Development

From the earliest writings of recorded history man has clearly taken an intense interest in the effects of the climate on his body. However, quantitative studies were not possible until Galileo had produced the first thermometer at the beginning of the seventeenth century; then it was to be more than another century before Fahrenheit and Celsius devised the more accurate instruments and convenient scales that we use today. This makes the work of Santorio Santorio in 1614 seem all the more remarkable, since he not only devised his own thermometer for recording body temperature, but also made a balance sensitive enough to demonstrate the presence of insensible perspiration (Foster, 1924).

It was almost two centuries later that De Sasaure devised his wet-bulb thermometer; the more accurate form of ventilated wet-bulb bearing Assmann's name was not freely available until the first decade of the twentieth century.

In spite of this lack of measuring equipment, some of the early writers from their observations made some penetrating deductions that still have a modern ring. Thus, Governor Ellis (1758), in his account of the heat† of the weather in Georgia, writes: "I have frequently walked an 100 yards under an umbrella, with a thermometer suspended from it by a thread to the height of my nostrils, when the mercury has risen to 105°; which is prodigious. At the same time I have confined this instrument close to the hottest parts of my body, and have been astonished to observe, that it has fallen several degrees. Indeed, I never could raise the mercury above 97° with the heat of my body." . . . "This same thermometer I have had thrice in the Equatorial parts of Africa; as often at Jamaica, and the West Indian Islands; and, upon examination of my journals, I do not find, that the quicksilver ever rose in those parts above the 87th degree, and to that but seldom: its general station was between the 79th and 86th degree; and yet I think I have felt those degrees, with a moist air, more disagreeable then what I now feel." . . . "What havoc must this make with an European constitution? Nevertheless, but few people die here out of the ordinary course; tho' indeed one can scarce call it living, merely to breathe, and trail about a vigourless body; yet such is generally our condition from the middle of June to the middle of September." Although he could not measure it, Governor Ellis had clearly appreciated the need to take the humidity of the air into account when trying to relate climatic stress to physiological strain, and in his last sentence he enunciates the age-old belief in the inferiority of Western man's physiological constitution for coping with hot climates, which persists to the present day and is still as unproven as it was in his day.

Fordyce's experiments, reported by Blagden (1775a, b), were among the earliest deliberate hot-room experiments for observing the effects of heat on man. They took place in the 1770's, and it is important to remember that there was then no wet-bulb thermometer, no simple and accurate method of measuring air speed and no measure of radiation. It was noted that when the room temperature was only 110° to 120°F (43° to 49°), but the air was almost saturated with moisture, body temperature and pulse rate rose rapidly, there was profuse sweating and a great increase in peripheral blood flow as evidenced by reddening of the skin and dilatation of the veins. On the other hand, when the air was dry the subjects could withstand much higher temperatures of up to 260°F (127°) for appreciable periods. This is well above the boiling-point of water, so it is perhaps hardly surprising to read that many of their contemporaries doubted whether it was possible. Recent work has fully confirmed the observation: individuals exposed to 266°F for 20 min survived unharmed (Blockley and Taylor, 1949); there have also been reports of exposure to much higher air temperatures for shorter periods. The limiting factor is the rise in skin temperature, and at about

† Temperatures quoted from Ellis are clearly in degrees Fahrenheit.

120°F it takes only a matter of seconds for the irreversible tissue damage of a burn to develop. In dry, still air at 260°F skin temperature does not reach this level, partly because air has a low specific heat, which means that a given volume of air carries only a small quantity of heat; partly because the latent heat of evaporation of sweat is highly effective in cooling the skin; and partly because the blood vessels in the skin dilate and heat is rapidly transported away from the hot skin into the body's cooler interior. If the exposure to such hot conditions is prolonged, the individual will indeed eventually collapse because of an excessive rise in the deep body temperature.

Blagden (1775a, b) and Fordyce only partly appreciated the importance of the evaporation of sweat in temperature regulation and believed in Stahl's phlogiston theory. This theory postulated that in all combustible compounds there was a substance that was released on combustion and that the process was reversible, so that heat could be consumed in the body in much the same way as it was liberated. Blagden (1775a) therefore concluded: "These experiments, therefore, prove in the clearest manner, that the body has a power of destroying heat. To speak justly on this subject we must call it a power of destroying a certain degree of heat communicated with a certain quickness. A powerful assistant evaporation must undoubtedly prove in keeping the body properly cool when exposed to great heats, but it can act only in a gross way . . . the finer balances of which are almost universally effected in that part of the body which is formed with the most subtle organisation." Almost three years later Lavoisier began finally to dispose of the phlogiston theory (Lavoisier, 1778, 1784a, 1785) and, with Pierre Simon de la Place, to construct his calorimeter and measure the respiratory quotient (Lavoisier, 1784b). It is interesting to speculate just how much scientific history would have been altered if the French Revolution had come twenty years earlier and Lavoisier's head had rolled then instead of in 1794.

The disposal of the phlogiston theory led to the gradual appreciation of the importance of the circulation in transporting heat from the interior of the body to the skin. Brodie (1812) experimented on decapitated animals and on animals with transections of the spinal cord and correctly concluded that heat loss depended on the control of the circulation from the brain. Aronsohn and Sachs (1885) finally located the heat loss centre in the anterior hypothalamus. During the early part of the nineteenth century much work on animal physiology contributed to the understanding of temperature regulation, but relatively little on human physiology. In tropical countries progress was greatly retarded by the almost complete confusion between illness caused by infective diseases of tropical origin and illness caused directly by the heat. This was the heyday for the belief in solar apoplexy, and it was widely thought that the sun's rays somehow penetrated the cranium—a belief that was apparently supported by the experiments of one John Davy, the brother of Sir Humphry, who wrote: "When the sun's rays are concentrated by a lens they penetrate through bone – such as a portion of the cranium. It is easy to ascertain the penetration through the cranium by a luminous point appearing on the inner surface. This circumstance may help to explain the effect of the sun on the brain in producing the malady called 'coup de soleil'." (Davy, 1839.)

The belief in special actinic rays capable of penetrating to the brain and spinal cord and causing sunstroke died hard. Sun helmets and spinal pads were the vogue, and Manson wrote: "The phenomenon connected with the Roentgen rays suggests the possibility that there may be solar rays other than ordinary heat rays which, although able to pass organic materials can, nevertheless, be arrested by metal. If this be true for the sun as for an electric spark, a useful addition to the sun hat would be a thin plate of light metal." (Manson, 1898.)

It required the experiments of Aaron (1911) and Shaklee (1917) to prove that ill effects from insolation in the tropics was due to a failure of temperature regulation. Even so, writing as recently as 1927, Sundstroem reviewed the position with extreme caution: "The trend in newer contributions to tropical physiology seems to be to minimise the action of the 'actinic' light and to focus attention on the existence in tropical regions of a continuous reduction of cooling power, which, moreover, unlike direct insolation, is omnipresent and requires more elaborate means of counteracting. One becomes inclined to reduce the importance of insolation in tropical physiological considerations and to consider it as a special case in connection with the aetiology of sunstroke." It is easy to forget how recently this belief in an unknown and sinisterly pejorative effect of the tropical sun was finally dispelled.

Progress during the twentieth century has rapidly gained momentum. The volume of new work published each year is now so large that it is extremely hard for workers to remain up to date in their knowledge of the whole field. In this chapter an attempt will be made to follow developments of two lines of enquiry. The first of these is one of the many attempts to evolve satisfactory indices of heat stress and strain. The second is the progress that has been made in understanding the thermoregulatory mechanisms of the human body and how these change during acclimatisation to heat.

C. Indices of Heat Stress and Physiological Effects

1. *Development of instruments*

The practical importance of finding some index or formula that would integrate all the factors contributing to the heat load into a single index expressing the heat stress or the physiological effect of a hot climate was quickly appreciated. Unfortunately this has not proved easy because of two fundamental difficulties; as a result, many indices have been devised and promoted. This is a source of considerable confusion in environmental physiology because it is often difficult to compare the work of two groups who have expressed their results in terms of different indices. One of these difficulties arises from the need to make such an index suitable for the most diverse applications. Thus, if a high degree of accuracy in prediction and universality of application is desired, it becomes necessary to take account of as many as possible of the variables contributing to the heat stress. Such an index is exceedingly complicated to formulate and cumbersome in use. Reducing the number of variables to the two that are usually the most important, i.e. dry-bulb and wet-bulb temperatures, is also unsatisfactory because in certain situations other variables, such as radiation level, wind speed and metabolic rate, not only become important but also are the very factors it is desirable to evaluate. The second difficulty has been in finding a satisfactory yardstick for measuring the effects of heat on man. The nature of these problems will become clearer as we follow the development of the indices themselves.

Early attempts were made to develop instruments that could in some degree mimic the effects of the climatic variables on the human body. Heberden (1826) produced a heated thermometer designed to give a combined measure of the effects of air temperature and air velocity. This was further developed in Hill's katathermometer (Hill, Flack, McIntosh, Rowlands and Walker, 1913) and was used for measuring the physiological effects of the environment. When the bulb of the instrument was covered with a wet silk finger stall, it also took some account of the cooling power of the environment in hot conditions.

One of the most elaborate of these instruments was the eupatheoscope developed by Dufton (1929, 1932, 1936). It consisted of a black cylinder 22 in (55 cm) high and $7\frac{1}{2}$ in (19 cm) in diameter containing a heater and a thermostat set at 78°F (25·5°). The surface of the cylinder therefore varied in temperature, depending on the environmental conditions of dry-bulb temperature, radiation and air velocity and mimicking to some degree the effects on the heat losses from the clothed body. The amount of current consumed by the heater was used to express the reading on a scale of Equivalent Temperature. This Equivalent Temperature index has been used by ventilating engineers and can also be derived from a nomogram based on the formula

$$\text{Equivalent temperature} = 0{\cdot}522ta + 0{\cdot}478w - 0{\cdot}0474\,v\,(100\text{-}ta),$$

if air temperature (ta), mean radiant temperature (tw) and air velocity (v) are known (Bedford, 1936).

It soon became clear that the thermal relationships between man and his environment were far too complicated to be reduced to a simple physical model, and that, even if an instrument could be devised to take proper cognisance of each of the climatic variables in determining the heat load, this still would not permit the prediction of the resultant physiological effect. It was also clear that, unless the index expressed the heat load of the environment in terms of the physiological effect on man, it would be of little practical value. The alternative approach to the problem was to use man as the measuring instrument and to construct formulae or nomograms relating his physiological responses to different combinations of the climatic variables. This is a largely empirical approach: to formulate such an index it is necessary to perform large numbers of experiments and test different combinations of the several climatic variables. There is also the further problem of what measure of physiological effect to adopt. The easily measured effects that occur when man enters a hot climate include changes in the pulse rate, skin and deep body temperatures, changes in body weight through sweat loss and subjective sensations of warmth and discomfort. Each of these variables shows a different pattern of change when man is exposed to heat, and none of them provides a perfect yardstick of physiological strain.

2. *Effective temperature*

This was the first scale of the empirical type; it was evolved by Houghten and Yagloglou (Houghten and Yagloglou, 1923; Houghten and Yagloglou, 1924; Yagloglou and Miller, 1925), who determined climates having widely differing combinations of air temperature, air speed and humidity, but having equivalent comfort as judged by the subjective impressions of groups of individuals. There are two scales, one to be applied if the individuals are stripped to the waist and the other for persons normally clad. The original scales contained no allowance for radiation, and it was quickly realised that this was a serious drawback. One of the reasons for this omission was the absence of any suitable measuring device in common usage. Aitken (1887) as early as the end of the last century proposed the use of a hollow metal sphere blackened on the outside with a thermometer at the centre, but it was not until Vernon (1930, 1932) reintroduced the blackened globe and, together with Bedford and Warner (1934), had worked out how to use it to measure mean radiant temperature that a relatively simple method of measuring radiation was available.

The introduction of the globe thermometer led to the elaboration by Vernon and Warner (1932) and Bedford (1946) of amended scales to include an allowance for radiation, called Corrected Effective Temperature. A further modification to the scales proposed by Smith (1955) includes an allowance for the level of energy expenditure.

The Effective Temperature scales have the great merit that they relate the thermal properties of the environment to an important physiological quality, i.e. the subjective sensation of warmth. They are of most value when investigating conditions near the optimum for comfort or only mildly hot, usually found in offices, homes and most factories.

3. *Wet-bulb globe temperature index*

The recently developed W.B.G.T. index was not based on the analysis of a new set of prime data, but was in effect derived from the Effective Temperature scales and represents a simplified form of them. It has rapidly found favour because of its simplicity; originally intended for use by the American Armed Forces (Yaglou and Minard, 1957), it combines the effects of dry-bulb (ta°F), wet-bulb (tw°F) and globe thermometer temperatures (tg°F) in the simple equation

$$\text{W.B.G.T.} = 0{\cdot}7\, tw + 0{\cdot}1\, ta + 0{\cdot}2\, tg.$$

Minard, Belding and Kingston (1957) have recommended that training of raw recruits should cease when the index reaches 85° and that all strenuous activity should be discontinued regardless of the state of acclimatisation when the index is 88° or more.

4. *Wet-bulb—dry-bulb index*

This index is similar to the W.B.G.T. index except that in calculating it no allowance is made for radiation and there is a small difference in the relative weightings of wet-bulb and dry-bulb temperatures. The index was developed by Lind, Hellon, Weiner, Jones and Fraser (1957) and Lind (1963) to relate tolerance times in saturated and non-saturated climates for mine rescue personnel.

The index is derived from the formula

$$\text{W.D.} = 0{\cdot}15\, d + 0{\cdot}85\, w$$

where d and w represent dry- and wet-bulb temperatures (°F), respectively.

5. *Predicted 4 hr sweat rate*

McArdle's P4SR nomogram (McArdle, Dunham, Holling, Ladell, Scott, Thomson and Weiner, 1947) was a direct result of the impetus given by the needs of war and was originally evolved for a specific purpose. In the early part of World War II the Board of Admiralty became concerned by the thermal conditions to which men in ships were exposed in tropical waters. This led to the formation of a group of workers at the National Hospital, London, charged with the task of examining the problems involved in predicting or assessing the stress of hot environments. The existing Effective Temperature scales were tested and found to be unsatisfactory in certain respects. It was thought that in a hot climate the sweat rate afforded a better yardstick of physiological effect than subjective sensations; accordingly the P4SR index was developed (Benson, Colver, Ladell, McArdle and Scott, 1945; Ladell and McArdle, 1945; Dunham, Holling, Ladell, McArdle, Scott, Thomson and Weiner, 1946). A direct correspondence between McArdle's P4SR index and the observed sweat rate of a group of individuals is only found when certain conditions are fulfilled. The individuals must be fully acclimatised to heat, the exposure to the hot climate must last 4 hr, and the conditions must not be so severe as to indicate a P4SR of above about 5·0. Above this level the observed sweat rate falls below the predicted value, because the maximum sweating capacity of the subjects has been exceeded.

The index takes account of various factors contributing to the stress of a heat exposure, namely, dry-bulb, wet-bulb and globe thermometer temperatures, air speed, metabolic rate and two levels of clothing.

The P4SR index was tested, and its predictions were compared with those of the Effective Temperature scales at the Tropical Research Unit at Singapore (Macpherson, 1960). In these exhaustive tests the reliability of the index was, in general, fully confirmed.

6. *Operative temperature*

We now come to indices of a rather different type, by means of which the aim has been to analyse the thermal exchanges between the human body and its environment and to base the assessment of the heat stress on the magnitude of heat flow.

The first attempts along this line were made at the J. B. Pierce Laboratory of Hygiene in America (Winslow, Herrington and Gagge, 1938; Winslow, 1941). They used the technique of partitional calorimetry to analyse the magnitude of heat flow through each of the avenues of thermal exchange and to derive constants for the coefficients of convection, radiation and evaporation. From the results they evolved a simplified heat stress index called Operative Temperature.

This index is somewhat similar to Equivalent Temperature and is derived from the formula

$$\text{Operative Temperature} = \frac{kv(tw) + kc(ta)}{kv + kc},$$

where tw and ta are the radiant wall and air temperatures (°C), respectively, and kv and kc are constants for radiation and convection, respectively. The index is applicable to lightly clothed or nude individuals; at operative temperatures below 29° and 31° (84° and 88°F) body temperature can be controlled by vasomotor regulation, and above 31° (88°F) evaporative cooling is required.

7. *The Belding and Hatch index*

To investigate the heat load imposed in a hot environment by analysing the pattern of thermal exchanges it imposes on man is clearly an attractive and rational approach to the whole problem. Haines and Hatch (1952) further developed the basic concepts of Winslow, Herrington and Gagge (1937, 1938) by showing how the principle of thermal exchange could be applied to the evaluation and control of industrial heat exposures; this was later amplified (Belding and Hatch, 1955) into the Belding and Hatch index (B.H.I.), which expresses the thermal stress of a hot climate as the ratio of the amount of sweat that must be evaporated to maintain the body in thermal equilibrium to the maximum evaporative capacity of the climate. To achieve such a mathematical approach in a sufficiently simple form for practical use required a number of assumptions and approximations. The index assumes that the individuals are of average build and dressed in shorts, with skin temperatures of 95°F and body surfaces uniformly wetted with sweat; it is further assumed that there is no storage of heat in the body and that the thermal exchanges by conduction and respiration can be ignored. The basic heat balance equation then becomes

$$\text{Evaporation required (E req)} = \text{Metabolic Heat (M)} + \text{Convective Heat Exchange (C)} \pm \text{Radiant Heat Exchange}$$

The Maximum Evaporative Capacity (E max) of the climate is also calculated, and the B.H.I. $= \dfrac{\text{E req}}{\text{E max}} \times 100$.

The Belding and Hatch index has the virtues of simplicity and directness, in that it expresses the heat stress quantitatively in terms of the stressing agent. It enables engineers to analyse the situation into its component parts and to determine rapidly how the stress can be diminished by altering one or another component. Unfortunately it has serious drawbacks. The approximations and assumptions introduced for the sake of simplicity reduce its accuracy; in treating man as a physical model one is, to some extent, ignoring his physiology. A more important criticism is that the index does not bear any simple relationship to physiological strain and that equivalent levels on the index produced by different combinations of levels of the climatic variables do not produce the same degree of physiological strain. This is easily demonstrated by choosing two climates with the same Heat Stress index, one a relatively cool but almost saturated climate and the other a much hotter and drier climate. The hotter drier climate is, in fact, much more stressful. In effect, the ratio $\dfrac{\text{E req}}{\text{E max}}$ only becomes an important determinant of strain when it is approaching unity; at lower levels it is primarily the size of E req itself that determines the strain. Belding and Hatch (1955) attempted to overcome this difficulty by stipulating that if E req exceeded 2400 B.T.U. (1 l of sweat per hour) the index should automatically become 100, whatever the value of $\frac{\text{E req}}{\text{E max}} \times 100$. This is a somewhat desperate compromise and throws some doubt on the value of the index for defining limiting conditions or for comparing the physiological strain imposed by different climates.

8. *Comparison of indices*

All the indices described above have defects and weaknesses, but each also has its advantages when used in a particular application.

The three most important are the Effective Temperature scales, the Heat Stress index and the P4SR index. Effective Temperature remains the best way of comparing and describing conditions at mild levels of heat stress and in the comfort zone; the Heat Stress index is valuable at higher levels of heat, because it enables the situation to be analysed so that the most appropriate remedy can be chosen; the P4SR index affords the most accurate way of relating heat stress and physiological strain.

By virtue of its simplicity the W.B.G.T. index also fills a useful place, but whenever possible an index giving a more accurate and complete description should be used. There seems no good reason for attempting to introduce variants of the W.B.G.T. index for general use, e.g. the wet-bulb—dry-bulb index; any benefits from increased accuracy seem likely to be more than offset by the confusion due to a multiplicity of indices.

9. *Unsolved problems*

It is nevertheless clear that the ideal index for assessing the heat load imposed on man in a hot situation and for predicting the resultant physiological strain is still lacking, and we may consider some of the problems involved.

The first is to decide exactly what such a new index is intended to do. From a practical point of view it might well be decided to combine the valuable attributes of both the rational approach of analysing the thermal exchange situation to determine the heat load with the ability to predict the resultant strain on the human body by using the empirical indices of physiological effect. This would at once raise two vital questions. (1) What measure of physiological effect affords the most accurate yardstick of physiological strain? (2) How should the thermal exchange analysis be formulated in order to relate the heat load to its physiological effect?

In the broadest terms it is clear that the physiological strain imposed by a particular thermal environment must be related to the body's difficulty in eliminating enough heat to maintain thermal equilibrium. To be more specific, it seems likely that the physiological strain is some function of the ratio of the level of thermoregulatory activity required by the particular thermal situation to the maximum thermoregulatory activity of which the body is capable. Sweat rate and peripheral blood flow are the two best indicators of the level of thermoregulatory activity; although one should not expect a simple linear relationship between the physiological effect and the resultant strain, they seem likely to afford the most consistent and predictable relationship within the limits imposed by biological variability.

The second question is much more difficult to answer. It seems rather unlikely that in the foreseeable future the characteristics of the human thermoregulatory system will prove sufficiently amenable to physical analysis for equations to be formulated that will relate changes in the environmental situation to changes in thermoregulatory activity.

If this is so, the most profitable line of attack may well be a further major empirical study designed to measure simultaneously the thermal exchanges and the physiological effects of exposure to a wide range of environments. Such a study could have several important aims: (1) to improve the thermal exchange coefficients; (2) to correlate the thermal exchange analyses with the physiological effects of body temperature, sweat rate and blood flow; (3) to measure skin temperatures, with a view to developing a nomogram or formula, perhaps similar to the P4SR nomogram, from which it would be possible to predict skin temperature for a wide range of conditions.

The first two aims are straightforward, but the third needs some explanation. The skin is the barrier between man and his environment, and its temperature is an important factor in determining the heat flow through all the avenues of thermal exchange. The assumption of a constant value is an important source of error in the B.H.I. Our present knowledge of skin temperature changes is extremely limited; in many earlier studies it was not measured because of technical difficulties. However, recent developments are overcoming these difficulties with improved techniques for scanning objects and recording their surface temperatures. Accurate measurements of skin temperature should undoubtedly form an important objective in any new study. Another essential preliminary step would be a re-examination of the techniques for measuring the climatic variables, especially radiant heat and air speed.

To develop a new index would be a large undertaking, and it would only be worth while if done thoroughly, but the use of electronic computers for analysing results could afford one great advantage not available in the earlier studies.

D. Theories and Measurement of Heat Acclimatisation

1. *Twentieth-century developments*

We must now return to the mainstream of physiological studies and the evolution of ideas on acclimatisation to heat, picking up the threads where we left them at the beginning of the twentieth century. In the early years of this century the ideas and hypotheses depended largely on ill-documented or anecdotal material of travellers' tales. The confusion between the effects of heat and tropical diseases and the belief in the sinister effects of insolation already described were great barriers to clear thinking and to progress. Writing on physiological responses to heat, Bazett (1927) described the confused state of knowledge: "The whole subject is unfortunately lamentably complicated and little understood; indeed we have advanced little beyond the views expressed by Claude Bernard in 1876 in his lectures on La Chaleur Animale." (*Leçons sur la Chaleur Animale*, Paris.)

It was realised that acclimatisation to heat should facilitate homoeothermy, but attempts to demonstrate such an effect were far from successful, so that Sundstroem (1927) wrote: "If it be true that the physiological response to a tropical climate is primarily due to the cooling power factor, the acclimatization process to such a climate should in the first hand consist of means to adjust the heat regulating mechanism to the highest attainable degree of efficiency in order to preserve a normal body temperature. It is a curious fact that, in spite of the numerous attempts to assay this heat-regulating efficiency by body temperature measurements, this simple point should still remain one of the most contested points in the whole field of tropical physiology."

Investigators had reported a rise in the resting body temperature when individuals went from temperate into tropical climates (Davy, 1850, Neuhass, 1893) and that the increase was more marked during the first weeks in the tropics (Rathay, 1870, 1871; Jousset, 1884). It was further shown that the rise in body temperature after muscular work was frequently much greater in the tropics (Young, 1915; Young, Breinl, Harris and Osborne, 1919). However, there were also many reports of no demonstrable influence of tropical living on body temperature (Boileau, 1878; Furnell, 1878; Thornley, 1878; Wick, 1910) and of little, if any, difference between the body temperatures of the indigenous races and white residents in the tropics (Jousset, 1884; Eijkman, 1895).

With hindsight it is now easy to see how these difficulties arose, but they had the unfortunate effect of diverting attention from the central problem of defining the changes in the temperature-regulating mechanisms to other indirect effects of repeated heat exposures.

2. *The controlled-climate laboratory*

In the 1920's it had become clear that further progress in the study of man's response to heat demanded a much closer control of the climatic variables contributing to the heat stress than could be readily obtained by observations made in the field; it was also inconvenient to have to go to hot regions to study the effects of heat. The solution was to mimic naturally occurring hot climates in the laboratory by building special rooms in which the climatic variables could be accurately controlled. The effects of repeatedly exposing individuals to a given type of hot climate could then be studied. This approach to the problem paid rich dividends; from it came the "classical" picture of man's adaptive responses to heat. Perhaps partly because it had proved so valuable initially, it ultimately became a brake on new ideas and further progress.

Hot-room experiments have usually followed a fairly well-defined pattern. A group of subjects is exposed to carefully controlled climatic conditions for a number of hours daily, during which they perform a known amount of physical work and their physiological responses are measured in terms of heart rate, body temperature, sweat loss and so on. After this initial test, the subjects continue to be exposed to hot conditions for a number of days, at the end of which the first test is repeated. The difference in response between the first and final tests shows the cumulative effect or "adaptation" induced by the intervening heat exposures.

3. *Features of acclimatisation*

From the many studies following this type of routine (Bean and Eichna, 1943; Henschel, Taylor and Keys, 1943; Robinson, Turrell, Belding and Horvath, 1943; Eichna, Bean, Ashe and Nelson, 1945; Horvath and Shelley, 1946; Eichna, Park, Nelson, Horvath and Palmes, 1950; Ladell, 1951; Bass, Kleeman, Quinn, Henschel and Hegnauer, 1955; Hellon, Jones, Macpherson and Weiner, 1956; Macpherson, 1960) the classical picture of acclimatisation to heat has emerged. The main features are a less marked increase in the heart rate while working, lower skin and deep body temperatures, a greater production of sweat and, subjectively, a lessened sense of discomfort. These changes clearly demonstrate that the organism becomes more efficient in coping with the imposed heat load, but they do not tell us the underlying mechanisms.

a. Endocrine system. There have been a number of theories about these underlying mechanisms; one of the earliest was an attempt to ascribe a major role to endocrine changes. It began with the observation by Dill, Jones, Edwards and Oberg (1933) of a decline in concentration of salt in the sweat of men who spent a period of days living in the desert. The fall in salt concentration was confirmed in other studies (Daley and Dill, 1937; Dill, Hall and Edwards, 1938) and led to the conclusion that salt conservation was an important mechanism in maintaining and improving the volume and composition of body fluids (Dill, 1938). It is certainly true that a pronounced deficiency of salt with a failure to replenish the losses in sweat and urine can increase the susceptibility to heat exhaustion (Taylor, Henschel, Michelsen and Keys, 1943). However, it was also found that the drop in salt excretion could be prevented by an adequate salt intake (McCance, 1938; Van Heyningen and Weiner, 1952), but it has never been proved that the administration of a high salt diet results in any speeding up of the acclimatisation process or alleviation of the effects of a hot climate. Indeed, too much salt and too little water can be dangerous.

The changes in salt concentration of sweat and urine led to another theory proposed by Conn, Johnston and Louis (Conn and Johnston, 1944; Conn, Johnston and Louis, 1946), who concluded "that in man the process of acclimatization to heat consists of an increased activity of pituitary adreno-corticotrophic hormone and that the resulting enhancement of the production and liberation of adrenal cortical steroids is responsible for bringing about the physiologic adjustments characteristic of the state of acclimatization." This is a pretty sweeping claim, which has not stood up to the test of time.

The injection of deoxycorticosterone acetate or adrenal cortical extracts produces the reduction in sweat and urinary chlorides that normally occurs during the first few days of heat exposure, but there is no evidence that this in any way facilitates or modifies the normal acclimatisation response (Moreira, Johnson, Forbes and Consolazio, 1945; Robinson, Kincaid and Rhamy, 1950).

Conn's theory was closely in line with the concept of the general adaptation syndrome developed by Selye (1946), who postulated that an organism exposed to any form of "stress" experiences a non-specific pattern of reaction, passing through the stages of alarm, resistance and finally, if the "stress" is overwhelming, exhaustion. The reactions described include loss of body weight and nitrogen, a rise in plasma potassium with a fall in chlorides and an increased production of adrenocorticotrophic hormones.

There is little doubt that the endocrine system has an important role to play in heat acclimatisation through the complex readjustments required in salt and water balances, and probably in many other ways, but the evidence seems to indicate that it is not the star performer.

b. Cardiovascular system. Changes in the cardiovascular system and, in particular, an improvement in the cardiovascular response to work in the heat have long been regarded as playing an essential role in heat adaptation. A decrease in pulse rates during the course of repeated heat exposures has been one of the most constant findings in the classical type of hot-room experiment (Bean and Eichna, 1943; Robinson *et al.*, 1943; Eichna *et al.*, 1945; Horvath and Shelley, 1946; Eichna, 1950). It has also been considered the most sensitive indication of acclimatisation to heat (Henschel *et al.*, 1943). It is obvious that the cardiovascular system must be important, for it has the vital task of transporting heat from the core of the body to the skin surface. Prominent symptoms of heat incapacitation in the unacclimatised subject, such as dizziness and fainting, are directly referable to the cardiovascular system. Nevertheless, there have been many differences of opinion as to the real nature of the changes underlying the improvement in performance.

It was early discovered that the blood volume expands when man goes into the heat (Barcroft, Binger, Bock, Doggert, Forbes, Harrop, Meakins and Redfield, 1922), and increases in blood volume of up to 25% after exposure to heat have been reported by a number of observers (Bazett, Sunderman, Doupe and Scott, 1940; Forbes, Hall and Dill, 1940; Conley and Nickerson, 1945; Bass *et al.*, 1955). The importance of this phenomenon has been generally accepted (Barcroft *et al.*, 1922), but recently Bass (1963) has argued that the expansion in blood volume cannot play an important role in heat acclimatisation. This is because it did not occur when subjects were acclimatised by short bouts of work in the heat (Bass, Buskirk, Lampietro and Mager, 1958), and in most of the studies in which it has been found the subjects were living a mainly sedentary existence. He also argues: "The cardiovascular system plays its thermoregulatory role – defined as convective transport of metabolic heat to the skin – quite efficiently on the very first exposure to work in the heat. Thus, skin blood flow is more than adequate (possibly maximal) on the first day in the heat, and does not increase as acclimatization progresses; if anything it decreases." (Bass, 1963.) This is hardly a tenable point of view; to explain why, we must interrupt the present train of thought and introduce two different ideas. First, it has long been thought necessary to work in the heat for acclimatisation to develop (Eichna *et al.*, 1945; Bean and Eichna, 1943). However, acclimatisation can be induced by passive elevation of body temperature (see below). Secondly, it has become clear that skin blood flow is neither adequate nor maximal on the first day of exposure, as is also described below. Altogether there seems no good reason to doubt that an expansion of blood volume normally occurs and is an important factor in improving the function of the cardiovascular system.

Increased venomotor tone is another cardiovascular adaptation that could achieve much the same purpose as an increase in circulating blood volume by helping to prevent pooling of the blood in the capacitance vessels (Bass and Henschel, 1956). This was tested by Wood and Bass (1960), and evidence of an increase in venomotor tone on the third and fourth days of exposure to heat was found, although the effect diminished again in succeeding days.

Unacclimatised subjects who collapse on exposure to heat stress often exhibit signs of cardiovascular insufficiency; this may take the form of postural hypotension or a complete fainting episode. Several observers have suggested that with acclimatisation the cardiovascular condition improves (Scott, Bazett and Mackie, 1939; Bean and Eichna, 1943; Taylor, Henschel and Keys, 1943).

c. Sweat rates. The great majority of the classical studies report an increase in the sweat rates as acclimatisation develops. The magnitude of the increase has varied; it has been small in some studies, large in others. A failure to show any increase has also been reported (Ladell, 1945). In spite of the great variability in results obtained, most observers have concluded that improvement in the efficiency of the sweating mechanism plays an important role in heat acclimatisation.

d. Summary. The summary of acclimatisation changes given by Eichna *et al.* (1950) seems particularly apt and to the point. "Mean skin temperature is adjusted to a level which permits thermal equilibrium between the body and the environment on the one hand, and on the other maintains an internal gradient which permits the transport of the deep heat to the surface without overtaxing the circulation. In the hot, dry environment of this study these conditions were attained almost wholly as a result of the increased evaporative cooling which an increased sweat secretion produced."

4. *Limitations of conventional hot-room approach*

We have adopted a rather different approach to the problem of investigating the nature of acclimatisation to heat (Fox, Goldsmith, Kidd and Lewis, 1963a). The aim has been to study the changes induced when the body temperature is maintained at a known level above normal for a period daily for a number of days, instead of studying the changes on exposure to a particular climatic stress and allowing body temperature to find its own level. There were a number of reasons for this departure from the conventional technique.

(*a*) In the conventional technique for studying heat acclimatisation the stimulus to heat adaptation has to be measured and expressed in terms of all the variables contributing to the heat load; these include the wet-bulb—dry-bulb temperatures, radiation flux, air speed, the amount of clothing worn, the physical characteristics of the subjects and their rate of energy expenditure. Each investigator has used different levels of these variables; when comparing the results of one study with another it is difficult to determine whether differences in response are related to the type of climate or the level of strain. It seemed essential to find some way of simplifying this confusion.

(*b*) It was argued that the fundamental basis of acclimatisation to heat must be a response by the body to the specific stress imposed, i.e. the response to heat. There seemed no reason to believe that the individual factors characterising the hot climate exert specific effects producing different types of adaptive response; if any such effects did occur, they would be additional to the main action of each factor in contributing to the total heat load. It was held that the stimulus to adaptation in a hot climate must be the activation of the thermoregulatory system as a result of the impediment to heat elimination from the body. Further, since elevation of deep body temperature appears to be the most powerful stimulus to the thermoregulatory system, it was argued that this should provide the best available yardstick for measuring the stimulus to adaptation.

One corollary to this line of reasoning was that the chosen situation should be as free as possible from other stresses, such as muscular work. The natural situation of going to live in a hot climate will almost certainly involve exposure to other such stresses, which may be expected to interact with the effects of heat and indeed may play a large role in the total response observed. However, if we wish to examine the adaptive response to heat alone it is important to begin by excluding other stresses as far as possible; the interactions between heat and other stresses should then prove easier to unravel. Heating the subjects passively rather than utilising the heat load imposed by raising metabolism with physical work had the additional advantage of raising the temperature as uniformly as possible throughout the body.

(*c*) The conventional technique for studying heat acclimatisation was thought to suffer from the disadvantage that, as the adaptations promoting the increased efficiency of thermoregulation developed, the presumptive stimulus to further adaptation progressively declined. Moreover, the rate of decline would be different in different types of hot climate. Thus in a hot humid climate the increased sweat may largely drip from the body without conferring the benefit of evaporative cooling, and the consequent fall in body temperature is small, whereas in a hot dry climate the same increase in sweat, evaporating completely on the body, confers a large benefit by substantially reducing body temperature. The presumptive stimulus also varies throughout each heat exposure, being low on entry into the hot climate and rising to a maximum towards the end. In other words, when using the conventional technique of a constant environmental stress to study heat adaptation, one is relating the observed response to a constantly changing and largely unpredictable baseline of stimulus strength.

(*d*) The methods used to measure the degree of acclimatisation produced by the heat exposure also seemed open to question. It has been customary to express the adaptation produced by a series of heat exposures in the form of an assessment based on the changes in at least three of the physiological responses, sweat loss, body temperature and pulse rate. It is necessary to do this because the degree of change in any one variable depends in some measure on the type of climate chosen. Thus, as the result of a series of exposures to a hot and humid climate, there is a large increase in sweating, but a comparatively small reduction in pulse rate and deep body temperature, whereas in a hot and dry climate the increase in sweating is small, but the changes in body temperature and pulse rate are large.

In effect the assessment has to be based on two entirely different types of response: direct evidence of an adaptive change, i.e. an increase in sweating, and evidence of the benefits conferred by all the adaptive changes, i.e. decreases in body temperature and pulse rate. This is already sufficiently complicated, but unfortunately the body temperature is also the presumptive stimulus for sweating, and there is a negative feed back from sweating to the body temperature, which varies according to the climate.

(*e*) Finally, the conventional technique is extremely arduous for subjects and experimenters alike. The subjects have to be coaxed and cajoled to perform their allotted work for several hours daily for a number of days. In order to achieve a constant thermal stress the amount of work performed must be accurately controlled and is therefore inevitably some form of simple repetitive task. It is hardly surprising that under these conditions tempers usually become frayed and relations between subjects and experimenters strained.

5. *Improved techniques*

Consideration of all these problems led to the search for an experimental situation in which the body temperature of a number of subjects could first be rapidly and passively raised to any desired level and then be maintained at the new level for as long as required.

The technique adopted was simple. The subjects were first exposed to a hot humid airstream that would raise body temperature 2° above normal in 10–15 min. As each subject approached the target temperature level he left the hot airstream, was quickly dressed in a vapour barrier suit and then reclined in a deck-chair in a room with an air temperature of about 38° Thc vapour barrier suit prevented the subject cooling by evaporating his sweat, and the room temperature kept heat exchanges by convection and radiation to a minimum. In this situation the body temperature tends to rise slowly because of the resting heat production, and it could therefore be controlled by blowing a variable, but usually small, amount of dry air into the vapour barrier suit. The subjects found this much more acceptable than the traditional type of heat exposure; the observers could be with the subjects and yet not be exposed to the stress; most important of all, accurate control of body temperature was possible.

The first step in using the technique was to find out whether it produced the characteristic changes of heat acclimatisation. Groups of subjects were given periods of controlled hyperthermia daily for a number of days, and once at the beginning and again at the end of the series their responses to a standard test of the conventional type were measured. It was shown that the controlled hyperthermia technique did produce the classical changes of heat acclimatisation and that in the second conventional tests the subjects had lower body temperatures and pulse rates and higher sweat losses and felt less distress. Although qualitatively the results were the same as with the traditional method, there was some evidence of quantitative differences in certain variables. For example, the reduction in pulse rate was comparatively small, and it seems probable that with the addition of physical work the combined blood-flow requirements of muscle and skin are a potent stimulus to improved cardiovascular efficiency. It is also possible that part of the large decreases reported in some conventional studies have been due to using the same exercise throughout, with a consequent improvement in efficiency from muscular training.

Having once established that controlled hyperthermia would induce acclimatisation, the next step was to compare the responses of the subjects during the first and last hyperthermia sessions. Sweat losses had increased markedly, and the change from day to day was surprisingly linear. Pulse rates either showed no change or, if anything, tended to be higher. The subjects complained that they felt more uncomfortable towards the end of the series of hyperthermia sessions than at the beginning. Thus the results indicated that, whereas the increase in sweating seen in the conventional test is a primary adaptive response to heat, the reductions in discomfort and lowered pulse rate are not.

The increase in sweating was found to behave rather like a training response, in that the magnitude of increase was determined by both the degree of body-temperature elevation and the time for which it was maintained each day. A relatively high correlation was found to exist between the total amount of sweat secreted during the series of hyperthermia sessions and the improvement remarked between the first and the last hyperthermia sessions when the subjects were tested in the conventional way. This increase in sweating capacity appears to be a local glandular change in response to repeated use, because repeated local heating of an area of skin will itself produce an increase in the local sweating capacity (Ito and Adachi, 1934); if sweating is suppressed in a local area of skin while the body as a whole is acclimatised to heat, the local area fails to develop an increase in sweating (Fox, Goldsmith, Hampton and Lewis, 1962). This observation has been confirmed in other ways (Brebner and Kerslake, 1963; Collins, Crockford and Weiner, 1963).

The changes in peripheral blood flow during adaptation to heat were also studied. At the time this was done the whole subject was in a state of great confusion. A number of observers had measured the peripheral blood flows of subjects in hot environments and had reported either a decrease or no obvious change with heat acclimatisation (Eichna *et al.*, 1950; Wyndham 1951; Whitney, 1954; Hellon and Lind, 1955). Only one group had reported an increase (Scott *et al.*, 1939; Bazett *et al.*, 1940). To reinvestigate this problem, the controlled hyperthermia technique was used to induce heat adaptation in groups of subjects, and their peripheral blood-flow responses to body heating were measured before and again after the period of heat adaptation (Fox, Goldsmith, Kidd and Lewis, 1963b). The blood flow through hand and forearm was measured, while the body temperature was being slowly raised from normal to 38·5°. It was found that after acclimatisation peripheral blood flow began increasing at a lower body temperature, and for a given rise in temperature the flows were up to 25% higher.

In another study, involving the same general approach and techniques, it was shown that a raised body temperature can impair the performance of one psychomotor task while improving another; here again there was no evidence that as acclimatisation developed there was any improvement in performance at a raised body temperature (Wilkinson, Fox, Goldsmith, Hampton and Lewis, 1964).

E. Primary and Secondary Adaptive Changes

If we accept the argument that the most essential physiological changes in heat adaptation are those affecting the thermoregulatory system, it becomes useful to classify the responses to heat exposure into two groups: the primary adaptive changes, i.e. those changes that directly help to protect the body from accumulating excessive heat; the secondary, or supporting, adaptive changes, i.e. those changes that make the primary adaptations possible or become necessary because of them.

It may be argued that this approach is simply to take refuge in teleology, and indeed so it is. Nevertheless, it has the important advantages of clarity and of indicating how the mechanisms involved in heat adaptation can be systematically studied. It is possible to begin by attempting to construct a list of all the theoretically possible primary adaptations that might be found in addition to those already known; an attempt to do this is shown in Table 1.

Those known to occur are marked with a dagger; the reader may be able to think of responses that should be added to complete the list without

Table I. *Adaptive Changes*

Mechanism	*Adaptation*
Sweating	a. Increased capacity†
	b. Quicker onset†
	c. Better distribution over body surface†
	d. Reduced salt content†
Cardiovascular	a. Greater skin blood flow†
	b. Quicker response†
	c. Blood flow closer to skin surface
	d. Better distribution over body surface
	e. Reduction in counter-current blood vessels
Metabolic	a. Lowered BMR
	b. Lowered energy cost for a given task
Respiratory	a. Hyperventilation (?Panting)†
Heat storage	a. Increased tolerance to higher body temperature
	b. A lower resting body temperature†
Anatomical	a. Change from short and stocky to long and thin†
Behavioural	

† Adaptations for which there is evidence.

resorting to such improbable things as the growth of a tail; although theoretically useful for man, this is something that we know does not happen. Changes in the endocrine system, in the composition of the tissue or even in blood volume have no place in this list by definition. Last, but certainly not least, come the behavioural changes, which are far too numerous to list. For the possible metabolic adaptations there is a conflict of evidence, some finding that adaptations do occur (Johnson and Kark, 1947; Quenouille, Boyne, Fisher and Leitch, 1951) and others concluding that they may not (Consolazio, Konishi, Ciccolini, Jamison, Sheen and Steffen, 1960). Finally, each of the primary adaptations must be dependent on secondary changes, and for each we can make some logical deductions as to the sort of mechanism we should be trying to identify and investigate.

F. Natural versus Artificial Acclimatisation

At this point it is desirable to step back for a moment and to examine how far the findings from these highly controlled laboratory studies are really relevant to what happens when man goes to live and work in a hot climate. It has always been recognised that the changes resulting from acute and usually short exposures to heat in a laboratory may not be the same as the effects of living for long periods in a hot climate. It has therefore been customary to refer to the laboratory-induced response as "artificial" acclimatisation to heat, to distinguish it from the "natural" acclimatisation acquired by living in the tropics. It became so obviously important to test the comparability of "natural" and "artificial" acclimatisation to heat that two identical experiments were undertaken, one in England and the other at Singapore, to compare the responses of unacclimatised and acclimatised subjects to a highly standardised hot-room test (Hellon *et al.*, 1956). The conclusion reached was that "The superior ability to withstand hot environments exhibited by those who live in the tropics involves physiological as well as behavioural adaptation, and the physiological basis of this 'natural acclimatization' is identical with that of the 'artificial acclimatization' produced in the laboratory."

This conclusion appeared to have far-reaching practical implications, since there are many circumstances in which it would be a great advantage to be able to acclimatise to heat artificially before going to a hot country. The problem is comparatively recent and is due to the change from leisurely travelling by sea to fast travelling by plane. An athlete who wishes to compete in a hot country would find it convenient to acclimatise while he trained at home, instead of having to go to the hot country some time before the competition in order to get acclimatised. For obvious reasons it could also have great military importance, and it was this that stimulated further work. Although it had been shown that the physiological basis of "natural acclimatisation" is identical with that of "artificial acclimatisation", the extent to which artificial acclimatisation would confer the benefits of natural acclimatisation with its added advantages of behavioural adaptation was by no means certain. Artificial acclimatisation had been successfully used to prepare African labourers for the heat stress of African mines (Wyndham, Strydom, Morrison, Du Toit and Kraan, 1954), but this was a highly specialised application. Accordingly it was decided to compare the physiological responses and military performance of naturally acclimatised, artificially acclimatised and unacclimatised troops carrying out a realistic programme of military tasks in the hot Aden climate (Edholm, Fox, Adam and Goldsmith, 1963). In this large-scale field experiment there were fifty-four volunteers divided into three groups of eighteen. After an initial laboratory test of their heat responses, one group was sent to Aden to become naturally acclimatised by training in the heat, one group was sent to a cool climate in Scotland, and the third group was artificially acclimatised to heat by twenty-three days of 4 hr heat exposures in a climatic chamber. The three groups were then reassembled in the laboratory to retest their reactions to heat, and both the naturally and artificially acclimatised groups were found to be fully acclimatised. Immediately afterwards all three groups were flown to Aden and took part in an arduous military exercise for twelve days in severe heat. In the field the artificially acclimatised subjects had lower temperatures than those in either of the other two groups, indicating that their thermoregulatory mechanism was more efficient. In spite of this physiological advantage their performance, as judged by a team of observers, was poorer, and they had a higher incidence of casualties than the naturally acclimatised subjects. Any advantage over the unacclimatised subjects was only obvious during the early part of this test. When discussing these results, the authors' comment that "the findings strongly indicate that natural acclimatization in the hot region in which the subjects will have to work confers additional benefits to those of simply training the thermoregulatory mechanisms. A greater freedom from minor ailments, such as skin conditions and diarrhoea, are definite examples encountered in this study; but less easily defined factors, such as familiarity with the terrain and how to do things and live in a particular hot climate with the minimum of strain, almost certainly play a part."

G. Acclimatisation by the Indigenes of Hot Countries

It is perhaps not entirely surprising to find that physiological acclimatisation to heat itself is only a part of the story of successful adjustment to tropical life. It certainly fits in well with the scanty knowledge that has been gleaned to date from the indigenous populations of hot places. Ladell (1957) has pointed out that the indigenous populations of hot countries are not usually fully acclimatised to heat, as judged by their responses to standard hot-room tests. This observation is supported by the work of many others (Weiner, 1950; Adam, Ellis and Lee, 1953; Hellon *et al.*, 1956; Strydom and Wyndham, 1963). Ladell therefore prefers to call the response produced by hot-room exposures "hyperacclimatisation". However, full acclimatisation can develop as the result of natural exposure, provided enough physical work is done in the heat (Edholm *et al.*, 1963), and it seems that, in his natural setting and left to his own devices, man does not normally choose to exert or expose himself sufficiently to develop his capacity for heat acclimatisation to the full, just as most of us do not choose a way of life that will develop our physical prowess or muscular strength to the full. There is one caveat that must be entered to this conclusion. It may be that there are essential differences between short-term and long-term response patterns of heat acclimatisation. Some of the primary adaptations, such as sweating, may develop more quickly than others on exposure to heat stress; in the acute heat exposures that have been studied in the laboratory these quickly developing components would appear the more important. The evidence that we now have makes it unlikely that these differences are large, but we certainly cannot exclude the possibility that they exist.

The problem of defining the natural state of heat acclimatisation of indigenous populations in the tropics has unfortunately become almost inextricably confused with the separate issue of whether there are true ethnic differences in the responses to heat stress.

So far most of the studies concerned with the problem of ethnic differences have depended on the conventional technique of exposing groups of individuals to some known hot climate and, from the sweat rate, body temperature and pulse rates observed, trying to interpret whether the groups differ (Robinson *et al.*, 1941; Ladell, 1950; Baker, 1958; Strydom and Wyndham, 1963). One of the difficulties inherent in the conventional technique – the problem of comparing results from different climates – has already led to arguments on interpretation (see *Fed. Proc.*, 1963, **22**, 808). Investigators in this field clearly have to be extremely careful to design experiments that will distinguish between the effects of the preceding pattern of energy expenditure and true ethnic differences in response. This is not going to be easy, and it may prove as difficult and baffling as trying to discover how it is that two individuals of the same race, age, occupation and physical build, living in the same climate, can differ so widely in their responses to heat.

REFERENCES

Aaron, H. (1911). *Philipp. J. Sci.* **6**, 101.
Adam, J. M., Ellis, F. P., and Lee, T. S. (1953). *M.R.C.* R.N.P. Rep. No. 53/749.
Aitken, J. (1887). *Proc. roy. Soc. Edinb.* **12**, 661.

Aronsohn, E., and Sachs, J. (1885). *Pflüg. Arch. ges. Physiol.* 37, 232.
Baker, P. (1958). *Amer. J. phys. Anthrop.* **16**, 207.
Barcroft, J., Binger, C. A., Bock, A. V., Doggert, J. H., Forbes, H. S., Harrop, G., Meakins, J. C., and Redfield, A. C. (1922). *Phil. Trans.* 211 : **13**, 351.
Bass, D. E. (1963). *In* "Temperature – Its Measurement and Control in Science and Industry", Vol. 3, Part 3, Reinhold Publishing Corp., New York.
Bass, D. E., Buskirk, E. R., Iampietro, P. F., and Mager, M. (1958). *J. appl. Physiol.* **12**, 186.
Bass, D. E., and Henschel, A. (1956). *Physiol. Rev.* **36**, 128.
Bass, D. E., Kleeman, C. R., Quinn, M., Henschel, A., and Hegnauer, A. H. (1955). *Medicine, Baltimore*, 34, 323.
Bazett, H. C. (1927). *Physiol. Rev.* **7**, 531.
Bazett, H. C., Sunderman, F. W., Doupe, J., and Scott, J. C. (1940). *Amer. J. Physiol.* **129**, 69.
Bean, W. B., and Eichna, L. W. (1943). *Fed. Proc.* **2**, 144.
Bedford, T. (1936). *Rep. industr. Hlth Res. Bd Lond.* No. 76.
Bedford, T. (1946). *M.R.C. (War) Memor.* No. 17. H.M.S.O., London.
Bedford, T., and Warner, C. G. (1934). *J. Hyg. Camb.* **33**, 330.
Belding, H. S., and Hatch, T. F. (1955). *Heat. Pip. Air Condit.* **27**, 129.
Benson, R. S., Colver, T., Ladell, W. S. S., McArdle, B., and Scott, J. W. (1945). *M.R.C. (War) Memor.* R.N.P. Rep. No. 45/205.
Blagden, C. (1775a). *Phil. Trans.* **65**, 111.
Blagden, C. (1775b). *Phil. Trans.* **65**, 484.
Blockley, W. V., and Taylor, C. L. (1949). *Heat. Pip. Air. Condit.* **21**, 111.
Boileau, J. P. (1878). *Lancet* i, 413.
Brebner, D. F., and Kerslake, D. McK. (1963). *J. Physiol.* **166**, 13P.
Brodie, B. (1812). *Phil. Trans.* **102**, 387.
Collins, K. J., Crockford, G. W., and Weiner, J. S. (1963). *J. Physiol.* **169**, 12P.
Conley, C. L., and Nickerson, J. L. (1945). *Amer. J. Physiol.* **143**, 373.
Conn, J. W., and Johnston, M. W. (1944). *J. clin. Invest.* **23**, 933.
Conn, J. W., Johnston, M. W., and Louis, L. H. (1946). *J. clin. Invest.* **25**, 912.
Consolazio, C. F., Konishi, F., Ciccolini, R. V., Jamison, J. M., Sheen, E. J., and Steffen, W. F. (1960). *Metabolism* **9**, 435.
Daley, C., and Dill, D. B. (1937). *Amer. J. Physiol.* **118**, 285.
Davy, J. (1839). *In* "Researches Physiological and Anatomical", 1, p. 161. Smith and Elder: London.
Davy, J. (1850). *Phil. Trans.* **140**, 437.
Dill, D. B. (1938). *In* "Life, Heat and Altitude". Harvard University Press, Cambridge, Mass.
Dill, D. B., Hall, F. G., and Edwards, H. T. (1938). *Amer. J. Physiol.* **123**, 412.
Dill, D. B., Jones, B. F., Edwards, H. T., and Oberg, S. A. (1933). *J. biol. Chem.* **100**, 755.
Dufton, A. F. (1929). *J. sci. Instrum.* **6**, 249.
Dufton, A. F. (1932). *Build. Res. Tech. Paper* No. 13.
Dufton, A. F. (1936). *J. Instn. Heat. Vent. Engrs.* **4**, 227.
Dunham, W., Holling, H. E., Ladell, W. S. S., McArdle, B., Scott, J. W., Thomson, M. L., and Weiner, J. S. (1946). *M.R.C. (War) Memor.* R.N.P. Rep. No. 46/316.
Edholm, O. G., Fox, R. H., Adam, J. M., and Goldsmith, R. (1963). *Fed. Proc.* **22**, 709.
Eichna, L. W., Bean, W. B., Ashe, W. F., and Nelson, N. (1945). *Johns Hopk. Hosp. Bull.* **76**, 25.
Eichna, L. W., Park, C. R., Nelson, N., Horvath, S. M., and Palmes, E. D. (1950). *Amer. J. Physiol.* **163**. 585.

Eijkman, C. (1895). *Virchows Arch.* **140**, 125.
Ellis, H. (1758). *Phil. Trans.* **50**, 755.
Forbes, W. H., Hall, F. G., and Dill, D. B. (1940). *Amer. J. Physiol.* **130**, 739.
Foster, M. (1924). *In* "Lectures on the History of Physiology during the 16th, 17th and 18th Centuries". Cambridge University Press, Cambridge.
Fox, R. H., Goldsmith, R., Hampton, I. F. G., and Lewis, H. E. (1962). *J. Physiol.* **162**, 59P.
Fox, R. H., Goldsmith, R., Kidd, D. J., and Lewis, H. E. (1963a). *J. Physiol.* **166**, 530.
Fox, R. H., Goldsmith, R., Kidd, D. J., and Lewis, H. E. (1963b). *J. Physiol.* **166**, 548.
Furnell, M. C. (1878). *Lancet* ii, 110.
Haines, G. F., and Hatch, T. F. (1952). *Heat. & Ventilating*, **49**, 93.
Hardy, J. D. (1961). *Physiol. Rev.* **41**, 521.
Heberden, W. (1826). *Phil. Trans.*, Ser. B. **116**, 69.
Hellon, R. F., Jones, R. M., Macpherson, R. K., and Weiner, J. S. (1956). *J. Physiol.* **132**, 559.
Hellon, R. F., and Lind, A. R. (1955). *J. Physiol.* **128**, 57P.
Henschel, A., Taylor, H. L., and Keys, A. (1943). *Amer. J. Physiol.* **140**, 321.
Hill, L., Flack, M., McIntosh, J., Rowlands, R. A., and Walker, H. B. (1913). *Smithson. misc. Coll.* **60**, No. 23, Publication No. 2170.
Horvath, S. M., and Shelley, W. B. (1946). *Amer. J. Physiol.* **146**, 336.
Houghten, F. C., and Yagloglou, C. P. (1923). *Trans. Amer. Soc. Heat. Vent. Engrs.* **29**, 163.
Houghten, F. C., and Yagloglou, C. P. (1924). *Trans. Amer. Soc. Heat. Vent. Engrs.* **30**, 193.
Ito, S., and Adachi, J. (1934). *J. orient. Med.* **21**, 93.
Johnson, R. E., and Kark, R. M. (1947). *Science* **105**, 378.
Jousset, A. *In* "Traite de l'acclimatement et l'acclimatation". Paris, 1884.
Ladell, W. S. S. (1945). *Brit. med. Bull.* **3**, 175.
Ladell, W. S. S. (1950). Abstract of communication of XVIII International Physiological Congress, Copenhagen.
Ladell, W. S. S. (1951). *J. Physiol.* **115**, 296.
Ladell, W. S. S. (1957). UNESCO Arid Zone Research VIII. Human and Animal Ecology. Reviews of Researches. Paris.
Ladell, W. S. S., and McArdle, B. (1945). *M.R.C. (War) Memor.* R.N.P. Rep. No. 45/229.
Lavoisier, A. L. (1778). *Hist. Acad. roy. Sci.* 520.
Lavoisier, A. L. (1784a). *Hist. Acad. roy. Sci.* 355.
Lavoisier, A. L. (1784b). *Hist. Acad. roy. Sci.* 448.
Lavoisier, A. L. (1785). *Hist. Acad. roy. Sci.* 530.
Lind. A. R., Hellon, R. F., Weiner, J. S., Jones, R. M., and Fraser, D. C. (1957). *National Coal Board M. R. Memor.* No. 1. N.C.B., London.
Lind, A. R. (1963). *Fed. Proc.* **22**, 891.
McArdle, B., Dunham, W., Holling, H. E., Ladell, W. S. S., Scott, J. W., Thomson, M. L., and Weiner, J. S. (1947). *M.R.C. (War) Memor.* R.N.P. Rep. No. 47/391.
McCance, R. A. (1938). *J. Physiol.* **92**, 208.
Macpherson, R. K. (1960). *M.R.C.* Spec. Rep. Ser. No. 298.
Manson, P. (1898). *In* "Tropical Diseases. A manual of the diseases of warm climates". Cassell, London.
Minard, D., Belding, H. W., and Kingston, J. R. (1957). *J. Amer. med. Ass.*, **165**, 1813.
Moreira, M., Johnson, R. E., Forbes, A. P., and Consolazio, F. (1945). *Amer. J. Physiol.* **143**, 169.
Neuhauss, R. (1893). *Virchows Arch*, **134**, 35.

Quenouille, M. H., Boyne, A. W., Fisher, W. B., and Leitch, I. (1951). *Tech. Commun. Bur. Anim. Nutr. Aberd.* No. 17.
Rathay, A. (1870). *Proc. roy. Soc.* 18, 513.
Rathay, A. (1871). *Proc. roy. Soc.* 19, 295.
Robinson, S., Dill, D. B., Wilson, J. W., and Nielsen, M. (1941). *Amer. J. trop. Med.* **21**, 261.
Robinson, S., Kincaid, R. K., and Rhamy, R. K. (1950). *J. Appl. Physiol.* **2**, 399.
Robinson, S., Turrell, E. S., Belding, H. S., and Horvath, S. M. (1943). *Amer. J. Physiol.* **140**, 168.
Scott, J. C., Bazett, H. C., and Mackie, G. C. (1939). *Amer. J. Physiol.* **129**, 102.
Shaklee, A. O. (1917). *Philipp. J. Sci.* **12**, 1.
Smith, F. E. (1955). *M.R.C.* Spec. Rep. No. 29. H.M.S.O. London.
Strydom, N. B., and Wyndham, C. H. (1963). *Fed. Proc.* **22**, 801.
Sundstroem, E. S. (1927). *Physiol. Rev.* **7**, 320.
Taylor, H. L., Henschel, A., and Keys, A. (1943). *Amer. J. Physiol.* **139**, 583.
Taylor, H. L., Henschel, A., Michelsen, O., and Keys, A. (1943). *Amer. J. Physiol.* **140**, 439.
Thornley, J. G. (1878). *Lancet* i, 554.
Van Heyningen, R., and Weiner, J. S. (1952). *J. Physiol.* **116**, 395.
Vernon, H. M. (1930). *J. Physiol.* **70**, 15P.
Vernon, H. M. (1932). *J. Industr. Hyg.* **14**, 95.
Vernon, H. M., and Warner, C. G. (1932). *J. Hyg., Camb.* **32**, 431.
Weiner, J. S. (1950). *Brit. J. industr. Med.* **7**, 17.
Whitney, R. J. (1954). *J. Physiol.* **125**, 1.
Wick. W. (1910). *Arch. Schiffs- u. Tropenhyg.* 14, 605.
Wilkinson, R. T., Fox, R. H., Goldsmith, R., Hampton, I. F. G., and Lewis, H. E. (1964). *J. appl. Physiol.* 19, 287.
Winslow, C.-E. A. (1941). *In* "Temperature, Its Measurement and Control in Science and Industry". Vol. 1, p. 509. Reinhold Publishing Corporation, New York.
Winslow, C.-E. A., Herrington, L. P., and Gagge, A. P. (1937). *Amer. J. Physiol.* **120**, 288.
Winslow, C.-E. A., Herrington, L. P., and Gagge, A. P. (1938). *Amer. J. Physiol.* **124**, 692.
Wood, J. E., and Bass, D. E. (1960). *J. clin. Invest.* **39**, 825.
Wyndham, C. H. (1951). *J. appl. Physiol.* 4, 383.
Wyndham, C. H., Strydom, N. B., Morrison, J. F., Du Toit, F. D., and Kraan, J. G. (1954). *Arbeitsphysiologie* **15**, 373.
Yagloglou, C. P., and Miller, W. E. (1925). *Trans. Amer. Soc. Heat. Vent. Engrs.* 31. 89.
Yaglou, C. P., and Minard, D. (1957). *Arch. Indust. Hlth.* 16, 302.
Young, W. J. (1915). *J. Physiol.* 49, 222.
Young, W. J., Breinl, A., Harris, J. J., and Osborne, W. A. (1919). *Proc. roy. Soc.* B, **91**, 111.

CHAPTER 4

Acute Anoxic Anoxia

J. DONALD HATCHER

A. Introductory

1. *Classification and terminology*

Anoxia is a term that indicates a deficiency of oxygen in the body. It can result from various causes, whose nature led Barcroft (1920) to recognise three main types – anoxic anoxia, anaemic anoxia, and stagnant anoxia. To this, a fourth kind, histotoxic anoxia, was added by Peters and Van Slyke (1931). Regardless of the aetiology of anoxia, the ultimate defect is a reduced oxygen tension in the tissues; when the condition is severe enough, there is an interference with the oxidative processes in the cells.

There are many who would insist that hypoxia is a more precise generic term, since they consider it to imply a reduction in the amount of oxygen rather than its absence. However, as Lambertsen (1961a) points out, the term hypoxia is less precise than anoxia, because the elision in the former's construction leaves it unclear whether hypo-oxia or hyper-oxia is meant. On these grounds, and on those of historical precedent, it is held that the term anoxia should be retained and used to refer to any deficiency of oxygen in the body: similarly, in referring to reduced oxygen in the blood, the term anoxaemia rather than hypoxaemia is used.

2. *Occurrence*

In anoxic anoxia, tissue oxygenation is reduced as a result of a primary reduction in arterial oxygen tension. It is encountered as a result of, first, the inhalation of air containing a reduced tension of oxygen. This may occur at high altitudes in mountaineers and in those living permanently in mountainous regions. It may be experienced by aviators if oxygen equipment fails or cabin pressure is lost. These situations can be simulated under controlled conditions in the laboratory by having experimental subjects breathe an oxygen-deficient gas mixture or by the use of a decompression chamber. Secondly, it may arise from abnormalities of the heart or pulmonary circulation in which, as a result of venous-arterial communications, venous blood is mixed with arterial blood. Thirdly, interference with normal respiratory function may occur, i.e. mechanical obstruction of the airway, paralysis of respiratory muscles, drowning and a variety of pathological processes that interfere with gas exchange at the alveloar capillary membranes. Anoxia due to any one of these causes may be associated with hypercapnia when a state of asphyxia exists. The abnormality must be severe before carbon dioxide retention occurs because of the hyperventilation associated with anoxia and the high diffusion constant of carbon dioxide.

It is recognised that in these various circumstances anoxia may be accompanied by additional stresses. The hyperventilation that ensues on sudden exposure to a low oxygen tension results in hypocapnia and alkalosis. The reverse change—hypercapnia and acidosis—can also occur, as described above. In respiratory disease fever may be present; this, by increasing the metabolic demand for oxygen, will increase the body's sensitivity to anoxia. At high altitudes, one must consider besides anoxia the effects of low environmental temperature. At the high speeds of modern aircraft and space vehicles, combinations of anoxia and high environmental temperature may be experienced. Explosive decompression of pressurised aircraft results in effects peculiar to the reduced total atmospheric pressure, although the degree of the associated anoxia is the major factor in determining survival.

The content of this chapter will be limited to a discussion of some of the effects of rapidly induced or acute anoxic anoxia in adult man. No attempt is made to deal with the effects of long-term or chronic anoxic anoxia; this is considered in Chapter 5. The conditions under examination are those associated with inhalation of air containing reduced tensions of oxygen, such as occur on sudden exposure to high altitudes and in respiration defects of rapid onset. Particular emphasis will be placed on the mechanisms that may be responsible for some of the major cardiovascular and respiratory compensatory changes essential if one is to survive sudden and severe anoxia. This, of necessity, will require a discussion of investigations carried out on species other than man. Some consideration will be given also to the modifying effects of such associated environmental factors as those mentioned above.

3. *Methods of quantitative expression*

In experiments carried out at high altitudes or in decompression chambers, the degree of anoxia is often expressed as altitude (in feet or metres), as total barometric pressure in mm Hg or as partial pressure of oxygen in mm Hg. The relationship between these three is shown in Fig. 1. In experiments at sea-level, anoxia is often produced by having subjects breathe gas mixtures deficient in oxygen, when the degree of anoxia is often expressed as the

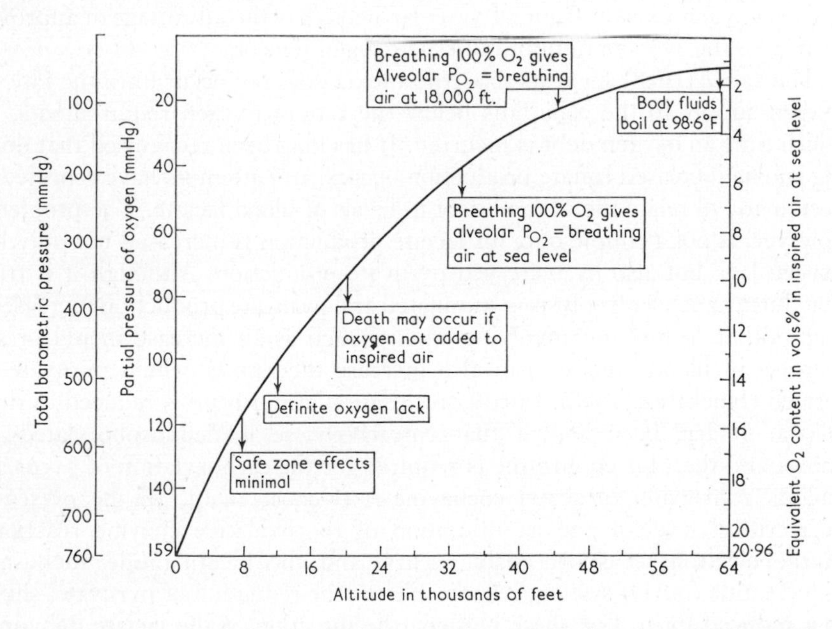

FIGURE 1. Curve relating the total barometric pressure (based on the International Commission for Air Navigation formula) with the partial pressure of oxygen at different altitudes between sea-level and 64,000 ft. Reductions in the percentage oxygen of the inspired air at sea-level equated with the oxygen tension observed at various altitudes.

percentage of oxygen in the mixture. It must be remembered that at high altitude the percentage amount of oxygen in the atmosphere is not altered, but only its partial pressure. At sea-level, reducing the percentage amount of oxygen in the air also alters its partial pressure, and in Fig. 1 the volume percentage of oxygen below the normal of 20·9 are equated with the partial pressure of oxygen and the total barometric pressures at altitudes between sea-level and 19,500 m (64,000 ft).†

There is a correlation between the partial pressure of oxygen in inspired air and the percentage saturation and tension of oxygen produced in arterial blood; however, largely owing to differences in respiratory response, the variation in blood oxygen tension between individuals at a given oxygen tension in inspired air is too great for reliable prediction (Houston, 1946; Dripps and Comroe, 1947). Thus use of the partial pressure of oxygen in inspired air or its equivalent should be confined to describing the conditions of the experiment.

A preferable measurement of the degree of anoxia is the percentage saturation or, better, the tension of oxygen in arterial or mixed venous blood. The oxygen tension in arterial blood approximates to the maximum tissue oxygen tension. The oxygen tension in mixed venous blood from the right heart or pulmonary artery, since it provides a physiologically integrated mean venous oxygen tension from all vascular beds, has the advantage of approximating to the average minimum tissue oxygen tension.

Huckabee (1958) suggests that true anoxia does not occur until the rate of oxygen supply to the cells falls below the rate of oxygen requirements, at which time an oxygen debt is incurred. It has long been recognised that during anoxia increased lactate production occurs, and attempts have repeatedly been made to relate the oxygen debt to levels of blood lactate. The problem, however, is not a simple one, for lactate production is increased not only by oxygen lack but also by increased pyruvate production. Although it is true that during anoxia glycolysis is stimulated and pyruvate production increased, many other factors not involving anoxia, such as an increase in pH or an increase in blood glucose, can also increase the rate at which pyruvate is formed (Huckabee, 1958). Further, utilisation of pyruvate is reduced during anoxia. It has been shown that co-carboxylase is dephosphorylated in anoxia and that this co-enzyme is required for the carboxylation of pyruvate and its conversion to acetyl coenzyme-A (Ochoa, 1939). In the presence of a critical level of anoxia, inhibition of the oxidative enzyme reactions in the electron transport chain occurs, and the nicotinamide adenosine dinucleotide (NAD) system, responsible for the reduction of pyruvate, shifts to a reduced form. For these reasons, the direction of the lactate dehydrogenase (LDH) system shifts to the right, with the production of increased lactate and a small amount of oxidised NAD (Eq. 1):

† See footnotes on p. 121.

$$\text{pyruvate} + \text{NADH} + \text{H}^+ \underset{}{\overset{\text{LDH}}{\rightleftarrows}} \text{lactate} + \text{NAD}^+ \quad (1)$$

In passing, it is worth pointing out that the amount of oxidised NAD formed in this way, although small in proportion to requirements, could play a role in prolonging life during severe anoxia.

On the basis of these considerations Huckabee (1958) states that any increase in lactate greater than that expected from pyruvate production alone can be considered to be the result of oxygen lack. He called such an increase, excess lactate (XL) and found that during anoxia excess lactate formation corresponds closely to oxygen debt. The calculation of excess lactate is based on the equation

$$XL = (L_n - L_o) - (P_n - P_o)\,\frac{(L_o)}{P_o}, \quad (2)$$

where L_o and P_o are basal control levels of lactate and pyruvate in arterial blood, and L_n and P_n are the levels of lactate and pyruvate in an arterial sample obtained at some specific time in the course of an experiment.

Huckabee (1958) has demonstrated that excess lactate, and thus true anoxia, does not appear in man and anaesthetised dogs until the arterial oxygen saturation is reduced to 60 to 74%, which is equivalent to an arterial oxygen tension of 26 to 32 mm Hg. Below this level there is a close relationship between the degree of reduction in arterial oxygen tension and excess lactate production. Huckabee (1958) cautions against ascribing symptoms and signs to anoxia in the absence of excess lactate production. It must however, be pointed out that many of the physiological manifestations of anoxia, including compensatory changes in cardiac and respiratory function, occur at levels of arterial oxygen tension above the threshold for excess lactate production. Measurements of excess lactate in arterial blood represent an average for the whole body, and it is important, as Huckabee (1958) remarks, to remember that the threshold level of anoxia required for excess lactate production no doubt varies among different organs and tissues. The concepts presented by Huckabee have been criticised (Krasnow, Neill, Messer and Gorlin, 1962; Olsen, 1963), but are important contributions to the subject; the application of his findings to such problems as the perception of anoxia in receptor areas and the initiation of the complex physiological changes occurring in the anoxic animal must await further study. The problem of assessing the degree of anoxia is dealt with in section C.2 below.

B. Cerebral Effects

The most striking subjective and objective signs of anoxia are ascribed to effects on the central nervous system, where alterations are found in sensory, motor and behavioural function. The extent of these changes and their duration and severity clearly depend on the rate, degree and duration of anoxia and on the efficiency of the cardiovascular and respiratory compensatory adjustments. With mild anoxia these changes include headache, dizziness, fatigue, sleepiness and loss of visual acuity. As the severity increases the behavioural changes observed range in different individuals from depression to euphoria. Auditory perception decreases, and sensitivity to other external stimuli, including pain, is diminished. These changes are associated with failure of judgement, muscle weakness and lack of co-ordination, and the subject is unable to carry out simple manual tasks. The person who is becoming anoxic is not a good witness and cannot be relied on to recognise his difficulty, because of impaired judgement. Throughout these stages compensatory alterations in cardiovascular and respiratory function are taking place. Finally unconsciousness occurs with clonic and tonic convulsions, collapse and death (Armstrong, 1943a).

There is some variation from subject to subject in the level of anoxia at which occur the phenomena related to altered cerebral function. It should be noted with reference to Fig. 1 that the effects of anoxia are minimal or absent on rapid ascent to an altitude of 2450 m (8000 ft), but that above this level signs of impairment appear; consequently above 3650 m (12,000 ft) oxygen should be supplied (or cabins pressurised) to prevent these cerebral manifestations.

It is generally accepted that unconsciousness occurs at an arterial oxygen saturation of about 55 to 60% (Armstrong, 1943b; Hoffman, Clark and Brown, 1945–6). Unconsciousness is preceded by a period of altered consciousness, but the arterial saturation at which this is reported to appear seems to depend on the criteria used to assess it. Thus, by using the first mistake made on sorting cards, Hoffman *et al.* (1945–6) found signs of altered consciousness at an average arterial oxygen saturation of 64%, but using other criteria others report signs of altered consciousness at arterial saturations as high as 85% (Armstrong, 1943a). Armstrong (1943b) studied the time after sudden removal of oxygen at various altitudes when individuals showed signs of mental and physical deterioration and also the time of onset of unconsciousness. The results are presented in Fig. 2 for simulated altitudes between 6100 m (20,000 ft) and 12,200 m (40,000 ft). At 9150 m (30,000 ft) full consciousness lasts for a matter of seconds, and within 2 min coma occurs. At 7600 m (25,000 ft) the period of consciousness is extended to about 3 min, and the interval between the onset of altered consciousness and unconsciousness is prolonged. The gradation in time intervals between low and high altitudes is a reflection of the time required for the arterial saturations to reach the threshold values mentioned above. If one breathes 100% oxygen, altitude tolerance is vastly improved. Anyone breathing pure oxygen at an altitude of about 10,000 m (33,500 ft) has an alveolar oxygen tension similar to that noted at sea-level in those breathing ambient air, and at about

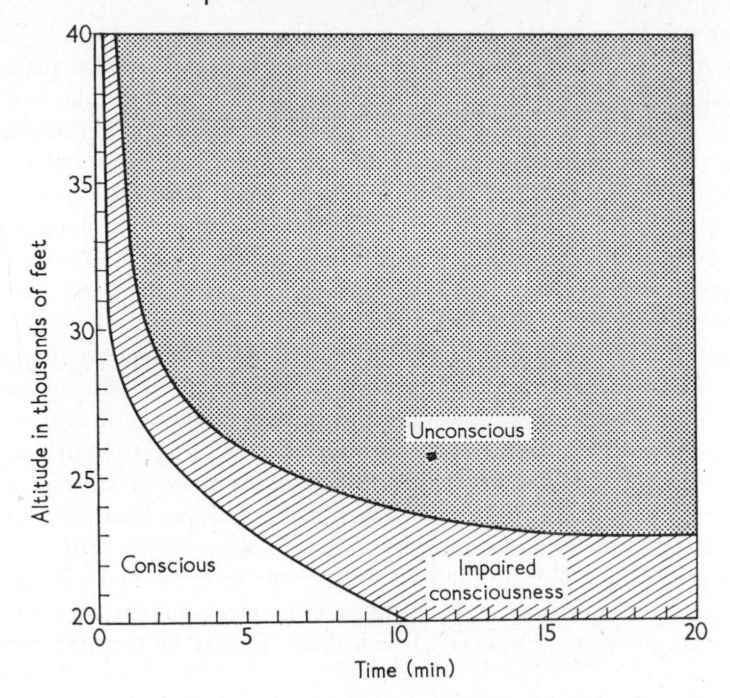

FIGURE 2. Duration of consciousness and of impaired consciousness at various "altitudes" between 20,000 and 40,000 ft. Redrawn from Armstrong (1943b).

13,500 m (44,000 ft) the alveolar oxygen tension is equivalent to that produced by breathing ambient air at 5,500 m (18,000 ft) (Fig. 1).

A relationship exists between the cerebral manifestations of anoxia and the electrical activity in the brain. During anoxia three phases of alteration in the electroencephalogram occur in man and laboratory animals. Although these phases have been recognised for a considerable time, their significance and mechanism were not appreciated until recently (Dell and Bonvallet, 1954; Hugelin, Bonvallet and Dell, 1959; Gastaut and Meyer, 1959). The sensory, motor and behavioural effects of oxygen-lack appear to involve the reticular and cortical diffuse systems rather than specific sensory and motor pathways (Gastaut and Meyer, 1959). The three phases of the changes noted in the electroencephalogram from the cortex during anoxia (Dell, Hugelin and Bonvallet, 1959) are described below.

Phase I: After a short latent period the electroencephalogram shows an arousal pattern indistinguishable from those produced by any alterations in sensory input. This arousal in anoxia is due to increased activity in the reticular formation, which reaches the cortex via ascending pathways. The increase in reticular activity is caused largely by increased afferent input from the chemoreceptors, for in their absence reticular activation is profoundly modified. It is possible that local effects of anoxia may play some role in this activation. The increased reticular activity coincides with the initiation of augmented respiratory activity and probably of certain cardiovascular changes. Activation of descending pathways from the reticular formation might be expected along with accompanying facilitation of motor phenomena, but these descending facilitatory effects are considered to be held in check at this stage by descending inhibitory impulses from the cortex, secondary to arousal. Thus at this stage there is a balanced negative feedback between the reticular and cortical systems.

Phase II: The cortex is much more susceptible to anoxia than lower parts of the brain, and signs of suppression of electrical activity appear first in the cortex as alterations of wave form and as δ-waves in the electroencephalogram. These electrical changes coincide with alterations in behaviour and in the level of consciousness. While cortical suppression occurs, reticular activity continues at the same or a higher level, but now, because of cortical suppression and the lack of inhibitory tone from these higher centres, alterations in motor function appear that may result in involuntary movements towards the end of this phase. If the onset of anoxia is sufficiently severe and sudden, the phenomena associated with Phase II may be the first alteration detected.

Phase III: At this stage the cortical electroencephalogram becomes silent, and only a flat trace is recorded. This indicates complete cortical suppression. The reticular formation, however, being less susceptible to oxygen deficiency, continues its activity, which, uninhibited now by the cortical system, results in marked motor facilitation via descending reticular pathways, and clonic and tonic convulsions appear.

Although all these changes described are referable to the decrease in oxygen tension, the hypocapnia occurring secondary to hyperventilation in anoxia facilitates these responses. Anoxia causes an increase in cerebral blood flow, but the beneficial effect of this in maintaining cerebral oxygen tension is modified by the extent of the hypocapnia, which has a vasoconstrictor action on cerebral vessels (see above). These interactions between the effects of oxygen and carbon dioxide on cerebral blood flow may explain the symptomatic improvement that can occur when carbon dioxide is added to air containing oxygen at a low tension (Lambertsen, 1961a).

C. Circulatory Changes

The major compensatory adjustments that occur in anoxia are related to the cardiovascular and respiratory systems. These are (*a*) increase in cardiac output; (*b*) redistribution of blood flow; (*c*) increases in rate and depth of respiration; and (*d*) increase in the oxygen-carrying capacity of the blood.

The extent to which these changes occur depends on a number of variables, the most important being the degree, rate of onset and duration of anoxia. However, size, age, metabolic rate, body temperature, environmental conditions and the presence of anaesthesia are also important in determining the extent of these adjustments.

1. *Cardiac output, heart rate and stroke volume*

In man, acute anoxia increases the cardiac output as determined by the Fick principle (Storstein, 1952; Goldring, Turino, Cohen, Jameson, Bass and Fishman, 1962), dye dilution technique (Asmussen and Nielsen, 1956; Fritts, Harris, Clauss, Odell and Cournand, 1958; Chidsey, Frye, Kahler and Braunwald, 1961; Goldring *et al.*, 1962), roentgen-kymograph technique (Keys, Stapp and Violante, 1943), acetylene method (Grollmann, 1930; Asmussen and Chiodi, 1941) and ballistocardiograph (Kety and Schmidt, 1948; Scarborough, Penneys, Thomas, Baker and Mason, 1951). The cardiac output rises when the arterial oxygen saturation falls below 80 to 85% (Korner, 1959), and the degree of rise in output appears directly related to the degree of fall in arterial oxygen saturation (Scarborough *et al.*, 1951; Korner, 1959). Rise of cardiac output in man is associated with a fall in arteriovenous oxygen difference (Storstein, 1952; Goldring *et al.*, 1962); the oxygen consumption on the average remains unchanged (Asmussen and Chiodi, 1941; Fishman, McClement, Himmelstein and Cournand, 1952).

In anaesthetised and unanaesthetised dogs the cardiac output also rises during acute anoxia (Harrison and Blalock, 1927; Lewis and Gorlin, 1952; Gorlin and Lewis, 1954; Nahas, Visscher, Mather, Haddy and Warner, 1954a; Baugh, Cornett and Hatcher, 1959). As in man, the rise is mainly due to an increase in heart rate, although an increase in stroke volume has also been observed (Harrison and Blalock, 1927; Hurlimann and Wiggers, 1953; Baugh *et al.*, 1959). The threshold of arterial oxygen saturation at which the increase in cardiac output occurs is about 70% saturation in the unanaesthetised dog and about 60% in the anaesthetised dog (Harrison and Blalock, 1927; Gorlin and Lewis, 1954). The response of cats appears similar to that of dogs (Doi, 1921).

Criticism has been expressed of the Fick technique when it is used for measuring cardiac output during acute anoxia in the absence of a steady state (Fishman *et al.*, 1952; Nahas *et al.*, 1954a; Fritts *et al.*, 1958). Criteria for a steady state include emotional stability and constancy of oxygen consumption and of respiratory exchange (R.Q.); in man this may not occur for 20 min or more (Fishman *et al.*, 1952). In anaesthetised dogs breathing 6% oxygen in nitrogen for 15 min we have found in our laboratory, from measurements of ventilation, blood R.Q. and oxygen consumption, that a steady state is established after 10 min and that the application of the Fick technique between 10 and 15 min is valid. Certainly, measurements of cardiac output made in the first 10 min of anoxia by the Fick method can result in errors (Nahas, Visscher and Haddy, 1953; Nahas *et al.*, 1954a). The importance of this problem can be readily appreciated by examining the changes in oxygen consumption during the first 15 min after an animal begins to inhale a gas mixture low in oxygen. The oxygen uptake regularly diminishes in dogs during the initial fall in blood-oxygen saturation and reaches a minimum, in our experience and that of others, within the first 5 min. After this the oxygen consumption gradually recovers and on the average reaches pre-anoxic control values at 10 min, by which time the saturation of the blood has levelled off at a new equilibrium (Hemingway and Nahas, 1952; Hemingway and Nahas, 1958; Huckabee, 1958). The degree of the initial transient fall in oxygen consumption appears to vary inversely with the percentage of oxygen in the inspired air (Hemingway and Nahas, 1952; Huckabee, 1958) and is considered to be due to the supply of oxygen coming from the desaturation of blood and body fluids (Huckabee, 1958).

2. *Influence of the metabolic rate and body temperature on the change in cardiac output*

Korner (1959) criticises experiments done on anaesthetised animals and points out that the degree of anoxia (generally below 50% arterial oxygen saturation) necessary to increase the cardiac output in the anaesthetised dog is rarely experienced by the unanaesthetised animal and is incompatible with prolonged survival. The level of metabolism is reduced by anaesthesia, and a greater degree of anoxaemia may be required in the anaesthetised than in the unanaesthetised dog to cause a level of tissue anoxia sufficient to evoke a rise in cardiac output.

In our experience not all dogs increase their cardiac output at the end of a 15 min period of breathing 6% oxygen. Indeed, in some the cardiac output may decrease to a slight or moderate degree, even though the saturation of arterial blood is reduced to or below the anoxaemic threshold at which an increase in cardiac output is normally elicited. None of these dogs showed signs of the acute cardiac crisis or acute cardiac failure occurring with extreme and prolonged anoxia, as described by Wiggers (1941). Dogs in which the cardiac output fell, or failed to increase, invariably showed a proportional reduction in oxygen consumption (steady state). In his recent review Korner (1959) points out that a decrease in cardiac output and oxygen consumption occurs in about one-third of rabbits exposed to 10% oxygen and that a marked decrease in oxygen consumption regularly occurs during anoxia in rats and guinea-pigs. It is recognised that anoxia alters temperature regulation and disturbs the ability of animals and man to maintain body temperature (Kottke, Phalen, Taylor, Visscher and Evans, 1948; Hemingway and Nahas,

1952, 1958). The decrease in rectal temperature during anoxia is more marked in small animals, such as mice, with a large ratio of surface area to body weight, than in larger animals, such as the dog and man. The decrease in body temperature is thought to involve an inhibition of heat production as well as increased heat loss (Hemingway and Nahas, 1952, 1958; Kottke *et al.*, 1948). Kottke and his colleagues conclude that the decrease in body temperature improves the tolerance of animals to anoxia, since at reduced body temperatures oxygen demands are also reduced. In dogs in which body temperature and metabolic rate were reduced by lowering environmental temperature, it was found that a given degree of anoxaemia was less effective in eliciting an increase in cardiac output and ventilation than when the same degree of anoxaemia was produced in euthermic dogs with a higher metabolic rate (Baumber and Hatcher, 1960; Hall and Salzano, 1963; Lim and Luft, 1963). At the opposite extreme are the observations of increased responsiveness to anoxia in dogs with an elevated metabolic rate secondary to an increase in body temperature (Baumber and Hatcher, 1963; Lim and Luft, 1963). Though the level of cardiac output and ventilation is then greater than that noted in euthermic anoxic dogs, the total load placed on the circulation is extreme, and acute cardiac failure is common (Baumber and Hatcher, 1963).

A common point in this discussion is the concept that the effectiveness of a given degree of anoxaemia as a stimulus to cardiovascular and respiratory compensatory changes should be viewed in the light of the metabolic rate of the animal. In an attempt to quantitate the degree of anoxic stress on these terms we have found it useful, particularly in studies of anoxia at different body temperatures, to calculate an anoxic index in arbitrary units as the ratio of oxygen consumption to the oxygen tension in mixed venous blood. Although oxygen consumption need not reflect the total metabolic requirement for oxygen in severe anoxia (Huckabee, 1958), such a ratio has a rough quantitative value. It would appear from the work of Huckabee that the accuracy and usefulness of such an index would be greatly improved if there were added to the oxygen uptake observed the oxygen equivalent of any excess lactate production, the latter providing a measure of the oxygen debt.

3. *Mechanisms responsible for elevation in cardiac output*

The mechanisms responsible for the increase in heart rate and stroke volume that lead to an increase in cardiac output have been considered by various investigators to be due to (*a*) a direct action of anoxia on the myocardium; (*b*) dilatation of the peripheral vessels secondary to the direct effects of anoxia on these vessels; (*c*) the Starling mechanism; (*d*) nervous mechanisms; and (*e*) a humoral mechanism.

a. Direct action on the myocardium. Harrison and Blalock (1927) and Strughold (1930) considered that anoxia increased the cardiac output by a direct action in the heart. This conclusion is contradicted by the results of experiments on the isolated mammalian heart and the heart-lung preparation, in which a depressant action of anoxaemia has been observed (Gremels and Starling, 1926; Bogue, Chang and Gregory, 1938; Lorber and Evans, 1943; Giotti and Beani, 1957; Kahler, Goldblatt and Braunwald, 1962).

b. Dilatation of the peripheral vessels. Several workers have postulated that the increase in cardiac output is secondary to a reduction in total systemic resistance (Harrison and Blalock, 1927; Feldman, Rodbard and Katz, 1948; Doyle, Wilson and Warren, 1952), the peripheral vasodilatation being caused by the direct effects of anoxia on peripheral vessels (Krogh, 1929). Gomori, Kovack, Takacs, Foldi, Szabo, Nagy, Wiltner and Kallay (1960a) accept this view, but suggest that the vasodilatation leads to an increase in the venous return and a consequent increase in cardiac output.

Baugh *et al.* (1959) found no decrease in total systemic resistance and no elevation in cardiac output in anoxic adrenalectomised dogs, despite a level of anoxia equivalent to that seen in intact anoxic dogs, in which a decrease in total systemic resistance and an increase in cardiac output occurred. They concluded that a decrease in resistance resulting from the direct effects of anoxia on peripheral vessels was not a primary mechanism. Further, some workers have noted an increase in cardiac output in anoxic dogs in the absence of any significant change or even with an increase in systemic resistance (Hurlimann and Wiggers, 1953; Stroud and Conn, 1954; Kahler *et al.*, 1962, Downing, Remensnyder and Mitchell, 1962; Penna, Soma and Aviado, 1962). It is significant that in these studies, with the exception of those by Stroud and Conn, artificially ventilated dogs were used, and the normal respiratory response and hypocapnia could not occur. Under these experimental conditions peripheral vasoconstriction is enhanced (Bernthal and Schwind, 1945; *v.* section C.4). In the experiments of Stroud and Conn (1954) the dogs inhaled 5% oxygen and an elevation in cardiac output of only 15% occurred, which is considerably less than is normally seen. The cardiac output was measured by use of ^{42}K in an isotope dilution technique, which could give low values because of loss of potassium from the circulation.

c. The Starling Mechanism. Gomori *et al.* (1960a) suggested that an increased venous return may be involved in determining the increase in cardiac output in anoxia, presumably by the Starling mechanism. There is no evidence for an increase of either right or left ventricular end-diastolic pressures or of effective pressure in the inferior and superior vena cavae and right atrium during anoxia; indeed, these pressures may even fall slightly (Sands and DeGraff, 1925; Lewis and Gorlin, 1952; Nahas, Visscher, Mather, Haddy and Warner, 1954a; Nahas, Josse and Muchow, 1954b, Baugh, Cornett and Hatcher, 1959). Keys, Stapp and Violante (1943), using the roentgen-kymograph technique, found no change in diastolic size of the heart during acute anoxia in healthy young men. An increase in the venule-vein pressure gradient and venoconstriction have been described which would favour venous return and an adequate level of cardiac filling for the increased cardiac output during anoxia (Nahas *et al.*, 1954b; Kahler *et al.*, 1962; Eckstein and Horsley, 1960). In the absence of an increase in central venous pressure the Starling mechanism was not considered by Nahas *et al.* (1954b) to be operative.

On the other hand, Downing *et al.* (1962) compared ventricular function curves by the technique of Sarnoff and Berglund (1954) before and during anoxia in open-chest dogs with controlled heart and respiratory rates. They found the ventricular function curve to be displaced to the left during anoxia, which means that for any given end-diastolic pressure a greater stroke work was attained. This change could be the result of nervous or humoral influences or both and would not necessarily be reflected by a change in cardiac filling pressures.

d. Nervous factors: (i) efferent cardiac nerves. As mentioned above, the increase in cardiac output is regularly associated with a tachycardia in both man and dogs, as in some instances is an increase in stroke volume. The role of the efferent nerves to the heart in these changes has been studied by various investigators.

Harrison, Blalock, Pilcher and Wilson (1927) found that sympathectomy (stellate to T-9) did not alter the increase in cardiac output or heart rate in dogs during anoxia, whereas Nahas, Mather, Wargo and Adams (1954c) noted that the increase in cardiac output and heart rate was smaller after sympathectomy (C-2 to T-8). Nahas (1956) also noted that vagotomy modified the increase in heart rate during anoxia, but did not affect the rise in cardiac output. It should be pointed out that the resting heart rate in these vagotomised animals was significantly elevated and that the absolute rate reached during anoxia was of the same order as in intact anoxic dogs. Alexander (1945) and Downing and Siegel (1963) detected an increased number of impulses in the inferior cardiac nerves during acute anoxia.

The sympathectomy procedures carried out by Harrison *et al.* (1927) and Nahas *et al.* (1954c) also denervate the adrenal gland. Parker, Baugh and Hatcher (1958) have shown that splanchnicectomy alone prevents the increase in cardiac output in acute anoxia. In this connection Woods and Richardson (1959) found that the increase in ventricular contractile force (strain-gauge arch) in open-chest bilaterally vagotomised dogs breathing 100% nitrogen for 2 min was reduced if the dogs were adrenalectomised and sympathectomised from stellate to T-7 before the induction of anoxia.

Chidsey, Frye, Kahler and Braunwald (1961) showed that pharmacological sympathetic denervation of the heart with the compound syrosingopine, which has a reserpine-like activity, did not prevent in man the increase in heart rate and cardiac output seen in acute anoxia. Harris, Bishop and Segel (1961) found that guanethedine, which blocks post-ganglionic sympathetic pathways, given in a dose sufficient to prevent the overshoot in blood pressure after the Valsalva manoeuvre, did not affect the rise in cardiac output and heart rate occurring in human subjects inhaling 13% oxygen.

Thus to date there seems to be no unanimity of opinion as to the role of the cardiac nerves in the cardiac output response to acute anoxia.

d. Nervous factors: (ii) the role of chemoreceptors. Many workers in the past have ascribed the tachycardia present during acute anoxia to activation of the chemoreceptor reflexes (Asmussen and Chiodi, 1941; von Euler and Liljestrand, 1942; Dripps and Comroe, 1947).

Perfusion of the isolated carotid bifurcation with venous blood in spontaneously breathing animals induces hypertension, hyperpnoea and tachycardia (Bernthal, Greene and Revzin, 1951; Daly and Scott, 1958, 1959). Though there is no doubt about the role of chemoreceptor stimulation in determining the hyperpnoea and the level of blood pressure in such an experiment, there is considerable evidence against involvement of a similar afferent mechanism in the tachycardia. When the respiration is artificially controlled during perfusion of the isolated carotid bifurcation with anoxic blood, a bradycardia is regularly seen (Bernthal *et al.*, 1951; Daly and Scott, 1958, 1959; Downing *et al.*, 1962) and is due to a reflex increase in efferent vagal activity and a decrease in efferent sympathetic activity to the heart (Downing *et al.*, 1962). Thus, as Downing and colleagues point out, with carotid body anoxia there is a dichotomous response of the sympathetic nervous system with increased sympathetic discharge to peripheral vessels, but decreased sympathetic discharge and increased vagal activity to the heart. This is in striking contrast with the increased sympathetic discharge to both heart and vessels that occurs with carotid sinus hypotension (Downing *et al.*, 1962).

The fact that tachycardia during perfusion of the isolated carotid bifurcations with anoxic blood is seen only in the presence of hyperpnoea led Bernthal *et al.* (1951) and Daly and Scott (1958, 1959) to suggest that the tachycardia in anoxia may be due to reflexes originating in the lungs, such as are described by Anrep, Pascual and Rossler (1935–6), which antagonise the primary reflex bradycardia caused by anoxia of the carotid bodies. Against this hypothesis are the observations that perfusion of the carotid body with oxygenated Ringer-Locke solution in anoxic cats (Neil, 1956) or with oxygenated blood in anoxic dogs (Daly and Scott, 1959) induces a reduction in respiration, but no change or even an increase in the level of tachycardia. Further, tachycardia and an increase in cardiac output occur during whole body anoxia when respiratory activity is mechanically controlled (Downing *et al.*, 1962; Penna *et al.*, 1962; Murray and Young, 1963).

Hyperventilation during anoxia results in hypocapnia, and the possibility that a reduced blood carbon dioxide level might play a role in the cardiac output elevation is suggested by the work of some investigators. Black and

Roddie (1958) observed that the increase in heart rate occurring in human subjects during the inhalation of 5 to 10% oxygen for up to 6 min could be modified by adding carbon dioxide to the anoxic gas mixture. An increase in heart rate and cardiac output is observed in artificially ventilated anoxic dogs, in which presumably no hypocapnia occurs (Penna, Soma and Aviado, 1962; Downing, Remensnyder and Mitchell, 1962; Murray and Young, 1963). However, the work of Keys, Stapp and Violante (1943) indicates that the rise in cardiac output during acute anoxia is less marked when the hypocapnia is prevented. There is a growing body of evidence based mainly on the work of Downing and colleagues to suggest that anoxia may initiate tachycardia and cardiac output elevation by a direct action on the central nervous system (Downing *et al.*, 1962; Downing, Mitchell and Wallace, 1963; Downing and Siegel, 1963). The modification of the tachycardia and rise in cardiac output seen when hypocapnia is prevented may possibly be related to an improved cerebral blood flow and oxygenation (see next section, C.6).

Recently some investigators (Woods and Richardson, 1959; Downing *et al.*, 1962; Kahler *et al.*, 1962) have examined the role of chemoreceptor reflexes in the increased cardiac contractility noted earlier by Wiggers (1941) in anoxic dogs. Kahler *et al.* (1962) studied right ventricular contractile force with a strain-gauge arch in dogs with cardiopulmonary by-pass prepared in such a way that the whole body and the isolated carotid bifurcations could be made anoxic independently. They found that contractile force and heart rate were increased during whole-body anoxia. Anoxia confined to the carotid bifurcations caused an increase in contractile force, but no consistent change in heart rate. A decrease in contractile force and a variable change in heart rate occurred when only the carotid circuit received oxygenated blood or when only the denervated carotid circuit received anoxic blood. They concluded that the augmentation of myocardial contractile force during anoxia is reflex in origin, with the chemoreceptor mechanism forming the afferent limb. The results obtained by Penna *et al.* (1962), although difficult to interpret, seem to support this contention.

There is a considerable body of evidence to challenge the conclusions reached by Kahler *et al.* (1962). First is the evidence of several groups showing that the primary chronotropic effect of anoxia of the chemoreceptors is a bradycardia (Bernthal *et al.*, 1951; Daly and Scott, 1958, 1959; Downing *et al.*, 1962). Second is the evidence that perfusion of the carotid body with oxygenated Ringer-Locke solution or oxygenated blood during systemic anoxia, though reducing the hyperpnoea, effects no change or increases the level of tachycardia (Neil, 1956; Daly and Scott, 1959). More recently Downing *et al.* (1962) assessed cardiac contractility by measuring ventricular function curves in open-chest dogs during whole-body anoxia and during anoxia confined to the isolated carotid circuit. During whole-body anoxia induced by the inhalation of 7% oxygen, cardiac output (electroturbinometer), heart rate and blood pressure rose, and an increase in ventricular contractility occurred, evidenced by a leftward displacement of the ventricular function curve. When anoxic blood was perfused through the carotid circuit there was a bradycardia, a small reduction in cardiac output and a reduced ventricular contractility, indicated by a shift to the right of the ventricular function curve. The results for cardiac contractility and change in heart rate when anoxia is confined to the carotid circuit appear to be in complete opposition to those described by Kahler *et al.* (1962). However, some caution should be exercised here, because of the great difference in methods. Indeed, one wonders whether myocardial contractile force and myocardial contractility as measured by these two groups can be equated. In considering the interpretation of these experiments in which myocardial contractility and myocardial contractile force are measured, it is important to realise, as Kahler *et al.* (1962) point out, that changes in cardiac output do not necessarily reflect changes in myocardial contractile force, since variations in venous return, systemic resistance and heart rate can also profoundly modify cardiac output.

Downing *et al.* (1962) conclude from their experiments that the chemoreceptors play no role in determining the cardiac response during anoxia. Subsequent experiments (Downing, Mitchell and Wallace, 1963) showed that the increase in cardiac output, cardiac contractility and heart rate, characteristic of whole-body anoxia, can be duplicated by anoxia confined to the cerebral circulation in artificially ventilated dogs with denervated aortic and carotid chemoreceptors and baroreceptors. Other studies (Downing and Siegel, 1963) showed an increased discharge in the left inferior cardiac nerves of cats during systemic anoxia after sectioning the chemoreceptor nerves. The authors consider these recent results to confirm their earlier suggestion that the cardiac effects of whole-body anoxia are due, at least in part, to direct hypoxic stimulation of the central nervous system with the efferent pathway in the cardiac sympathetic nerves. Similar conclusions were reached by Alexander (1945) on the basis of similar experiments. These results are not in agreement with those of Gomori *et al.* (1960a), who found no increase in cardiac output when anoxia was confined to the cerebral circulation in spontaneously breathing dogs, but noted the usual rise in cardiac output when anoxia was confined to the trunk. The reason for these discrepancies is not apparent.

In summary, the weight of evidence indicates that anoxic stimulation of the carotid chemoreceptors plays no role in determining the increased cardiac output and heart rate in anoxia. There is evidence that direct anoxic stimulation of the central nervous system may induce these changes, possibly via the cardiac sympathetic nerves. However, there are sufficient conflicts in results obtained by various groups to indicate clearly the need for more investigation of these possible mechanisms. The above review is based on experiments involving carotid chemoreceptors almost exclusively, presumably because the carotid body is so much more accessible than the aortic chemoreceptor sites. The inference has generally been made (by either the authors or those reading the paper) that both carotid and aortic chemoreceptors elicit similar reflex effects. Recently, Comroe and Mortimer (1964), in confirmation of earlier work by Comroe (1939), showed that stimulation of the aortic chemoreceptors induced a tachycardia and hypertension, whereas stimulation of carotid chemoreceptors usually initiated a bradycardia and hypotension; the degree of hyperpnoea was greatest with carotid body stimulation. Similar observations have been reported by Soma, Penna and Aviado (1965).

e. Humoral mechanisms: (i) adrenal gland. Baugh, Cornett and Hatcher (1959) showed that acute adrenalectomy prevented the rise in cardiac output, heart rate and stroke volume normally seen in intact anaesthetised dogs made anoxic by breathing 6% oxygen for 15 min (Fig. 3). In these studies the right adrenal gland was removed several days before experiment, and a loose ligature was placed about the left adrenal so that when it was tightened all vessels to and from the gland were cut off from the circulation. Ligature control studies indicated that the absence of cardiac output elevation in the adrenalectomised anoxic dogs could in no way be attributed to the effects of ligation itself (Fig. 3). Nahas, Mather, Wargo and Adams (1954c) investigated the effect of acute anoxia two weeks after adrenalectomy in five unanaesthetised dogs maintained on replacement therapy with deoxycorticosterone acetate. After breathing 8% oxygen for 3 min, these animals were unable to increase their cardiac output or heart rate, as did intact anoxic dogs; their general condition deteriorated if anoxia was prolonged beyond this time.

Harrison, Blalock, Pilcher and Wilson (1927) had earlier reported the effects of anoxia in dogs a few hours after acute surgical removal of the adrenals and in dogs in which the lumbo-adrenal vein was occluded by twisting a ligature placed about the vein proximal to the gland. When "post-adrenalectomy shock" was severe, no rise or a fall in cardiac output was observed. When there was little or no shock (presumably those in which ligation was involved, although this is not stated), a rise in cardiac output occurred. Ligation confined to the lumbo-adrenal vein proximal to the gland was shown by Cannon (1919) to be inadequate to prevent adrenal secretions from entering the general circulation. The more complete ligation technique employed by Baugh *et al.* (1959) is not subject to this criticism. Gomori, Kovack, Takacs, Foldi, Szabo, Nagy, Wiltner and Kallay (1960a) observed a rise in cardiac output with anoxia in anaesthetised dogs adrenalectomised one week before experiment and maintained with deoxycorticosterone. Their results are questionable, since in both adrenalectomised and intact groups significant elevations in cardiac output, of the order of 50%, were noted at arterial oxygen saturations of 80 to 90%. Such levels of arterial oxygen saturation are well above the "anoxaemic threshold"

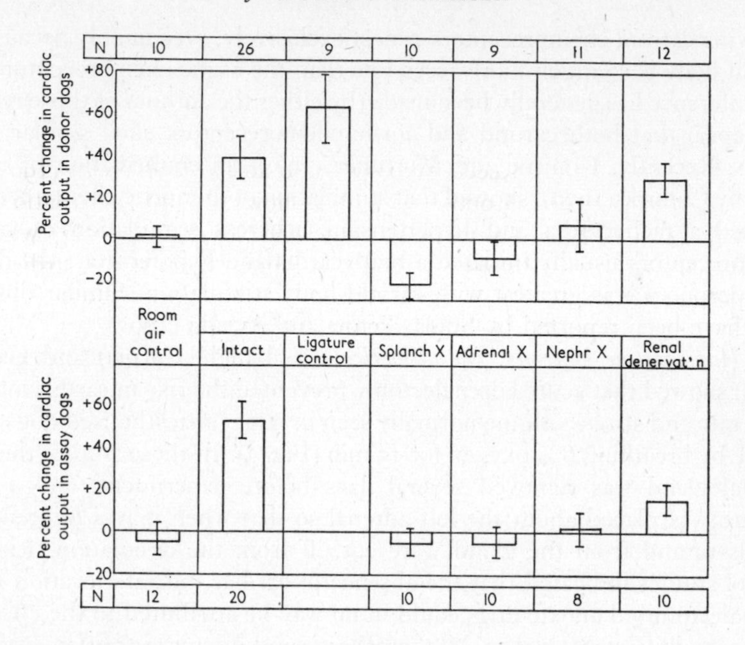

FIGURE 3. Percentage changes in cardiac output in various groups of dogs after 15 min breathing 6% oxygen shown in the panels in upper half. Change in cardiac output in intact donor dogs breathing room air for similar periods shown in first panel.

In the panels at bottom of figure are changes in cardiac output in assay dogs infused with blood from corresponding groups of anoxic donor dogs. For technique of assay, see text. In the first panel are the effects on cardiac output of similar infusions of blood from normal dogs breathing room air.

Results for anoxic donor and assay groups, illustrated in panels 1, 2, 3 and 5, are from Baugh, Cornett and Hatcher (1959); in panel 4 from Parker, Baugh and Hatcher (1958); in panels 6 and 7 from Binnion, Ackles and Hatcher (1962).

Each bar in the different panels shows the mean ± its standard error.

Abbreviations used are: splanch X = splanchnicectomy
adrenal X = adrenalectomy
nephr X = nephrectomy
renal denervat'n = renal denervation.

established for the anaesthetised dog (Harrison and Blalock, 1927; Lewis and Gorlin, 1952; Gorlin and Lewis, 1954).

Parker, Baugh and Hatcher (1958) splanchnicectomised dogs a few days before subjecting them to acute anoxia. This procedure, which effectively denervates the adrenal gland, prevented the usual rise in cardiac output, stroke volume and heart rate noted in intact anoxic animals (Fig. 3). Woods and Richardson (1959) and Penna *et al.* (1962) found that epidural sympathetic blockade prevented or modified the usual increase in contractile force and cardiac output seen in intact anoxic dogs.

There is no doubt that adrenal cortical secretion is enchanced in chronic anoxia, but there is no evidence indicating an adrenal cortical response during short periods of severe anoxia in either man (Hale, Sayers, Sydnor, Sweat and Van Fossan, 1957; Thorn, Jenkins and Laidlaw, 1953) or dogs (Lewis, Thorn, Koepf and Dorrance, 1942; Fowler, Shabetai and Holmes, 1961).

There is good evidence for the secretion of adrenaline in anoxia, based on measurements of adrenaline levels in the blood by various techniques (Kellaway, 1919; Houssay and Molinelli, 1926; Surtshin, Rodbard and Katz, 1948; Bülbring, Burn and DeElio, 1948; Ludemann, Filbert and Cornblath, 1956; Manger, Wakim and Bollman, 1959; Fowler, Shabetai and Holmes, 1961). However, some have failed to find evidence of increased catecholamine secretion (Toyooka and Blake, 1961; Goldring *et al.*, 1962).

There is evidence that the increased secretion of adrenaline in acute anoxia is due to sympathetic nerve stimulation of the adrenal (Kellaway, 1919; Houssay and Molinelli, 1926; Parker *et al.*, 1958). There is also evidence that adrenaline can be released by a direct action of anoxia on the gland (Malmejac, Chardon and Gross, 1950; Bülbring *et al.*, 1948), but the degree of anoxia required was severe.

Rapela and Houssay (1952) found catecholamines coming from the left adrenal vein of asphyxiated anaesthetised dogs at a rate of 0·44 μg/kg per min, of which 86% was adrenaline. Fowler *et al.* (1961) reported that the secretion rate of "norepinephrine-like substances" in the left adrenal vein of dogs breathing 7% oxygen was 0·14 μg/kg per min. In the experiments of Rapela and Houssay (1952) and Fowler *et al.* (1961) a bioassay technique was used. In both studies the abdominal cavity was opened before induction of anoxia for the purpose of blood collection; this operative procedure alone could affect the secretion of adrenaline (Walker, Zileti, Reutter, Shoemaker and Moore, 1959). It is known that an infusion of adrenaline at 0·2 μg/kg/min will increase cardiac output in the anaesthetised dog to levels similar to those occurring in acute anoxia (Binnion and Hatcher, 1963). Thus it appears, on the basis of the work of Rapela and Houssay (1952) and Fowler *et al.* (1961), that there may be enough adrenaline secreted in acute anoxia to produce the increased cardiac output seen in the anoxic dog; however, when it is remembered that the action of catecholamines is reduced in the presence of anoxia (Stravraky, 1942; Grandpierre and Franck, 1943; Van Loo, Surtshin and Katz, 1948; Giotti and Beani, 1957; Duner and von Euler, 1959), this conclusion should perhaps be questioned.

A transmissible humoral agent that increases the cardiac output of normal dogs can be demonstrated in the blood of intact anoxic dogs (Baugh *et al.*, 1959; Fig. 3). In the assay procedure used by them, blood withdrawn from peripheral arteries of anoxic dogs was infused into normal assay dogs at 0·5 ml/kg min for 30 min, cardiovascular measurements being made before and at the end of the infusion. This cardiotonic humoral agent is not present in the blood of normal dogs breathing room air or in blood obtained from splanchnicectomised or adrenalectomised anoxic dogs (Parker *et al.*, 1958; Baugh *et al.*, 1959; Fig. 3). These workers suggested that the humoral agent is adrenaline and that its release is due to a nervous reflex, the efferent arm being the splanchnic nerves. The average rise in cardiac output in assay dogs infused with blood from intact anoxic dogs was 53%. If this were due exclusively to adrenaline it would have to be infused at a rate of about 0·2 μg/kg per min (Binnion and Hatcher, 1963). Since blood from the anoxic donor dog is infused into assay dogs at 0·5 ml/kg per min this would mean that the adrenaline level in the peripheral blood of the anoxic dog required to produce the effect noted in the assay animal would be of the order of 0·4 μg/ml. This level is considerably in excess of those reported for the dog under a variety of stressful situations, including anoxia (Ludemann, Filbert and Cornblath, 1956; Manger *et al.*, 1959). Thus the nature and origin of the transferrable humoral agent demonstrated by Baugh *et al.* (1959) remains uncertain.

e. Humoral mechanisms: (ii) the kidney. Adrenaline may not be the only humoral agent with cardiovascular activity released in acute anoxia. Marshall and Kolls (1919) and Kriss, Futcher and Goldman (1948) found the renal nerves damaged by adrenalectomy in the dog. It is likely from the work in our laboratory on splanchnicectomised and adrenalectomised dogs (Parker *et al.*, 1958; Baugh *et al.*, 1959) that the nerves to the kidney were damaged or destroyed by the operative procedures.

The autonomic nerve supply to the kidney has a part to play in the renal vasoconstriction seen in anoxia (Franklin, McGee and Ullman, 1951; Gomori, Kovack, Takacs, Foldi, Szabo, Nagy and Wiltner, 1960b, 1960c) and may be required for optimal production of renal humoral agents, such as renin and vaso-excitor material (VEM), that may be released during anoxia (Shorr, 1948; Taquini, 1950; Edelman, Zweifach, Escher, Grossman, Mokotoff, Weston, Leiter and Shorr, 1950). VEM, which is released by the kidney under conditions of reduced oxygen tension, potentiates the action of adrenaline on vascular smooth muscle of the precapillary sphincters and metarterioles (Shorr, Zweifach and Furchgott, 1948; Edelman *et al.*, 1950; Shorr, 1955). It has not been possible to assess the full action of this material, since it has not been isolated in pure form. In so far as is known, VEM has no cardiac effect. Mylon and Heller (1948) demonstrated that natural angiotensin, which was inactive alone, produced intense vasoconstriction in the isolated rabbit ear when administered with subthreshold amounts of adrenaline. It was shown recently that infusions of angiotensin and adrenaline, in doses that separately do not raise cardiac output, do so when given together (Binnion and Hatcher, 1963).

Most investigators agree that the primary stimulus for renin release and consequent angiotensin production is a reduction in intrarenal blood pressure

(Huidobro and Braun-Menendez, 1942; Mikasa and Masson, 1961) acting on the juxtaglomerular cells (Goormaghtigh, 1939; Hartroft and Hartroft, 1961). It is possible that in anoxia the renal vasoconstriction leads to a fall in intrarenal blood pressure, with a subsequent liberation of humoral material, possibly renin, and that after renal denervation the stimulus for liberation of humoral substances during anoxia is reduced. Although renal anoxia is considered a stimulus for the production of VEM (Edelman *et al.*, 1950), it is not seriously considered by most investigators to be a stimulus for renin release, although there is some unwillingness to rule out this possibility (Huidobro and Braun-Menendez, 1942; Taquini, 1950). Analysis of the work of Divry (1951) bears directly on these points. She perfused, with a mechanical heart, the homotransplanted kidneys of dogs with arterial and venous blood at normal and reduced pressures and, besides assessing renin discharge, measured the renal blood flow and calculated the oxygen consumption of the kidney from the blood flow and the renal arteriovenous oxygen difference. On changing the perfusion from arterial to venous blood at the same pressure, there was no renin release detected. In these experiments there was a sixfold reduction in available oxygen, with a concomitant fivefold fall in oxygen consumption; thus, when the state of oxygenation of the renal tissue is considered in the light of its metabolism, little or no tissue anoxia appears to have been produced. While the kidney was still being perfused with venous blood, a reduction in perfusion pressure induced a small (10%) reduction in available oxygen, but an increase of 115% in oxygen consumption. Considering the oxygen supply in relation to renal metabolism, anoxia would be present, and in these circumstances a release of renin was noted. Although Divry (1951) considered that anoxaemia is not a cause of renin production, this analysis of her results suggests that anoxia may, in fact, be a stimulus to renin release. Recently, Skinner, McCubbin and Page (1963) carried out similar experiments on three dogs subjected to a milder degree of hypoxia than Divry's and concluded that hypoxia is not a factor in renin release.

These various observations led us to investigate the role of the kidney in acute anoxia. It has recently been shown (Binnion, Ackles and Hatcher, 1962) that acute nephrectomy by a ligation technique carried out just at the moment dogs begin to breathe 6% oxygen in nitrogen prevents the rise in cardiac output, heart rate and stroke volume normally seen at the end of 15 min of anoxia in intact dogs. A few days before these experiments, unilateral nephrectomy was performed and loose ligatures placed about the pedicle of the other kidney. The ends of the ligatures were brought out through the abdominal wall and, when tightened, effectively removed the kidney from the circulation. By the assay technique previously described (Baugh *et al.*, 1959) it was shown that blood from nephrectomised dogs did not raise the cardiac output of normal assay dogs, indicating absence of a cardiotonic agent. It therefore appears that both adrenal gland and kidney are essential for the increase in cardiac output normally seen in acute anoxia of anaesthetised dogs. It appears also that the cardiotonic agent has both adrenal and renal components.

In a subsequent series of experiments it was shown that, in dogs in which one kidney was removed and the other denervated, the rises in cardiac output, heart rate and stroke volume, though slightly reduced, were not significantly different from those seen in intact dogs (Binnion *et al.*, 1962). On the other hand, assays of blood from anoxic renal denervated dogs produced a rise in cardiac output that, although significant in comparison with the effects of infusions of normal blood, was of smaller magnitude than that noted with infusions of blood from intact anoxic dogs (Binnion *et al.*, 1962; Fig. 3). If the change in cardiac output in the assay dog can be considered a quantitative measure of the amount of cardiotonic agent present in the blood of anoxic dogs, then it appears that the renal nerves are necessary for maximal production of the cardiotonic material.

Summarising the work done in our laboratory on the role of humoral factors in the cardiovascular responses to anoxia, it appears that both adrenal gland and kidney are required for producing the cardiotonic agent and that in their absence there is no cardiotonic agent present and the usual increase in cardiac output seen in intact anoxic dogs does not occur. This suggests that the cardiotonic agent plays at least some part in determining the rise of cardiac output in acute anoxia. Removal of the renal nerves reduces the amount of cardiotonic agent produced, although the rise in cardiac output seen in the anoxic renal denervated dogs is not significantly different from that noted in intact dogs. It is possible that the agent plays a permissive role in the elevation of cardiac output in acute anoxia.

Using an assay procedure similar to that outlined in studies of acute anoxic anoxia (Baugh *et al.*, 1959), we have also found a cardiotonic agent in the blood of dogs with severe anaemia, one that increases the cardiac output of normal dogs (Justus, Cornett and Hatcher, 1957; Hatcher, Jennings, Parker and Garvock, 1963). Studies have been carried out to determine the length of time for which the increase in cardiac output is sustained in the assay dog after completion of the 30 min infusion with blood obtained from donor dogs with each type of anoxia. The results are shown in Fig. 4, along with the change in cardiac output noted on infusion of normal blood into assay dogs. It is apparent that the increase in cardiac output observed at the end of the infusion of anaemic blood is sustained for periods up to 2 hrs after completion of the infusion. In contrast, the increase in cardiac output noted with infusions of blood from dogs with acute anoxic anoxia is not sustained and reaches control levels by 30 min after the completion of the infusion and possibly sooner. Since the rise in cardiac output is essentially similar in both series at the end of the infusion period, these differences in duration of action cannot be explained on the basis of dose; they suggest that the cardiotonic agent is different in the two types of anoxia.

The nature of the cardiotonic agent found in the blood of dogs with acute anoxic anoxia is not known; our results indicate that it is not a single substance, but rather several agents from the adrenal gland and kidney. Evidence indicating that small amounts of adrenaline are secreted in anoxia has been reviewed, and the conclusion has been reached that insufficient amounts are released to account for the cardiac output rise in anoxic dogs and in assay dogs infused with blood from these animals. The identity of the renal component is not known, but there is presumptive evidence in favour of both angiotensin and VEM being produced in anoxia. Of these two, only

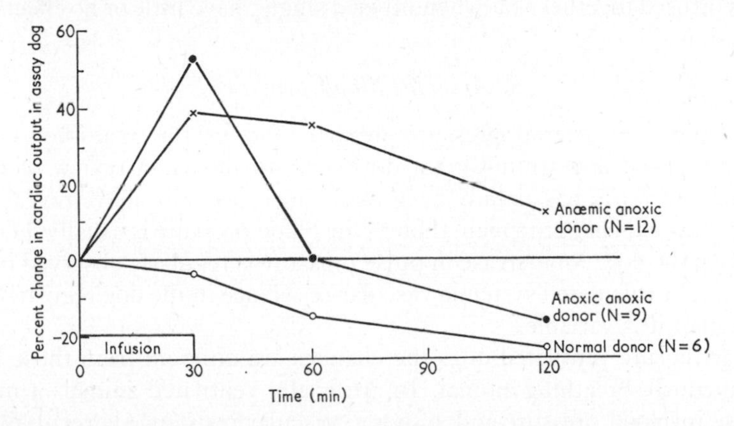

FIGURE 4. Comparison of changes in cardiac output induced in normal assay dogs by infusions of blood from anaemic donor dogs (X), from anoxic donor dogs (●) and from normal donor dogs breathing room air (O). The assay results are for anaemic and normal donor dogs from Hatcher, Jennings, Parker and Garvock (1963) and for anoxic donors from Binnion, Ackles and Hatcher (1962).

angiotensin has been isolated in a pure form. Accordingly we have studied the cardiovascular effects of adrenaline and angiotensin infused singly and together for 30 min (Binnion and Hatcher, 1963). Infusions of adrenaline (0·1 μg/kg per min), which alone did not change cardiac output, and angiotensin (0·05 μg/kg per min), which only reduced cardiac output slightly, when given together produced a rise in cardiac output of approximately 20% after 15 min of infusion. This effect was not sustained, for after 30 min of infusion the cardiac output had returned to control levels. It was suggested that this transient potentiation might involve a release of aldosterone by the angiotensin (Carpenter, Davis and Ayers, 1961; Genest, 1961) and an interaction of all three hormones. Recently this hypothesis was tested; when aldosterone (0·05 μg/kg per min), which alone had no effect on the circulation, was added to the other two hormones and the three were infused together, an increase in cardiac output averaging 65% was observed at 15 min of infusion. Again the effect was not sustained, and at 30 min after beginning infusion the increase was about one-third of that noted at 15 min (Binnion and Hilton, 1963). Unfortunately, in the assays of blood from anoxic dogs the cardiac output was not measured at the 15 min point of the 30 min infusion. However, the absence of a sustained effect, despite continued infusions of the hormones, suggests that some other combination of agents is involved in the anoxic animal. These infusion experiments do illustrate two important principles, namely (*a*) that the net cardiovascular action of these hormones, all of which may be released in certain stresses, cannot be predicted from an assessment of their individual action and (*b*) that profound effects can be produced with doses of these hormones infused together that when infused singly, have little or no effect on the circulation.

4. *Arterial blood pressure*

The changes in arterial pressure in man and in the anaesthetised and unanaesthetised dog are not great during acute anoxia. A review of papers to which reference has already been made in section C.1 above reveals little or no consistent change in man, though the blood pressure is usually increased slightly in the dog. An increase in pulse pressure is regularly observed in both species. The calculated systemic vascular resistance in the dog regularly falls, but in man it is variable.

In artificially ventilated dogs the changes are different from those in the spontaneously breathing animal. In artificially ventilated animals a marked increase in blood pressure and systemic vascular resistance is regularly seen (Bernthal and Schwind, 1945; Bernthal, Motley, Schwind and Weeks, 1945; Hurlimann and Wiggers, 1953; Stroud and Conn, 1954; Downing, Remensnyder and Mitchell, 1962; Penna, Soma and Aviado, 1962; Downing, Mitchell and Wallace, 1963). This difference in response has been the subject of comment by Bernthal and Schwind (1945), who pointed out that the intensity of peripheral vasoconstriction during stimulation of the chemoreceptors may be augmented by artificial control of the pulmonary ventilation. This increased vasoconstriction and systemic resistance during acute anoxia is due to stimulation of the chemoreceptors and is mediated by the efferent sympathetic vasoconstrictor fibres (Bernthal and Schwind, 1945; Bernthal *et al.*, 1945; March and Van Liere, 1948; Bernthal and Woodcock, 1951; Heymans and Neil, 1958a).

In the spontaneously breathing anoxic dog, denervation of the chemoreceptors results in a fall in arterial pressure, indicating that the chemoreceptor reflexes play a part in the maintenance of blood pressure at relatively normal levels in these animals (Heymans and Neil, 1958a). Thus the difference between spontaneously breathing dogs and those with controlled ventilation is one of degree. This difference may be accounted for on the basis of the hypocapnia in spontaneously breathing dogs, which could result in reduced vasomotor tone (Dale and Evans, 1922).

More recently Downing *et al.* (1963) showed in artificially ventilated dogs with chemoreceptor denervation that anoxia confined to the central nervous system resulted in an increase in arterial pressure and total systemic resistance. They also demonstrated (Downing and Siegel, 1963) that the impulse traffic in the left inferior cardiac nerve was enhanced by systemic anoxia sustained after chemoreceptor denervation. They suggest, as had Alexander (1945) previously on the basis of similar experiments, that anoxia may directly stimulate centres in the central nervous system. Gomori *et al.* (1960a) found no change in total systemic resistance during anoxia confined to the head in spontaneously breathing dogs, but from other experiments Gomori *et al.* (1960b, c) conclude that the renal vasoconstriction in spontaneously breathing anoxic dogs is due to increased sympathetic vasoconstriction resulting from the direct effects of anoxia on centres in the brain. These investigators are unwilling to rule out the possibility that the vasoconstriction may involve humoral mechanisms. In this connection Parker, Baugh and Hatcher (1958) and Baugh, Cornett and Hatcher (1959) found the rise in blood pressure during anoxia to be modified by adrenalectomy and splanchnicectomy; instead of falling, the total systemic resistance tended to show a rise. Nahas, Visscher, Mather, Haddy and Warner (1954a) and Nahas, Mather, Wargo and Adams (1954c) noted that the blood pressure during anoxia was lower in adrenalectomised dogs than in intact dogs. However, the control arterial pressure in the adrenalectomised group was 64 mm Hg lower than in the intact group, which suggests that the corticoid replacement therapy they used was inadequate. This criticism does not pertain to the experiments of Baugh *et al.* (1959). Binnion *et al.* (1962) noted that neither nephrectomy nor renal denervation altered the blood pressure response to anoxia in spontaneously breathing dogs.

In summary, the level of arterial pressure during anoxia is largely determined by the chemoreflexes. There is evidence that sympathetic vasoconstrictor activity may also be enhanced by a direct action of anoxia on the central nervous system and reduced by the hypocapnia that occurs as a result of hyperventilation. There is some evidence of humoral regulation through the adrenal gland.

5. *Pulmonary vascular system*

There is general agreement that the pulmonary arterial pressure rises in man and animals during acute anoxia (Korner, 1959; Fowler, 1960; Harris and Heath, 1962). Some investigators conclude that the major factor accounting for this rise is the increase in pulmonary blood flow (cardiac output) seen in acute anoxia (Hurlimann and Wiggers, 1953; Nahas *et al.*, 1954a). Most workers find as well an increase in calculated pulmonary resistance which they consider to play a significant role (Korner, 1959; Fowler, 1960). It is not certain what part of the pulmonary vascular tree is involved in this resistance change, but there is some evidence to support a post-arteriolar site (Hall, 1953; Rivera-Estrada, Saltzman, Singer and Katz, 1958). The mechanism of the increased pulmonary vascular resistance has not been settled (Fowler, 1960; Harris and Heath, 1962). There is evidence in favour of a direct vasoconstrictor effect of anoxia (von Euler and Liljestrand, 1946; Hall, 1953; Duke, 1951) and of increased sympathetic vasoconstrictor tone (Stroud and Rahn, 1953; Aviado, Ling and Schmidt, 1957) mediated by the chemoreceptor reflexes (Aviado *et al.*, 1957). Humoral factors have been considered, in particular adrenaline (Heemstra, 1954; Aviado *et al.*, 1957), but there is no good evidence in support of this suggestion (Duke, 1951; Goldring *et al.*, 1962).

6. *Regional systemic circulation*

The myocardial blood flow is increased by anoxia, coronary resistance decreases in the anaesthetised dog (Hackel, Goodale and Kleinerman, 1954; Hackel and Clowes, 1956; Feinberg, Gerola and Katz, 1958; Gomori and Takacs, 1960; Scott, Finkelstein and Croll, 1962), and the coronary fraction of the cardiac output increases (Takacs, 1957a). Hackel and Clowes (1956) compared the coronary blood flow response to anoxia in intact dogs and in dogs that had been both adrenalectomised and sympathectomised (stellate to T-10). They concluded that the increased coronary blood flow during anoxia was due to local metabolic products. Attempts to characterise these substances are being made (Berne, 1963). Feinberg *et al.* (1958) point out that the changing oxygen requirement of the myocardium at normal and reduced arterial oxygen contents is primarily met by changes in coronary blood flow and thus by oxygen availability and not by changing the oxygen extraction. Others report increased extraction as well (Hackel *et al.*, 1954). Feinberg *et al.* (1958) suggest that oxygen extraction increases only when coronary flow fails to maintain the oxygen requirements.

The cerebral blood flow is increased during acute anoxia, and cerebral vascular resistance falls (Lennox and Gibbs, 1932; Kety and Schmidt, 1948; Kety, 1958; Lassen, 1959). These changes occur when the arterial oxygen tension is reduced to about 50 mm Hg and are generally considered to be due to the direct effects of anoxia on cerebral vessels (Kety and Schmidt, 1948; Lassen, 1959; Lambertsen, 1961b). Consciousness is lost when the oxygen tension in cerebral venous blood reaches 15 to 20 mm Hg, and at this level abnormal cortical electrical activity begins abruptly (Lassen, 1959). The effectiveness of the increase in cerebral blood flow in maintaining oxygen tension in the brain during anoxia is moderated by the hypocapnia that occurs secondary to the hyperventilation. If the hypocapnia is prevented, the increase in cerebral blood flow is greater (Kety, 1958; Lassen, 1959; Lambertsen, 1961b).

Anoxia decreases the blood flow through the hand (Schneider and Truesdell, 1924; Abramson, Landt and Benjamin, 1943), but increases it through skeletal muscle (Lennox and Gibbs, 1932; Abramson *et al.*, 1943; Anderson, Allen, Barcroft, Edholm and Manning, 1946; Black and Roddie, 1958). Abramson *et al.* (1943) consider the increased muscle blood flow to be due to the increased cardiac output and the direct vasodilator effect of anoxia on peripheral vessels. Black and Roddie (1958) have investigated this problem further and point out that forearm blood flow increases and vascular resistance falls in both voluntary hyperventilation and anoxia. In both states these vascular changes were considered to involve a humoral mechanism rather than vasomotor nerves, for similar changes occurred in the nerve-blocked limb. When hypocapnia was prevented by the addition of carbon dioxide to the air breathed by hyperventilating subjects or to the low oxygen gas mixture, no increase in blood flow occurred. These investigators suggest that hypocapnia leads to some biochemical change or to the release of some humoral agent, such as adrenaline, that causes the vasodilatation in muscle vessels. Clarke (1952) made similar observations, but pointed out that the increase in forearm blood flow was greater in the first minute of hypoxia than in the next 2 min, when the alveolar carbon dioxide tension had fallen even farther.

Crawford, Fairchild and Guyton (1959) perfused the hind limb of the anaesthetised dog with the animal's own venous blood and demonstrated an increase in blood flow nearly proportional to the degree of desaturation of the blood. They concluded that the vasodilatation was due to the direct effects of anoxia on the limb vessels. In anaesthetised anoxic dogs there is an increase in hind limb blood flow (Takacs, 1957b), but vasoconstriction is observed when ventilation is controlled and is considered to be due to increased sympathetic vasomotor tone and initiated by the chemoreceptor reflexes (Bernthal *et al.*, 1945; Bernthal and Schwind, 1945; Litwin, Dil and Aviado, 1960).

Granberg (1962) has recently reviewed the literature of the effects of acute anoxia on renal haemodynamics and comments on the extreme variability of results both in man and in anaesthetised and unanaesthetised dogs. He points out that there is a great weight of evidence to show that the renal blood flow and glomerular filtration rate are increased or unchanged in man, whereas in the anaesthetised dog a decrease in renal blood flow and glomerular filtration rate are found more frequently. In both species antidiuresis is more commonly reported than diuresis or an unchanged urine flow. The variability is ascribed by Granberg (1962) to differences in methods and conditions of experiment.

Renal denervation modifies or prevents the decrease in renal blood flow and urine output in anaesthetised anoxic dogs (Toth, 1940; Gomori *et al.*, 1960c). The Gomori group also showed that perfusion of the head of a dog with anoxic blood while the rest of the body was normally oxygenated resulted in an increase in renal vascular resistance and oliguria similar to that seen in anoxia of the whole body. These changes occurred in many of the dogs after denervation of the carotid sinus, and they suggest that anoxia stimulates the central nervous system directly, with a subsequent increase in renal vasoconstrictor nerve activity. Adrenaline has been considered by some to play a part, but these renal vascular changes are still observed in anaesthetised dogs after adrenalectomy (Toth, 1940; Gomori *et al.*, 1960c).

Selkurt (1953) studied dogs whose kidneys were perfused with either arterial or venous blood supplied from the carotid artery and jugular vein, respectively. He found that perfusion with venous blood increased renal plasma flow and urine volume. Thus the variability in results mentioned by Granberg (1962) may be due in part to the predominance of one of two opposing factors, neurogenic vasoconstriction and the direct vasodilatory action of hypoxia on renal vessels.

The antidiuresis commonly seen in acute hypoxia is considered to be due to a decrease in glomerular filtration rate and possibly an increase in secretion of antidiuretic hormone (Granberg, 1962).

7. *Arterial oxygen capacity*

A rapid increase in haematocrit is a characteristic feature of acute anoxia in animals and man (Gregg, Lutz and Schneider, 1919; Izquierdo, 1928; Hurtado, Merino and Delgado, 1945; Baugh *et al.*, 1959). This increase is considered to be due mainly to splenic contraction (Barcroft, Harris, Orahovats and Weiss, 1925) and is not seen after splenectomy, but the usual rise occurs after splenic denervation (Izquierdo, 1928). Splenic contraction is not observed during anoxia of the head alone, but is observed in anoxia of the trunk and whole body (Gomori, Takacs and Kallay, 1960). The rise in haematocrit is not observed after removal of the adrenal medulla from cats (Izquierdo, 1928) or after total adrenalectomy in dogs (Nahas *et al.*, 1954c; Baugh *et al.*, 1959). The role of the spleen in these changes is more important in dogs than in man (Ebert and Stead, 1941; Reeve, Gregersen, Allen and Sear, 1953). In severe anoxia an increase in capillary permeability may contribute to the rise in haematocrit (Henry, Goodman and Meehan, 1947).

D. Respiratory Adjustments

1. *The Mechanisms of the ventilatory response*

Survival from a sudden exposure to an anoxic environment depends on both cardiovascular and respiratory adjustments. Indeed, an increase in ventilation spares the cardiac response in that the rate of blood flow required to maintain a given oxygen tension is less in the presence of a full ventilatory response.

It is generally agreed that the increased ventilation, which involves an increase in both rate and depth of breathing, is due to stimulation of the chemoreceptors by the reduced oxygen tension of the blood. Some of the details of experimental evidence has been reviewed in section C.3 d (ii) above. In the absence of chemoreceptors, anoxia depresses the central respiratory mechanisms, ventilation decreases, and respiratory failure ultimately occurs (Heymans and Neil, 1958b).

By recording action potentials in the sinus nerve of animals, it has been established that the chemoreceptor threshold for oxygen is above an arterial oxygen tension of 100 to 115 mm Hg and that, as the oxygen tension falls, the impulse traffic in the nerve increases progressively (von Euler, Liljestrand and Zotterman, 1939; Witzleb, Bartels, Budde and Mochizucki, 1955; Hornbein, Griffo and Roos, 1961a; Eyzaguirre and Lewin, 1961). In Fig. 5 the impulse activity recorded by various investigators from chemoreceptor fibres in the sinus nerve of anaesthetised cats is shown at different levels of anoxia. One of the striking features is the difference in the shape of response curve based on the work of Witzleb *et al.* (1955) and on that of other workers. Witzleb and colleagues found the greatest increase in impulse activity occurring with minimal changes in arterial oxygen tension between 110 and 100 mm Hg, and for each decrease of 10 mm Hg oxygen tension below this the increase in impulse acitivity was progressively less. In contrast to these findings are those of von Euler *et al.* (1939) and Hornbein *et al.* (1961a) (curves H_1, H_2 and E in Fig. 5), which show smaller increases in impulse activity occurring with changes in arterial oxygen tension around the normal level and the greatest increases at the more extreme reductions in arterial oxygen tension. Witzleb *et al.* (1955), in contrasting their results with those of von Euler *et al.* (1939), were unable to account for the differences, but there is general agreement that changes in temperature, stability of the preparation, the associated level of carbon dioxide and pH are all factors that could influence the slope and possibly the shape of the curve (Witzleb *et al.*, 1955; Hornbein *et al.*, 1961a; Eyzaguirre and Lewin, 1961; Dejours, 1962). Further, the degree of activity recorded after a change in oxygen tension appears to depend on whether or not the readings are taken during a steady state (Eyzaguirre and Lewin, 1961; Hornbein *et al.*, 1961a).

As Dejours (1962) points out, these relationships between chemoreceptor nerve activity and change in arterial oxygen tension do not mean that the same relationship should necessarily hold between oxygen tension and the ventilatory response. Moreover, work involving the recording of impulse activity has all been done on animals, and the projection of these findings to the human respiratory changes in acute anoxia should be made with caution.

The fact that firing of chemoreceptors occurs at and above the arterial

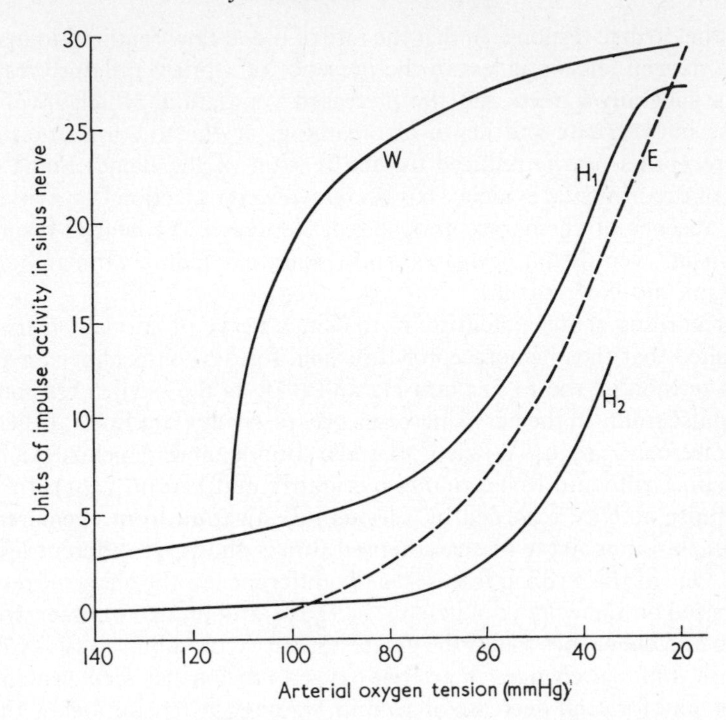

FIGURE 5. Impulse activity in carotid sinus nerves of anaesthetised cats at various arterial oxygen tensions. Curves designated W based on results from Witzleb, Bartels, Budde and Mochizucki (1955). Curve E based on results obtained by von Euler, Liljestrand and Zotterman (1939). Impulse activity measured by Witzleb *et al.* (1955) and by von Euler *et al.* (1939) as impulses per second (accurate units for their experiments can be obtained by adding a zero to each number on the Y ordinate). Impulse activity in the experiments of Hornbein, Griffo and Roos (1961a), shown as H_1 and H_2, determined with an integrating device and expressed in arbitrary units of chemoreceptor activity. Units on the Y ordinate, as given, are accurate for their values. For curve H_1, pH 7·21 and arterial P_{CO_2}, 40 mm Hg. For curve H_2, pH 7·55, and arterial P_{CO_2} 16·5.

oxygen tension noted in the resting animal is an argument in favour of there being a tonic oxygen-sensitive chemoreceptor influence on respiration (Dejours, 1962). Support for this is seen in experiments of other types. The inhalation of 100% and 33% oxygen results, in the first few minutes, in a transient decrease in ventilation (Dripps and Comroe, 1947; Loeschcke, 1953). Prolonged inhalation of 100% oxygen may produce an increase in ventilation, but this is due to associated metabolic changes, which are minimal and out of phase with the changes in oxygen tension in the short-term oxygen breathing studies (Dejours, 1962). Single or double breaths of oxygen ("oxygen tests") result in a transient slight decrease in ventilation, and this effect can be amplified if the subject is first made mildly anoxic, but to a degree insufficient to affect the resting ventilation (Dejours, Labrousse, Raynaud, Girard and Teillac, 1958; Dejours, 1962). The opposite experiment, a two-breath anoxia test, done by Hornbein, Roos and Griffo (1961b), has shown a transient slight increase in ventilation (Fig. 6).

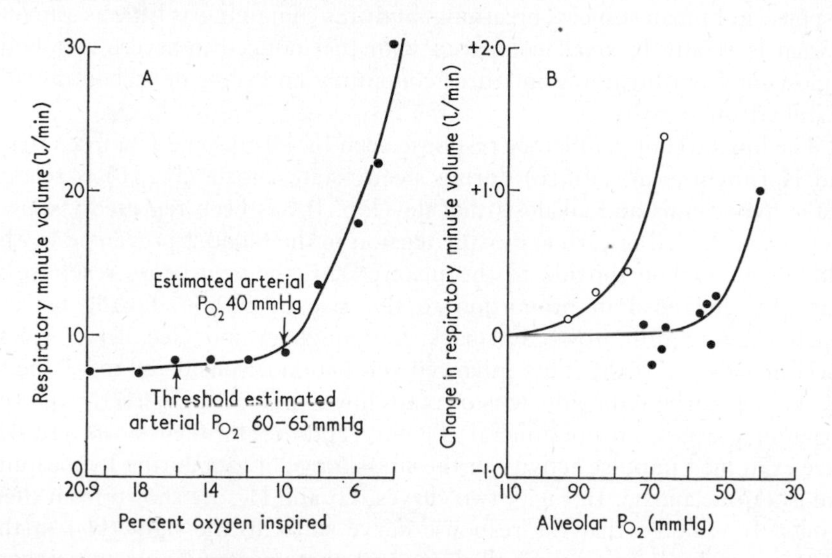

FIGURE 6. A. Respiratory minute volume observed in human subjects at different percentage oxygen contents in inspired air, based on results of Dripps and Comroe (1947). B. Open circles (O), the transient ventilatory response in a human subject at different alveolar oxygen tensions achieved by a "two-breath anoxia test". Closed circles (●), the "steady-state" response of ventilation at the same and lower alveolar oxygen tensions. Based on results of Hornbein, Roos and Griffo (1961b).

These various experiments indicate that at rest there is tonic chemoreceptor activity, which under certain conditions may be a positive drive to ventilation (Dejours, 1962). This conclusion is supported by the presence of impulses in chemoreceptor nerves when the oxygen tension is in the normal range or above (Fig. 5).

The changes in ventilation that occur when human subjects inhale low oxygen mixtures for prolonged periods (several minutes) and thus reach a steady state are totally different from those seen after a one- or two-breath anoxic test when a steady state is not present (Fig. 6). Dripps and Comroe (1947) found under steady-state conditions no consistent change in ventilation in human subjects breathing 18% oxygen, but a slight increase was observed in most subjects when 16% oxygen was inhaled (average increase 0·5 l/min). At this time the oxygen saturation of arterial blood was 91%, equivalent to about 60 to 65 mm Hg tension. Though this percentage of oxygen in inspired air may indicate the threshold, it was necessary to reduce the oxygen in inspired air to 10% before an increase in ventilation occurred in all subjects. At this time the increase in ventilation was only 1·3 l/min, despite a reduction in arterial oxygen tension to approximately 40 mm Hg. Similar findings were obtained by Hornbein *et al.* (1961b). It is recognised that the ventilatory response in human subjects breathing mixtures containing as little as 4 to 6% oxygen is relatively small compared with that noticed in severe metabolic acidosis or from breathing mixtures containing an excess of carbon dioxide (Lambertsen, 1961b).

The low level of ventilatory response seen by Dripps and Comroe (1947) and Hornbein *et al.* (1961b) during steady-state anoxia (Fig. 6) is related to the hypocapnia and alkalosis that develop. It has been repeatedly shown that, when the fall in carbon dioxide tension in the blood is prevented by the addition of carbon dioxide to the inspired air, the ventilatory response is markedly enhanced in proportion to the rise in carbon dioxide tension (Nielsen and Smith, 1951; Cormack, Cunningham and Gee, 1957; Loeschcke and Gertz, 1958). This enhanced ventilation is considered to be due to the higher carbon dioxide tension and lower pH acting directly on the respiratory centre and peripheral chemoreceptors. Hornbein *et al.* (1961a) have examined impulse activity in the sinus nerve of cats during hypocapnic and isocapnic anoxia. In Fig. 5 two curves, H_1 and H_2, are shown from their results. It is noted that the response curve shifts to the right (H_2) in the presence of hypocapnia and alkalosis and that no significant increase in impulse activity occurs until the arterial oxygen tension is reduced to below 60 mm Hg. When the hypocapnia and alkalosis are prevented, the curve is displaced to the left (H_1), and significant impulse activity occurs at arterial oxygen tensions above 60 mm Hg. These results provide electrophysiological evidence that fits reasonably well with the ventilatory responses observed during steady-state anoxia with and without hypocapnia. The slight increase in ventilation obtained in the two-breath anoxic test by Hornbein *et al.* (1961b), in which the carbon dioxide tension is essentially unaltered (Fig. 6), can be explained by alterations in impulse activity, such as that shown in curve H_1 in Fig. 5. The impulse activity measured by Witzleb *et al.* (1955), although possibly fitting the ventilatory response of unsteady states associated with the oxygen and anoxic breath tests, does not give results that relate to the ventilatory changes of steady-state anoxia. Why the results of Witzleb and associates are so different from others is not clear, but clearly studies relating ventilation response with impulse activity in the chemoreceptor nerves during both unsteady states and steady states of acute anoxia would be profitable.

In relation to the influence of carbon dioxide tension on the ventilatory response to anoxia, the experiments of Nielsen and Smith (1951) are of particular interest. They determined first the increase in ventilation of a trained subject to anoxia (alveolar oxygen tension approximately 37 mm Hg) during which the alveolar carbon dioxide tension was reduced from normal to below 20 mm Hg. The alveolar carbon dioxide tension was then elevated in steps by adding carbon dioxide to the inspired gas mixture while adjusting the oxygen content of the anoxic gas mixture to keep the alveolar oxygen tension constant. They found that the increase in ventilation with anoxia was unaffected by increments in alveolar carbon dioxide tension up to a threshold value of 30 to 33 mm Hg (Fig. 7). With further elevations in alveolar carbon dioxide tension, an abrupt and linear increase in ventilation occurred, and the slope of the increase was greater in acute anoxia than when the subject breathed room air with similar increments of carbon dioxide (Fig. 7). The multiple factor theory for the control of ventilation proposed by Gray (1950) states that the alterations in pH and tensions of oxygen and carbon dioxide

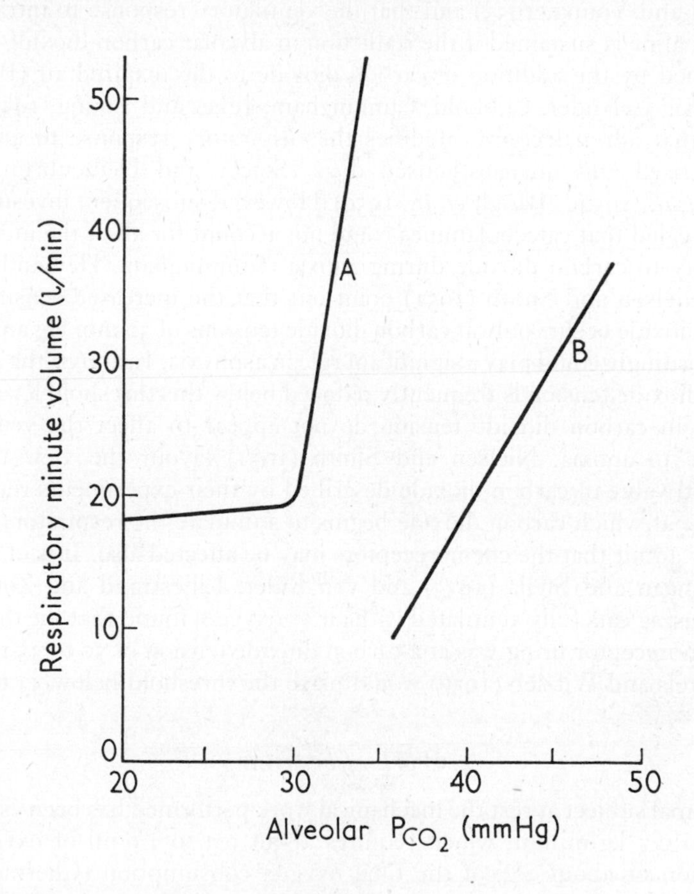

FIGURE 7. Ventilatory response in human subject at different levels of alveolar carbon dioxide tension when alveolar oxygen tension in A is kept constant at approximately 37 mm Hg and in B at 110 mm Hg. Results from Nielsen and Smith (1951).

during anoxia determine the ventilatory response on the basis of their additive effects. Nielsen and Smith (1951) point out that, if this were so, the two slopes shown in Fig. 7 should be the same. The fact that the slope was steepest during anoxia indicates a greater sensitivity of the respiratory control system to carbon dioxide during acute anoxia than during the inhalation of room air and is consistent with a positive interaction of stimuli rather than a simple additive effect. The studies of Cormack *et al.* (1957), Lloyd, Jukes and Cunningham (1958), Loeschcke and Gertz (1958), Tenney, Remmers and Mithoefer (1963), though designed differently, lead to the same conclusion. Cormack *et al.* (1957) suggested that catecholamines might be involved in the increased sensitivity to carbon dioxide during anoxia. This suggestion was based on the knowledge that catecholamines stimulate respiration (Whelan and Young, 1953) and that the ventilatory response to intravenous noradrenaline is sustained if the reduction in alveolar carbon dioxide tension is modified by the addition of carbon dioxide to the inspired air (Barcroft, Basnayake, Celander, Cobbold, Cunningham, Jukes and Young, 1957). It is known that adrenalectomy modifies the respiratory response to anoxia in anaesthetised and unanaesthetised dogs (Stacey and Demenbrun, 1950; Nahas *et al.*, 1954c; Baugh *et al.*, 1959). However, subsequent investigations have revealed that catecholamines could not account for all of the increase in sensitivity to carbon dioxide during anoxia (Cunningham, Hey and Lloyd, 1958). Nielsen and Smith (1951) point out that the increased sensitivity to carbon dioxide occurs only at carbon dioxide tensions of 30 mm Hg and above and accordingly could play a significant role in asphyxia. In anoxia the alveolar carbon dioxide tension is frequently reduced below this threshold level, when changes in carbon dioxide tension do not appear to affect the ventilatory response to anoxia. Nielsen and Smith (1951) favour the view that the threshold value of carbon dioxide described by their experiments represents the value at which carbon dioxide begins to stimulate the respiratory centre, but they admit that the chemoreceptors may be affected also. It is of interest that Samaan and Stella (1935) and von Euler, Liljestrand and Zotterman (1939), using cats fully ventilated with air or oxygen, found that the threshold for chemoreceptor firing was at a carbon dioxide tension of 30 to 35 mm Hg, but Bartels and Witzleb (1956) would place the threshold below 33 mm Hg.

2. *Work of breathing*

In a normal subject at rest the mechanical work performed has been estimated at 0·3 to 0·7 kg/m/min, which requires about 0·3 to 1·0 ml of oxygen/l of ventilation or about 2% of the total oxygen consumption (Cherniack and Cherniack, 1961). During anoxia certain evidence indicates that the work of breathing may rise disproportionately to the moderate ventilatory response evoked (Fig. 6). Widdicombe (1963) in a review of the regulation of tracheobronchial smooth muscle concludes that anoxia increases the tone of these muscles, which to a large degree is due to stimulation of chemoreflexes with the efferent pathway in the vagus nerve. Such an effect would increase the work of breathing. Though there is reasonable evidence for this in animals, there is a paucity of results for human subjects. An increase in carbon dioxide acting locally on bronchial smooth muscle has a dilator action, but in the intact animal it induces bronchoconstriction. This effect is independent of the chemoreceptors, but is dependent on vagal efferent fibres, which may be stimulated by a direct action of the carbon dioxide in the central nervous system (Daly, Lambertsen and Schweitzer, 1953). Widdicombe (1963) points out that the effects of hypocapnia on bronchial muscle have not been studied. After the period encompassed by Widdicombe's review, McGregor and his colleagues (McGregor, Donevan and Anderson, 1962; Newhouse, Becklake, Macklem and McGregor, 1963) found that, during voluntary hyperventilation of air at a minute volume of 30 l/min, the mean inspiratory flow resistance was 133% and the mean respiratory work of ventilating the lungs was 68% greater when the alveolar carbon dioxide tension fell to 20 to 25 mm Hg than when this tension was maintained at 45 to 50 mm Hg. End respiratory oesophageal pressure and compliance were unaffected. The increased work of ventilation could be modified by administration of atropine or isoprenaline; when these drugs were given together no increase in respiratory work was observed (Newhouse *et al.*, 1963). Thus in acute anoxia the work of breathing may be increased out of proportion to the ventilatory response as a result of the effects of anoxia and hypocapnia on resistance to air flow.

REFERENCES

Abramson, D. I., Landt, H., and Benjamin, J. E. (1943). *Arch. intern. Med.* **71**, 583.

Alexander, R. S. (1945). *Amer. J. Physiol.* **143**, 698.

Anderson, D. P., Allen, W. J., Barcroft, H., Edholm, O. G., and Manning, G. W. (1946). *J. Physiol.* **104**, 426.

Anrep, G. V., Pascual, W., and Rossler, R. (1935–6). *Proc. roy. Soc.* **B119**, 191.

Armstrong, H. G. (1943a). "Principles and Practice of Aviation Medicine", p. 274. 2nd Edition. Williams and Wilkins Company, Baltimore.

Armstrong, H. G. (1943b). "Principles and Practice of Aviation Medicine", p. 334. 2nd Edition. Williams and Wilkins Company, Baltimore.

Asmussen, E., and Chiodi, H. (1941). *Amer. J. Physiol.* **132**, 426,

Asmussen, E., and Nielsen, M. (1956). *Acta physiol. scand.* **35**, 73.

Aviado, D. M., Ling, J. S., and Schmidt, C. F. (1957). *Amer. J. Physiol.* **189**, 253.

Barcroft, H., Basnayake, V., Celander, O., Cobbold, A. F., Cunningham, D. J. C., Jukes, M. G. M., and Young, I. M. (1957). *J. Physiol.* **137**, 365.

Barcroft, J. (1920). *Lancet* ii, 485.

Barcroft, J., Harris, H. A., Orahovats, D., and Weiss, R. (1925). *J. Physiol.* **60**, 443.

Bartels, H., and Witzleb, E. (1956). *Pflüg. Arch. ges. Physiol.* **262**, 466.

Baugh, C. W., Cornett, R. W., and Hatcher, J. D. (1959). *Circulation Res.* **7**, 513.
Baumber, J. S., and Hatcher, J. D. (1960). *Proc. Canad. Fed. Biol. Soc.* **3**, 9.
Baumber, J. S., and Hatcher, J. D. (1963). *Proc. Canad. Fed. Biol. Soc.* **6**, 7.
Berne, R. M. (1963). *Amer. J. Physiol.* **204**, 317.
Bernthal, T., Greene, W., Jr., and Revzin, A. M. (1951). *Proc. Soc. exp. Biol. N.Y.* **76**, 121
Bernthal, T., Motley, H. E., Schwind, F. J., and Weeks, W. F. (1945). *Amer. J. Physiol.* **143**, 220.
Bernthal, T., and Schwind, F. J. (1945). *Amer. J. Physiol.* **143**, 361.
Bernthal, T., and Woodcock, C. C. (1951). *Amer. J. Physiol.* **166**, 45.
Binnion, P. F., Ackles, K. N., and Hatcher, J. D. (1962). *Proc. Canad. Fed. Biol. Soc.* **5**, 13.
Binnion, P. F., and Hatcher, J. D. (1963). *Circulation Res.* **12**, 393.
Binnion, P. F., and Hilton, D. (1963). *Proc. Canad. Fed. Biol. Soc.* **6**, 10.
Black, J. E., and Roddie, I. C. (1958). *J, Physiol.* **143**, 226.
Bogue, J. Y., Chang, I., and Gregory, R. A. (1938). *Quart. J. exp. Physiol.* **27**, 319.
Bülbring, E., Burn, J. H., and DeElio, F. J. (1948). *J. Physiol.* **107**, 222.
Cannon, W. B. (1919). *Amer. J. Physiol.* **50**, 399.
Carpenter, C. C. J., Davis, J. O., and Ayers, C. R. (1961). *J. clin. Invest.* **40**, 2026.
Cherniack, R. M., and Cherniack, L. (1961). "Respiration in Health and Disease", p. 28. W. B. Saunders Company, Philadelphia.
Chidsey, C. A., Frye, R. L., Kahler, R. L., and Braunwald, E. (1961). *Circulation Res.* **9**, 989.
Clarke, R. S. J. (1952). *J. Physiol.* **118**, 537.
Comroe, J. H. (1939). *Amer. J. Physiol.* **127**, 176.
Comroe, J. H. and Mortimer, L. (1964). *J. Pharm. exp. Therap.* **146**, 33.
Cormack, R. S., Cunningham, D. J. C., and Gee, J. B. L. (1957). *Quart. J. exp. Physiol.* **42**, 303.
Crawford, D. G., Fairchild, H. M., and Guyton, A. C. (1959). *Amer. J. Physiol.* **197**, 613.
Cunningham, D. J. C., Hey, E. N., and Lloyd, B. B. (1958). *Quart. J. exp. Physiol.* **43**, 394.
Dale, H. H., and Evans, C. L. (1922). *J. Physiol.* **56**, 125.
Daly, M. de Burgh, Lambertsen, C. J., and Schweitzer, A. (1953). *J. Physiol.* **119**, 292.
Daly, M. de Burgh, and Scott, M. J. (1958). *J. Physiol.* **144**, 148.
Daly, M. de Burgh, and Scott, M. J. (1959). *J. Physiol.* **145**, 440.
Dejours, P. (1962). *Physiol. Rev.* **42**, 335.
Dejours, P., Labrousse, Y., Raynaud, J., Girard, F., and Teillac, A. (1958). *Rev. franc. étud. clin. biol.* **3**, 105.
Dell, P., and Bonvallet, M. (1954). *C. R. Soc. Biol., Paris* **148**, 885.
Dell, P., Hugelin, A., and Bonvallet, M. (1959). "Cerebral Anoxia and the Electroencephalogram", p. 46. Charles C. Thomas, Springfield, Illinois.
Divry, A. (1951). *Arch. int. Physiol.* **59**, 211.
Doi, Y. (1921). *J. Physiol.* **55**, 43.
Downing, S. E., Mitchell, J. H., and Wallace, A. G. (1963). *Amer. J. Physiol.* **204**, 881.
Downing, S. E., Remensnyder, J. P., and Mitchell, J. H. (1961). *Circulation Res.* **10**, 676.
Downing, S. E., and Siegel, J. H. (1963). *Amer. J. Physiol.* **204**, 471.
Doyle, J. T., Wilson, J. S., and Warren, J. R. (1952). *Circulation* **5**, 263.
Dripps, R. D., and Comroe, J. H. (1947). *Amer. J. Physiol.* **149**, 277.
Duke, H. (1951). *Quart. J. exp. Physiol.* **36**, 75.
Duner, H., and von Euler, U. S. (1959). *Acta physiol. scand.* **46**, 201.
Ebert, R. V., and Stead, E. A. (1941). *Amer. J. med. Sci.* **201**, 655.
Eckstein, J. W., and Horsley, A. W. (1960). *J. Lab. clin. Med.* **56**, 847.
Edelman, I. S., Zweifach, B. W., Escher, D. J. W., Grossman, J., Mokotoff, R., Weston, R. E., Leiter, L., and Shorr, E. (1950). *J. clin. Invest.*, **29**, 925.
Eyzaguirre, C., and Lewin, J. (1961). *J. Physiol.* **159**, 222.

Feinberg, H., Gerola, A., and Katz, L. N. (1958). *Amer. J. Physiol.* **195**, 593.
Feldman, M., Jr., Rodbard, S., and Katz, L. N. (1948). *Amer. J. Physiol.* **154**, 391.
Fishman, A. P., Himmelstein, A., Fritts, H. W., and Cournand, A. (1955). *J. clin. Invest.* **34**, 637.
Fishman, A. P., McClement, J., Himmelstein, A., and Cournand, A. (1952). *J. clin. Invest.* **31**, 770.
Fowler, N. O. (1960). *Amer. J. Med.* **28**, 927.
Fowler, N. O., Shabetai, R., and Holmes, J. C. (1961). *Circulation Res.* **9**, 427.
Franklin, K. J., McGee, L. E., and Ullman, E. A. (1951). *J. Physiol.* **112**, 43.
Fritts, H. W., Jr., Harris, P., Clauss, R. H., Odell, J. E., and Cournand, A. (1958). *J. clin. Invest.* **37**, 99.
Gastaut, H., and Meyer, J. S. (editors) (1959). "Cerebral Anoxia and the Electroencephalogram". *Proceedings of the Marseille Colloquium.* Charles C. Thomas, Springfield, Illinois.
Genest, J. (1961). *Canad. med. Ass. J.* **84**, 403.
Giotti, A., and Beani, L. (1957). *Sci. med. Italica* **6**, 274.
Goldring, R. M., Turino, G. M., Cohen, G., Jameson, A. G., Bass, B. G., and Fishman, A. P. (1962). *J. clin. Invest.* **41**, 1211.
Gomori, P., Kovack, A. G. B., Takacs, L., Foldi, M., Szabo, Gy., Nagy, Z., Wiltner, W., and Kallay, K. (1960a). *Acta med. hung.* **16**, 93.
Gomori, P., Kovack, A. G. B., Takacs, L., Foldi, M., Szabo, Gy., Nagy, Z., and Wiltner, W. (1960b). *Acta med. hung.* **16**, 37.
Gomori, P., Kovack, A. G. B., Takacs, L., Foldi, M., Szabo, Gy., Nagy, Z., and Wiltner, W. (1960c). *Acta med. hung.* **16**, 43.
Gomori, P., and Takacs, L. (1960). *Amer. Heart J.* **59**, 161.
Gomori, P., Takacs, L., and Kallay, K. (1960). *Acta med. hung.* **16**, 75.
Goormaghtigh, N. (1939). *Proc. Soc. exp. Biol., N. Y.* **42**, 688.
Gorlin, R., and Lewis, B. M. (1954). *J. appl. Physiol.* **7**, 180.
Grandpierre, R., and Franck, C. (1943). *C. R. Soc. Biol., Paris* **137**, 743.
Granberg, P. (1962). *Scand. J. clin. Lab. Invest.* **14**, Supplementum 63.
Gray, J. S. (1950). "Pulmonary Ventilation and its Physiological Regulation", p. 51. Thomas, Springfield, Illinois.
Gregg, H. W., Lutz, B. R., and Schneider, E. C. (1919). *Amer. J. Physiol.* **50**, 216.
Gremels, H., and Starling, E. H. (1926). *J. Physiol.* **61**, 297.
Grollmann, A. (1930). *Amer. J. Physiol.* **93**, 19.
Hackel, D. B., and Clowes, G. H. A. (1956). *Amer. J. Physiol.* **186**, 111.
Hackel, D. B., Goodale, W. T., and Kleinerman, J. (1954). *Circulation Res.* **2**, 169.
Hale, H. B., Sayers, G., Sydnor, K. L., Sweat, M. L., and Van Fossan, D. D. (1957). *J. clin. Invest.* **36**, 1642.
Hall, F. G., and Salzano, J. V. (1963). Technical documentary report No. AMRL-TDR-63-19; Project No. 7163; Contract No. AF 33(616)-6803). Biomedical Laboratory, Air Force Systems Command, Wright-Patterson Air Force Base, Ohio.
Hall, P. W. (1953). *Circulation Res.* **1**, 238.
Harris, P., Bishop, J. M., and Segel, N. (1961). *Clin. Sci.* **21**, 295.
Harris, P., and Heath, D. (1962). "The Human Pulmonary Circulation: Its Form and Function in Health and Disease", p. 130. 1st Edition. E. S. Livingston Ltd., Edinburgh and London.
Harrison, T. R. and Blalock, A. (1927). *Amer. J. Physiol.* **80**, 169.
Harrison, T. R., Blalock, A., Pilcher, C., and Wilson, C. P. (1927). *Amer. J. Physiol.* **83**, 284.
Hartroft, W. S., and Hartroft, P. M. (1961). *Fed. Proc.* **20**, 845.

Hatcher, J. D., Jennings, D. B., Parker, J. O., and Garvock, W. B. (1963). *Canad. J. Biochem. Physiol.* 41, 1887.

Heemstra, H. (1954). *Quart. J. exp. Physiol.* 39, 83.

Henry, J. Goodman, J., and Meehan, J. (1947). *J. clin. Invest.* 26, 1119.

Hemingway, A., and Nahas, G. G. (1952). *Amer. J. Physiol.* 170, 426.

Hemingway, A., and Nahas, G. G. (1958). *J. appl. Physiol.* 13, 267.

Heymans, C., and Neil, E. (1958a). "Reflexogenic Areas of the Cardiovascular System", p. 176. 1st Edition. J. and A. Churchill Limited, London.

Heymans, C., and Neil, E. (1958b). "Reflexogenic Areas of the Cardiovascular System", p. 135. 1st Edition. J. and A. Churchill Limited, London.

Hoffman, C. E., Clark, R. T., and Brown, E. B. (1945–6). *Amer. J. Physiol.* **145**, 685.

Hornbein, T. F., Griffo, Z. J., and Roos, A. (1961a). *J. Neurophysiol.* **24**, 561.

Hornbein, T. F., Roos, A., and Griffo, Z. J. (1961b). *J. appl. Physiol.* **16**, 11.

Houssay, B. A., and Molinelli, E. A. (1926). *Amer. J. Physiol.* 76, 538.

Houston, C. S. (1946). *Amer. J. Physiol.* **146**, 613.

Huckabee, W. E. (1958). *J. clin. Invest.* 37, 264.

Hugelin, A., Bonvallet, N., and Dell, P. (1959). *Electroenceph. clin. Neurophysiol.* **11**, 325.

Huidobro, F., and Braun-Menendez, E. (1942). *Amer. J. Physiol.* 137, 47.

Hurlimann, A., and Wiggers, C. J. (1953). *Circulation Res.* **1**, 230.

Hurtado, A., Merino, C., and Delgado, E. (1945). *Arch. intern. Med.* 75, 284.

Izquierdo, J. J. (1928). *Amer. J. Physiol.* 86, 145.

Justus, D. W., Cornett, R. W., and Hatcher, J. D. (1957). *Circulation Res.* 5, 207.

Kahler, R. L., Goldblatt, A., and Braunwald, E. (1962). *J. clin. Invest.* 41, 1553.

Kellaway, C. H. (1919). *J. Physiol.* 53, 211.

Kety, S. S. (1958). "Physiology of Cerebral Circulation in Man." *In Circulation* (Proceedings of the Harvey Tercentenary Congress), p. 331. Blackwell, Oxford.

Kety, S. S., and Schmidt, C. F. (1948). *J. clin. Invest.* 27, 484.

Keys, A., Stapp, J. P., and Violante, A. (1943). *Amer. J. Physiol.* 138, 763.

Korner, P. I. (1959). *Physiol. Rev.* 39, 687.

Kottke, F. J., Phalen, J. S., Taylor, C. B., Visscher, M. B., and Evans, G. T. (1948). *Amer. J. Physiol.* **153**, 10.

Krasnow, W., Neill, W. A., Messer, J. V., and Gorlin, R. (1962). *J. clin. Invest.* 41, 2075.

Kriss, J. P., Futcher, P. M., and Goldman, M. L. (1948). *Amer. J. Physiol.* **154**, 229.

Krogh, A. (1929). "The Anatomy and Physiology of Capillaries", p. 133. Revised edition. Yale University Press, New Haven, Connecticut.

Lambertsen, C. J. (1961a). "Anoxia Altitude and Acclimatization." *Medical Physiology Bard*, p. 691. 11th Edition. C. V. Mosby Company, St Louis, Missouri.

Lambertsen, C. J. (1961b). "Chemical Factors in Respiratory Control." *Medical Physiology Bard*, p. 633. 11th Edition. C. V. Mosby Company, St Louis, Missouri.

Lassen, N. A. (1959). *Physiol. Rev.* 39, 183.

Lennox, W. G., and Gibbs, E. L. (1932). *J. clin. Invest.* **11**, 1155.

Lewis, B. M., and Gorlin, R. (1952). *Amer. J. Physiol.* **170**, 574.

Lewis, R. A., Thorn, G. W., Koepf, G. F., and Dorrance, S. S. (1942). *J. clin. Invest.* 21, 33.

Lim, T. P. K., and Luft, U. C. (1963). Technical document report No. AAL-TDR-62-19; Project No. 8238-20; Contract No. AF41(657)-330. Arctic Aeromedical Laboratory, Air Force Systems Command, Fort Wainwright, Alaska.

Litwin, J., Dil, A. H., and Aviado, D. M. (1960). *Circulation Res.* 8, 585.

Lloyd, B. B., Jukes, M. G. M., and Cunningham, D. J. C. (1958). *Quart. J. exp. Physiol.* 43, 214.

Loeschcke, G. C. (1953). *Pflüg. Arch. ges. Physiol.* **257**, 349.

Loeschcke, H. H., and Gertz, K. H. (1958). *Pflüg. Arch. ges. Physiol.* **267**, 460.
Lorber, V., and Evans, G. T. (1943). *Proc. Soc. exp. Biol., N. Y.* **54**, 1.
Ludemann, H. H., Filbert, M. G., and Cornblath, M. (1956). *J. appl. Physiol.* **8**, 59.
Malmejac, J., Chardon, G., and Gross, A. (1950). *C. R. Soc. Biol., Paris* **144**, 522.
Manger, W. M., Wakim, K. G., and Bollman, J. L. (1959). "Chemical Quantitation of Epinephrine and Norepinephrine in Plasma", p. 182. Charles C. Thomas, Springfield, Illinois.
March, D., and Van Liere, E. J. (1948). *J. Pharmacol.* **94**, 221.
Marshall, E. K., and Kolls, A. C. (1919). *Amer. J. Physiol.* **49**, 302.
McGregor, M., Donevan, R. E., and Anderson, N. M. (1962). *J. appl. Physiol.* **17**, 933.
Mikasa, A., and Masson, G. M. C. (1961). *Proc. Soc. exp. Biol., N. Y.* **106**, 315.
Murray, J. F., and Young, I. M. (1963). *Amer. J. Physiol.* **204**, 963.
Mylon, E., and Heller, J. H. (1948). *Proc. Soc. exp. Biol., N. Y.* **67**, 62.
Nahas, G. G. (1956). *J. appl. Physiol.* **9**, 65.
Nahas, G. G., Visscher, M. B., and Haddy, F. J. (1953). *J. appl. Physiol.* **6**, 292.
Nahas, G. G., Visscher, M. B., Mather, G. W., Haddy, F. J., and Warner, H. R. (1954a). *J. appl. Physiol.* **6**, 467.
Nahas, G. G., Josse, J. W., and Muchow, G. C. (1954b). *Amer. J. Physiol.* **177**, 315.
Nahas, G. G., Mather, G. W., Wargo, J. D. M., and Adams, W. L. (1954c). *Amer. J. Physiol.* **177**, 13.
Neil, E. (1956). *Arch. int. Pharmacodyn.* **105**, 468.
Newhouse, M. T., Becklake, M. R., Macklem, P. T., and McGregor, M. (1963). *J. appl. Physiol.* (In press.)
Nielsen, M., and Smith, H. (1951). *Acta physiol. scand.* **24**, 293.
Ochoa, S. (1939). *Biochem. J.* **33**, 1262.
Olsen, R. E. (1963). *Ann. int. Med.* **59**, 960.
Parker, J. O., Baugh, C. W., and Hatcher, J. D. (1958). *Proc. Canad. Fed. Biol. Soc.* **1**, 38.
Penna, M., Soma, L., and Aviado, D. M. (1962). *Amer. J. Physiol.* **203**, 133.
Peters, J. P., and Van Slyke, D. D. (1931). "Quantitative Clinical Chemistry", Vol. I, p. 597. 1st Edition. Williams and Wilkins Company, Baltimore.
Rapela, C. E., and Houssay, B. A. (1952). *C. R. Soc. Biol., Paris* **146**, 1977.
Reeve, E. B., Gregersen, M. I., Allen, T. H., and Sear, H. (1953). *Amer. J. Physiol.* **175**, 195.
Rivera-Estrada, C., Saltzman, P. W., Singer, D., and Katz, L. N. (1958). *Circulation Res.* **6**, 10.
Samaan, A., and Stella, G. (1935). *J. Physiol.* **86**, 309.
Sands, J., and DeGraff, A. C. (1925). *Amer. J. Physiol.* **74**, 416.
Sarnoff, S. J., and Berglund, E. (1954). *Circulation* **9**, 706.
Scarborough, W. R., Penneys, R., Thomas, C. B., Baker, B. M., and Mason, R. E. (1951). *Circulation* **4**, 190.
Schneider, E. C., and Truesdell, D. (1924). *Amer. J. Physiol.* **71**, 90.
Scott, J. C., Finkelstein, L. J., and Croll, M. N. (1962). *Amer. J. Cardiol.* **10**, 840.
Selkurt, E. E. (1953). *Amer. J. Physiol.* **172**, 700.
Shorr, E. (1948). *Amer. J. Med.* **4**, 120.
Shorr, E. (1955). "Polypeptides Which Stimulate Smooth Muscle", p. 120. (J. H. Gaddum, ed.). E. and S. Livingston Limited, London.
Shorr, E., Zweifach, B. W., and Furchgott, R. F. (1948). *Ann. N. Y. Acad. Sci.* **49**, 571.
Skinner, S. L., McCubbin, J. W., and Page, I. H. (1963). *Circulation Res.* **15**, 64.
Soma, L. R., Penna, M., and Aviado, D. M. (1965). *Pflüg. Areh. ges. Physiol.* **282**, 209.
Stacey, R. W., and Demenbrun, D. O. (1950). *Amer. J. Physiol.* **161**, 51.
Stavraky, G. W. (1942). *Amer. J. Physiol.* **137**, 485.

Storstein, O. (1952). *Acta med. scand.* **143**, supplement 269.
Stroud, R. C., and Conn, H. L. (1954). *Amer. J. Physiol.* **179**, 119.
Stroud, R. C., and Rahn, H. (1953). *Amer. J. Physiol.* **172**, 211.
Strughold, H. (1930). *Amer. J. Physiol.* **94**, 641.
Surtshin, A., Rodbard, S., and Katz, L. N. (1948). *Amer. J. Physiol.* **152**, 623.
Takacs, L. (1957a). *Acta physiol. hung.*, **11**, 55.
Takacs, L. (1957b). *Magy. belorv. arch.* **10**, 74.
Taquini, A. C. (1950). "Factors Regulating Blood Pressure." *Transactions of the Fourth Conference*, p. 209. Josiah Macy, Mr., Foundation, New York.
Tenney, S. M., Remmers, J. E., and Mithoefer, J. C. (1963). *Quart. J. exp. Physiol.* 48, 192.
Thorn, G. W., Jenkins, D., and Laidlaw, J. C. (1953). "Recent Progress in Hormone Research", Vol. 8, p. 205. Academic Press, New York and London.
Toth, L. A. (1940). *Amer. J. Physiol.* **129**, 532.
Toyooka, E. T., and Blake, W. D. (1961). *Amer. J. Physiol.* **201**, 448.
Van Loo, A., Surtshin, A., and Katz, L. N. (1948). *Amer. J. Physiol.* **154**, 397.
von Euler, U. S., and Liljestrand, G. (1942). *Acta physiol. scand.* **4**, 34.
von Euler, U. S., and Liljestrand, G. (1946). *Acta physiol. scand.* **12**, 301.
von Euler, U. S., Liljestrand, G., and Zotterman, Y. (1939). *Skand. Arch. Physiol.* 83, 132.
Walker, W. F., Zileli, M. S., Reutter, F. W., Shoemaker, W. C., and Moore, F. D. (1959). *Amer. J. Physiol.* **197**, 765.
Whelan, R. F., and Young, I. M. (1953). *Brit. J. Pharmacol.* 8, 98.
Widdicombe, J. G. (1963). *Physiol. Rev.* 43, 1.
Wiggers, C. J. (1941). *Ann. intern. Med.* 14, 1237.
Witzleb, E., Bartels, H., Budde, H., and Mochizucki, M. (1955). *Pflüg. Arch. ges. Physiol.* **261**, 211.
Woods, E. F., and Richardson, J. A. (1959). *Amer. J. Physiol.* **196**, 203.

CHAPTER 5

High Altitudes

L. G. C. E. PUGH

A. High Altitude Populations†

1. *Indigenous peoples*

It has been estimated that at least 10 million people live permanently at heights between 12,000 ft and 13,000 ft (3600 m and 4000 m).‡ Of these, four-fifths live in the Andes and the rest in Tibet and surrounding areas. Climate and ecology rather than reduced oxygen pressure determine the altitude limit for human habitation in these regions. Three-quarters of the

† A table for converting in to ft and ft to in is on p. xiii.

‡ In this chapter, as elsewhere in the book, heights are given to the nearest 50 metres and the nearest 100 ft.

area of Tibet has an elevation of 16,000 ft (4900 m) or more, with 8 to 10 in (20 to 25 cm) of precipitation a year, winter temperatures extending down to −27°F (−33°) and summer temperatures rising only to 45° to 58°F (7° to 14·5°). It is a region of grass, scrub and desert, where only pastoral nomads can exist. The bulk of the population is found in south-eastern Tibet at altitudes of 9000 ft (2700 m) to 15,000 ft (4600 m) along the valleys of the great rivers that run into India, Burma and China and their tributaries. This region has a milder climate and supports an agricultural population, cultivating barley, wheat and other crops (Bell, 1924; Richardson, 1962).

The pastoral and agricultural communities of the Andean plateau live at altitudes ranging from 12,000 ft to 14,500 ft (3600 m to 4400 m). In southern Peru 13,000 ft (4000 m) of altitude marks a significant line in subsistence activity. Above this altitude agriculture is seriously hampered by temperatures below freezing-point in every month of the year. Despite these frosts, some crops, such as the native grains and certain varieties of potato, will grow. Thus the area above 13,000 ft (4000 m) is predominantly pastoral with some agriculture, whereas below it is agricultural with some pasture (Baker, 1964).

The communities living highest in the Andes are the mining communities that depend for their subsistence on food brought up from below. In Chile, Bolivia and Peru it is not uncommon to find mines at altitudes as high as 16,000 ft to 17,000 ft (4900 m to 5200 m) up (McFarland, 1936). There are several tin and silver mines near Potosí in Bolivia at 17,000 ft (5200 m) and numerous copper mines in the Cerro de Pasco region in Peru around 16,000 ft (4900 m). In northern Chile many of the sulphur deposits are found on the summits of extinct volcanoes at 18,000 ft to 20,000 ft (5500 m to 6100 m). Probably the highest permanently occupied settlement in the world is the mining camp situated at 17,500 ft (5300 m) on Mount Aucanquilcha, which was visited by the International Physiological High Altitude Expedition of 1935 (McFarland, 1937; Dill, 1938). At that time there were 150 people living there. Many had been there for years, although the labour turnover was large. The high wages brought a continual stream of employees, but many of them could not become adjusted to 17,500 ft (5300 m) or 19,000 ft (5800 m). Hence the group was a highly selected one. Surprisingly, many of the workers were Chileans or Bolivians born and bred near sea-level, who were competing successfully with natives of 12,000 ft to 14,00 ft (3600 m to 4300 m) after some months of acclimatisation. Some of the miners had their families living with them, the women and children faring as well as the men. The women went down to Ollagüe (12,000 ft, 3600 m) to give birth to their children, but returned to the camp a few weeks later. The miners climbed daily 1500 ft (450 m) to work in the sulphur mines at 19,000 ft (5800 m). An attempt on the part of the mine authorities to induce them to live nearer their work was a failure. The miners refused to occupy a camp built for them at 18,500 ft

(5600 m), the reasons being loss of weight and appetite and inability to sleep. On this evidence 17,500 ft (5300 m) has come to be regarded as the limit of altitude to which man can become permanently adjusted.

The principal difficulties encountered by incoming peoples at heights over 12,000 ft (3600 m) have been (1) cold and (2) infertility. Many references to both occur in the historical records of the Incas and Spanish colonists of Peru (Monge, 1948).

The Incas moved the populations of towns and villages from one region to another for political reasons. In doing so they were careful to move them to places at the same altitude or "air temper", because it was found that the mortality of lowlanders moved to high altitudes was high, as was that of highlanders moved to low altitudes. When the Spaniards founded the imperial city of Potosí at 13,000 ft (4000 m) in the sixteenth century, 100,000 natives and 20,000 Spanish colonists were settled there. The fertility and fecundity of the natives was unimpaired, but the Spanish women had to descend to lower altitudes to rear their families. The birth in Potosí of the first Spaniard to survive infancy took place fifty-three years after the founding of the city and was known as the miracle of St Nicholas Tolentino. The child grew to maturity and married in Potosí, but none of his six children survived infancy. Eventually the Spanish population recovered its fertility, largely through interbreeding with the Indians. But the capital was finally moved to Lima, because of the relative infertility of cattle and horses at Potosí. The Spaniards made wide use of negro slaves in their colonial empire, but they were unable to use them for working the mines at high altitude because of the high mortality from cold and exhaustion.

Adaptation to cold is largely achieved by clothing and shelter, and it seems surprising that incoming populations should have so much difficulty over this. It would seem that native populations are slow to change their customs and habits of dress, even when the climate demands it. This is not always a matter of prejudice or conservatism: often economic or logistic reasons make it impossible, as for refugees or armies in the field.

2. *Visitors*

By ascending in slow stages a man may reach an altitude of 18,000 ft (5500 m) without adverse symptoms other than shortness of breath, a certain lassitude and Cheyne-Stokes breathing at night. If the ascent to altitude is too rapid, as by car or plane, symptoms of mountain sickness will appear some hours after arrival. The critical height for the average young adult is about 11,000 ft (3300 m). Susceptible persons may have symptoms as low as 6000 to 8000 ft (1800 to 2400 m), whereas some exceptional individuals can tolerate 15,000 ft (4600 m). At White Mountain in California about 50% of visitors to the lower station at 12,000 ft (3600 m) have symptoms of malaise, headache and insomnia, having come up by car from sea-level (N. Pace—personal communication). It has been claimed that the critical height for mountain sickness varies in different parts of the world, but differences are most likely related to the manner of ascent.

There has been some discussion about the greatest height to which men born and bred at sea-level can become permanently acclimatised. This question was studied by the Himalayan Scientific and Mountaineering Expedition of 1960–1 (Pugh, 1962a, b), which spent eight months at heights over 15,000 ft (4600 m) and more than three months continuously at 19,000 ft (5800 m) before attempting to climb Mount Makalu (27,900 ft, 8500 m) without the aid of oxygen equipment. It was found that a fresh party of climbers who joined the expedition for the ascent of Makalu were, after two months in the field, fitter if anything and more energetic than those who had been at 19,000 ft (5800 m) all the winter.

The wintering party felt reasonably well and did six hours or more of physiological work a day, with one or two hours of outdoor physical exercise, yet they lost weight continuously at the rate of 0·5 to 1·5 kg a week, and two out of eight members of the party had to descend at intervals to rest at lower altitudes.

It was concluded that for most people 17,500 ft (5300 m) would be the limit and that for some people even 15,000 ft (4600 m) would be too high for permanent acclimatisation. This conclusion is in line with the findings of the International High Altitude Expedition of 1935 (Dill, 1938) and is on the whole supported by the experiences of the Indian Army in the defence of their Himalayan border. The Indians are said to have kept troops for up to two years at altitudes as high as 18,000 ft (5500 m), but precise information on this point is lacking.

Our knowledge of life at heights over 19,000 ft (5800 m) depends mainly on the recorded experiences of mountaineering expeditions and on the observations made by doctors and physiologists who have accompanied them.

The Duke of Abruzzi's expedition to Bride's Peak in the Karakorum in 1909 (Filippi, 1912) first demonstrated that it is possible for a man to ascend by his own efforts to 25,000 ft (7600 m), contrary to the predictions of physiologists. Climbers on Mount Everest (Norton, 1925; Bruce, 1923; Ruttledge, 1934) were the first to emphasise the need for careful acclimatisation and to describe the signs and symptoms of high altitude deterioration. By 1933 it had become clear that, if the mountain was to be successfully climbed, the time spent above 23,000 (7000 m) would have to be kept to a minimum. Although there was initial improvement in well-being and climbing performance at heights up to 23,000 ft (7000 m), there was an underlying process of physical deterioration, which was the more rapid and severe the higher they went above 20,000 ft (6100 m). These expeditions also experimented with oxygen equipment, but failed to derive much benefit from its use, because of insufficient flow rates of oxygen and inadequate breathing apparatus. Two climbers in 1924 and four in 1933 succeeded in reaching an altitude of about 28,200 ft (8600 m), being defeated on each occasion by several factors, of which oxygen lack was only one.

These expeditions had approached Everest from the north, which entailed a six-week march through Tibet, all of it at altitudes over 13,000 ft (4000 m). They were therefore comparatively well acclimatised by the time they reached their base camp. In 1932 a party led by Smythe succeeded in climbing Mount Karmet (25,600 ft, 7700 m) within three weeks of leaving the plains (Smythe, 1932). This achievement started a controversy about the time required for acclimatisation that lasted virtually until the first ascent of Everest in 1953.

After World War II, Nepal opened its frontiers to foreigners for the first time since 1873, and in 1951 Shipton discovered a possible route to Mount Everest from the south. Two Swiss expeditions (Kurz, 1953) attempted to climb the mountain by this route in 1952. On the first attempt a party of five climbers reached the South Col (26,000 ft, 7900 m) after an exhausting climb of 3500 ft (1000 m) from the Western Cwm. They remained there five days with little food and less than 1 l of fluid a day. Only Lambert and Tensing were strong enough to continue. After camping at 27,400 ft (8300 m), these two climbers reached an altitude of about 28,000 ft (8600 m), climbing at a rate of only 150 ft an hour. Although they had oxygen equipment with them, like others before them, they were unable to use it successfully and were forced to turn back at the same altitude and for the same reasons as previous parties. The second Swiss expedition, which set out in the autumn of 1952, was defeated mainly by the intense cold at and above the South Col (25,800 ft (7800 m)).

The mountain was finally climbed in 1953 by a British expedition led by Sir John Hunt (1953). This expedition benefited by the experience of the Swiss who had pioneered the route to the South Col (25,800 ft, 7800 m); they were also greatly helped by the results of investigations carried out on a training expedition to Mount Cho Oyu (26,800 ft, 8100 m), led by Eric Shipton in 1952. Shipton's party had studied the problems of acclimatisation and high altitude deterioration (Pugh, 1954a), as well as nutrition and protective equipment, but above all they had shown that oxygen equipment could be used successfully, given adequate flow rates of oxygen and apparatus that did not restrict breathing at high ventilation rates.

The 1953 expedition took with them 198,000 l of oxygen, and the assault parties used it for climbing all the way up from the West Cwm (22,000 ft, 6700 m). Bourdillon and Evans used closed circuit oxygen equipment in their attempt to climb from the South Col to the summit in a single day. They achieved a rate of ascent of nearly 1000 ft an hour (Bourdillon, 1954);

but the apparatus was giving trouble by the time they reached the South Summit (28,700 ft, 8750 m), and they turned back. Hillary and Tensing set out on the next day, using open-circuit equipment at a flow rate of 3 l O_2 STP/min, and were able to maintain a climbing rate of 600 ft an hour. They camped at 27,900 ft (8500 m) and next day reached the summit.

Since 1953 all the 28,000 ft (8500 m) Himalayan peaks have been climbed with the aid of oxygen. Everest was climbed again by the Swiss in 1956 (Schmied, 1957) and by an American expedition in 1963 (Dyhrenfurth, 1963).† But the highest peak to be conquered so far without supplementary oxygen has been Mount Dawlaghiri (27,000 ft, 8200 m), which was climbed by a Swiss expedition in 1959 (Eiselin and Forrer, 1961).

Mountaineers at great heights, like armies and intrusive populations at more moderate altitudes, have suffered as much from cold as from oxygen lack. Almost every party pinned down by illness or accident at heights over 24,000 ft (7300 m) has suffered severely from frostbite. Notable examples are Herzog's party on Mount Annapurna (Herzog, 1952) and the 1961 expedition to Mount Makalu (Hillary, 1962). In 1963 four Americans, after ascending to the summit of Mount Everest, were forced to spend a night in the open at about 28,000 ft (8500 m). To their great good fortune it was a calm night, and they escaped with relatively minor injuries.

3. *Mental effects*

Up to about 20,000 ft (6100 m) mountaineers have found remarkably little impairment of mental function other than lassitude, slowness and a tendency to forgetfulness (Pugh and Ward, 1956). Despatches written by expedition leaders and the poetry and painting produced by artistically gifted climbers have been of high quality. Complicated tasks, such as gas analysis, have been carried out without loss of accuracy and with few mistakes in calculation. On the scientific and mountaineering expedition of 1960–1 only minor impairment was revealed by card sorting and other psychological tests during a winter spent at 19,000 ft (5800 m) (Gill, Poulton, Carpenter, Woodhead and Gregory, 1964).

At 25,850 ft (7800 m) on Mount Everest most climbers have been rather severely affected by oxygen lack and have described blunting of emotional response, lack of insight, mental inertia, and slowing of thought processes (Pugh and Ward, 1956)—a state, in fact, resembling a psychiatric depressive illness at sea-level. However, although the ability to trace faults in equipment or meet unforeseen situations has been impaired, carefully laid plans have been correctly followed out, and sound judgement has been displayed in mountaineering matters.

None of the climbers who have ascended without oxygen equipment to

† And by other expeditions since then (Ed.).

28,200 ft (8600 m) have suffered any permanent after effects, and all have had remarkably successful careers in occupations demanding high standards of mental ability. Their number includes three university professors, a general, a former colonial governor and a prominent surgeon. One member of the 1933 Everest expedition, however, who spent five nights at 25,700 ft (7800 m), found difficulty in remembering the names of guests at social functions when he returned to his post as aide-de-camp to the Viceroy of India.

4. *Hypoxia*

Fig. 1 shows the partial pressure of oxygen in inspired air and alveolar gas and the percentage saturation of arterial blood with oxygen in acclimatised persons at various altitudes. The alveolar gas values were obtained by analysis of Haldane end-expiratory gas samples taken from recumbent subjects on mountaineering expeditions (Table 1). The results for blood are those reported

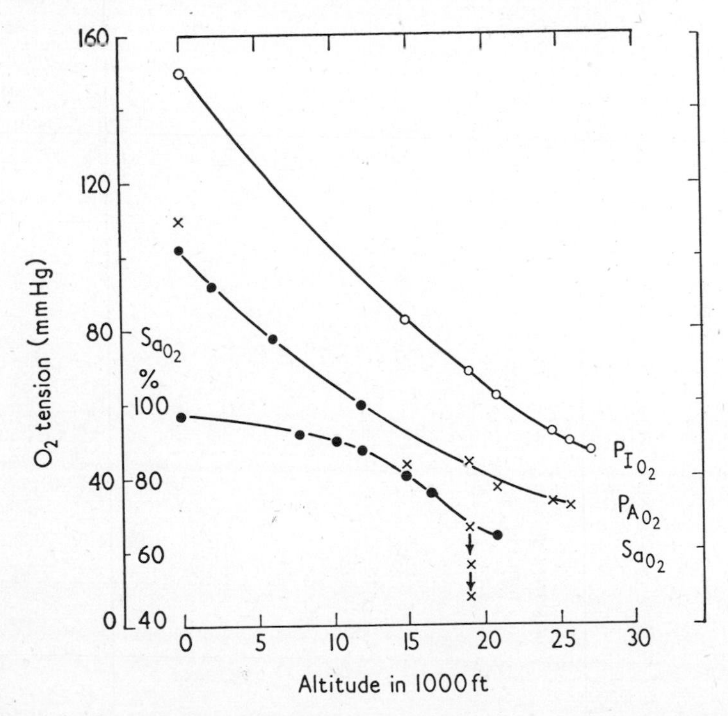

FIGURE 1. Partial pressure of oxygen in inspired air ($P_{I\,O_2}$) and alveolar gas ($P_{A\,O_2}$), and percentage saturation of arterial blood with oxygen ($S_{a\,O_2}$) in acclimatised or partially acclimatised visitors at various altitudes up to 25,700 ft (7800 m) (Bar press 288 mm Hg) (Pugh, 1962b). The crosses represent mean values reported by the Himalayan Scientific and Mountaineering Expedition 1960–1. Filled circles represent published values from other sources. The arrows point to results obtained during exercise at 19,000 ft (5800 m) (Gill, Milledge, Pugh and West, 1962).

Table I. *Composition of Inspired and Alveolar Gas at Various Altitudes from Sea-level to 7830 m (25,700 ft)*

Altitude	*Barometric pressure* mm Hg	P_{IO_2} mm Hg	P_{AO_2} mm Hg	P_{ACO_2} mm Hg	$P_I - P_{AO_2}$ mm Hg	*Number of samples*
Sea-level	750	150	110	38	40	
5800 m (19,000 ft)	380	69	45	22	24	21
6400 m (21,000 ft)	344	62	38	21	24	5
7440 m (24,400 ft)	300	53	34	16	19	8
7830 m (25,700 ft)	288	50	33	14	17	8

by Houston and Riley (1947) from their chamber experiment, Operation Everest, and by West, Gill, Lahiri, Milledge, Pugh and Ward (1962) from the Himalayan Scientific and Mountaineering Expedition 1960–61.

At 25,700 ft (7800 m) on Mount Makalu (pressure 288 mm Hg) the alveolar P_{O_2} was only 33 mm Hg. The corresponding value for arterial P_{O_2} would be about 31 mm Hg, and at pH 7·5 the arterial oxygen saturation would be 66%. The mean values for resting arterial oxygen saturation observed by the Himalayan Scientific and Mountaineering Expedition during a winter spent at 19,000 ft (5800 m) was also 67% and similar values were reported at somewhat higher simulated altitudes by Houston and Riley (1947). It is likely that the resting arterial oxygen saturation does not go much below 65% at extreme altitude owing to hypocapnia (low CO_2). During exercise, however, arterial oxygen saturation may fall to extremely low levels because of diffusion limitation: values as low as 45% were observed in short periods of maximum exercise at 19,000 ft (5800 m) by the 1960–1 Himalayan expedition (West *et al*, 1962).

B. Physiological Adjustments

1. *Respiration*

The complex and finely integrated responses of the organism to reduced atmospheric oxygen pressure are generally classified according to the duration of exposure. One reason for this is that the different types of exposure are important in different situations, e.g. acute exposure in aviation, chronic exposure in terrestrial regions. Native populations that have become adjusted to high altitude over many generations show certain differences from acclimatised visitors, and some physiologists seek to restrict the word adaptation to the special changes found in this group.

The possible mechanisms available to the body for transporting oxygen to the tissues in normal amounts in the presence of reduced atmospheric oxygen pressure gradient are:

(i) Increased ventilation,
(ii) Shift of the arterio-venous points to the steep part of the Hb_{O_2} dissociation curve,
(iii) Circulatory increase,
(iv) Elevation of blood oxygen capacity,
(v) Changes at tissue level, viz. increased concentration of respiratory pigments and enzymes, increased vascularity and increased regional blood flow.

Of these mechanisms, some are immediately available, for example the shift in the oxygen transport to the steep portion of the dissociation curve and the circulatory increase. Others take time to develop, either because they involve structural changes or because they are partially inhibited by opposing forces.

a. The P_{O_2} gradient. Fig. 2 illustrates the gradient of oxygen pressure at various stages in the transport of oxygen between the atmosphere and the mixed venous blood, (*a*) for persons at sea-level and (*b*) for acclimatised persons living at 19,000 ft (5800 m). The slope of the O_2 gradient is much

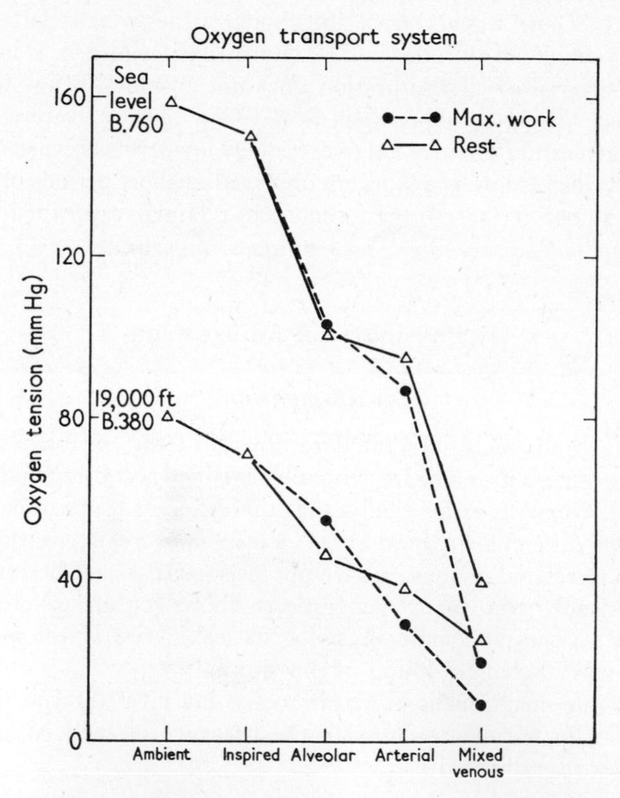

FIGURE 2. Oxygen tension at various stages in the oxygen transport system in rest and maximum exercise for subjects at sea-level and at 19,000 ft (5800 m) and in the acclimatised state (Pugh, 1964c).

smaller at altitude than at sea-level, and the mixed venous P_{O_2} at altitude is within 10 mm Hg of the sea-level value both in rest and exercise, in spite of 80 mm Hg difference in atmospheric P_{O_2}. Economy in P_{O_2} is achieved mainly by two adjustments; one of these is increased ventilation, which is responsible for the reduction in P_{O_2} gradient between inspired and alveolar gas; the other is the displacement of the zone of O_2 uptake and unloading by the blood to the steep portion of the Hbo_2 dissociation curve, where the change in blood o_2 content per mm Hg change in o_2 pressure is large. This, being an inherent physicochemical property of the haemoglobin molecule, is not an adaptive response in the ordinary sense of the word, but it is nevertheless one of the principal mechanisms upon which man's tolerance of altitude depends.

b. Lung ventilation. The respiratory response to altitude begins within a few hours of arrival and increases rapidly during the first week. A slow further increase takes place over the ensuing three to six weeks, but there is much individual variation. In the early stage of acclimatisation respiration is poorly controlled and tends to "hunt". The commonest abnormality is Cheyne-Stokes breathing, which may be observed in persons at rest or asleep, at heights over 10,000 ft (3000 m). The respiratory cycle consists typically of three to four respirations followed by a pause lasting up to 10 sec. At great heights Cheyne-Stokes breathing tends to persist, and some people suffer considerable discomfort from it at night, because they fall asleep in the apnoeic phases and wake with a feeling of suffocation each time breathing begins again.

In acclimatised visitors the resting minute volume at any given altitude tends to be constant and independent of altitude, so that the reduced density of the air is completely compensated for (Pugh, 1957); again, however, there is considerable individual variation (Gill and Pugh, 1964). Chiodi (1957, 1963) found that permanent residents at heights around 13,000 to 15,000 ft (4000 to 4600 m) ventilated less than acclimatised or partly acclimatised visitors. Some of his subjects had CO_2 response curves similar to those of persons at sea-level and showed reduced sensitivity to hypoxia.

c. The CO_2/O_2 diagram. The effect of high altitude hyperpnoea in reducing the gradient of oxygen pressure between the atmosphere and the lung alveoli, and the difference between acutely exposed unacclimatised and chronically exposed acclimatised persons are conveniently illustrated by the Rahn-Otis CO_2/O_2 diagram, in which alveolar P_{CO_2} ($P_{A\ CO_2}$) is plotted against alveolar P_{O_2} ($P_{A\ O_2}$).

Figure 3 shows the curves of Rahn and Otis (1949) with the addition of results obtained at extreme altitude on Himalayan expeditions. The iso-altitude lines are lines along which the alveolar gas points must lie at any given altitude and at the specified alveolar exchange ratio of 0·85. The intercept on the abscissa gives the inspired P_{O_2} at the particular altitude. As acclimatisation proceeds, the alveolar gas point moves downwards along the iso-altitude line from the point of intersection with the curve for acute exposure to the point of intersection with the curve for acclimatised persons. $P_{A\ CO_2}$ is a measure of the increase in ventilation, and alveolar $P_{A\ O_2}$ indicates the degree of compensation for reduced atmospheric oxygen pressure. It is seen that the advantage in $P_{A\ O_2}$ gained by acclimatisation is greatest in the range of altitudes between 9000 ft (2700 m) and 15,000 ft (460 m). In this range

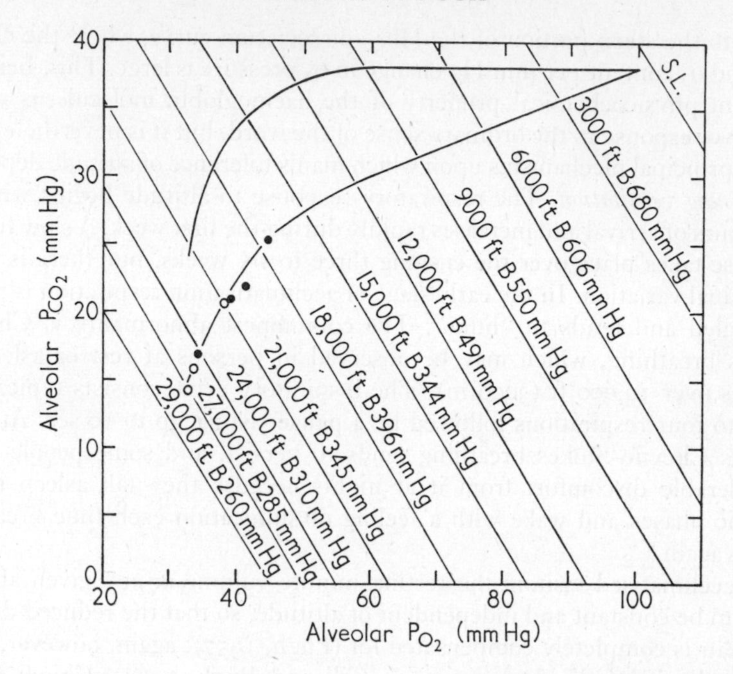

FIGURE 3. The $CO_2 - O_2$ diagram (Rahn and Otis, 1949) showing alveolar P_{CO_2} and P_{O_2} in unacclimatised persons acutely exposed to altitude (upper curve) and in acclimatised persons chronically exposed to altitude (lower curve). The plotted points represent mean values of Haldane end expiratory samples taken on Himalayan expeditions. The oblique lines are lines along which the alveolar values at any given altitude must lie at a given alveolar exchange ratio of 0·85.

the gain in $P_{A\,O_2}$ in acclimatised persons amounts to 8 to 9 mm Hg and corresponds to an altitude difference of 3000 to 4000 ft (900 to 1200 m).

Above 15,000 ft (4600 m) the gain in $P_{A\,O_2}$ with acclimatisation becomes smaller; at 21,000 ft (6400 m) it is only 5 mm Hg, and above 27,000 ft (8200 m) mountaineers are in much the same situation as acutely exposed persons would be.

d. The alveolar-arterial O_2 gradient. In addition to its effect on $P_{A\,O_2}$, increased ventilation reduces the P_{O_2} gradient between the lung alveoli and the arterial blood. Most observers have found that the aveolar-arterial P_{O_2} gradient in resting subjects at heights above 14,000 ft (4300 m) is about 2 mm Hg, to be compared with the normal value of 10 mm Hg at sea-level (Dill, 1938; Houston and Riley, 1947). This is explained by the raised ventilation-perfusion ratio secondary to increased ventilation (Hurtado, 1964), the improved distribution of ventilation is underventilated parts of the lung and the reduction at low P_{O_2} of the O_2 gradient associated with venous admixture.

e. Respiratory alkalosis. At heights over 12,000 ft (3600 m) blood pH rises within two to three days to values in the region of 7·50 (Houston and Riley, 1947; Severinghaus, Mitchell, Richardson and Singer, 1963) and it may be many weeks before the normal value is regained.

The compensating adjustment consists in reduction of plasma bicarbonate by the action of the kidney, bicarbonate being replaced by chloride. There is comparatively little information on the time course or completeness of these adjustments at various altitudes. The International High Altitude Expedition of 1935 found their blood pH still elevated after six weeks at 17,500 ft (5300 m), the mean value being 7·45 compared with 7·35 for native residents (Dill, 1938), and the Himalayan party in 1960–61 obtained similar results at 19,000 ft (5800 m).

f. Respiratory regulation. Our understanding of this complex and incompletely understood subject has been greatly clarified by experiments in which ventilation was measured on subjects breathing varying concentrations of CO_2 while P_{O_2} was kept constant at different levels.

The changes observed during acclimatisation to altitude are schematically illustrated in Fig. 4, which is based on values collected by Milledge (1963a)

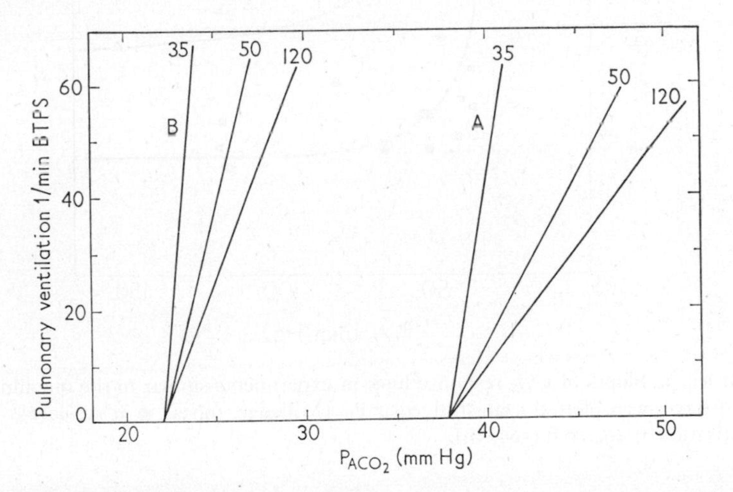

FIGURE 4. Ventilatory response to CO_2 at constant O_2 tension in a subject A at sea-level and B after acclimatisation at 19,000 ft (5800 m). The numbers above each line represent P_{O_2} in mm Hg (Milledge, 1963a).

at 19,000 ft (5800 m) on the Himalayan Scientific and Mountaineering Expedition 1960–1. (1) The fan of lines representing the ventilatory response to CO_2 at different oxygen pressures is displaced to the left, and the extrapolated value of P_{ACO_2} at zero ventilation (or apnoea point), which represents the response threshold to CO_2, was reduced to 24 mm Hg compared with 38 mm Hg at sea-level. This change was virtually complete one week after arrival at 19,000 ft (5800 m). (2) The slope of the $V/P_{A\,CO_2}$ line at sea-level P_{O_2} was considerably steeper after acclimatisation, indicating increased CO_2 sensitivity. This was a more gradual change, taking place over several weeks. (3) The point representing resting ventilation with the subjects breathing atmospheric air (not shown in Fig. 4) was always slightly to the left of the apnoea point, showing that even after many weeks at 19,000 ft (5800 m) the subjects were still responding primarily to the hypoxic drive from the carotid body. (4) Fig. 5 shows the curves obtained by plotting the slope of $\dot{V}/P_{A\,CO_2}$ lines against P_{O_2} at sea-level and at altitude. Although the

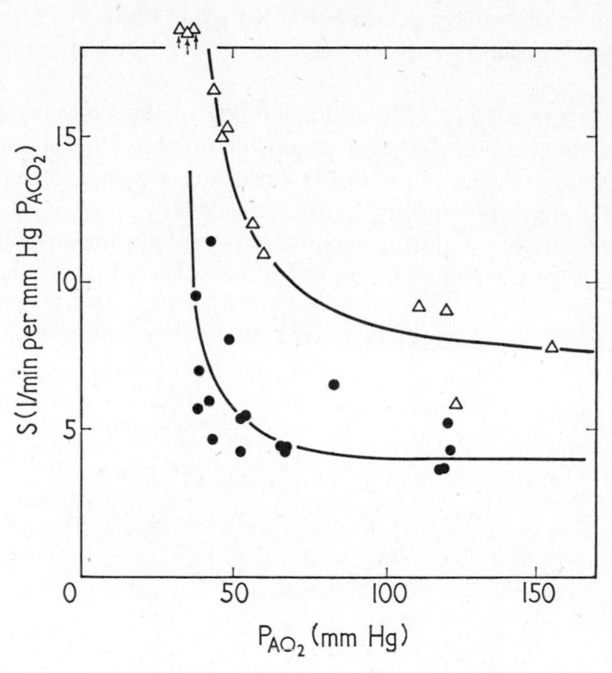

FIGURE 5. Slopes of CO_2 response lines in experiments similar to the one illustrated in Fig. 4 have been plotted against alveolar P_{O_2} (Milledge, 1963a). ● at sea-level, △ after acclimatisation at 19,000 ft (5800 m).

CO_2 sensitivity at a given P_{O_2} is greater at altitude, the P_{O_2} at which the curves become vertical is unchanged at altitude (i.e. 30 mm Hg). Lloyd, Jukes and Cunningham (1958) have called this the point of infinite P_{O_2} sensitivity and regard it as a measure of O_2 sensitivity. On this and other evidence it was concluded that the sensitivity to hypoxia was not changed after six to twelve weeks at 19,000 ft (5800 m). It is worth pointing out here that the $P_{A\,CO_2}/P_{A\,O_2}$ curves on the Rahn-Otis CO_2/O_2 diagram also become vertical at a P_{O_2} of about 30 mm Hg (Fig. 3).

Until recently the shift of the CO_2 response curve during acclimatisation was attributed to renal reduction of plasma bicarbonate and restoration of blood pH towards normal. However, Severinghaus, Mitchell, Richardson and Singer (1963) have produced evidence that a major role in this mechanism is played by chemoreceptor areas in the fourth ventricle, which respond to the pH of the cerebrospinal fluid. Their hypothesis is that the pH of the cerebrospinal fluid is actively regulated by the choroid plexus and glia by accepting or rejecting HCO_3 ions. On ascending to altitude, hypoxic stimulation of the carotid body results in some increase in ventilation and a fall in P_{CO_2}. The resulting mild alkaline shift of cerebrospinal fluid (H^+) decreases CO_2 chemoreceptor activity, thus restraining the respiratory response to the carotid body stimulation. The alkalinity extends to all brain extracellular fluid and cerebrospinal fluid. As new cerebrospinal fluid of normal pH reaches the CO_2 chemoreceptors, their pH is restored towards normal, permitting ventilation to increase in response to carotid body stimulation. Severinghaus *et al.* (1963) obtained evidence supporting their hypothesis on human subjects during a visit to 12,400 ft (3600 m). They found that blood pH rose to an average value of 7·48 within a day of arrival and remained near that through the eight-day period of observation. The pH of the cerebrospinal fluid, on the other hand, showed only minor changes. Ventilation showed the expected increase associated with acclimatisation.

2. *Changes in the blood*

Elevation of the red cell and haemoglobin contents of the blood was the first adaptive adjustment to be clearly recognised in persons ascending to high altitude (Viault, 1891). Figures for residents at high altitudes show a linear relation between haemoglobin concentration and barometric pressure at altitudes up to 12,000 ft (3600 m), with a steeper increase at higher altitudes (Fig. 6). In acclimatised visitors the haemoglobin curve levels off at around 20·5 gm % (Pugh, 1964a, b). Among native residents the O_2 capacity of the blood is raised to such a degree that the arterial blood contains more O_2 than the arterial blood of persons at sea-level. In visitors the arterial O_2 content is the same as or rather lower than that at sea-level (Table 2).

The significance of increased O_2 capacity in acclimatisation to altitude lies in the fact that it increases the volume of O_2 transported per litre of blood flow when the available P_{O_2} gradient is restricted by reduction of arterial P_{O_2}. However, the steep increase in the apparent viscosity of the blood at haematocrits above 45% (Whittaker and Winton, 1933), particularly in the presence of vasoconstriction (Pappenheimer and Maes, 1942), makes it seem likely that special adaptive changes must take place in the vascular system before haemoglobin levels above 20 gm% become wholly advantageous. It is interesting to note in this connection that no correlation has been observed on Himalayan expeditions between haemoglobin concentration and physical

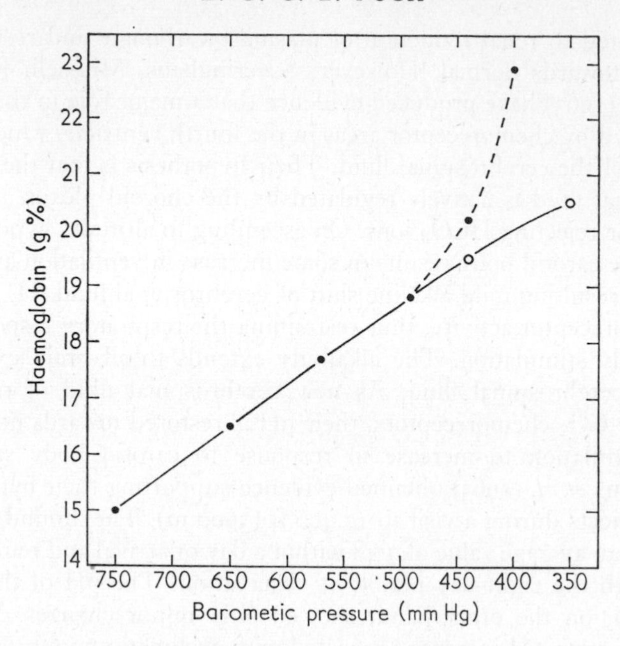

FIGURE 6. Relation between haemoglobin concentration and barometric pressure in native residents and acclimatised visitors at various altitudes. ●—●- - -●, mean values for residents at Denver (5000 ft, 1500 m), Mexico City (7500 ft, 2200 m), Oroya (12,000 ft, 3600 m), Morococha (14,900 ft, 4600 m) and 'Quilcha (17,500 ft, 5300 m), published by various authors. ○—○ Values from Himalayan expedition (Pugh, 1964a).

performance among either climbers or Sherpa porters (Pugh, 1954b). Others have pointed out that among high altitude populations occasional individuals are found who have sea-level haemoglobin values without obvious limitation of work capacity (Dill, 1938; Hurtado, 1964).

a. Blood morphology. The mean corpuscular volume, in both visitors (Pugh, 1962b and 1964a and b) and natives of high altitude (Hurtado, 1964), is somewhat higher than at sea-level (MCV 97 μ^3), but the cells are normally filled with haemoglobin (MCHC 32 to 34%).

According to Hurtado (1964), the number and distribution of leucocytes is normal among permanent residents at 14,900 ft (4550 m), but the reticulocyte count is raised (to 46 thousand/cu mm, compared with 18 thousand/cu mm at sea-level).

b. Erythrokinetics. Many recent advances have been made in our knowledge of the development and maintenance of polycythaemia at altitude. Barcroft (1925) pointed out that an increase of as much as 10 to 15% may occur in the haemoglobin concentration in persons ascending rapidly to about 14,000 ft (4300 m). This finding was confirmed by Assmussen and

Table II. *Haemoglobin Concentration, O_2 Capacity, Arterial O_2 Content ($C_a O_2$) and Percentage Saturation ($S_a O_2$) of Native Residents and Visitors at Various Altitudes.* (1. – Dill, 1958; 2. – Hurtado, 1964; 3. – Dill, Talbot and Consolazio, 1937b; 4 and 5 – Reynafarje, 1964; 6 and 7 – Pugh, 1962b).

Place	*Altitude* km	*Bar. press* mm Hg	Hb g %	O_2 cap. ml/100	C_{aO_2} ml/100	$S_a O_2$ %	*Time spent at altitude*
			Permanent residents				
1. Mexico City	2·3	570	17·7	23·8	21·3	91	
2. Morococha	4·5	440	20·1	26·9	22·6	80	
3. 'Quilcha	5·3	400	22·9	30·5	23·2	76	
			Acclimatised visitors				
4. Morococha	4·5	440	19·5	24·8	19·7	80	6 months
5. 'Quilcha	5·3	400	18·9	25·1	18·9	75	6 weeks
6. Ming Bo	5·8	380	19·6	26·2	17·5	67	5 months
7. Five Himalayan expeditions	5·3–6·5	400–350	20·5	27·4	17·8–19·2	65–70	
			Sea-level U.K.				
London	S.L.	750	14·5	19·5	18·9	97	

Consolazio (1941) on Mount Evans; they showed that fluid shifts and reduction of plasma volume were responsible. The slow subsequent rise, which continues for about two months, was formerly attributed entirely to increased erythropoiesis. Recent investigations have revealed a more complicated picture. Fig. 7 shows the changes in haemoglobin concentration, blood

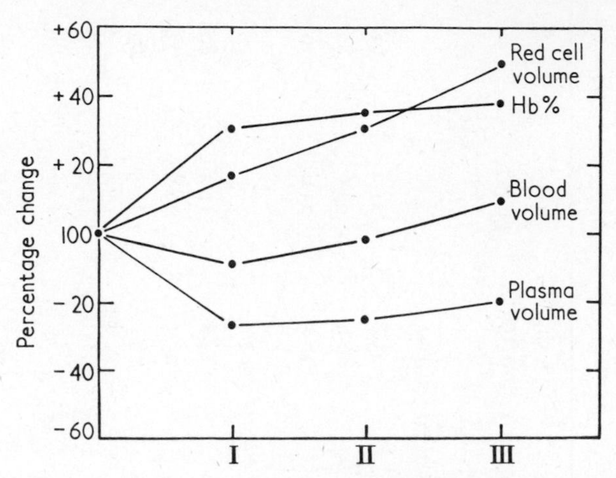

FIGURE 7. Average changes in haemoglobin concentration, red cell volume, plasma volume and blood volume in four subjects during eight months on Himalayan expedition: (I) after eighteen weeks at altitudes between 13,000 ft (4000 m) and 19,000 ft (5800 m); (II) after three to six weeks at 19,000 ft; and (III) after nine to fourteen weeks at or above 19,000 ft.

volume, plasma volume and red cell mass observed by the 1960–1 Himalayan expedition during eight months spent at heights above 15,000 ft (4600 m) and three months at 19,000 ft (5800 m). Similar results have been reported by Reynafarje (1964) from Morococha (14,900 ft, 4550 m). Red cell mass rose continuously over the whole period of observation, although the haemoglobin concentration was nearly constant after two to three months. Blood volume and plasma volume fell during the first two to four months, but subsequently rose, blood volumes reaching values 10 to 20% above the sea-level control values and plasma volumes remaining somewhat below the sea-level values. It is evident from these results that the haemoglobin concentration depends to a considerable extent on changes in plasma volume.

Reynafarje (1964) and his colleagues measured the turnover rate of haemoglobin iron in visitors to Morococha (14,900 ft, 4600 m). They found that red cell formation exceeded red cell destruction within two hours of arrival at Morococha and in seven to fourteen days had risen to three times the sea-level rate. Even eight months was insufficient for equilibrium to be reached between red cell formation and destruction. In native residents haemoglobin and blood volume were higher than among visitors even after one year's residence at Morococha, and red cell production was 30% higher than in normal subjects at sea-level. The survival time of red cells by the glycine ^{14}C or chromium 51 techniques, both in residents and newcomers, was within the normal range.

There is animal evidence that the polycythaemia of altitude is regulated by renal secretion of erythropoietin. Merino (1956) found some evidence of this in man. Plasma obtained from high altitude residents was injected into healthy subjects at sea-level and produced a moderate but constant reticulocytosis, in contrast with plasma from sea-level control subjects.

c. HbO_2 dissociation curve. Dissociation curves obtained by Barcroft during an expedition to Mount Teneriffe (12,200 ft, 3600 m) in 1910 led him to suppose that the affinity of blood for oxygen increases during acclimatisation to altitude (for references, cf. Haldane and Priestley, 1935), but others failed to confirm his findings. The matter was considered to have been finally settled by the International High Altitude Expedition of 1935, who found no evidence of any shift of the HbO_2 dissociation at a given pH either in native residents or in visitors at 17,500 ft (Hall, 1936). Hurtado (1964), however, has since reported a small but unmistakable shift to the right in the curve of native residents of Morococha (14,900 ft, 4550 m). It seems clear from these findings that increase in the affinity of blood for O_2 plays no important part in man's adaptation to altitude, as it does in some mammals, notably the llama (Dill, 1938; Meschia, Prystowsky, Hellegers, Huckabee, Metcalfe and Barron, 1960).

d. CO_2 dissociation curve. The changes taking place in the blood in response to low O_2 and CO_2 tension can be illustrated by considering the CO_2 dissociation curves for whole blood. Figure 8 shows CO_2 dissociation curves for fully oxygenated blood, plotted on log-log paper. The upper and lower curves were constructed from the nomograms of Dill, Edwards and Consolazio (1937a) and Dill, Talbot and Consolazio (1937b); they represent the blood of normal adults at sea-level and of South American miners living at 17,500 ft (5300 m) and working daily at 19,000 ft (5800 m). The middle curve represents the blood of members of the Himalayan Scientific and Mountaineering Expedition 1960–1 at 19,000 ft (5,800 m). The pH isopleths were constructed by joining the pH points, which were obtained from the nomograms. This empirical chart differs in two ways from the more familiar HCO_3^+/P_{CO_2} chart for plasma based on the Henderson-Hasselbalch equation. (1) The pH lines are not parallel, but slightly convergent in an upward direction; (2) the dissociation curves for whole blood are slightly non-linear at low P_{CO_2}. The usefulness of the chart lies in the fact that, given the P_{CO_2} and total CO_2 content of a sample of blood, its pH can be obtained without measuring the O_2 capacity of percentage saturation.

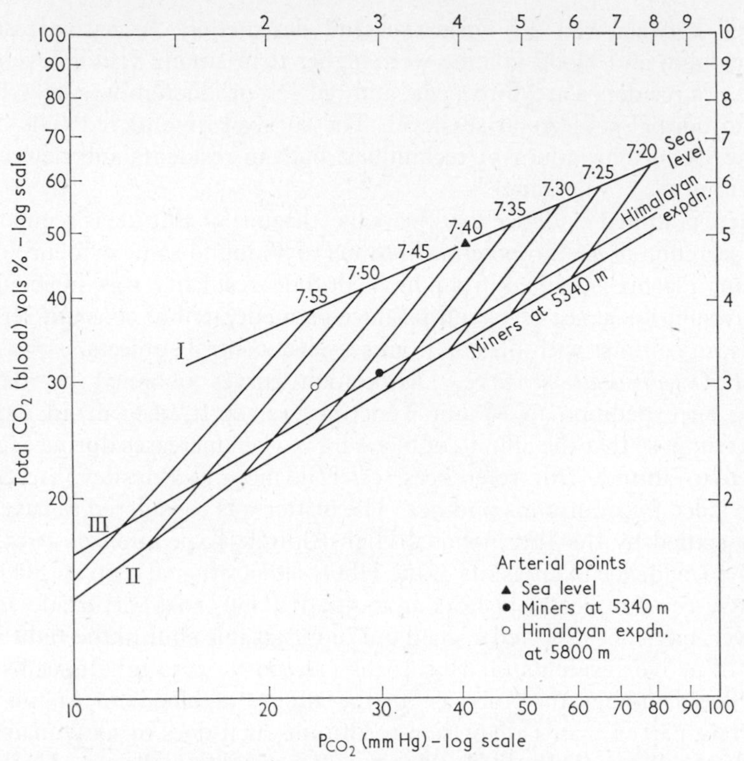

FIGURE 8. Carbon dioxide dissociation curves for fully oxygenated blood at sea-level and at high altitude (log-log scale). Curve I constructed from the nomogram of Dill, Edwards and Consolazio (1937a) for blood at sea-level; Curve II from the nomogram of Dill, Talbot and Consolazio (1937b) for the blood of South American miners living at 5340 m (17,500 ft) and working at 5800 m (19,000 ft); Curve III from experimental results obtained by the Himalayan Scientific and Mountaineering Expedition 1960–1 on blood samples from five subjects living at 5800 m (19,000 ft).

Reference to Fig. 8 shows that the blood of acclimatised or partially acclimatised persons at altitude contains less CO_2 at any given P_{CO_2} than the blood of persons at sea-level. This is the result of the renal compensation for respiratory alkalosis. The curve for miners at altitude 5340 m (17,500 ft) lies at a lower level than that for visitors, and the bicarbonate values at a given P_{CO_2} are correspondingly lower, suggesting a greater degree of renal compensation. This is confirmed by the arterial points, which show a pH of 7·37 for the miners, compared with 7·45 for the visitors. The arterial points, also show lower values for P_{CO_2}, evidence of lower ventilation. Both arterial points lie above the respective dissociation curves, owing to partial unsaturation of the arterial blood; mean saturation values were 76% for the miners (pressure 400 mm Hg) compared with 67% for the visitors (pressure 380 mm Hg). Another matter illustrated by this chart is the slightly steeper slope of the

CO_2 dissociation curves at altitude. A given rise of P_{CO_2} causes a slightly smaller fall in pH in the blood of acclimatised persons at altitude than in the blood of persons at sea-level. Thus, in spite of the lower plasma HCO_3^+, the buffering power of whole blood for P_{CO_2} is somewhat greater at altitude than at sea-level. This is explained by the increase in haemoglobin concentration.

3. *Circulatory adjustment*

a. Cardiac output. Elevation of resting heart rate and cardiac output is an immediate response to hypoxia, with a threshold at around 8000 ft (2400 m). The mechanism would seem to be an indirect one, depending on reduced peripheral resistance in certain vascular beds, since no direct control mechanism has been demonstrated. As other adjustments are established, namely increased ventilation and blood O_2 capacity, the resting heart rate and the resting cardiac output revert to the normal sea-level values, at least at altitudes up to 15,000 ft (4600 m). At still higher altitude, there is probably some persistent circulatory increase, since resting heart rates remain elevated (Gill and Pugh, 1964), but we have no unequivocal information about this. (For references, see Korner (1959).

b. Pulmonary hypertension. Elevation of pulmonary arterial pressure in response to low O_2 tension was first demonstrated in patients undergoing cardiac catheterisation at sea-level. Rotta, Canepa, Hurtado, Velasquez and Chavez (1956) reported pulmonary hypertension in healthy adults after one year of residence in Morococha, and this finding was confirmed in persons who had resided at 14,900 ft (4550 m) from childhood upwards (Penaloza, Sime, Banchero, Gamboa, Cruz and Marticorena, 1963b). They found that the pulmonary arterial pressure was higher in native residents than in visitors and was associated with right ventricular hypertrophy, increased pulmonary vascular resistance and increased right ventricular work. Pulmonary wedge pressure, left ventricular work and resting cardiac output were the same as in normal subjects at sea-level. Arias-Stella and Seldaña (1964) have described a thickening of the muscular layer of the pulmonary arteries and muscularisation of the arterioles, somewhat resembling the pulmonary vasculature of the foetus. The critical range of altitude for the development of pulmonary hypertension is 8000 to 10,000 ft, according to Vogel, Weaver, Rose, Blount and Grover (1963). Their results on school children at Leadville, Colorado, (2400 m, 8000 ft) showed large increases in pulmonary arterial pressure during exercise, but little change in resting values.

Electrocardiograph studies in the Himalaya (Jackson and Davis, 1960; Milledge, 1963b) have shown right axis deviation of increasing degree with increase of altitude up to 24,400 ft (7400 m) in acclimatised or partly acclimatised visitors, the changes being only incompletely reversed by breathing sea-level oxygen mixtures.

Whether pulmonary hypertension at altitude has an adaptive significance or whether it should be regarded as one of the penalties of living at high altitude is an open question. Vogel and his colleagues (1963) believe that the elevation of pulmonary arterial pressure improves the perfusion of all the pulmonary areas and increases the effectiveness of the alveolar blood gas interface.

c. Anatomical changes. Studies on the native residents of Morococha (14,900 ft, 4550 m) have revealed larger lung volume (Hurtado, 1964) and greater heart size than in comparable subjects at sea-level (Theilen, Gregg and Rotta, 1955), especially when differences in body size are taken into account. Post-mortem specimens have shown permanent dilatation of the capillary bed of the lungs and greater size of the alveoli (Campos and Iglesias, 1957). These changes are associated with greater diffusing capacity during exercise (Velasquez, 1956) and higher cardiac output during submaximal exercise (Theilen, Gregg, and Rotta, 1955). Whether such differences depend on residence at high altitude from childhood or whether they are associated with racial factors awaits investigation, although on general grounds the former seems more likely.

d. Regional adjustments. It is logical to assume that adjustments at tissue level take place and favour the delivery of O_2 at normal rates in the presence of reduced blood-tissue P_{O_2} gradients. Such changes would include increased regional blood flow, increases in the capillary bed and changes in respiratory enzymes.

Animal experiments have shown increased capillary density in muscles (for reference, see Hurtado, 1964), higher concentration of myoglobin and alterations in respiratory enzymes. Reynafarje (1962a, b) has reported evidence of the last two changes in biopsy specimens of human muscle.

No direct information is available on the regional distribution of blood flow to different organs in man at high altitude, but certain inferences can be drawn from studies of acute hypoxia and other evidence.

e. Cerebral circulation. In acute hypoxia, the cerebral blood flow begins to increase when the arterial P_{O_2} falls to 50 mm Hg and rises steeply when arterial P_{O_2} falls below 40 mm Hg (Fig. 9). Hyperventilation hypocapnia reduces the cerebral blood flow at a given arterial P_{O_2} and displaces the curve relating cerebral blood flow to arterial P_{O_2} downwards and to the left in the manner illustrated in Fig. 8. Thus the cerebral blood flow in acute hypoxia depends on both $P_{A\,O_2}$ and $P_{A\,CO_2}$.

This has implications for acclimatisation to altitude. Hyperventilation hypocapnia may prevent or reduce the rise in cerebral blood flow in response to hypoxia during the early stages of acclimatisation and be partly responsible for the onset of mountain sickness some hours after arrival at altitude. Unfortunately no information is available about whether the effect is a permanent

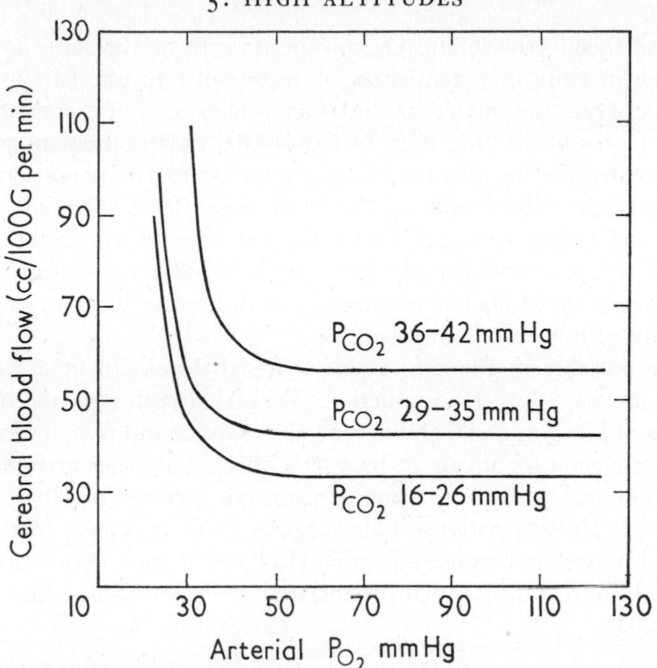

FIGURE 9. Relation of cerebral blood flow to arterial P_{O_2} and P_{CO_2} in persons acutely exposed to hypoxia. Redrawn from a figure published by Lambertsen (1958).

one or whether the cerebral blood vessels become adjusted to reduced P_{CO_2} levels, and if so what mechanism is involved.

f. Muscle circulation. Indirect evidence that the blood flow through active muscles in maximal or submaximal exercise is greater in natives of high altitude than in acclimatised visitors is provided by the observation that natives of high altitude have notably higher cardiac outputs than acclimatised visitors when working at the same intensity (Theilen, Gregg, and Rotta, 1955; Pugh, Gill, Lahiri, Milledge, Ward and West, 1964).

According to Hurtado (1964), natives of higher altitude have lower O_2 debts and blood lactate values after submaximal exercise than acclimatised athletes at the same work intensity. This could be regarded as evidence that their muscles have a greater diffusing capacity for O_2, associated with a greater number of open capillaries or higher blood flow or both. For the theoretical aspects of this subject the reader is referred to papers by Kety (1957) and Otis (1963).

4. *Fertility*

Barron, Metcalfe, Meschia, Huckabee, Hellegers and Prystowsky (1964) studied pregnant ewes and their foetuses at Morococha (14,900 ft, 4500 m).

They found foetal growth and O_2 consumption to be the same as in sheep at sea-level, in spite of a reduction of approximately one-half in the P_{O_2} gradient between the maternal and foetal bloods. They concluded that the sheep foetus at altitude lives in an internal environment in which the oxygen pressure and the quantity of oxygen available to it are not significantly different from the values found in the foetus at sea-level. The rate of oxygen supply to the foetus appeared to be regulated by adjusting the diffusing capacity of the placental barrier, either by increasing the diffusion surface or by reducing the diffusion resistance per unit placental barrier, or by a combination of both mechanisms.

Another possible mechanism assisting foetal development at altitude is increased haemoglobin F production. Barker (1957) reported increased production of Hb F in foetal and neo-natal rats, mice and puppies exposed to low oxygen tension for up to 20 hr a day; sheep and other mammals apparently do not make use of this mechanism. No increase in Hb F has been found in high altitude natives. This suggests that the human foetus is well supplied with oxygen, because increased Hb F synthesis occurs in a variety of complicated human pregnancies associated with low blood-flow hypoxia (Barker, 1964).

C. Muscular Exercise

1. *General*

In the resting state the oxygen transport system is fully adequate to supply the needs of the body for oxygen, even at very great altitude. For example, on the second ascent of Everest in 1956 (Eggler, 1956), two parties on the summit (pressure c. 250 mm Hg) remained for 2 hr without their oxygen equipment, and Shipton and Smythe in 1933 spent three nights at 27,400 ft (8300 m) and about 270 mm Hg. It is only during muscular exercise that the system is fully extended and its limitations can be adequately studied. Studies on well-acclimatised mountaineers over a wide range of altitudes from sea-level to 24,500 ft (7400 m) have shown that maximum oxygen intake declines progressively with increase of altitude from sea-level upwards (Fig. 10). By 20,000 ft (6100 m) maximum oxygen intake has fallen to about 2 l/min, and by 24,500 ft (7400 m) to about 1·5 l/min (Pugh *et al.*, 1964). Mountaineers on Everest before World War II used to claim that after suitable acclimatisation they could climb nearly as fast at 20,000 ft (6100 m) as at much lower altitudes in the European Alps. Although this claim was hardly acceptable to physiologists of the day, it was later substantiated by measurements of O_2 intake of men walking uphill at their habitual pace on Himalayan expeditions (Pugh, 1958). The results showed that climbers do, in fact, maintain surprisingly high oxygen intakes up to about 20,000 ft (6100 m). They accomplish this by working nearer to capacity (Fig. 10), and for this reason the hours

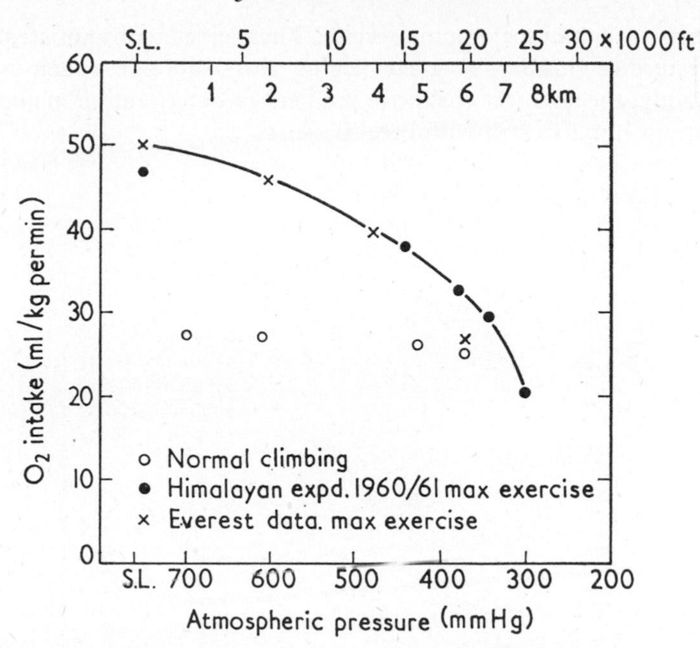

FIGURE 10. Maximum oxygen intake at various altitudes and oxygen intake of men walking uphill at their habitual pace.

of climbing decline progressively with increase of altitude. Thus (Pugh, 1958) a man capable of a 10 to 14 hr day at alpine altitude will climb for, say, 5 to 7 hr at heights around 20,000 ft (6100 m). At heights above about 25,000 ft (7600 m) continuous progress is no longer possible, and the climber has to adopt the device of working intermittently so that much of the work can be done anaerobically (Pugh, 1958). On the higher slopes of Everest climbers have reported (Pugh, 1958) having to stop every twelve paces in order to rest and recover their breath. Astrand, Astrand, Christensen and Hedman (1960) in studies carried out at sea-level demonstrated that the device of working intermittently permits a man to work for an hour or more with an oxygen intake that would exhaust him in 10 min if carried on continuously. Himalayan porters who carry loads exceeding half their body weight up to 20,000 ft (6100 m) use the same method, the work periods becoming shorter and the rests longer as the height increases.

The respiratory aspects of muscular exercise at high altitude are illustrated in Fig. 11, which is based on results collected at 19,000 ft (5800 m) by the Himalayan Scientific and Mountaineering Expedition of 1960–1 (Pugh *et al*, 1964). Ventilation, measured in l/min at body temperature, at observed barometric pressure and saturated with water vapour (BTPS) increased progressively with increase of altitude over the whole range of work loads from mild exercise to 6 min maximum exercise. The increase substantiates a well-known feature of climbing at great heights, that, although a man can keep going steadily, the slightest change in gradient or deterioration in snow conditions brings him to a halt with breathlessness.

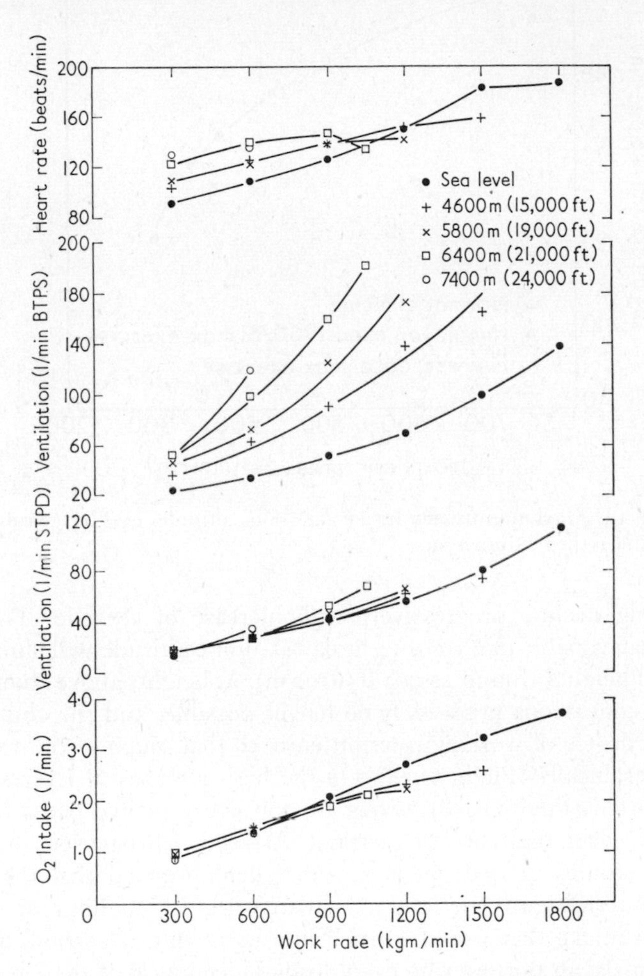

FIGURE 11. Oxygen intake, ventilation, and heart rate during ergometer exercise at various altitudes, from sea-level to 7400 m (24,400 ft).

Ventilation BTPS with maximum work loads at heights of 19,000 ft (5800 m) and over sometimes reached values of over 200 l/min. These extreme rates, although of value in assisting in diffusion of oxygen across the lung, led to considerable respiratory alkalosis, with pH levels as high as 7·55. Since the work of breathing is known to increase steeply at high minute volumes, one may infer that these high rates were not wholly advantageous, but were the result of extreme hypoxic stimulation associated with falling arterial oxygen saturation. In this connection it was interesting to find that the only Sherpa studied at 19,000 ft (5800 m) was able to reach a higher oxygen intake with a much lower minute volume than the rest of the party. Ventilation expressed as l/min of dry gas at 760 mm Hg and 0°C (STPD) for a given work rate was virtually independent of altitude, except at work rates approaching the maximum for a given altitude, when there was relative hyperventilation. This means that in light and moderate work, therefore, ventilation was so regulated that the mass of oxygen inhaled per min was approximately constant and independent of altitude, as in resting subjects.

The arterial oxygen during exercise at 19,000 ft (580 m) fell to an average value of 56% in 5 min maximum exercise, compared with an average resting value of 67% at this altitude (West *et al.*, 1962). In shorter periods of work at a higher intensity, 1200 kg/min, arterial oxygen saturation fell progressively throughout the work period, reaching values between 40 and 50%. This clearly implicates the diffusing capacity for oxygen as a limiting factor in exercise at extreme altitude (West, 1962), although the work of breathing may be another, because each of the subjects was brought to a halt by extreme dyspnoea.

For the cardiovascular system in work at extreme altitude, the findings of the party at 19,000 ft (5800 m) by a modified acetylene method showed that the outputs for a given work load were similar to the values observed at sea-level, but the maximum values were reduced in parallel with maximum oxygen intake, the mean being 17·3 l/min, compared with 23 l/min at sea-level (Pugh, 1964a). The heart rate during mild and moderate exercise at 19,000 ft (5800 m) was higher than at sea-level but the maximum rates were only 140 to 150 beats/min, compared with 180 to 190 beats/min at sea-level (Fig. 11).

2. *Effects of oxygen*

The administration of oxygen at sea-level pressure to subjects working on the ergometer at 19,000 ft (5800 m) caused ventilation (BTPS) for a given work rate to fall to values half-way between the sea-level control value and the values observed when the subjects were breathing atmospheric air. Heart rate fell to the sea-level control value, or below it, and the maximum rate sometimes increased to the maximum value observed in exercise at sea-level (Pugh *et al.*, 1964). The maximum work load tolerated for 5 min was restored almost to the sea-level value. From the cardiac output/work-rate relation (which appeared to be unchanged at altitude) and the observed changes in heart rate it was inferred that the stroke volume of the heart and its maximum output also increased, although probably not to the sea-level value, because of pulmonary hypertension and increased blood viscocity.

These results contradict the impression created from reports by mountaineers that the work capacity of acclimatised men at great altitude cannot be restored to the sea-level value by breathing oxygen (Houston and Riley, 1947). The same thing is shown by the finding that the party using closed-circuit oxygen equipment on Mount Everest in 1953 climbed at an oxygen consumption of 2 l/min, which would be normal for similar conditions at sea-level (Pugh, 1958).

D. Failure of Adjustment

1. *Mountain sickness*

The symptoms of mountain sickness are headache, insomnia, malaise, irritability, weakness, anorexia, vomiting, disturbance of breathing, cyanosis and tachycardia, in varying degrees, and combinations of these, depending on the individual and the altitude stress. The symptoms appear characteristically some hours after arrival and may last from one to five days.

Many years ago Barcroft (1925) had a controversy with the Italian physiologist Mosso (1898) over the causes of mountain sickness. Mosso ascribed it to hypocapnia, but Barcroft claimed that low arterial oxygen tension was responsible. No one has demonstrated any close correlation between arterial oxygen saturation and acute mountain sickness in individual subjects. In fact, the symptoms are worst at the stage when hyperpnoea is developing and P_{AO_2} is rising. The observation that reduction of arterial P_{CO_2} associated with hyperventilation has a powerful vasoconstrictor effect on the cerebral blood vessels suggests that hypocapnia may after all be implicated as a cause of mountain sickness by restricting the rise in cerebral blood flow in response to low oxygen tension.

2. *Acute pulmonary oedema*

An important hazard involved in ascending over-rapidly to heights above 12,000 ft (3600 m) is acute pulmonary oedema. The condition comes on usually at night and may or may not be associated with respiratory infection. The patient becomes progressively cyanosed and dyspnoeic, with cough and frothy sputum, and may die within a few hours unless oxygen is administered. The physiological pathology of the condition has been much discussed: it is thought to be connected either directly or indirectly with pulmonary hypertension, which is a normal cardiovascular response to altitude (Penaloza, Sime, Banchero and Gamboa, 1963a; Houston, 1963; Cournand, 1963).

3. *Chronic mountain sickness (Monge's disease)*

South American investigators have described a condition affecting residents at altitude around 14,000 ft (4300 m), in which acclimatisation to altitude is gradually lost (Monge, 1937, 1948). Recent evidence suggests that the cause is loss of chemoreceptor sensitivity to hypoxia (Hurtado, 1964; Severinghaus *et al.*, 1964). The condition is characterised by extreme polycythaemia with red cell counts up to $8 \times 10^6/mm^3$ and haematocrit readings of 70 to 80%. Blood volume and resting cardiac output are increased, and there is severe pulmonary hypertension. Occasionally congestive heart failure follows. Recovery takes place on descending to low altitude. The condition has to be distinguished from emphysema, which is common among mining populations at high altitude.

4. *High altitude deterioration*

The symptoms of high altitude deterioration are loss of weight, increasing lethargy and weakness. Sometimes symptoms of acute mountain sickness, such as vomiting, reappear. There is evidence that dehydration and starvation were important contributory factors on mountaineering expeditions before 1953. Climbers on pre-war Everest expeditions returned grossly emaciated, and some members of assault parties passed no urine for up to twenty-four hours (Finch, G. I., personal communication). Warren (1937) reported that urinary pH never exceeded pH 5·4 at heights above 18,000 ft (5500 m), which could be related to low food intake. Shipton (1938) kept dietary records on the 1935 expedition, and analysis showed intakes of less than 1500 cal/day for all the time spent over 18,000 ft (5500 m); other descriptions suggested low fluid intake. The high water requirement at high altitudes was demonstrated by Pugh (1954a), who showed that an intake of 3 to 4 l/day was necessary to secure normal urinary volumes even in climbers at altitudes around 18,000 ft (5500 m). This was attributed to increased water loss from the respiratory tract, due to hyperpnoea and the dry atmosphere and also to sweating associated with the intense solar radiation on snow-covered glaciers (Pugh, 1962c).

On the Mount Everest expedition of 1953 the importance of this problem was recognised, and measures were taken to ensure a 4 to 5 l fluid intake together with a palatable diet of suitable calorific value (Pugh, 1954a). These measures, along with use of oxygen for climbing and sleeping, adequately solved the problem of high altitude deterioration on mountaineering expeditions and within the next five years all the peaks over 28,000 ft (8500 m) in altitude were to be climbed.

REFERENCES

Arias-Stella, J., and Seldaña, M. (1964). *Med. Thoracalis.* (In press.)

Assmussen, E., and Consolazio, F. C. (1941). *Amer. J. Physiol.* **132**, 555.

Astrand, I., Astrand, P. O., Christensen, E. H., and Hedman, R. (1960). *Acta physiol. scand.* 48, 448.

Baker, P. T. (1964). "The biology of population of anthropological importance." Wenner Gren Foundation Burg Wartenstein Symposium Paper No. 23.

Barker, J. N. (1957). *Amer. J. Physiol.* 189, 281.
Barker, J. N. (1964). *In* "The Physiological Effect of High Altitude", p. 125, (W. H. Weihe, ed.). Pergamon Press, Oxford.
Barcroft, J. (1925). "The Respiratory Function of the Blood, Part I", Lessons from High Altitude, Chapter 1. University Press, Cambridge.
Barron, D. H., Metcalfe, J., Meschia, G., Huckabee, W., Hellegers, A., Prystowsky, H. (1964). *In* "Physiological Effect of High Altitude", p. 115 (W. H. Weihe, ed.). Pergamon Press, Oxford.
Bell, Charles, (1924). "Tibet Past and Present." Oxford University Press, Oxford.
Bourdillon, T. D. (1954). *Proc. roy. Soc.* B. 143, 24.
Bruce, C. G. (1923). "The Assault on Mount Everest 1922." E. Arnold, London.
Campos, J. and Iglesias, B. (1957). Cited by Hurtado (1964).
Chiodi, H. (1957). *J. appl. Physiol.* 10, 81.
Chiodi, H. (1963). *In* "The Regulation of Human Respiration", p. 363 (D. J. C. Cunningham and B. B. Lloyd, eds.). Blackwell, Oxford.
Cournand, A. (1963). *In* "Progress in Research in Emphysema and Chronic Bronchitis," Vol. 1, p. 257 (R. F. Grover, ed.). S. Karger, Basel and New York.
Dill, D. B. (1938). *In* "Heat, Life and Altitude", Chapter VII. Harvard University Press, Cambridge, Massachusetts.
Dill, C. B. (1958). "Handbook of Respiratory Data", p. 107 (P. L. Altman, G. F. Gibson, and C. C. Wang, eds.). W. B. Saunders, Philadelphia.
Dill, D. B., Edwards, H. T., and Consolazio, W. V. (1937a). *J. biol. Chem.* 118, 635.
Dill, D. B., Talbot, J. H. and Consolazio, W. V. (1937b). *J. biol. Chem.* 118, 649.
Dyrenfurth, N. G. (1963). *Nat. geogr. Mag.* 124, 460.
Eggler, A. (1956). *Alpine J.* 61, 239.
Eiselin, M., and Forrer, E. (1961). "The Mountain World 1960/61", p. 131. Swiss Foundation for Alpine Research. Allen and Unwin, London.
Filippi, Fileppo de. (1912). "Karakoram and Western Himalaya", 1909. Constable, London.
Gill, M. B., Milledge, J. S., Pugh, L. G. C. E., and West, J. W. (1962). *J. Physiol.* 163, 373.
Gill, M. B., Poulton, E. C., Carpenter, A., Woodhead, M. M., and Gregory, M. H. R. (1964). *Nature, Lond.* 203, 436.
Gill, M. B., and Pugh, L. G. C. E. (1964). *J. appl. Physiol.* 19, 949.
Haldane, J. S., and Priestley, J. G. (1935). "Respiration", p. 85. Clarendon Press, Oxford.
Hall, F. G. (1936). *J. biol. Chem.* 115, 485.
Herzog, M. (1952). "Annapurna". Jonathan Cape, London.
Hillary, E. P. (1962). "High in the Thin Cold Air". Hodder and Stoughton, London.
Houston, C. S. (1963). *In* "Progress in Research in Emphysema and Chronic Bronchitis", Vol. 1, p. 313 (R. F. Grover, ed.). S. Karger, Basel and New York.
Houston, C. S., and Riley, R. L. (1947). *Amer. J. Physiol.* 149, 565.
Hunt, J. (1953). "Ascent of Everest." Hodder and Stoughton, London.
Hurtado, A. (1964). *In* "Handbook of Physiology; Adaptation to the Environment", Section 4, p. 843 (D. B. Dill, E. F. Adolph and C. G. Wilber, eds.). American Physiological Society, Washington, D.C.
Jackson, F., and Davis, H. (1960). *Brit. Heart J.* 22, 671.
Korner, P. I. (1959). *Physiol. Rev.* 39, 687.
Kety, S. S. (1957). *Fed. Proc.* 16, 666.
Kurz, M. (1953). "The Mountain World". Swiss Foundation for Alpine Research. Allen and Unwin, London.
Lambertsen, C. J. (1958). *In* "Man's Dependence on the Earthly Atmosphere", p. 234 (K. E. Schaefer, ed.). Macmillan, New York.

Lloyd, B. B., Jukes, M. G. M., and Cunningham, D. J. C. (1958). *Quart. J. exp. Physiol*, XLIII, 214.

McFarland, R. A. (1937). *J. comp. physiol. Psychol.* 24, 189.

Merino, C. F. (1956). Report 56–103. U.S. School of Aviation Medicine, Randolph Field, Texas.

Meschia, G., Prystowsky, H., Hellegers, A., Huckabee, W., Metcalfe, J., and Barron, D. H. (1960). *Quart. J. exp. Physiol.* 45, 284.

Milledge, J. S. (1963a). *In* "The Regulation of Human Respiration", p. 397 (D. J. C. Cunningham and B. B. Lloyd, eds.). Blackwell, Oxford.

Milledge, J. S. (1963b). *Brit. Heart J.* 25, 291.

Monge, C. (1937). *Arch intern. Med.* 59, 32.

Monge, C. (1948). "Acclimatization in the Andes: Historical Conformation of Climatic Aggression in the Development of Andean Man." (Translated by D. F. Brown.) Johns Hopkins Press, Baltimore.

Mosso, A. (1898). "Life of Man in the High Alps", Ch. XXII. Unwin, London.

Norton, E. F. (1925). "Fight for Everest." E. Arnold, London.

Otis, A. B. (1963). *In* "The Regulation of Human Respiration", p. 111 (D. J. C. Cunningham and B. B. Lloyd, eds.). Blackwell, Oxford.

Pappenheimer, J. R., and Maes, J. P. (1942). *Amer. J. Physiol.* 126, 38–39P.

Penaloza, D., Sime, F., Banchero, N., and Gamboa, R. (1963a). *In* "Progress in Research in Emphysema and Chronic Bronchitis", Vol. 1, p. 257 (R. F. Grover, ed.). S. Karger, Basel and New York.

Penaloza, D., Sime, F., Banchero, N., Gamboa, R., Cruz, J., and Marticorena, E. (1963b). *Amer. J. Cardiol.* 11, 150.

Pugh, L. G. C. E. (1954a). *Geogr. J.* 120, 183.

Pugh, L. G. C. E. (1954b). *J. Physiol.* 126, 38–39P.

Pugh, L. G. C. E. (1957). *J. Physiol.* 135, 590.

Pugh, L. G. C. E. (1958). *J. Physiol.* 141, 233.

Pugh, L. G. C. E. (1962a). *Geogr. J.* 128, 447.

Pugh, L. G. C. E. (1962b). *Brit. med. J.* ii, 621.

Pugh, L. G. C. E. (1962c). Paper read at Unesco Symposium in Lucknow on "Environmental Physiology and Psychology in Arid Conditions", p. 325. Lucknow.

Pugh, L. G. C. E. (1964a). *J. appl. Physiol.* 19, 441.

Pugh, L. G. C. E. (1964b). *J. Physiol.* 170, 344.

Pugh, L. G. C. E. (1964c). "The Scientific Basis of Medicine", Annual Reviews, Ch. III, p. 32.

Pugh, L. G. C. E., Gill, M. B., Lahiri, S., Milledge, J. S., Ward, M. P., and West, J. B. (1964). *J. appl. Physiol.* 19, 431.

Pugh, L. G. C. E., and Ward, M. P. (1956). *Lancet* ii, 1115.

Rahn, H., and Otis, A. B. (1949). *Amer. J. Physiol.* 157, 445.

Reynafarje, B. (1962a). *J. appl. Physiol.* 17, 301.

Reynafarje, B. (1962b). Report No. SAM-TDR-62-89, U.S.A.F. School of Aerospace Medicine, Brooks Air Force Base, Texas.

Reynafarje, B. (1964). *In* "The Physiological Effect of High Altitude", p. 73 (W. H. Weihe, ed.). Pergamon Press, Oxford.

Richardson, L. E. (1962). "Tibet and its History." Oxford University Press, Oxford.

Rotta, A., Canepa, A., Hurtado, A., Velasquez, T., and Chavez, R. (1956). *J. appl. Physiol.* 9, 328.

Ruttledge, H. (1934). "Everest: the Unfinished Adventure 1933." Hodder and Stoughton, London.

Schmied, E. (1957). "Mountain World 1956–1957", p. 157. Allen and Unwin, London.

Severinghaus, J. W., Mitchell, R. A., Richardson, B., and Singer, M. M. (1963). *J. appl. Physiol.* 18, 1155.
Severinghaus, J. W., and Mitchell, R. A. (1964). *In* "The Physiological Effects of High Altitude", p. 273. (W. H. Weihe, ed.). Pergamon Press, Oxford.
Severinghaus, J. W., Bainton, C. R., and Carcelen, A. (1964). *Proc. phys. Soc. Lond.* 6–7 Nov., C. 20.
Shipton, E. (1938). *Chem. & Ind.* 57, 1231.
Smythe, F. S. (1932). "Kamet Conquered." Gollancz, London.
Theilen, G. O., Gregg, D. E., and Rotta, A. (1955). *Circulation* 12, 383.
Velasquez, T. (1956). Report No. 56–108. U.S.A.F. School of Aviation Medicine, Randolph Field, Texas.
Viault, E. (1891) *C.R. Acad. Sci., Paris* 112, 295.
Vogel, J. H. R., Weaver, W. F., Rose, R. L., Blount, S. G., and Grover, R. F. (1963). *In* "Progress in Research in Emphysema and Chronic Bronchitis", Vol. 1, p. 269 (R. F. Grover, ed.). S. Karger, Basel and New York.
Warren, C. B. (1937). *Geogr. J.* 90, 127.
West, J. B. (1962). *J. appl. Physiol.* 17, 421.
West, J. B., Gill, M. B., Lahiri, S., Milledge, J. S., Pugh, L. G. C. E., and Ward, M. P. (1962). *J. appl. Physiol.* 17, 617.
Whittaker, S. R. F., and Winton, F. R. (1933). *J. Physiol.* 78, 339.

CHAPTER 6

High Pressures

PART I. GENERAL

H. JOHN TAYLOR

A. Introductory

The effects of increased barometric pressure can be divided into three distinct phases: (1) during compression; (2) when under pressure; (3) during and after decompression.

B. Compression

A man is subjected during compression not only to the physical effects of changing increased pressure, but also to the effects of breathing increased partial pressures of the gases in the environment. This second effect is discussed in Part II of this chapter. The physical effects may be noticed on the various cavities in the body, such as the middle-ear spaces and the nasal accessory sinuses. When a man goes under pressure the outer surface of the ear drum is exposed; unless the pressure on the inner surface is also raised, the drum will be subjected to a one-sided pressure, severe pain will result, and the drum will eventually rupture. In order to equalise the pressure on both sides of the drum, the Eustachian tube must be opened by swallowing, shouting, holding the nose and blowing, and so on. If the Eustachian tube is blocked by mucus or tissue, this opening cannot be accomplished. Similarly, it is essential that pressure in sinuses such as the nasal accessory sinus should be equalised.

The most important cavities in the body are the lungs. By breathing, as the pressure rises the pressures are equalised. This, under water, supposes that the diver is breathing from a set or other source. If a man dives without a set, he cannot breathe and he must hold his breath, and so the air in his lungs is compressed. Instructors in the Submarine Escape Training Tank and also pearl divers are examples of those who are subjected to such conditions. At a depth of 100 ft, for example, the volume of air in the diver's lungs is compressed to one-fourth of its original volume at atmospheric pressure. This amount approximates to the volume of the residual air in the lungs after the most forceful expiration. Should the diver now descend deeper than 100 ft the additional pressure, as he is unable to diminish the lung volume by compressing the chest wall or pushing up the diaphragm, will force blood and tissue fluids into the lung spaces, bringing about what is known as a "squeeze". If severe enough, this can prove fatal. In the example quoted, when diving to 100 ft holding the breath, the partial pressure of the gases in the lungs can be increased to approximately four times the original volume. However, it is to be noted that the partial pressure of carbon dioxide in the alveoli, measured on reaching 100 ft, differs little from the value at atmospheric pressure, showing how rapidly the process of elimination of carbon dioxide can be reversed.

C. Pressure

Problems are raised by the respiration of oxygen and inert gases at increased partial pressures.

1. *Oxygen*

Two effects are known. The first is the Paul Bert or acute effect, characterised by a convulsion, epileptiform in character, produced after breathing oxygen at pressures greater than 2 atm of oxygen. The second, the chronic, effect of raised oxygen pressure is inflammation of the pulmonary alveoli, also known as the Lorrain Smith effect (Smith, 1899). Exposures to oxygen at much higher pressures can clearly also produce this effect if the exposure is not terminated by the onset of a convulsion. Intermittent exposure can produce some degree of protection against this chronic effect (Taylor, 1949b); the mechanism is somewhat obscure, but is probably connected with thickening of the alveolar lining. The acute convulsive effect is produced by breathing oxygen at partial pressures of over 2 atm† of oxygen. In some sensitive subjects the critical value may be somewhat less; if carbon dioxide is present, the time to onset of convulsions can be considerably reduced (Taylor, 1949a). It has been found, in animals at any rate, possible by giving various drugs to increase greatly the time of exposure to high-pressure oxygen before convulsions intervene. There is one effect, however, that should be mentioned here. Bannister and Cunningham (1954) quote instances when it was more beneficial, under exercise conditions, to breathe 66% oxygen rather than 100% oxygen. Excellent reviews on the subject of oxygen toxicity have been given by Bean (1945) and Stadie, Riggs and Haugaard (1944). Papers by Donald (1947) and Lambertsen and his colleagues (1953a, b, c, d) should also be consulted.

2. *Inert gases*

The narcotic effect of breathing air under pressure is well known to divers. Inert gases all show to greater or lesser extent the properties of anaesthetic agents. Argon and the denser inert gases are more narcotic than nitrogen, but helium is less so, which is why helium is now used as a diluent for oxygen in nearly all deep-diving work.

Many workers, particularly Bean (1950), have claimed that the cause of the narcosis is a retention of carbon dioxide in tissues, due to increased alveolar carbon dioxide when compression takes place, and difficulties of diffusion of this gas, due to the increased density of gases in the lungs. Rashbass (1955b) at the Royal Naval Physiological Laboratory (R.N.P.L.) gave multiplication tests to men subjected to pressure. The scores were lower than those for performance at atmospheric pressure. Hyperventilation made practically no difference to the score obtained while under pressure. Alveolar CO_2 measurements showed an extremely small increase under pressure. It follows that CO_2 is not an important factor in this narcosis, although all experimenters agree that the effect is enhanced by breathing CO_2 at the same time as the inert gas.

Research into the cause and prevention of the narcosis found when men are subjected to pressures over 4 atm absolute of air has progressed over a number of years at the R.N.P.L. Bennett and Glass (1959) have demonstrated changes in the electroencephalograms of men at raised pressures of air that correlate with the pressures to which the men are subjected. By far the most work has, however, been carried out on animals. Having developed an electrical stimulus technique for the quantitative measurement of narcosis in small mammals, Bennett, Dossett and Kidd (1960) have used it to investigate the effect of rate of compression, the anti-narcotic action of azacyclonol and the comparative narcotic properties of a number of inert gases. The technique has also been used to investigate the effects of some eleven drugs on both nitrogen

† See Appendix A for notes on atmospheric pressure, gauge pressure and absolute pressure and for conversion tables (lb/in^2 to dynes/cm^2, g/cm^2 and Kg/cm^2).

narcosis and oxygen poisoning. Drugs that control the one condition apparently affect the other in a similar manner. Further, the drugs that control these conditions are antipyretics, hypnotics or depressants, and those that enhance are stimulants, suggesting an active process for the mechanisms of narcosis, such as a histotoxic hypoxia. Preliminary experiments have suggested that azacyclonol has a similar protective action in man.

Barnard, Hempleman and Trotter (1962) carried out experiments to compare the relative effects of nitrogen partial pressure, gas density of the mixture breathed and carbon dioxide on the production of narcosis at raised ambient pressures. Between performance at simple arithmetic and nitrogen partial pressure they found an inverse relationship approximately linear over the range studied (down to 330 ft).†

Experiments carried out recently by Bennett (1963) on the in vivo measurement of transmission of rat spinal synapses and sciatic nerve have confirmed the finding of Carpenter (1956) and Marshall (1951) that the synapse is the principal site of action of inert gases at raised pressures. Further, it was demonstrated that the synaptic block is similar to that produced by asphyxia or hypoxia.

The previous EEG experiments implied involvement of the multisynaptic reticular activating system. As the rat experiments further implied synaptic involvement, measurements were made of auditory evoked potentials in the cat's reticular formation and cortex. The results showed the primary action of inert gases, like that of many gaseous anaesthetics, to be on cortical apical dendrites and the multisynaptic reticular formation. The potentials evoked in the brainstem were found to be depressed some 50% more than those in the cortex.

Two principle theories have been suggested for the cause of the narcosis. Most workers believe the cause is the high partial pressure of inert gas. A number have suggested that retained carbon dioxide might be the cause. This latter theory has received some criticisms, and experiments by Bennett on measurement of cortical P_{CO_2} and P_{O_2} in chloralosed cats have established that carbon dioxide retention is certainly not the cause, although it has a marked potentiating action, as suggested earlier by Case and Haldane (1941). The results suggest that whether narcosis occurs or not depends on the summed result of three factors, the partial pressures of inert gas and of oxygen and the density of the gas mixture breathed.

Current and future experiments are designed to investigate further the potentiating nature of carbon dioxide narcosis by controlled variation of P_{CO_2} in the presence of high cerebral oxygen tensions, together with measurements of cortical pNa, pK and pH and respiratory variables. The results should decide the correctness of the hypothesis that the mechanism in narcosis is a consequence of a histotoxic hypoxia.

† For convenience in this chapter the metric equivalents of the values for ft have been collected in Appendix B and have not been inserted in the text.

D. Decompression

1. *Decompression ratios*

Paul Bert (1878) found that bends, after exposure to increased air pressure, could usually be prevented by slow uniform decompression to atmospheric pressure. This method was developed by other workers, notably in this country by Hill and Macleod (1903) and Hill and Greenwood (1907); it was the Admiralty Diving Committee (1905–7) that introduced the method of stage decompression. The research on which this method was based was carried out by Haldane, Damant and Boycott (1908) and the original British Diving Tables were the result. These authors when calculating the stage decompression tables assumed the existence of 5, 10, 20, 40 and 75 min tissues (i.e. those that reach half-saturation in these times). They introduced the concept of a safe decompression ratio; thus, with a 2 : 1 ratio it is safe to decompress from a partial pressure of inert gas in the tissue of x to one of $x/2$ whatever value is given to x. They usually used a decompression ratio of 2 : 1, but on occasions used 2·1 : 1 or even 2·3 : 1. However, it was soon discovered that the permitted tissue partial pressure decompression ratio of 2 : 1 was dangerous when long periods of time elapsed at depths exceeding 120 ft of sea water; accordingly the permitted ratio of the controlling tissue (a slowly saturating one) was modified to a decompression ratio of 1·75 : 1. The British tables were universally adopted for diving operations, but were subsequently modified by U.S. workers Hawkins, Shilling and Hansen (1935). They showed that 5 and 10 min tissues had no bearing on the production of bends. New tables were now calculated allowing a ratio of 2·8 : 1 for the 20 min tissue and 2·0 : 1 for the 40 and 75 min ones. The American tables are so different from the British ones in many respects that it was thought worth while to reinvestigate the whole subject at the R.N.P.L. The critical point is the difference in the decompression ratios for British and American methods. The original theory of Haldane of a fixed decompression ratio could in no way predict the possibility of such findings of variable decompression ratios in any quantitative and useful way; it is clear that a different approach must be made to the problem of inert gas exchange in the tissues. The first objective, therefore, was to try to obtain the correct ratio for the specified half-time tissues. The work, carried out on goats, has been described by Davidson, Sutton and Taylor (1950); let it suffice to say that the variation in these ratios was so great from animal to animal that no reasonable decompression schedule could be compiled.

2. *Decompression tables*

Hempleman (1952) considered the pathological, physiological and mathematical conditions involved in decompression sickness and came to the conclusion that the quantity of nitrogen present in a tissue (Q) is proportional to $t^{\frac{1}{2}}$, where t is the exposure time, K the depth in feet, so that

$$Q = K\, t^{\frac{1}{2}}.$$

The units are arbitrary, incorporating a constant.

Crocker and Taylor (1952) produced decompression tables based on these ideas without, however, much success, in so far as they produced an unacceptably high incidence of bends.

However, the so-called square root law does give results in remarkable agreement with experiment, when the safe time of exposure at depth before direct surfacing is found. For example, the safe exposure time at 100 ft of sea water, before immediate surfacing, is about 22 min. In arbitrary units therefore $Q = 100\sqrt{22} = 475$ units. When this figure is used to calculate the safe time at 200 ft, the time of exposure is $5\frac{1}{2}$ min. When all the experimental work in this country, in the United States and in other countries on the question of direct surfacing with safety is considered, it is remarkable how it fits the proposed formula. However, it soon became obvious that for long dives the relationship is not valid (Crellin, Crocker and Hollis, 1954).

Rashbass (1954) calculated the quantity of gas entering an avascular slab of tissue as a result of partial pressure changes in the nearby blood supply. The equations used were derived from Fick's law of diffusion, with the appropriate boundary conditions inserted. The tissue was presumed to be able to withstand an excess quantity of gas before a "bend" situation developed. This quantity was measured in arbitrary units called "footsworth", one "footsworth" being the quantity of gas that would dissolve in the tissue as a result of a pressure of 1 ft being applied for an indefinitely long time. Rashbass assumed that an excess of 30 footsworth was critical. For complex pressure-time courses the footsworth present in the tissue is impossible to calculate by any formal methods. Rashbass therefore used a graphical solution involving the superimposition of positive and negative pressure changes. This has been extended and rendered more accurate by numerical methods.

Rashbass (1955a) then calculated decompression tables based on such ideas, and these were well tested in a pressure chamber, but gave an unacceptable incidence of bends when trials in the sea were carried out under operational conditions (Crocker, 1957). In the meantime calculations were made in the light of knowledge then available with the ultimate object of putting a diver down to 600 ft. These dives were carried out in 1956 (Crocker and Hempleman, 1957). Dives to 300, 450 and 600 ft were carried out in the open water, the first in Loch Linnhe near Fort William in Scotland and the others in Sjo Fjörd near Bergen. The idea was to carry out these dives at different depths, altering the mixture breathed and the time on the bottom, so that the theoretical tissue saturation of the 300 and 450 ft dives was of the same order as that predicted for the 600 ft dive. As it turned out, this was not entirely successful, but the 600 ft dive was accomplished with only minor incidents. During the descent oxygen-helium mixtures were breathed from 40 ft downwards, but in the ascent the change back to air was at 150 ft for the 300 ft dives and at 200 ft for the 400 and 600 ft ones. Oxygen was also breathed at the shallow depths on the ascent.

In response to operational requirements, Crocker (1958) modified the previous Rashbass tables, without a great deal of theoretical background, but as a result of practical experience. New tables for diving to 200 ft were produced with success and were incorporated in the Diving Manual.

3. *Bubble formation*

It was naturally realised at the R.N.P.L. that, without a much more fundamental knowledge than was then available, it was going to be an almost impossible undertaking to dive deep with an acceptable degree of safety. It would appear possible to achieve equally satisfactory decompression schedules for most diving programmes by using either decompression ratios or a fixed head of excess pressure as the criterion for bends production. Clearly, one of these must be false. Haldane's original idea that it was as safe to come from 2 atm to 1 atm absolute as from 6 atm to 3 atm absolute and that a 2 : 1 decompression ratio applied safely to all tissues of the body, was insecurely established experimentally. Theoretical treatments of bubble formation from liquids and tissues (Harvey, 1951) led to the conclusion that the fundamental equation is $\Delta P = t - T$ (where t = gas tension in the liquid, T = hydrostatic pressure and ΔP = excess gas pressure). From this it might be concluded that a certain excess gas tension can be tolerated in the tissues and that a greater excess than this critical pressure will lead to bubble formation and hence "bends". Thus, if a prolonged dive to a pressure P_1 was carried out and decompression was then to a pressure P_2, just escaping a bend, on Haldane's theory the ratio $\frac{P_1}{P_2}$ is the controlling factor, whereas $P_1 - P_2$ controls on the other approach. No evidence is available at present to distinguish between these two possibilities. It has been predicted (Bateman, 1951) that, assuming Haldane's theory is correct, then the decompression ratio must itself be pressure dependent, and present air-diving research by the U.S.N. is indeed based on this concept. There is no practical evidence that ratios do alter with pressure. All that had been established was that prolonged deep dives were unsafe, whatever tables were used (British or U.S.). Hempleman (1957), using goats, made an attempt to solve this problem. He concluded that (*a*) P_1/P_2 is a constant ratio for any fixed dive time, provided P_2 is greater than atmospheric pressure, and (*b*) that the value of P_1/P_2 is a maximum for short dives and a minimum for long dive times. This work also suggested that the process of uptake and elimination of gases by the body may not be reversible when similar pressure changes are involved. Hempleman (1962a), in continuation of this line of approach and with experiments on goats, showed conclusively that the uptake and elimination of inert gases under those conditions were certainly not strictly reversible. Thus it seems that a discontinuity in bodily physics takes place when large and rapid drops in pressure occur. There can therefore be no unified theory; there will be one dealing with the uptake of gas and another dealing with the elimination. Hempleman suggests that the so-called silent bubble idea will play a central part in any theory dealing with the elimination of gas. In retrospect, it is interesting to see how the idea of "silent bubbles" was readily accepted as the only feasible explanation of certain phenomena connected with decompression sickness, and yet no one, apart from Hempleman, appears to have thought that "silent bubbles" could be found in tissues under conditions of so-called safe diving routines and that this formation would considerably modify the kinetics of gas elimination from those tissues.

4. *Deep diving*

In the meantime, in view of the great strides in the development of diving outfits, notably the replacement of the old helmeted diving by selfcontained equipment, the decision was taken to resume experimental work and trials for deep diving, the suggested depth to be attained eventually being 1000 to 1200 ft. The first series of experimental dives were made to 300 ft, the stay at depth being 10 min and the decompression schedule for such a dive not to exceed 3 hr.

The calculations were based (Hempleman, 1962b) on certain facts:

(1) The U.S. Navy have conducted a number of no-stop dives from which it is concluded that for a given duration at depth it seems possible to go deeper on helium (20% O_2 + 80% He) than on air.

(2) The shape of the no-stop curve for air is well established, and this is taken as representing the gas uptake curve for the situation that produces decompression sickness on air.

(3) There is a safe decompression ratio to be used in conjunction with the quantity of gas obtained as in (2). This ratio varies with the depth of the dive, being 1·9 : 1 at low pressures and reducing to 1·5 at deeper depths (goats to 200 ft). Experimental work on men and animals has verified this variation.

(4) Oxygen plays a part in the occurrence of decompression sickness. This was first shown by Donald (1955). For deep dives, therefore, the breathing mixture was treated as one material without discriminating between oxygen with helium and oxygen with nitrogen.

The dives were carried out with air and 5% oxygen/95% helium, the change over occurring at 100 ft on the descent and at 90 ft on the ascent. Reasonably successful diving to 300 ft was carried out, but rather inexplicable differences between diving in a chamber and in the sea were noticed. These were the first oxygen-helium dives carried out in the British Navy for many years.

Further experimental work was carried out by Hempleman and Trotter, and as a result it became obvious that there were two lines of approach to the problems of decompression.

It is possible to decompress rapidly to a comparatively shallow stop. This initial large pressure drop means that a large helium "extraction gradient" is created and the rate of loss of helium is rapid. Bubble formation and growth is initiated during this opening phase, as shown in this series of experiments. If the rate of loss of gas from the lungs is high enough, it could prevent the growth of the bubble, and nothing untoward would happen. If, however, this does not occur, then bubble growth reaches a stage at which it succeeds in impeding the removal of gas and the effectiveness of the "extraction gradient" is much reduced. A process of bubble growth has been initiated and can only be terminated by recompression.

The second alternative is to attempt to prevent bubble formation altogether. For this purpose the initial fast phase of decompression must be of short duration, and the slow phase must be prolonged. The results of the series of experimental dives described by Hempleman and Trotter seem to support more and more the second approach. The deeper and longer the dive, the more important the second approach becomes. This possibly means abandonment of the fixed ratio ideas. Besides this problem, we have the fact that the results on goats at the R.N.P.L. for no-stop dives conflicted with U.S. ideas (Duffner, Snyder and Smith, 1959), in so far as they indicate that dives on helium are distinctly more hazardous than corresponding dives on air. Up to the present, as mentioned previously, the U.S. values have been accepted in calculating the necessary decompression schedule. However, a schedule for a dive to 400 ft for 10 min was found to be successful. The next series of experiments by Hempleman and Trotter was begun to decide between the alternatives already mentioned for direct surfacing dives in the U.S. results and our own. The human experiments showed results clearly not in agreement with those of the U.S. but closely resembling the previous results obtained here on goats. As a result of this work, new schedules were calculated for dives of 16 min to 300, 400 and 500 ft, and exhaustively tested in the pressure chamber. These were all carried out under exercise conditions, on a rowing machine, and gave satisfactory results, so much so that they were considered to be suitable for sea-testing. Much more work has been done, and experiments at deeper depths have been carried out, but their results have not yet been reported.

Another hazard of decompression is the overdistension of the lung when the ambient pressure is reduced by non-equalisation of the gaseous pressure in the lung. This might be caused by breath-holding, laryngeal spasm or the blocking off of portions of the inflated lung by, for instance, a plug of mucus acting as a non-return valve. Work on this problem at the R.N.P.L. has been carried out by Dr H. C. Wright. He showed that, if this over-extension of the lung does occur and if small blood vessels are ruptured in the process, then this will allow air to be forced into the pulmonary circulation. Air emboli are formed with the well-known dangerous consequences. Malhotra and Wright (1960a,b), using rabbits, confirmed the previous work on dogs that the dangers of producing air embolism were greatly reduced if the chest and abdomen were firmly bound, thus limiting the degree of expansion and stretching in the lungs.

Malhotra and Wright (1960c) also carried out an attempt to detect the site of entry of the air into the circulation. Further work on this subject is in progress.

REFERENCES

Bannister, R. G., and Cunningham, D. J. C. (1954). *J. Physiol.* **125**, 118.
Barnard, E. E. P., Hempleman, H. V., and Trotter, C. (1962). *M.R.C. Memor.* R.N.P. Rep. No. 62/1025.
Bateman, J. B. (1951). In "Decompression Sickness" (J. F. Fulton, ed.), pp. 242–77. W. B. Saunders, Philadelphia.
Bean, J. W. (1945). *Physiol. Rev.* **25**, 1.
Bean, J. W. (1950). *Amer. J. Physiol.* **161**, 417.
Bennett, P. B. (1963). "Investigations into the Aetiology of Inert Gas Narcosis." Ph.D. Thesis, University of Southampton.
Bennett, P. B., Dossett, A. N., and Kidd, D. J. (1960). *M.R.C. Memor.* R.N.P. Rep. No. 60/1001.
Bennett, P. B., and Glass, A. (1959). *M.R.C. Memor.* R.N.P. Rep. No. 59/937.
Bert, P. (1878). "La Pression Barometrique" (translators M. A. Hitchcock and F. A. Hitchcock). College Book Co., Columbus, Ohio, 1943.
Carpenter, F. G. (1956). *Amer. J. Physiol.* **187**, 573.
Case, E. M., and Haldane, J. B. S. (1941). *J. Hyg., Camb.* **41**, 225.
Crellin, R. Q., Crocker, W. E., and Hollis, H. (1954). *M.R.C. Memor.* R.N.P. Rep. No. 54/791.
Crocker, W. E. (1957). *M.R.C. Memor.* R.N.P. Rep. No. 57/885.
Crocker, W. E. (1958). *M.R.C. Memor.* R.N.P. Rep. No. 58/902.
Crocker, W. E., and Hempleman, H. V. (1957). *M.R.C. Memor.* R.N.P. Rep. No. 57/887.
Crocker, W. E., and Taylor, H. J. (1952). *M.R.C. Memor.* R.N.P. Rep. No. 52/708, Pt. B.
Davidson, W. M., Sutton, B. M., and Taylor, H. J. (1950). *M.R.C. Memor.* R.N.P. Rep. No. 50/582.
Donald, K. W. (1947). *Brit. med. J.* **1**, 172.
Donald, K. W. (1955). *J. appl. Physiol.* **7**, 639.
Duffner, G. J., Snyder, J. F., and Smith, L. L. (1959). "Adaptation of helium-oxygen to mixed gas SCUBA." U.S.N. Experimental Diving Unit Research Report No. 3–59.
Haldane, J. S., Damant, G. C. C., and Boycott, A. E. (1908). *J. Hyg., Camb.* **8**, 342.
Harvey, E. N. (1951). *In* "Decompression Sickness" (J. F. Fulton, ed.), pp. 90–114. W. B. Saunders, Philadelphia.

Hawkins, J. H., Shilling, C. W., and Hansen, R. A. (1935). *U.S. Nav. Med. Bull.*, Washington, 327.
Hempleman, H. V. (1952). *M.R.C. Memor.* R.N.P. Rep. No. 52/708, Pt. A.
Hempleman, H. V. (1957). *M.R.C. Memor.* R.N.P. Rep. No. 57/896.
Hempleman, H. V. (1962a). *M.R.C. Memor.* R.N.P. Rep. No. 62/1019.
Hempleman, H. V. (1962b). *M.R.C. Memor.* R.N.P. Rep. No. 62/1020.
Hill, L., and Macleod, J. J. (1903). *J. Hyg., Camb.* 3, 401.
Hill, L., and Greenwood, M. (1907). *Proc. roy. Soc.* B, 79, 21.
Lambertsen, C. J., Kough, R. H., Cooper, D. Y., Emmel, G. L., Loeschcke, H. H., and Schmidt, C. F. (1953a). *J. appl. Physiol.* 5, 471.
Lambertsen, C. J., Kough, R. H., Cooper, D. Y., Emmel, G. L., Loeschcke, H. H., and Schmidt, C. F. (1953b). *J. appl. Physiol.* 5, 803.
Lambertsen, C. J., Stroud, M. W., Ewing, J. H., and Mack, C. (1953c). *J. appl. Physiol.* 6, 358.
Lambertsen, C. J., Stroud, M. W., Gould, R. A., Kough, R. H., Ewing, J. H., and Schmidt, C. F. (1953d). *J. appl. Physiol.* 5, 587.
Malhotra, M. S., and Wright, H. C. (1960a). *M.R.C. Memor.* R.N.P. Rep. No. 60/996.
Malhotra, M. S., and Wright, H. C. (1960b). *M.R.C. Memor.* R.N.P. Rep. No. 60/997.
Malhotra, M. S., and Wright, H. C. (1960c). *Proc. roy. Soc.* B, 154, 418.
Marshall, Jean M. (1951). *Amer. J. Physiol.* **166**, 699.
Rashbass, C. (1954). *M.R.C. Memor.* R.N.P. Rep. No. 54/789.
Rashbass, C. (1955a). *M.R.C. Memor.* R.N.P. Rep. No. 55/847.
Rashbass, C. (1955b). *M.R.C. Lond. R.N.P. Rep.* No. 55/854.
Smith, J. L. (1899). *J. Physiol.* 24, 19.
Stadie, W. C., Riggs, B. C., and Haugaard, N. (1944). *Amer. J. med. Sci.* **207**, 84.
Taylor, H. J. (1949a). *J. Physiol.* **108**, 264.
Taylor, H. J. (1949b). *J. Physiol.* **109**, 272.

PART II. NARCOTIC ACTION OF INERT GASES

PETER B. BENNETT

A. Introductory

With the increasing interest in underwater diving shown by the navies of the world, its growth as a sport and the widening commercial interest taken in it by, for example, the big oil companies, there is an awareness of how much we do not know about the physiology of man under increased air pressure. The three major problems, decompression sickness, nitrogen narcosis and oxygen toxicity, remain mysteries, in spite of considerable research. Here the problem of nitrogen or inert gas narcosis will be considered. As with the other problems, a complete solution is not at hand, though a part of the mystery has been unravelled. In this, research at the R.N.P.L. over the last seven to eight years has played a prominent part, and especial attention will be given to the work of this Laboratory. A more general appreciation of the problem may be found elsewhere (Rinfret and Doebbler, 1961; Featherstone and Muehlbaecher, 1963; Bennett, 1964a).

That compressed air does have narcotic properties, if breathed at sufficiently increased pressures, has been realised for over 100 years. One of the first reports is that of Green in the U.S.A., in 1861, who maintained that at 160 ft (50 m) a diver becomes somnolent and his judgment is impaired, the condition being serious enough to merit his immediate return to the surface.

Many others have since then attested to the signs and symptoms of narcosis, attaching various degrees of importance to them and various depths or pressures for their onset (Hill and Macleod, 1903; Damant, 1930; Hill and Phillips, 1932; Behnke, Thomson and Motley, 1935; Shilling and Willgrube, 1937; Behnke and Yarbrough, 1939; Case and Haldane, 1941; Carpenter, 1955; Bennett, 1963a).

The signs and symptoms of compressed air narcosis are similar to those of oxygen lack and alcoholic intoxication. They are minimally present at 100 ft (30 m), equivalent to 4 atm absolute, and become increasingly severe with increasing depth or pressure. If the latter is sufficient, loss of consciousness may result. The signs and symptoms of the condition have been effectively summarised by Behnke *et al* (1935) as "euphoria, retardment of the higher mental processes and impaired neuromuscular co-ordination" and are believed to be caused by the nitrogen of air. They appear in man and animals almost immediately on reaching the limiting pressure. Hard work (Adolfson, 1964) and increased oxygen and carbon dioxide partial pressures (Case and Haldane, 1941; Bennett, 1964b, c) considerably enhance the narcosis. Frequent exposure produces some acclimatisation to the condition. Decompression results in rapid recovery.

The narcosis is not restricted solely to air. Behnke and Yarbrough (1939), Lawrence, Loomis, Tobias and Turpin (1946), Cullen and Gross (1951) and Pittinger (1962), besides many of the workers already mentioned, were quick to note that the rare, inert or noble gases—xenon, krypton, argon, neon and helium—are capable of causing similar signs and symptoms of narcosis. The potency of any gas may be related to physical constants, such as molecular weight (Behnke and Yarbrough, 1939), absorption coefficients (Case and Haldane, 1941), lipid solubility (Meyer, 1899; Overton 1901; Behnke and Yarbrough, 1939), thermodynamic activity (Ferguson, 1939; Brink and Posternak, 1948), van der Waals constants (Wulf and Featherstone, 1957) and the formation of clathrates (Miller, 1961; Pauling, 1961).

Up to World War II most workers were content to note the signs and symptoms of the narcosis and to measure the decrement in performance by means of various psychometric tests of memory, intellectual ability and neuromuscular control. In the early post-war years Carpenter (1953, 1954, 1955, 1956) and Marshall (1951), in particular, made special efforts to clarify the cause and mechanisms of the narcosis produced by air and these seemingly inert gases, whose atomic structures suggest that they will not readily enter into biochemical reactions.

B. The Electroencephalogram

Early studies in our laboratory made use of the electroencephalograph and assorted mental and neuromuscular tests (Bennett and Glass, 1957a, b, 1961) to examine the psychological changes and the electrical activity of the cortex of the brain in men exposed to increased pressures of air and to compare these with the effects of isonarcotic concentrations (Carpenter, 1955) of inert gas anaesthetics such as nitrous oxide. Electroencephalogram measurements were also made in rabbits in similar circumstances.

At an air pressure equivalent to 250 ft, rabbits seemed more excitable. There was an increase in the percentage low voltage fast electroencephalogram activity, which agreed with observations by Jullien, Roger and Chatrian (1953) on cats at 165 ft (50 m). Albano, Criscuoli and Ciulla (1962) and Albano and Criscuoli (1962), on the basis of human experiments at 300 ft (90 m) with various oxygen: nitrogen partial pressures, maintain that this hyperexcitability is a function of the associated increased oxygen partial pressure masking the narcotic action of nitrogen. In more recent work (Bennett, 1964c) the effect of increased pressures of argon, nitrogen, helium and oxygen on auditory-induced evoked potentials in cats was studied.

The results supported the presence of a period of increased cortical excitability during the initial 10 min or so at pressure. However, with increasing time at pressure this is replaced by a cortical depression (see Section E below).

In man the subjective and electroencephalographic effects of nitrous oxide narcosis are similar to those caused by isonarcotic concentrations of air. However, the theoretical isonarcotic concentration of 20% nitrous oxide is rather more narcotic than 200 ft (60 m) of air. With the latter, abolition of blocking of the α-rhythm (8 to 13 c/s) to a mental stimulus occurred as discussed in greater detail below. Nitrous oxide induced a similar abolition of α-blocking, but the narcosis proceeded to a deeper plane, causing a "faraway state" when the external environment receded, accompanied in the electroencephalogram by long silent periods and "reversed blocking", when an external stimulus causes a burst of α-activity to reappear from an almost isoelectric record.

C. The Nitrogen Threshold

The most interesting observation in this work may be thought the effect of compressed air on the α-blocking response in man (Bennett and Glass, 1957a, b, 1961). In some 47% of subjects asked to do mental sums, the α-activity disappeared while the individual was thinking, to return on the solution of the problem (Fig. 1A). Exposure to compressed air at 100 ft (30 m) abolished this phenomenon in 5 to 25 min, depending on the subject (Fig. 1C). On the return to atmospheric pressure the blocking response returned (Fig. 1D).

The time to the abolition of the α-blocking response is a function of both the individual and the pressure to which he is subjected. For a given indi-

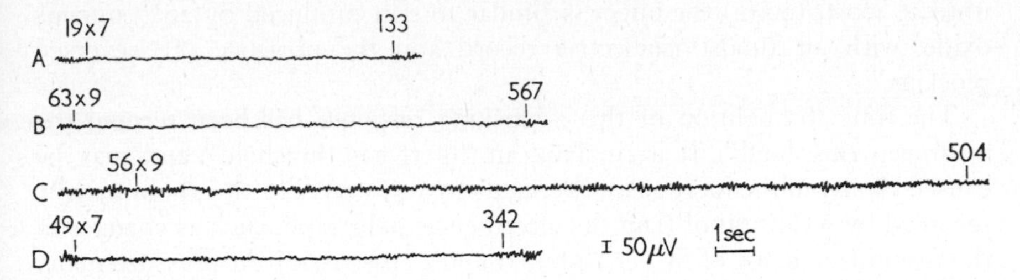

FIGURE 1. Effect of 4 atm absolute air (100 ft) on the human EEG during calculation.
A. At surface.
B. After 3 min at pressure.
C. After 15 min at pressure. Blocking abolished.
D. One minute after returning to the surface. Blocking returning (Bennett and Glass, 1957).

vidual there is a constant relationship between the reciprocal of the square root of the time to blocking abolition and the depth or pressure at which it occurs (Fig. 2). There is a tendency after the abolition of blocking for individuals to make more errors in mental arithmetic, and it was concluded that a "critical threshold change has occurred in the neurophysiological state of the brain, possibly in the ascending reticular formation, due to its progressively increasing saturation by nitrogen under pressure".

At pressures above 7 atm absolute (200 ft, 60 m) abolition of blocking may occur during compression. Between 200 and 300 ft (60 and 90 m) the electroencephalogram shows a progressive diminution in electrical activity,

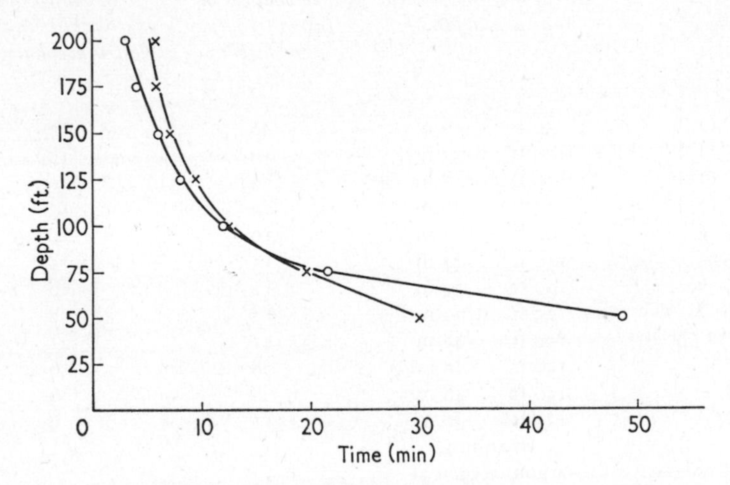

FIGURE 2. Variation of time to abolition of α-blocking (the "nitrogen threshold") with depth or pressure to which subjects are exposed. Inter-individual variation is illustrated (Bennett and Glass, 1961).

until at 300 ft (90 m) the effect is similar to that produced by 20% nitrous oxide, with an almost isoelectric record and the presence of "reversed blocking".

The time to abolition of the α-blocking response has been termed the "nitrogen threshold". It is, in fact, an "inert gas threshold" and may be caused by any of the inert gases (Bennett, 1958, 1960). The threshold may be recorded by means other than the electroencephalograph, such as changes in the fusion frequency of flicker (Bennett and Cross, 1960). At pressure many individuals show a change of at least one cycle per second in the frequency at which a flickering neon light is perceived as a steady light. A comparison of the time to this change with the time to cause abolition of α-blocking (Table I) emphasises that there is a fundamental change occurring in the brain. Unfortunately, neither of these techniques may be readily used for determining the "inert gas threshold". Not all subjects have an α-blocking response, and many are unable to differentiate between the small changes in fusion frequency with sufficient reliability. It is to be hoped that some other more widely applicable technique will be found to measure the "inert

Table I. *Comparison of times to a maintained change in the fusion frequency of flicker and abolition of α-blocking in a number of subjects exposed to various equivalent depths breathing compressed air and in one subject (P.B.) breathing* 80% *argon* + 20% *oxygen as an alternative (Bennett* 1958).

Subject	*Breathing compressed air at a depth of*		*Times in minutes to changes in fusion frequency of flicker*	*Times in minutes to changes in α-blocking in the electroencephalogram*
C.P.O. F	50 ft	16 m	38	37
C.P.O. F	100 ft	32 m	19	18
C.P.O. F	150 ft	48 m	9	9
C.P.O. F	175 ft	56 m	12	12
L/S. C.	100 ft	32 m	19	19
L/S. C.	150 ft	48 m	15	15
P.O. N	100 ft	32 m	12	10
P.O. N.	125 ft	40 m	8	8
A/B U.	150 ft	48 m	21	21
Lt. N.	150 ft	48 m	8	6
P.B.	150 ft	48 m	8	6
P.B.	125 ft	40 m	11	9
	Breathing argon/oxygen at			
P.B.	75 ft	24 m	9	9
P.B.	100 ft	32 m	6	8

gas threshold" in the future, for it seems to provide a quantitative index of the rate of saturation of brain tissue, which could be of value in dealing with decompression sickness as well as inert gas narcosis.

D. Rate of Compression

The widely held view that the faster the rate of compression the more severe the signs and symptoms of narcosis is based mostly on practical experience; the effect is thought to relate to an increase in alveolar and cerebral carbon dioxide, for this gas markedly potentiates inert gas narcosis (Case and Haldane, 1941; Marshall, 1951; Bennett, 1964b, c).

The increase in carbon dioxide is believed to be caused by the increased oxygen partial pressure blocking haemoglobin transport and by respiratory embarrassment (Bean, 1950; Albano, 1962; Adolfson, 1964). However, this is not entirely correct. There are many variables that can affect the issue and fast compression (1500 ft/min; 450 m/min) can be beneficial in specific circumstances, as described below.

The effect of three different rates of compression on various nitrogen and oxygen mixtures was examined in 114 Wistar rats (Bennett, Dossett and Kidd, 1960). Narcosis in the rats was measured by finding the voltage of a square wave stimulus required to effect a twitch of the caudal muscles. On compression with air or inert gas mixtures the voltage had to be increased to effect the same caudal twitch as that observed at atmospheric pressure (Bennett, 1963b). This increase is expressed as a percentage, called the "narcotic level" = 100 (voltage at increased pressure − voltage at atmospheric pressure) divided by the voltage at atmospheric pressure. Results are shown in Fig. 3. In rats it is evident that with mixtures of low oxygen partial pressure fast compression is beneficial. Only if the oxygen partial pressure exceeds 35 lb/in^2 does fast compression increase the degree of narcosis.

In man fast compression may also be beneficial. This is emphasised by experiments carried out recently (Bennett, Dossett and Ray, 1964) for the purpose of improving submarine escape techniques. At 400 to 500 ft (125 to 170 m) men breathing compressed air would normally be unconscious or stupefied. Eight were compressed to 400 ft and 500 ft in 20 sec, and the degree of narcosis from the period 20 sec after leaving atmospheric pressure to 60 sec, when they were decompressed back to the surface, was measured by means of a two-choice reaction time test.

At 400 ft there was no significant difference between the mean reaction times of 37·8 and 37·2 milliseconds at the surface and at pressure, respectively. At 500 ft there was a significant ($t = 0{\cdot}02$, for 13 degrees of freedom) 14 to 15% increase in the reaction time, compared with control values. This suggests that the level of narcosis is equivalent to that produced by some 90 to 100 ft (30 to 35 m) of air (Kiessling and Maag, 1961). The reason why narcosis

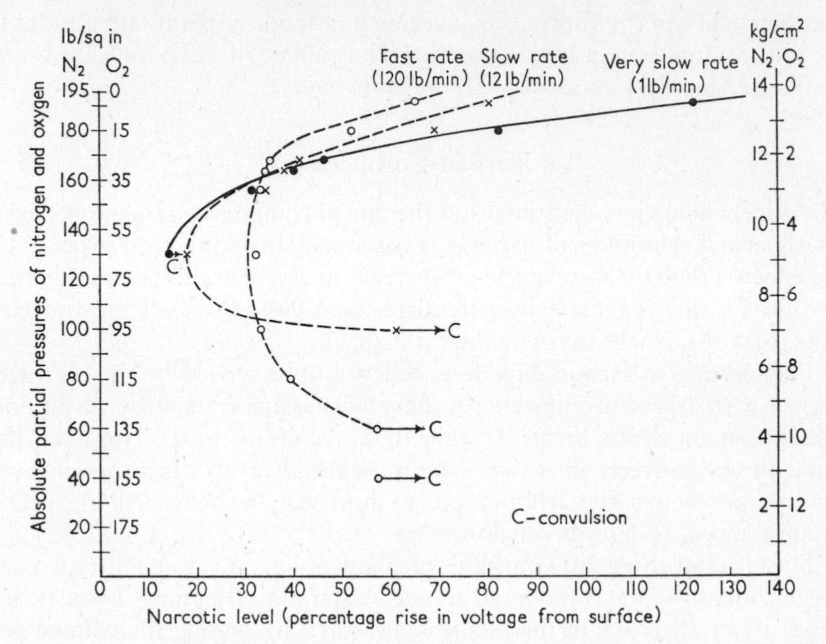

FIGURE 3. Effect of different rates of compression on the narcotic effect of rats exposed to different nitrogen/oxygen partial pressures at an absolute pressure of 195 lb/in² (Bennett, Dossett and Kidd, 1960).

is not severe is probably the inability of the inert gas to reach specific sites in the brain at a sufficient concentration before decompression begins.

E. Neurophysiological Studies

That these specific sites of action are at synapses has been inferred from studies in vitro on isolated frog reflex preparations (Marshall, 1951), isolated rat sciatic nerve, electroconvulsive shock experiments on mice (Carpenter, 1953, 1954, 1955, 1956), in vivo electromyograms of decerebrate cats (Chun, 1959) and in vivo peripheral nerve and spinal synapse potentials in rats (Bennett, 1963a, c).

In the last-mentioned, experiments on the effect of 220, 200 and 180 lb/in² argon, nitrogen and helium in the presence of 15 lb/in² oxygen were done on sixty-three rats lightly anaesthetised with pentobarbitone sodium. As by Carpenter, no change was found in peripheral nerve conduction. Spinal synapses, however, were depressed by the argon and to a less extent by the nitrogen mixtures. Helium had no effect on the potentials. With argon and nitrogen the potentials showed an initial augmentation and then a progressive depression, similar to that produced by asphyxia. Evidence was also obtained to suggest that the post-synaptic cell was depressed before the pre-synaptic (Fig. 4).

Other experiments indicated that brain synapses are depressed in a similar manner, but that they are more sensitive to inert gases than are spinal synapses (Bennett, 1964c). Under chloralose anaesthesia (50 to 60 mg/kg) the effect of inert gases on auditory-evoked potentials in twenty-seven cats was examined when the animals were exposed to 150 lb/in^2 argon, nitrogen or helium in the presence of 35 lb/in^2 oxygen absolute for one hour. Control experiments were carried out at atmospheric pressure with the animals breathing air, and the effect was studied of exposure to an absolute pressure of 75 lb/in^2 nitrogen and 18 lb/in^2 oxygen (i.e. 63 lb/in^2 nitrogen, 15 lb/in^2 oxygen gauge pressures plus atmospheric partial pressures of 12 lb/in^2 nitrogen and 3 lb/in^2 oxygen).

Exposure to atmospheric air or to the helium mixtures caused no significant change in evoked potentials. With the argon and nitrogen mixtures the positive cortical potential, believed to be pre-synaptic and to originate from the soma of deep cortical cells, was first augmented and then progressively depressed (Fig. 5). The argon mixture required about twice as long as the nitrogen, 30 as against 15 min, to produce a stable level of depression. This was regarded as further evidence to support the hypothesis of a critical gas concentration. On the other hand, the reticular formation potentials and the negative cortical potential, believed to be of post-synaptic origin from superficial soma and apical dendrites, were steadily and markedly depressed without previous augmentation.

The reasons for this augmentation of the presynaptic potential are not clear. It may be the result of a "release" effect from inhibitory control of the cortex by the ascending reticular formation of the brain stem or a more general depression of inhibitory mechanisms (Chun, 1959), or asphyxial depolarisation as a result of a histotoxic hypoxia (Ebert, Hornsey and Howard, 1958; Schreiner, 1962; Bennett, 1964a, b), interference with Na and K ion exchange (Sears, 1962; Bennett, 1964a, b) or some other unknown factor.

Further support for the critical concentration theory is provided by the fact that the 75 lb/in^2 nitrogen mixture required about twice the time of the 150 lb/in^2 nitrogen mixture to reach a maximum level of depression.

It was concluded from these experiments that inert gases at raised pressures induce narcosis by depressing transmission in polysynaptic systems of the central nervous system, such as are found in the cortical mantle and the reticular formation of the brain stem. The synaptic depression may be due to interference with one or all of a number of factors (Bennett, 1964a), such as subcellular hypoxia (Ebert *et al.*, 1958; Schreiner, 1962), interference with the release of chemical transmitters (Sears, 1962) and prevention of Na/K ion exchange. Some recent unpublished work of my own shows that extracellular

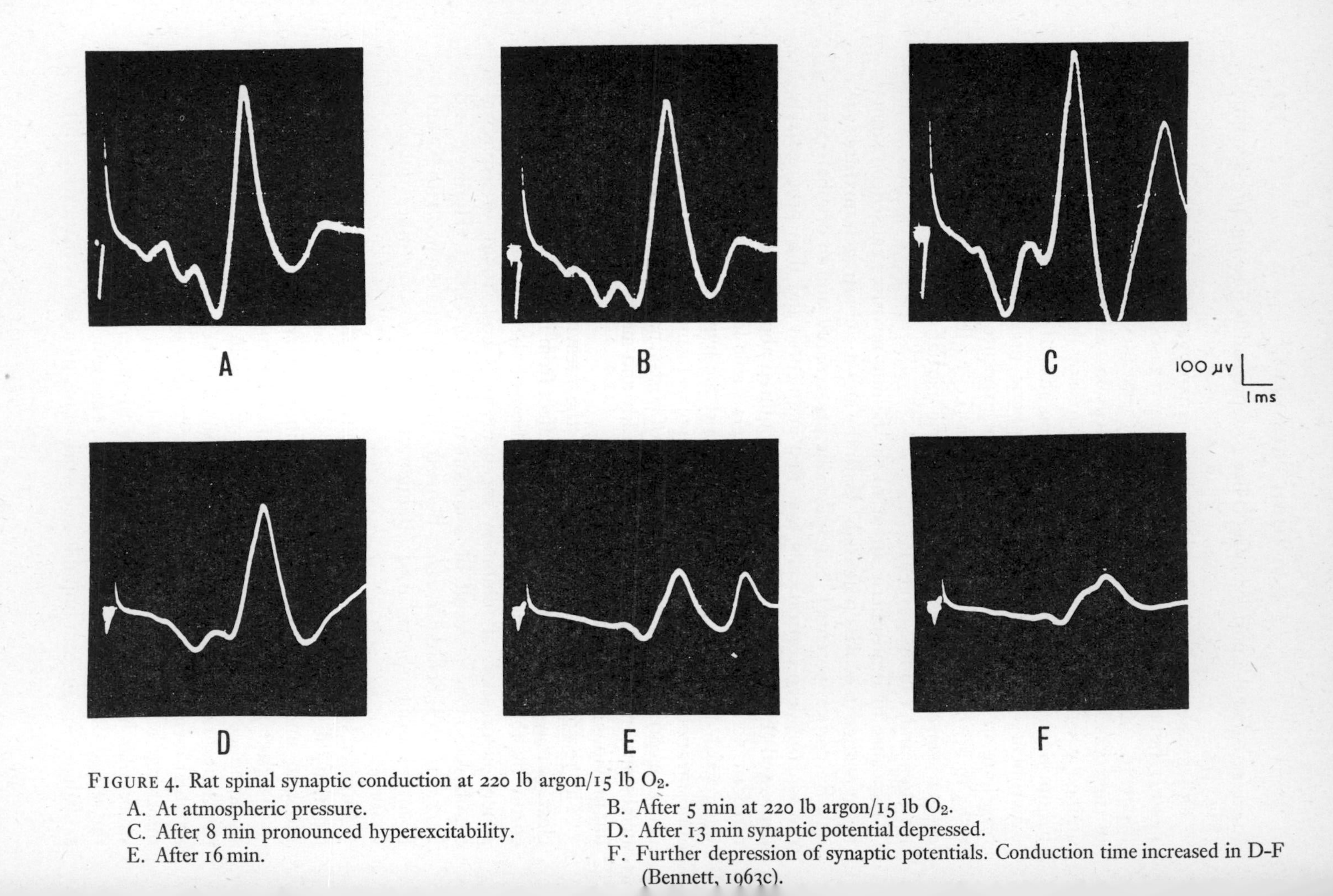

FIGURE 4. Rat spinal synaptic conduction at 220 lb argon/15 lb O_2.

A. At atmospheric pressure.
B. After 5 min at 220 lb argon/15 lb O_2.
C. After 8 min pronounced hyperexcitability.
D. After 13 min synaptic potential depressed.
E. After 16 min.
F. Further depression of synaptic potentials. Conduction time increased in D-F (Bennett, 1963c).

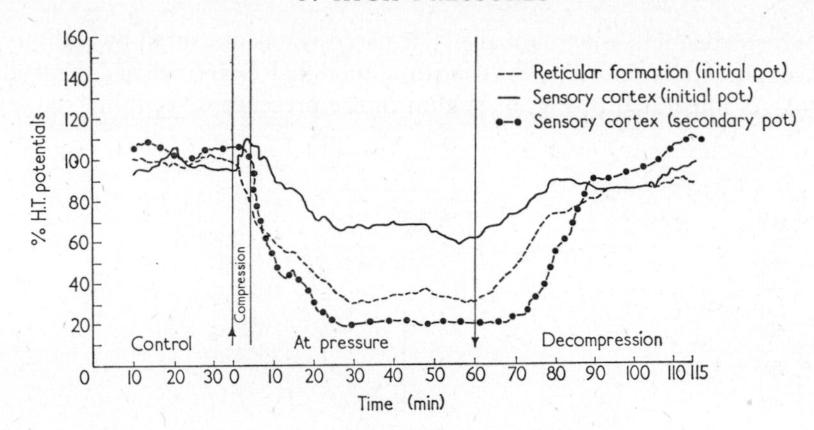

FIGURE 5. Advancing means of changes in the amplitude of potentials evoked by auditory stimuli in the sensory cortex and reticular formation of cats exposed to 150 lb/in^2 argon and 35 lb/in^2 oxygen absolute (Bennett, 1964c).

cortical Na ions increase markedly during inert gas narcosis. This tends to support the theory of ion prevention: a subcellular hypoxia would seem less likely. Polarographic measurements of cortical oxygen tension in cats (Bennett, 1963c, 1964b) shows that exposure to compressed air or inert gas+oxygen mixtures causes an increased cerebral oxygen tension. Attempts to increase this to levels 1000 to 2000% greater than normal failed to show any marked decrease in the narcosis, as might be expected with a histotoxic hypoxia. Further, hypoxia normally produces a failure of the sodium extrusion pump, with a consequent decrease in extracellular Na ions (Meyer, Gotoh and Tazaki, 1961), which is the opposite to what has been found with inert gas narcosis. The explanation of the true mechanism awaits more research.

F. The Carbon Dioxide Theory

Whereas the mechanism must therefore remain controversial, the cause of the narcosis is clearer. From time to time, and especially in recent years (Bean, 1950; Seusing and Drube, 1960; Buhlman, 1961), the theory has arisen that inert gas narcosis is really carbon dioxide narcosis due to a raised cerebral carbon dioxide tension attributable to respiratory embarrassment because of the increased density of the mixtures breathed at pressure and the prevention of carbon dioxide transport by the haemoglobin as a result of its being fully saturated owing to the high oxygen partial pressure.

That this theory is not correct has been indicated by experiments on thirty-five cats under chloralose anaesthesia exposed to increased pressures of argon, nitrogen and helium in the presence of either 35 lb/in^2 oxygen or 3 lb/

in² oxygen (Bennett, 1963c, 1964b). The narcosis was measured by auditory-evoked potentials and the P_{CO_2} with a modified Severinghaus electrode (1959). As illustrated in Fig. 6, helium in the presence of 35 lb/in² oxygen

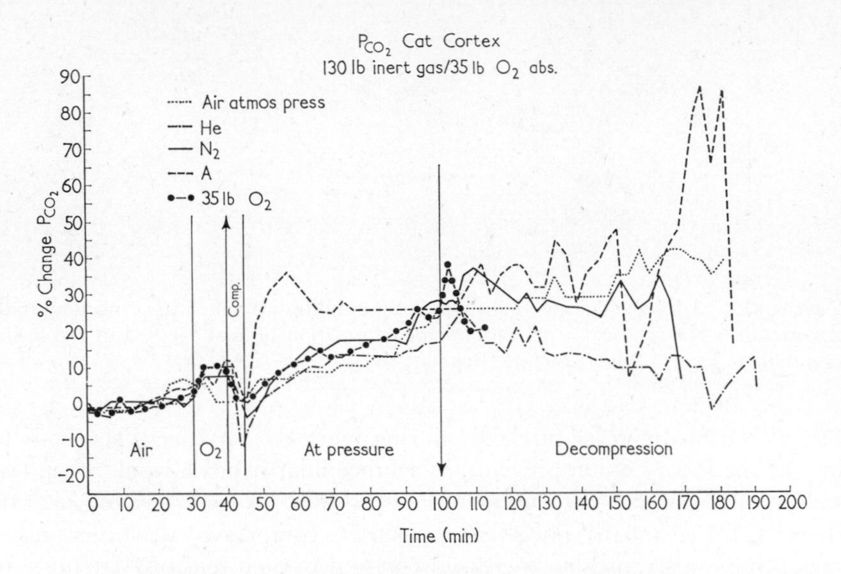

FIGURE 6. Cortical P_{CO_2} of chloralosed cats exposed to increased pressures of argon, nitrogen and helium in the presence of 35 lb/in² oxygen (Bennett, 1963a).

caused no significant increase in carbon dioxide. Both nitrogen and argon, however, did induce a significant increase in cortical P_{O_2}. Argon, with its greater density, caused the largest percentage increase. Exposure to 35 lb/in² oxygen alone caused an increase in P_{CO_2} similar to that found with 130 lb/in² nitrogen and 35 lb/in² oxygen. The evoked potentials in the presence of nitrogen were depressed, indicating narcosis, but were unaffected by the exposure to oxygen alone. It is therefore evident that carbon dioxide is unlikely to be the cause of the narcosis, which is due to the increased partial pressure of the inert gas. Further, it is confirmed that a high oxygen tension is an important factor in raising cortical P_{CO_2}, as are increases in density of the breathing mixtures. The effect of exposure to mixtures with a low oxygen partial pressure (Fig. 7) confirms this hypothesis. Although the cats exposed to the nitrogen mixture were narcotised, the P_{CO_2} was not significantly different from that of controls at atmospheric pressure. Argon with its greater density does still induce an increased P_{CO_2}, whereas helium causes a reduction in cerebral P_{CO_2}. The experiments also supported the evidence for the potentiating action of an increased P_{CO_2} on inert gas narcosis (Case and Haldane, 1941; Hesser, 1963). In this connection carefully controlled hyperventilation

(Bennett, 1964a), which lowers the P_{CO_2}, may be used to decrease the level of narcosis. Care must, however, be taken that the work involved in hyperventilating at increased pressures does not increase the P_{CO_2} and produce a more severe rather than a milder narcosis. Some training is required to achieve the correct technique.

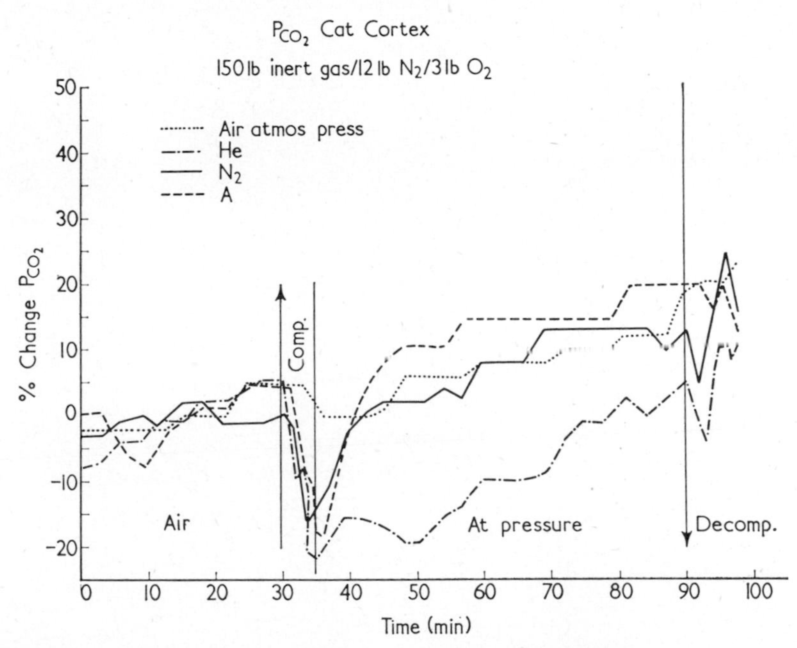

FIGURE 7. Cortical P_{CO_2} of chloralosed cats exposed to increased pressures of argon, nitrogen and helium in the presence of 3 lb/in^2 oxygen (Bennett, 1963a).

G. The Effects of Drugs

Reference to Fig. 3 and the work of Bennett, Dossett and Kidd (1960) shows that it is possible to measure an impairment of the twitch of caudal muscles to a square wave stimulus in rats exposed to high oxygen partial pressures as well as to high nitrogen partial pressures. This suggests that there may be some similar mechanisms involved in oxygen toxicity and inert gas narcosis. Accordingly, eleven drugs were selected, known either to protect, to have no effect on or to enhance oxygen convulsions in rats, for comparison with their effects on nitrogen narcosis. The level of narcosis was measured by the caudal twitch electro-shock technique described previously. The results of these experiments (Bennett, 1962) are shown in Table II. It is clear that drugs do have a similar action in the two conditions.

One drug in particular, azacyclonol, was selected for a more thorough

Table II. *Comparative Effects of Eleven Drugs on Nitrogen Narcosis and Oxygen Toxicity in Rats (Bennett 1962)*

	No drugs	*Carbachol* 1 mg/kg	*Azacyclonol* 80 mg/kg	*Glutethimide* 30 mg/kg	*Phenacetin* 200 mg/kg	*Acetylsalicylic acid* 100 mg/kg	*Physostigmine* 0·1 mg/kg	*Adrenaline* 1 mg/kg	*Hyoscine* 1 mg/kg	*Methylamphetamine* 5 mg/kg	*Bemegride* 5 mg/kg	*Leptazol* 25 mg/kg
Individual	53	0	0	0	7·5	28	52	58	57	61	46	97
percentage rise in	51	0	2	4	5	32	50	42	77	87	66	97
voltage of stimulus	57	0	5	5	5	19	48	54	57	87	88	75
to effect a caudal	57	0	5	5	0	30	51	56	52†	65	92	90
twitch in rats ex-	51	0	2	0	5	33	46	56	57	75	95	74
posed to 180 lb in²	49											
nitrogen/15 lb in²	47											
oxygen (12·66 kg	52											
cm² nitrogen/1·05	44											
kg cm² oxygen)	47											
	48											
	65											
Mean percentage	52	0	3	3	5	28	49	53	60	75	77	87
	22	79	33	28	48	31	21	21†	25	18	7	20
Individual	29	55	33	37	52	32	20	13†	37	12	36	6
time in minutes	24	50	35	45	38	36	12	30	42†	8	6	9
from start of	20	61	33	34	54	28	31	37	24	20	9	6
compression to	23	44	39	32	46	31	21	9	20	18	15	5
a convulsion	15			34			27					
in rats exposed	24			36			11					
to 80 lb in²	28			31								
(5·62 kg cm²)	15			33								
oxygen	24			34								
	27			42								
	27			30								
	16											
	23											
Mean time (minutes)	22	58	35	33	48	32	20	22	30	15	15	9

† Animal died. Means to nearest whole numbers.

study of its action in preventing inert gas narcosis, owing to its lack of side effects. Some forty-six rats were exposed in 102 experiments to increased pressures of argon, nitrogen and helium at a partial pressure of 180 lb/in^2 in the presence of 15 lb/in^2 oxygen before and after oral, intraperitoneal and intravenous administration of azacyclonol (Bennett, 1963b). The electroshock technique was used to evaluate the narcosis. It was found that the effective oral or intraperitoneal dose was 40 mg, regardless of the inert gas to which the animals were exposed, suggesting a mechanical barrier effect. The protection to rats afforded by 40 mg intraperitoneal azacyclonol when they

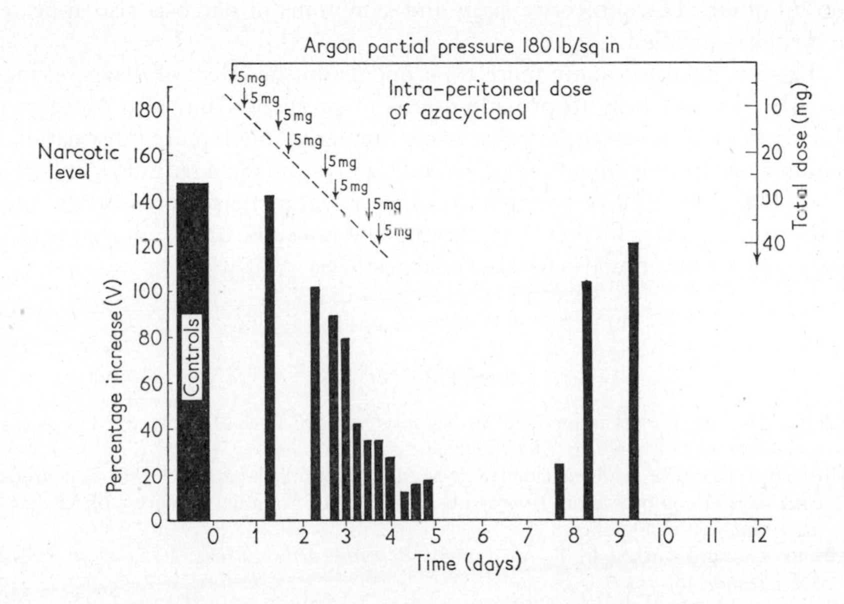

FIGURE 8. Effect on mean response index of 40 mg azacyclonol (intraperitoneal) on four rats exposed to an argon partial pressure of 180 lb/in^2 compared with seven controls at atmospheric pressure (Bennett, 1963b).

are exposed to the argon mixture is illustrated in Fig. 8. The narcotic threshold of twelve rats exposed to the nitrogen mixture was a 50·2% ± 4% increase in the volts needed to effect a stimulus. The percentage increase 48 hr after oral administration of azacyclonol was 1·21% ± 2·45%, a value not significantly different from that for the controls. The slow action of azacyclonol is also clear from Fig. 8.

Preliminary studies of the action of azacyclonol in men exposed to 300 ft (90 m) of compressed air have also been made (Bennett, 1961). Tests of the narcotic effect were made by means of simple arithmetic, letter cancellation, a visual co-ordination test and critical flicker fusion (c.f.f.). A narcotic index, based on the results of the tests, showed a mean value of 11·5 at 300 ft (90 m) compared with 17·5 at atmospheric pressure, giving a 34% decrement in average performance. Oral administration of 900 mg (daily dose of 300 × 3) azacyclonol for seven days provided a partial protection, the decrement in performance being only 23%. By use of the c.f.f. test alone to measure the narcosis, a 1200 mg (600 × 2) oral dose for three days was found more effective. At 300 ft (90 m) of air the c.f.f. increased significantly by 1·5 cycles over atmospheric pressure. After administration of the drug there was no significant difference between values at atmospheric pressure and 300 ft (90 m) of air. The subjective signs and symptoms of narcosis also appeared to be well controlled.

There is need for many more experiments on the effect of drugs on inert gas narcosis, not only to provide means of protection but also to assist in elucidating its causes and mechanisms. Similarly, much more information is required on the movement of Na, K and Ca ions and changes in P_{O_2} and P_{CO_2} in cerebral tissue and on synaptic transmission at increased pressures of inert gases in particular, for here may lie the key not only to the mechanisms of inert gas narcosis, but also to those of anaesthesia.

REFERENCES

Adolfson, J. (1964). *In* "Compressed Air Narcosis". M.Sc. Thesis, Institute of Psychology, University of Gothenburg, Stockholm.

Albano, G. (1962). *In* "Influenza della velocita di discesa sulla latenza dei disturbi neuropsichici da aria compressa nel lavoro subacqueo". 25th National Congress of Medicine, Taormina, 15–18 October.

Albano, G., and Criscuoli, P. M. (1962). *Bolletino della Società Italiana di Biologia Sperimentale* 38, 754.

Albano, G., Criscuoli, P. M., and Ciulla, C. (1962). *Lav. Um.* **14**, 351, 396.

Bean, J. W. (1950). *Amer. J. Physiol.* **161**, 417.

Behnke, A. R., Thomson, R. M., and Motley, E. P. (1935). *Amer. J. Physiol.* **112**, 554.

Behnke, A. R., and Yarbrough, O. D. (1939). *Amer. J. Physiol.* **126**, 409.

Bennett, P. B. (1958). *M.R.C. Memor.* R.N.P. Rep. No. 59/935.

Bennett, P. B. (1960). *Ergonomics*, 3, 273.

Bennett, P. B. (1961). *M.R.C. Memor.* R.N.P. Rep. U.P.S. 196.

Bennett, P. B. (1962). *Life Sciences* No. **12**, 721.

Bennett, P. B. (1963a). *In* "Neurophysiologic and neuropharmacologic changes in inert gas narcosis". Proceedings 2nd Symposium on Underwater Physiology. Publication 1181, 209. National Academy of Sciences, Washington.

Bennett, P. B. (1963b). *Amer. J. Physiol.* **205**, 1013.

Bennett, P. B. (1963c). *In* "Investigations into the aetiology of inert gas narcosis". PhD. Thesis, University of Southampton.

Bennett, P. B. (1964a). *In* "The Aetiology of Compressed Air Intoxication and Inert Gas Narcosis". Pergamon Press, London and Oxford. (In Press.)

Bennett, P. B. (1964b). *J. appl. Physiol.* (In Press.)

Bennett, P. B. (1964c). *Electroenceph. clin. Neurophysiol.* **17**, 388.
Bennett, P. B., and Cross, A. V. C. (1960). *J. Physiol.* **151**, 28–29P.
Bennett, P. B., Dossett, A. N., and Kidd, D. J. (1960). *M.R.C. Memor.* R.N.P. Rep. No. 60/1001.
Bennett, P. B., Dossett, A. N., and Ray, P. (1964). *M.R.C. Memor. R.N.P.* (In press.)
Bennett, P. B., and Glass, A. (1957a). *J. Physiol.* 138, 18–19.
Bennett, P. B., and Glass, A. (1957b). *M.R.C. Memor.* R.N.P. Rep. No. 59/937.
Bennett, P. B., and Glass, A. (1961). *Electroenceph. clin. Neurophysiol.* 13, 91.
Brink, F., and Posternak, J. M. (1948). *J. cell. comp. Physiol.* **32**, 211.
Buhlmann, A. A. (1961). *Schweiz. med. Wschr.* **19**, 774.
Carpenter, F. G. (1953). *Amer. J. Physiol.* **172**, 471.
Carpenter, F. G. (1954). *Amer. J. Physiol.* **178**, 505.
Carpenter, F. G. (1955). *In* "Inert gas narcosis". Proceedings of the 1st Symposium on Underwater Physiology, Publication 377. National Academy of Sciences, Washington.
Carpenter, F. G. (1956). *Amer. J. Physiol.* **187**, 573.
Case, E. M., and Haldane, J. B. S. (1941). *J. Hyg., Camb.* **41**, 225.
Chun, C. (1959). *Fiziol. Zh.* **45**, 605.
Cullen, S. C., and Gross, E. G. (1951). *Science* **113**, 580.
Damant, G. C. C. (1930). *Nature, Lond.* **126**, 606.
Ebert, M., Hornsey, S., and Howard, A. (1958). *Nature, Lond.* **181**, 613.
Featherstone, R. M., and Muehlbaecher, C. (1963). *Pharmacol. Rev.* **15**, 97.
Ferguson, J. (1939). *Proc. roy. Soc.* B. **197**, 387.
Green, J. B. (1861). "Diving with and without Armour", cited by Unsworth, I. P. *In* "Nitrogen narcosis: some aspects". *St Mary's Hospital Gazette, London*, 66, 272, (1960).
Hesser, C. M. (1963). *In* "Measurement of inert gas narcosis in man". Proceedings of the 2nd Symposium on Underwater Physiology. Publication 1181. National Academy of Sciences, Washington.
Hill, L., and Macleod, J. J. (1903). *J. Physiol.* **29**, 492.
Hill, L., and Phillips, A. E. (1932). *J. roy. nav. med. Serv.* 18, 157.
Jullien, G., Roger, A., and Chatrian, G. E. (1953). *Riv. Neurol.* **23**, 357.
Kiessling, R. J., and Maag, C. H. (1961). *J. appl. Psychol.* 46, 91.
Lawrence, J. H., Loomis, W. F., Tobias, C. A., and Turpin, F. H. (1946). *J. Physiol.* **105**, 197.
Marshall, J. M. (1951). *Amer. J. Physiol.* **166**, 699.
Meyer, H. H. (1899). *Arch. exp. Path. Pharmak.* **42**, 109.
Meyer, J. S., Gotoh, F., and Tazaki, Y. (1961). *Electroenceph. clin. Neurophsiol.* 13, 762.
Miller, S. L. (1961). *Proc. nat. Acad. Sci. Wash.* **47**, 1515.
Overton, E. (1901). *In* "Studien über die Narkose". G. Fisher, Jena.
Pauling, L. (1961). *Science* **134**, 15.
Pittinger, C. B. (1962). *In* "Mechanisms of anesthesia. Xenon as an anesthetic". Proceedings of 22nd International Congress of Physiological Sciences, Leiden, Holland, Vol. 1, 531. Exerpta Medica Foundation, London.
Rinfret, A. P., and Doebbler, G. F. (1961). *In* "Physiological and biochemical effects and applications. Argon, helium and the rare gases", (G. A. Cook, ed.), Vol. 2. Interscience, London.
Schreiner, H. R. (1962). *In* "Mechanisms of anesthesia. III. Role of lipid molecules in anesthesia and narcosis". Proceedings of 22nd International congress of Physiological Sciences, Leiden, Holland, Vol. II (2), 535. Exerpta Medica Foundation, London.
Sears (1962). *In* "Mechanisms of anesthesia. III. Role of lipid molecules in anesthesia and narcosis". Proceedings of 22nd International Congress of Physiological Sciences, Leiden, Holland, Vol. V (2), p. 540. Exerpta Medica Foundation, London.

Seusing, J., and Drube, H. (1960). *Klin. Wschr.* 38, 1088.
Severinghaus, J. W. (1959). *In* "A symposium on pH and Blood Gas Measurement", (R. F. Woolmer, ed.), p. 126. Churchill, London.
Shilling, C. W., and Willgrube, W. W. (1937). *Nav. med. Bull., Wash.* 35, 373.
Wulf, R. J., and Featherstone, R. M. (1957). *Anesthesiology* 18, 97.

APPENDIX A

1 lb/in² = 68,497·6 dyne/cm² = 70·3067 g/cm²

A

lb/in²		(approximate values) Kg/cm²
1	*equivalent to*	0·07
5	*equivalent to*	0·35
10	*equivalent to*	0·70
15	*equivalent to*	1·05
20	*equivalent to*	1·40
25	*equivalent to*	1·76
50	*equivalent to*	3·52
60	*equivalent to*	4·22
70	*equivalent to*	4·92
80	*equivalent to*	5·62
90	*equivalent to*	6·33
100	*equivalent to*	7·03
150	*equivalent to*	10·55
200	*equivalent to*	14·06
250	*equivalent to*	17·58

B

Kg/cm²		lb/in²
1	*equivalent to*	14·2
2	*equivalent to*	28·4
3	*equivalent to*	42·7
4	*equivalent to*	56·9
5	*equivalent to*	71·1
10	*equivalent to*	142·2
15	*equivalent to*	213·4
20	*equivalent to*	284·47

Notes

Atmospheric pressure: At sea-level normally 1,013,250 dynes/cm² = 1,033, 227 g/cm² = 14·69 lb/in² (often taken as 14·7 lb or even 15 lb in²).

Gauge pressure: Atmospheric pressure is not normally shown in gauges, which are graduated to read zero at 14·7 lb/in², so that they register pressures in excess of atmospheric.

Absolute pressure: This is the sum of gauge and atmospheric pressures, often written simply "absolute".

APPENDIX B

feet		metres
1	*equivalent to*	0.3
10	*equivalent to*	3
90	*equivalent to*	27
100	*equivalent to*	30
150	*equivalent to*	45
165	*equivalent to*	50
200	*equivalent to*	60
250	*equivalent to*	75
300	*equivalent to*	90
330	*equivalent to*	100
400	*equivalent to*	120
450	*equivalent to*	135
500	*equivalent to*	155
600	*equivalent to*	185
1,000	*equivalent to*	300
1,200	*equivalent to*	360
1,500	*equivalent to*	450

CHAPTER 7

High and Low Gravitational Forces

PETER HOWARD

A. Introductory

Variations in gravitational force are, strictly speaking, only a part of normal human experience to the extent that the force exerted by the earth varies with geography and with altitude. Regional differences in the gravitational constant G are minute, and the inverse square relationship means that only when the height above sea-level is a significant fraction of the earth's radius will the changes from this cause be important. However, the Principle of Equivalence formulated by Einstein states that gravity cannot be distinguished in its effects from an acceleration of appropriate magnitude. It is, therefore, legitimate to discuss the physiological consequences of acceleration under the heading of this chapter.

Although acceleration is a familiar concept, everyday acquaintance with it is limited, in terms either of magnitude or of duration. The force that an accelerating motor car exerts upon the mass of its occupants may last for an appreciable time, but it is usually gentle enough to evoke no physiological disturbances. On the other hand, the forces generated when the same car collides with a cliff may well be large enough to cause fatal injury, but they act only for a few milliseconds. Accelerations that last for more than 1 sec, and yet are sufficiently great to affect human function, are encountered almost exclusively in aircraft, in space vehicles and in some of the more bizarre fair-ground devices. In all these they may reasonably be described as "self-sought".

In this chapter some effects of various types of acceleration upon the major systems of the body will be examined, with certain necessary restrictions. First, only forces of long duration will be considered, and the mechanical consequences of crashes, catapulting and the like will be excluded. Secondly, attention will be given more to those findings of which interpretation is difficult than to those about which there is no dispute. Thirdly, the low gravitational force of the title will be discussed only briefly, because the peculiar circumstances of its occurrence have severely limited the study of its physiological effects.

B. Terminology

Acceleration is the rate of change of velocity. Velocity is a vector quantity, having direction as well as magnitude, and an alteration in either property will result in acceleration, which is also a vector quantity. Changes in speed produce linear accelerations, whereas motion in a curved path at a constant speed gives rise to a centripetal acceleration directed along the radius of the curve. Angular acceleration results from a change in the speed of rotation. The dimensions of velocity are length and time, and it follows that those of acceleration are l/t^2, i.e. centimetres (or feet)/second/second. It is more convenient, however, to express the rate of change of velocity in terms of the earth's gravitational constant, *G*, which has a mean value of 980·7 cm/sec^2 or 32·2 ft/sec^2. It is also common practice to use the words "acceleration" and "force" as though they were synonymous. In this sense, accelerations can be interpreted as changes in weight; indeed, most accelerometers measure the change in weight of a suspended mass, but their scales may be calibrated in units of acceleration or of force or merely as multiples of gravity.

From a physiological standpoint, accelerations are best classified according to the direction in which they act upon the body. The nomenclature is confused because effects are produced by an inertial or reactive force acting, according to Newton's Third Law of Motion, in a direction opposite to the acceleration. For example, the forward acceleration of a vehicle tends to push the driver back in his seat, and centripetal accelerations produce a centrifugal displacement, away from the centre of rotation. Many classifications have been suggested in an attempt to overcome these difficulties, but most of them have found favour only with their own authors. They range from the mathematical, in which the acceleration is described according to rectangular co-ordinates of positive or negative sign, to the unequivocal terms used by American astronauts for indicating the direction in which the eyeballs tend to move under the influence of the applied force. In this country, the terms "positive", "negative" and "transverse" are firmly established, and they will be used here. They refer, respectively, to the vernacular descriptions "eyeballs down", "eyeballs up" and "eyeballs in or out"; "eyeballs left (or right)" corresponds to lateral accelerations.

Apart from the difference in basic terminology, there are two other sources of confusion between the aviation physiologist and the engineer. The latter usually begins his reckoning from the baseline imposed upon him by the earth; to him an acceleration of $+1g$ implies a doubling of the normal weight. Similarly, he interprets an acceleration of $-1g$ as a cancellation of gravity, resulting in weightlessness. To the physiologist, who works in both directions from zero, $+1g$ represents the usual state of a man in the upright posture, and $-1g$ suggests that the man is standing on his head. Moreover, the physiological effects of acceleration and of deceleration are the same if the force acts in the same direction on the body, and it is not uncommon to read of "an acceleration of $11g$ sustained during the re-entry of a space capsule". In engineering parlance, this is a deceleration, and the phrase is yet another example of the loose thinking that bedevils many biological sciences.

C. Positive Acceleration

Most of the physiological effects of positive and of negative acceleration can be ascribed to disturbances of the cardiovascular system. This is because the applied force acts along the long axis of the body, and hence in the line of the major blood vessels, producing fluid shifts and alterations in the normal pressure gradients between the different parts of the circulatory tree. The disturbances themselves evoke compensatory reflexes; although these tend to restore conditions to normal, they introduce further complications.

1. *Pressure gradients*

The hydrostatic pressure exerted by a vertical column of fluid depends upon the height of the column above the point of measurement, upon the density of the liquid and upon the acceleration due to gravity, g. If the first two factors are assumed to be constant (which is nearly true for a rigid, open-ended column), the pressure at a given point in the fluid will vary directly with the applied acceleration. As a model of the circulation, such a simple tube is clearly inadequate, but it is possible to construct more sophisticated models to imitate the behaviour of the cardiovascular pressures during gravitational stress. The first step is to close the ends of the tube by elastic diaphragms, representing the compliance of the blood vessels. There will now be a point within the column at which the pressure is zero in relation to the atmosphere. Pressures above this point will be sub-atmospheric, and those in the lower part of the tube will be greater than the external reference pressure (Fig. 1).

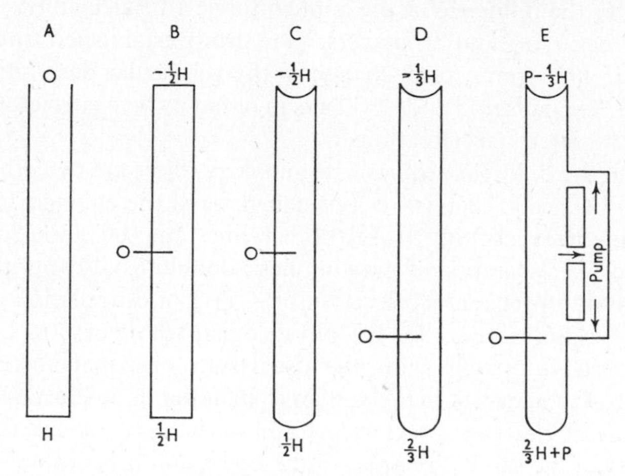

FIGURE 1. The position of the hydrostatic indifference point in models of the circulation.

A. Open-ended rigid tube.
B. Closed rigid tube.
C. Tube closed by membranes of equal compliance.
D. Tube closed by membranes of different compliance.
E. Addition of a pump, delivering a pressure P. (Modified from Clark, Hooker and Weed, 1934.)

If the two diaphragms are equally stiff, the zero level will be at the middle of the column, but it may be shifted up or down by strengthening either membrane, by forcing more fluid into the model or by providing the walls with "vasomotor tone". As a further refinement, a pump may be added, but it will not alter the general distribution of pressure within the system. The total pressure gradient between the two ends of the column will remain unchanged, but an evenly distributed increment (which may be steady or pulsating) will be added throughout the fluid.

These considerations imply that there exists within the cardiovascular system a "hydrostatic indifference point" (Wagner, 1886) at which the absolute pressure remains unaltered during changes of posture or centrifugal acceleration. The pressure at all points above this level will be decreased by positive acceleration, whereas pressures below the indifference point will increase in proportion to the distance from it. It also follows that, if the capacitance of the vessels is reduced, the indifference point will be displaced upwards and the decline in the pressure at head level will be reduced.

The site of this theoretical point has been the subject of some confusion. On the basis of alterations of posture, and of measurements from the venous side of the circulation, it has been placed about 9 cm below the diaphragm (Gauer, 1961) or just headward of the atria (Wilkins, Bradley and Friedland, 1950; Gauer and Hull, 1954). Gauer has also urged that the vessels in the lower part of the body form a U-tube and that the indifference point must therefore lie in the same plane in both the arteries and the veins. As evidence for this view he cites the studies of Lawton, Greene, Kydd, Peterson and Crosbie (1958), who found a point of zero change in the arterial pressure well below the diaphragm of monkeys exposed to positive acceleration, and concludes that it is the distensibility of the venous system that determines the reference point for both sides of the circulation. The results of Lawton *et al.* (1958) were, however, obtained during the onset of acceleration, whereas the measurements of Gauer and Sieker (unpublished) were made during a steady state after a change in posture. It is not possible to equate these two pieces of evidence, for vasoconstriction, which is known to appear 6 to 10 sec after the acceleration is applied, will tend to raise the indifference point in the arterial system. Equivalence of the two sides of the circulation can only be assumed if the resistance that connects them responds in a linear fashion to alterations of driving pressure; despite the observations of Salzman and Leverett (1956) on venoconstriction, this assumption is unlikely to be justified.

There is, in fact, respectable evidence that the arterial point of hydrostatic indifference lies above the diaphragm. From measurements of the brachial arterial pressure with the arm supported at different heights (Howard, unpublished results) the level of the second costal interspace appears to represent a reference plane at 2, 3 and 4*g*. A similar location can be deduced from the measurements of pressure in the aortic arch made by Lindberg, Wood, Sutterer, Marshall and Headley (1961). The results of Lambert and Wood (1946), who studied the fall in blood pressure produced at eye level by positive acceleration, suggest that the site of zero change is at a rather lower point, situated approximately 40 cm below the eye. Its precise position is not of great practical importance. If it is known, however, a reasonable estimate can be made of individual tolerance to acceleration. Moreover, both the efficiency of the heart and the activity of the reflex compensatory mechanisms will depend to some extent upon the site of the hydrostatic indifference point.

2. *Black-out and unconsciousness*

Apart from the sensation of increased weight, the earliest effects of positive acceleration are related to vision. A generalised "veiling", with apparent loss of contrast, leads to a progressive impairment in the extent of the visual field, beginning at the periphery. This stage, in all its ill-defined degrees, is known as "grey-out". Higher accelerations, above about 4*g*, produce further narrowing of the field, until central vision is also lost, a state referred to as "black-out" or "amaurosis fugax". If the applied force is again increased by 0·5 to 1 *g*, consciousness is lost.

All these symptoms can reasonably be explained in terms of the arterial blood pressure in the upper part of the body. Peripheral vision is affected first, because of the pattern of distribution of the central retinal artery, whose end-branches are at a comparatively low pressure. Further, the presence of an intra-ocular tension of 15 to 25 mm Hg opposes the flow of blood into the eye by reducing the transmural pressure of the retinal vessels, with the result that the circulation may be severely affected by a relatively small decline in the systemic pressure. Lambert and Wood (1946) showed that black-out was produced by an acceleration that reduced the arterial blood pressure at eye level to about 20 mm Hg, and Lambert (1945) demonstrated that vision could be restored at this stage by applying suction to the eye. The considerable individual variations in the stress required to produce loss of vision depend upon hydrostatic factors; these in turn are influenced by the heart-to-eye distance, the degree of muscular relaxation, the resting value of the systolic pressure and the efficacy of the cardiovascular compensation that can be achieved. Similar factors determine the point at which consciousness is lost, an event precipitated by the fall of the systolic pressure to zero.

Although this interpretation of the symptoms is broadly correct, it needs some qualification. In the first place, black-out is not the clear-cut entity usually assumed. It is possible to demonstrate an impairment of central vision at accelerations as low as 1·4*g* (Howard and Byford, 1956), and a sufficiently bright stimulus can be seen up to the onset of unconsciousness (Howard, 1964). It seems that the curve of target brightness against what may loosely be called "black-out threshold" at first rises steeply, then forms a plateau and finally rises steeply once more. At the plateau, vision for a wide range of brightnesses disappears over a small band of accelerations, and it happens that this condition corresponds to the reduction in systolic blood pressure required by theory and found by Lambert and Wood (1946).

There have been many attempts to implicate the central nervous system in the mechanism of black-out, both on mechanical and on haemodynamic grounds. It has been argued that descent of the brain through the less dense cerebrospinal fluid might cause pressure on the optic chiasma or on the occipital cortex (Gauer, 1950) or that the blood flow in the vertebral vessels might be reduced below a critical level (Young, 1946). Others have considered that the fall in systolic pressure associated with black-out cannot be without effect upon the entire cerebral circulation. The most recent evidence on this point is that of Beckman, Duane and Coburn (1962), who claimed that the limitation of voluntary movements of the eyes, which they observed when vision had been lost, indicated central involvement. However, this phenomenon, and every other manifestation of black-out, can be accurately reproduced by external pressure on the eyeball, and it is difficult to believe that pressure blindness is other than a purely retinal lesion. All the evidence so far available suggests that black-out and pressure blindness are both due to an anoxic failure in the ganglion-cell layer of the retina (Howard, 1964).

Unconsciousness is to be expected when the blood pressure at the base of the brain falls to zero, and Lambert and Wood (1946) showed that this indeed occurs. In the interval between black-out and loss of consciousness the other senses remain intact, and mental efficiency appears to be normal. The question then arises as to how cerebral function is maintained in the presence of a blood pressure that may be only a few mm of Hg. Akesson (1948) pointed out that the cerebral blood and cerebrospinal fluid are in equilibrium within the closed box of the skull, so that the vessels remain patent during high accelerations. The reduced pressure in the jugular veins acts as a siphon, drawing blood from the arterial tree through the effectively rigid cerebral vasculature, and collapse will only occur when the siphon is broken by a failure of flow in the carotid arteries. Henry, Gauer, Kety and Kramer (1950) produced direct evidence for the siphon by demonstrating that the pressure in the jugular bulb falls to sub-atmospheric levels, so that the arterio-venous pressure difference across the brain is maintained. Further indications as to the normality of the cerebral circulation at 4·5*g* were given by the unchanged venous oxygen saturation and the normal uptake of glucose by the brain. This, combined with the apparent insensitivity of cerebral blood flow to systemic hypotension even when the venous pressure does not fall (Slack and Walter, 1963), can explain how consciousness is adequately sustained during positive acceleration. Some even more recent experiments have shown that the total cerebral blood flow is reduced at 2, 3 and 4*g*, but that the reduction is confined to the white matter, cortical flow remaining unchanged (Howard and Glaister, unpublished observations, 1963).

Unconsciousness in the air or on the centrifuge usually occurs a few seconds after the onset of a high acceleration, as a form of acute cerebral anoxia. However, collapse may also take place after a longer time at a lower acceleration, which may even not be of sufficient magnitude to produce black-out. In the latter form there is often a prodromal period of progressive increase in pulse rate, with a slow decline in blood pressure, but both these may remain steady until unconsciousness supervenes. The collapse is accompanied by all the changes characteristic of a vaso-vagal attack; a precipitous fall of blood pressure, a sudden slowing of the heart, peripheral vasodilation, pallor and sweating. The mechanism is clearly the same as that of postural fainting, and one factor in its development is shown in Fig. 2. The volume of the legs rises steeply when positive acceleration is applied, because of the displacement of blood into the capacity vessels, but it does not reach a steady value. The volume curve continues to rise with time at a rate directly proportional to the applied force. The slope represents the filtration of fluid from the blood into the soft tissues, secondary to the high intravascular pressure, and this leads to a reduction in the effective circulating volume. Fainting normally succeeds the withdrawal or trapping of 750 to 1000 ml of fluid, but in the presence of a lowered blood pressure at head level a smaller loss may suffice to trigger the collapse.

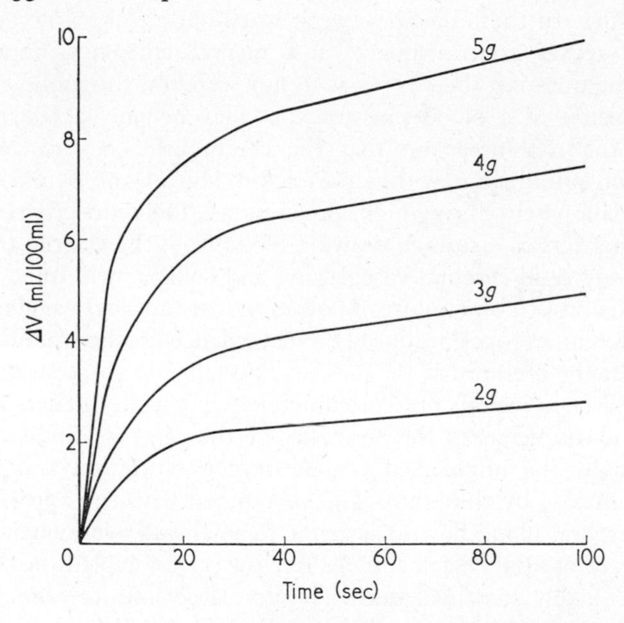

FIGURE 2. Changes in the volume of the calf during positive acceleration. The initial steep increase, caused by pooling, is followed by a more gradual rise as fluid is filtered into the soft tissues.

In more than 75% of the incidence of sudden unconsciousness of the first type (i.e. those due to cerebral anaemia) epileptiform convulsions occur during the recovery phase. They usually coincide with the removal of the acceleration and appear to be associated with the restoration of an adequate blood supply to the brain. Similar fits are sometimes seen after the administration of oxygen to anoxic subjects, but they are an infrequent sequel to circulatory collapse.

3. *The heart*

Early studies of the electrocardiogram during positive acceleration (Gauer, 1950) demonstrated changes in the form of the complexes entirely unlike those found with other varieties of tachycardia. The total amplitude was reduced, with alterations in the R and S waves, suggesting right ventricular preponderance. The T waves became flattened or diphasic and were sometimes inverted. Depression or elevation of the S-T segment, with loss of identity of the T wave, was also noted at the higher accelerations. The findings were at first interpreted as evidence of coronary insufficiency associated with an increased work load on the heart, but this explanation does not bear close examination. Although the weight of each unit of blood ejected is increased by a factor x at x g, the stroke volume is reduced, and the work of the heart is therefore less severe than the magnitude of the acceleration might indicate. Moreover, the mass of blood is ejected into an aorta, in which the pressure is sensibly invariant, and the supply pressure to the coronary vessels is also unchanged. Later work has shown that all the changes can be explained by the rotation and downward shift of the electrical axis of the heart (Pryor, Sieker and MacWhorter, 1952) and that they are simple exaggerations of the normal respiratory variations in the E.C.G. (Browne and Fitzsimons, 1957). The question of myocardial ischaemia has been raised again by the report of Zuidema, Cohen, Silverman and Riley (1956), who found serious or potentially serious arhythmias in four out of five subjects exposed to moderate accelerations for several minutes. Ventricular extrasystoles and premature beats of nodal or auricular origin were the most common abnormalities, but one man complained of severe substernal pain. The ECG changes were all equivocal, but ischaemia was not excluded. A similar study by Miller, Riley, Bondurant and Hiatt (1959) failed to show any cardiac irregularities, although the exposures were equally prolonged. In more than 2000 experiments on the R.A.F. centrifuge in which the ECG or arterial blood pressure were recorded, only two types of arhythmia have been observed. The first consists of ventricular extrasystoles, usually seen in subjects who have an occasional ectopic beat even at 1g. The other irregularity was seen in one man, who, after a slow ascent to 8g developed coupled beats just before he fainted. These continued for some seconds into the recovery period, when the heart rapidly reverted to normal rhythm (Fig. 3).

When the acceleration is removed, there is usually a transient bradycardia, often with AV nodal rhythm, and sometimes with partial AV dissociation. These changes are apparently due to vagal stimulation caused by the restoration of pressure to the carotid sinus receptors. They are abolished by vagotomy or by carotid sinus denervation (Jongbloed and Noyons, 1934).

The cardiac output is decreased by positive acceleration, a result predictable from measurements made after alterations of posture, but only recently obtained on the centrifuge (Howard, 1959; Lindberg, Sutterer, Marshall, Headley and Wood, 1960). Because of the concomitant tachycardia, the stroke volume is more severely restricted, and at the same time the pattern of filling and emptying of the heart is altered (Gauer, 1950). The diastolic

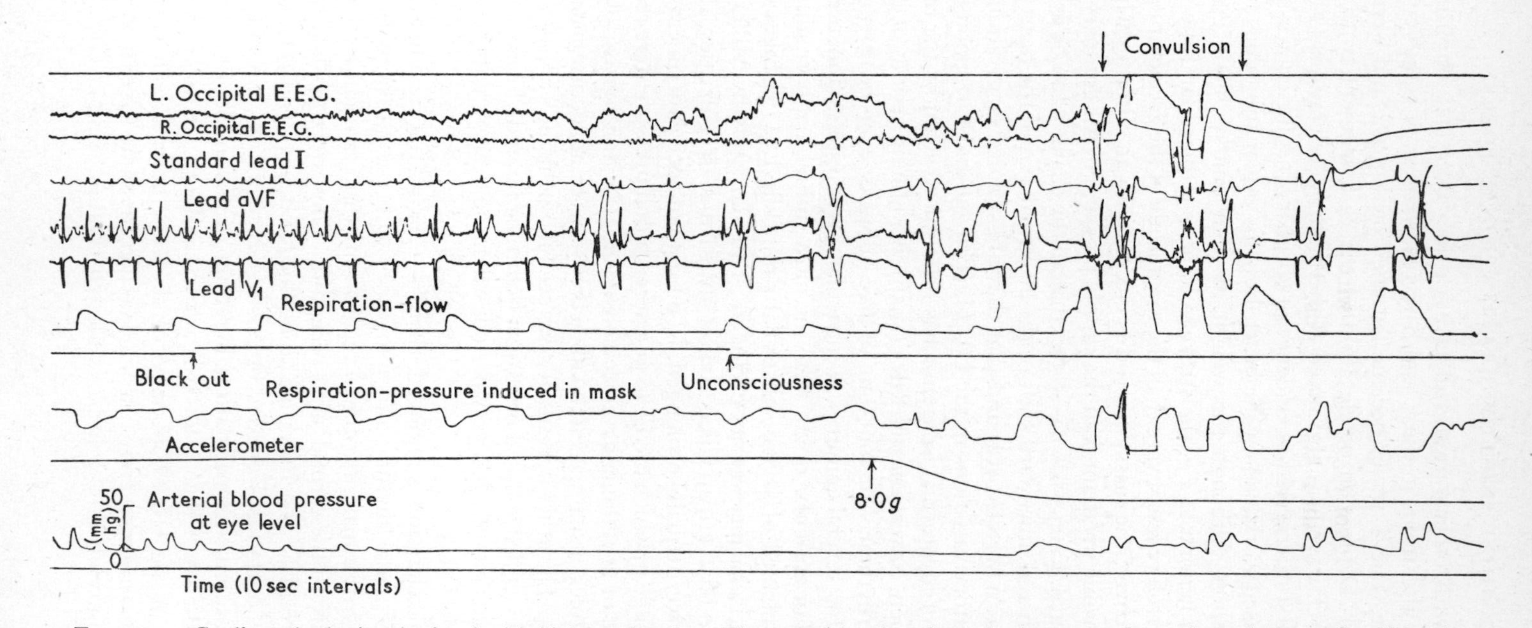

FIGURE 3. Cardiac arhythmia associated with high acceleration. The record is taken from the last part of a prolonged run to 8*g*. Note the appearance of α-rhythm during black-out and the convulsion accompanying recovery.

volume of both auricles and ventricles is reduced, and in systole the ventricular cavity is no more than a slit. During exceptionally high accelerations, the strong contractions of an almost empty heart may produce subendocardial haemorrhages (Henry, 1950), and it has been suggested that the Jarisch-Bezold reflex may be induced (Gauer, 1950). Gauer has also shown that the cardiac silhouette remains of almost constant size throughout the cycle and that the movement of blood is accomplished by piston-like alterations in the level of the AV valves.

4. *Circulation*

The alterations in the pressure gradients throughout the cardiovascular system, induced by postural changes or by acceleration, evoke reflex responses that tend to restore the blood pressure at head level. The compensatory mechanisms do not become fully effective for at least 5 sec after the application of the stress, unless the rate of onset is so slow that they can keep pace with the acceleration. Visual symptoms related to the initial fall in the arterial blood pressure usually remit at a later stage as the pressure at eye level is restored, but they may inexplicably reappear after compensation has been established.

The two major responses to positive acceleration are tachycardia and vasoconstriction. The heart rate begins to rise almost immediately (Lindberg and Wood, 1963), but the pressor response is delayed up to 10 sec. When it comes into effect, there may be some decrease in pulse rate. When the steady state has been achieved, the tachycardia is a linear function of the acceleration, at least for rates up to about 160/min. The increase in the vascular resistance of the forearm and hand is also related in simple fashion to the applied force (Howard, 1964; Fig. 4).

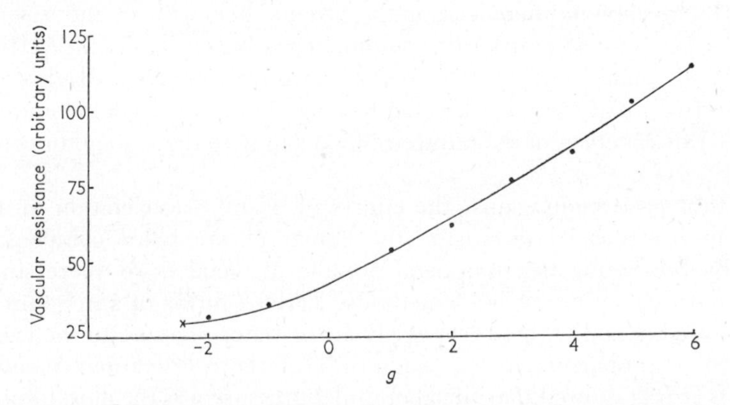

FIGURE 4. Changes in the vascular resistance of the forearm during positive acceleration.

It is usual to ascribe both types of response to the stimulation of the carotid sinus baroreceptors brought about by the fall in blood pressure. Jongbloed and Noyons (1934) found that the reflex responses could be abolished by section of the carotid sinus nerves in animals exposed to positive acceleration, and Greenfield (1945) obtained similar results. There are, however, certain objections to this simple explanation, and direct stimulation of the carotid sinus region has given conflicting results in both animals and man. It is, at first sight, surprising that the latencies of the tachycardia and of the vasoconstriction should be so different if both are mediated by the same mechanism, but the extent of the pressor response is perhaps more surprising still. If it is assumed that the purpose of the reflex is to preserve general homoeostasis by restoring the blood pressure within the sinus to normal (and so to safeguard the cerebral blood supply), a greater response would be expected than is in fact observed. On this theory the vasoconstriction should be maximal at a relatively low acceleration and remain constant thereafter. Fig. 4 shows that this does not happen, but that the response continues to increase at higher accelerations, although the heart rate has by then reached a plateau. These results suggest that, in man at least, the carotid sinus mechanism is not vitally concerned in vasopressor responses, although it undoubtedly participates in vasodilator reflexes (Ernsting and Parry, 1957). This idea is supported by the work of Roddie and Shepherd (1957), who found that direct stimulation of the carotid sinus in man produced the expected changes in the heart rate but had little effect on peripheral blood flow.

Salzman and Leverett (1956) demonstrated that positive acceleration caused peripheral venoconstriction, and the relationship between the increase in venous tone and the recovery of the blood pressure at head level led Gauer (1961) to attribute the venoconstriction to carotid sinus stimulation. There are, however, other reflex areas that are stimulated by gravitational stress. The aortic arch receptors, for example, probably experience a fall of pressure, for the hydrostatic indifference point is situated below them, and vasomotor impulses may also arise from the pulmonary vessels. The local reactions of arterial and venous walls to increases of transmural pressure (Bayliss, 1902; Folkow, 1953) may also be expected to contribute to the compensatory response. A direct effect of the reduced blood suppy to the vasomotor centre is possible.

Practical protection against the effects of positive acceleration in the air relies upon artificially increasing the extent of the reflex compensation, either by preventing the peripheral pooling of blood or by increasing the vascular resistance or by both methods. Various forms of anti-*g* suit have been devised (Franks, 1961) that apply air or water pressure to the legs and abdomen in proportion to the magnitude of the acceleration. Wood and Lambert (1952) showed that the abdominal pressure was the most important determinant of the protection afforded, but that the best results could be obtained if the abdominal bladder was combined with arterial occlusion cuffs on the thighs. Immersing the body in water, up to the level of the sternum, has a beneficial effect for similar reasons, but it is not without danger. Thus, Jasper and Cipriani (1943) produced right-sided heart failure, with distension of the right auricle and pulmonary engorgement, in monkeys exposed to prolonged acceleration with hydrostatic protection.

Other methods of delaying or preventing black-out include voluntary tensing of the muscles, which both promotes venous return and increases the resistance to flow, and shouting or straining. The blood pressure is increased by the rise of intra-thoracic pressure induced by the latter; however, if the effort is prolonged, the blood pressure falls steeply, as the return of venous blood from the peripheral vessels is impaired.

5. *Respiration*

The effects of positive acceleration on the mechanics of breathing and on gas exchange received little attention until recently. By analogy with the effects of alterations of posture, both these functions should be disturbed by acceleration, for it is known that unevenness of ventilation and of perfusion occur in the erect position (West, 1963). Gauer (1938) reported an increase of respiratory rate, which he attributed partly to hyperventilation, but it may also be due in part to stretching of the lungs as the diaphragm is pulled down. The same mechanism affects the end-expiratory level, and the residual capacity is increased. The tidal volume remains within normal limits at low accelerations, but it is decreased by greater forces.

The disturbance of the ventilation: perfusion ratio has been demonstrated by Glaister (1963), who found that the CO_2 content of the expired gas does not continue to climb slowly throughout a prolonged expiration, but that it declines slowly after about a litre of air has been expelled. This finding indicates the late emptying of ventilated but underperfused areas of the lung. At the same time, the oxygen consumption falls, and the saturation of arterial blood with oxygen is decreased, suggesting that other parts are underventilated but well perfused. Arterial desaturation has also been observed by others (Henry, 1950; Barr, Bjurstedt and Coleridge, 1958; Barr, 1963), all of whom attributed it to the opening of AV anastomoses. It seems, therefore, that only in the middle zones of the lungs is the balance of ventilation and perfusion preserved.

Within the last few years complaints of chest pain have become increasingly common in pilots. Typically, the pain appears on taking a deep breath or on moving from the cockpit after a flight involving exposure to moderate acceleration. A deep inspiration provokes coughing, which usually eases the symptoms. The pain, which is of a sharp or stabbing character, may be substernal or in the region of the lower ribs. The vital capacity is reduced, often to one-half, and X-rays taken while the symptoms persist show the appearances of basal atelectasis (Green and Burgess, 1962). The exact mechanism of the syndrome is not yet clear, but acceleration, oxygen breathing and the use of the anti-*g* suit have all been shown to be implicated, although limitation of full inspiration may occur in susceptible subjects after positive *g*, without a suit and breathing air. It is suggested that the lower parts of the lungs are compressed and distorted, especially as the anti-*g* suit prevents the descent of the diaphragm, and may splint the lower ribs. This leads to blockage or closure of the finer air passages, and the alveolar gas is then absorbed by a blood supply that is greater than normal. The exacerbating effect of oxygen is due to the more rapid rate of absorption of this gas from a closed space. Deep breathing or coughing forces the blocked air passages to open, and the collapse is relieved, but residual symptoms and X-ray signs may persist for as long as 48 hrs. The long duration, together with the fact that both the arterial desaturation and the pain may become progressively more severe after the return to normal conditions if no deep breaths are taken, make it improbable that oedema plays an important part, as Lindberg and Wood (1963) have suggested. It is possible, however, that some exudation into the alveoli may occur from hydrostatic effects.

D. Negative Acceleration

The signs and symptoms of negative acceleration are much more dramatic than those of positive *g*, and manoeuvres producing it are usually shunned and often prohibited. Tolerance is so low that negative accelerations do not, in any event, confer much tactical advantage. For these reasons they have received scant attention, but their physiological effects are at least as intriguing as those produced by forces in the opposite direction. Like the latter, they arise primarily in the cardiovascular system, because the acceleration is applied in the long axis of the great vessels.

With negative *g*, vascular pressures in the neck and head are increased; those below the heart (anatomically) are reduced. According to Henry (1950), the rise of pressure in the jugular bulb amounts to about 75 mm Hg at 3*g*, and a similar increase occurs in the carotid artery. However, Lindberg and Wood (1963) claim that the arterial pressure rises to a greater extent than the venous pressure, so that the AV gradient across the brain is maintained or even increased. On the venous side, at least, the point of hydrostatic indifference is apparently a little headward of the heart—a conclusion that explains the fall of right auricular pressure noted by Wilkins, Bradley and Friedland (1950) during passive tilting to the head-down position. Gauer and Hull (1954) found a similar decline in the left auricular pressure and argued that the indifference point of the arterial system also lies above the heart.

The symptoms of negative acceleration arise primarily from the venous congestion in the upper part of the body. The face feels (and looks) swollen and plethoric. The eyes are painful and vision is blurred; diplopia may occur. Headache, which often persists after the return to $1g$, is a common complaint. At accelerations greater than about $2g$ breathing becomes difficult, with a sensation of choking. Above $2\frac{1}{2}\,g$ petechial haemorrhages may appear in the soft tissues of the face and forehead, and with still greater forces frank haemorrhages may occur into the conjunctivae. Epistaxis is also common. "Red-out", a state analogous to black-out with blindness and a red mist over the field of view, is the classical accompaniment of negative *g*, but its existence is uncertain. Dropping of the lower lid over the cornea and the presence of blood-stained tears have both been suggested as causes.

Early experiments with negative *g* appeared to indicate that the high venous and arterial pressures in the head might lead to massive cerebral haemorrhage from rupture of branches of the Circle of Willis (Armstrong and Heim, 1938). There is no reason why such lesions should occur, for the vessels are protected (as in positive *g*) by simultaneous and almost equal changes in the cerebrospinal fluid pressure (Rushmer, Beckman and Lee, 1947). No evidence of intracranial haemorrhage could be found in animals exposed to $40g$ (Beckman and Ratcliffe, 1956), and the gross bleeding observed by Armstrong and Heim (1938) was probably a post-mortem artefact. The commonest cause of death was found to be asphyxia secondary to oedema of the airways and tongue; bandaging of the head and tracheotomy markedly increased the tolerance of animals for 12 to $15g$ (Henry, 1950).

The reflex responses to negative acceleration consist in the main of bradycardia and vasodilatation, although it is arguable whether the former confers much benefit. The high pressure in the carotid sinus stimulates the baroreceptors to such an extent that asystole may develop for periods of as long as 9 sec (Jasper and Cipriani, 1943). The occurrence and duration of the arrest depends in part upon the rate at which the acceleration is applied, and it is uncommon if the onset is slow (Howard, unpublished observations). When contractions reappear, they are usually arhythmic. Nodal rhythm, with absent P-waves in the ECG, is the mildest form, but extrasystoles from various abnormal foci are more common. In a minority of subjects these take the form of regular ventricular ectopic beats, each coupled to a preceding normal pulse. The rhythm returns to normal as soon as the stress is removed. The slowing of the heart and the accompanying decrease in the cardiac output lead to a secondary fall in the blood pressure at head level, and in this sense the bradycardia is a compensatory mechanism. However, the high venous pressure is not reduced, and the arteriovenous pressure difference across the brain accordingly falls. The severe restriction of cerebral blood flow that this must entail is probably responsible for the reported loss of consciousness at $4g$ (Henry, 1950), but it has also been suggested that irradiation within the central nervous system may contribute by affecting muscle tone (Gauer,

1961). The bradycardia can be abolished by cutting the vagi or reduced by the administration of atropine. With both these treatments the arterial pressure remains high, and the AV difference is maintained. As with positive *g*, the alterations in heart rate appear immediately, but at least 6 sec are required for the full development of the vasomotor reactions. These results provide further support for the view that the carotid sinus is not primarily responsible for the changes of tone.

Although little or no investigation into the respiratory effects of negative acceleration has been carried out, Henry (1950) reported that it produced a fall in the arterial oxygen saturation in both animals and man. A disturbance of the balance between ventilation and perfusion may be postulated, with the upper lobes as the site of the overperfusion. Overventilation of the relatively anaemic parts of the lung is, however, less likely than in positive acceleration, for with negative *g* these regions are in the lower lobes. The headward movement of the diaphragm and abdominal viscera will tend to compress the bases, and it is probable that their aeration, as well as their blood supply, is reduced.

E. Transverse Acceleration

Tolerance to transverse acceleration has long been known to be high (Bührlen, 1937). The prospect, and later the advent, of manned spaceflight led to exploitation of this fact, because the launching and re-entry of such vehicles entails the use of high accelerations for relatively long periods of time; about 820 *g*-seconds are required to achieve orbital speed. Most attention has been given to forward acceleration, partly because tolerance for it is slightly higher and partly because support is given by the seat rather than by the restraining harness. Lateral forces, in which the acceleration acts from side to side, are of little practical importance, and they have not been extensively studied.

1. *Circulation*

The cardiovascular effects of transverse acceleration are slight, because the vector lies at right-angles to the major vessels. Some of the smaller vessels must obviously be in line with the applied force, and the high pressure gradient in them can lead to capillary rupture and petechiae in dependent parts of the body.

In practice, most of the preferred sitting or lying positions involve some degree of flexion of the trunk in relation to the acceleration vector, and the hips and knees are also usually bent. A small component of positive or negative *g* is, therefore, present, and the appropriate cardiovascular disturbances become apparent at high accelerations. Figure 5 shows the fall in blood pressure at head level produced by an acceleration of 6*g*, with the trunk inclined forwards by about 10 degrees. The decrease is less marked than hydrostatic considerations would suggest, because the thighs and knees were flexed at 90 degrees, thus shortening the effective column length. After some minutes of exposure in this position, the feet and legs become numb, from a combination of pressure on the calves and a diminished blood supply. Reactive hyperaemia of the feet is a result of return to normal conditions.

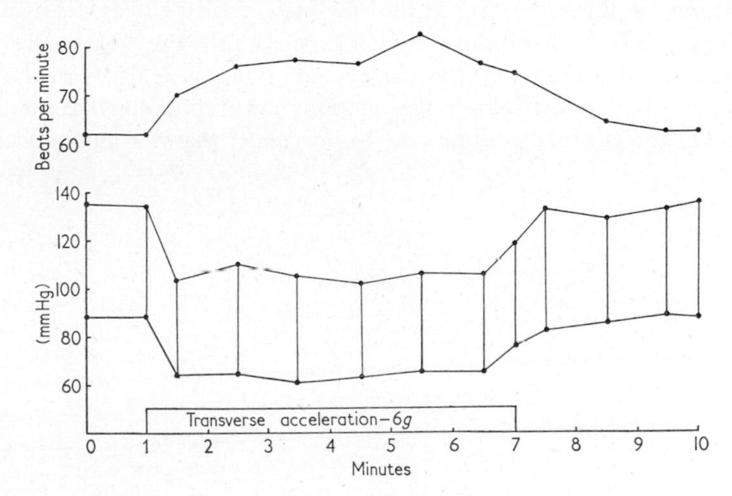

FIGURE 5. Changes in arterial pressure (radial artery supported at head level) and in pulse rate during exposure to 6*g* transverse acceleration.

The pressure in the right auricle increases greatly and may rise by as much as 12 mm Hg at 5*g* (Lindberg, Wood, Sutterer, Marshall and Headley, 1961). Compression of the chest under the weight of the anterior wall is probably partly responsible, but the decline of auricular pressure towards the control value if the exposure is prolonged indicates that changes in the thoracic blood volume may also be involved. In spite of the increased filling pressure of the auricle, the stroke volume remains unchanged at 5*g*, although the total cardiac output may rise slightly as a result of concomitant tachycardia.

2. *Respiration*

Difficulty in breathing is the main symptom of transverse acceleration and determines the ultimate limit of tolerance. The increased effort associated with inspiration is noticeable at 3*g*, and thoracic respiration may become impossible at 6*g*. With each expiration there is a further encroachment upon the residual volume, until a deliberately large breath must be taken to restore the "lost" gas. The respiratory rate rises; the tidal volume is at first increased and later decreased. The vital capacity tends to fall progressively throughout the exposure, especially if the inspired gas is 100% oxygen (Fig. 6). At the same time the oxygen consumption falls and the CO_2 content of the expired air is reduced, changes occur similar to those accompanying positive *g* (Section C.5 above) and due to the alteration in ventilation: perfusion ratio, which results here from congestion of the posterior zones of the lungs. The arterial oxygen saturation is reduced to a greater extent than with positive acceleration of the same magnitude and may reach 75% at 5*g* (Lindberg *et al.*, 1961). The reduction of the vital capacity probably owes more to the effort of breathing than to the occurrence of atelectasis, although the latter may occur. At extremely high accelerations even abdominal respiration is ineffective, and the tidal volume may be no greater than the anatomical dead space at 12*g*.

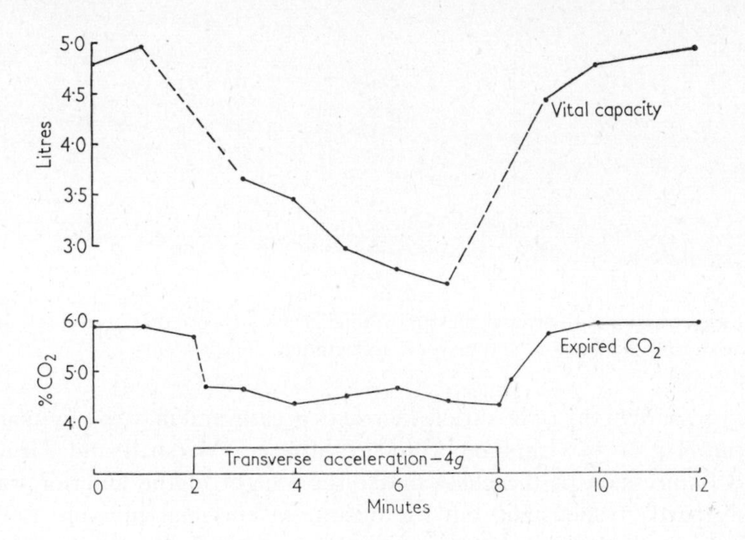

FIGURE 6. Progressive diminution of vital capacity and changes in end-tidal CO_2 during an exposure to 4*g* (transverse) for 6 min.

With accelerations in excess of about 4*g*, chest pain may be a prominent feature. It is usually referred to the lower sternum and epigastrium and is of a dull sickening character. Some of the limitation of vital capacity is due to the exacerbation of the pain that a full inspiration produces. Electrocardiograms show no abnormality (Bondurant and Finney, 1958), and the symptom is probably related to stretching and distortion of the mediastinum, which can be demonstrated by X-rays (Hershgold, 1960).

One other effect of transverse acceleration may be noted here, although it is unrelated to respiration. After prolonged exposure to high values of transverse *g*, severe giddiness occurs in most subjects. Its intensity is related both to the level of the acceleration and to its duration, and it is entirely unlike the vertigo that occurs in susceptible subjects after any type of centrifuge run. This makes it improbable that the angular accelerations of the centrifuge are responsible, as Clarke and Bondurant (1958) suggest. The disturbance is almost specific for movements of the head in a fore-and-aft direction, and it is associated with a distorted sense of the vertical. On rising suddenly from a sitting position, the subject feels unnaturally tall, and ceilings seem close to the head. Climbing a flight of stairs is difficult, because of their steep apparent slope, and walls and doors appear to slope sharply forwards. All these effects are probably of otolithic origin, but the lesion producing them is not known. After an exposure to 6*g* for 20 min, one subject experienced dizziness and vomiting for 4 hrs; he still had residual symptoms on moving his head 24 hrs later. It is likely, therefore, that high linear accelerations cause actual, if transitory, damage to the otolith organs.

F. Weightlessness

The condition of weightlessness, which is often and incorrectly termed zero-gravity, can only be experienced on earth for a small fraction of a second – during, for example, a jump from a wall. In aircraft performing certain well-defined manoeuvres (Haber and Haber, 1950) the duration of weightlessness can be extended to, at most, a minute. Long-term experience of the condition can be gained only in spaceflight, and experimental findings on its effects are accordingly sparse. Much of the available information has been based on theoretical considerations, from simple tests in aircraft and from simulation that is admittedly imperfect. The success of the American and Russian manned space programmes is evidence of the ability to adapt to the weightless state, but neither side has, apparently, gleaned any significant physiological insight into the mechanism of the adaptation. In the circumstances, this is understandable, if regrettable.

1. *Circulation*

The most obvious consequence of weightlessness is the abolition of all hydrostatic pressure gradients. There is no reason why this should have any adverse effect; an essentially similar state is achieved by lying down in a 1*g* environment. Certain predictions can be made about the changes occurring, and some of these have been fulfilled during exposures in flight. The heart rate would be expected to decrease, for example, in the absence of postural redistribution of blood. Henry, Ballinger, Maher and Simons (1952) found this to be true for animals during rocket flights, and Hawkins, Young and Roman (1961) demonstrated a relative bradycardia in man. An interesting observation made by these and other workers is that the decline in the heart rate is much slower during the transition from transverse acceleration to weightlessness than it is during the change from acceleration to normal 1g conditions. A small increase in the arterial pressure and a decrease in the peripheral resistance are other predictable results of the absence of pressure gradients and have been shown to occur, at least in animals (Yasdovsky, 1958). Gauer (personal communications, 1962) suggested that the redistribution of blood, compared with the normal, might affect the volume receptors of the left auricle and pulmonary veins and evoke a diuresis by the so-called Gauer-Henry reflex (Gauer and Henry, 1956). There has been no indication in any of the spaceflights to date of such a diuresis.

Although the cardiovascular system behaves normally during the period of weightlessness, its function appears to be disturbed when normal conditions are restored. It is well known that the compensatory reflexes are impaired after prolonged immobility in bed and that return to the erect posture may lead to fainting. Similar intolerance of positive acceleration has been shown after immersion in water, used in simulation of weightlessness. Whereas a night in bed has no significant effect on tolerance for acceleration, however, as little as 4 hrs immersion produces demonstrable impairment of postural responses (Graveline and Barnard, 1961), which may not return to normal for two days after the exposure (Graybiel and Clark, 1961). The decompensation caused by prolonged confinement to bed is associated with a loss of muscle strength (Dietrich, Whedon and Shorr, 1948), but this cannot be a factor in the asthenia produced by simulated weightlessness. Graveline (1963) has shown that, although the postural hypotension and decreased tolerance for stress can be reduced by binding the limbs with bandages after removal from the bath, attempts to maintain muscular tone by graded exercises during immersion have no effect on the subsequent weakness. Arguing that sympathetic vasomotor control is normally kept alert by hydrostatic changes in the peripheral venous pressure, with accompanying reduction of venous return, the same author applied pneumatic cuffs to the limbs of immersed subjects and inflated the tourniquets periodically to produce venous congestion. Full protection was afforded by this method in most subjects, and in some the reaction to passive tilting was better than that observed during the pre-immersion control period. The implications of these findings for the control of venomotor tone are interesting.

There is some evidence that the true weightlessness occurring in spaceflight has a similar adverse effect on the cardiovascular responses. At least two of the American astronauts are known to have had some degree of postural hypotension after their return to earth, but the severity and duration of their symptoms have not been revealed. It is probable that the more prolonged weightlessness experienced by the Russian cosmonauts had a similar effect. Beckh (1959) showed that the symptoms of positive and transverse acceleration were more severe when exposure was preceded by the weightless state, and it would be of interest to know whether the disturbed circulatory reflexes are restored more quickly to normal by accelerations greater than 1 *g* than by a simple return to standard terrestrial conditions. It might well be that the acceleration experienced during re-entry through the earth's atmosphere mitigates the severity of the postural response.

2. *Central nervous system*

Most of the published studies of central nervous function during weightlessness have, in fact, been concerned with performance. For example, Beckh (1954) described the inability of weightless subjects accurately to place crosses in a series of diagonal squares and the inaccuracy of turtles in striking for food in the weightless state. A tendency to overshoot was noted in both tests. Gerathewohl and Stallings (1958) found that after-images imprinted upon the retina appeared to move upwards during parabolic flight, a result that complements the studies of Whiteside, Graybiel and Niven (1963) during increased acceleration. Investigations of the tendon reflexes have given conflicting results. Matthews (1956) reported that the ankle jerk was abolished during 140 msec of weightlessness, but Hawkins (1963) claimed that the knee jerk was markedly exaggerated.

The importance of these and other observations lies in the ill-appreciated fact that they represent the behaviour of the central nervous system under open-loop conditions. Gerathewohl (1958) has pointed out that vision provides the only sensory cue during weigthlessness, and many investigators have found that the subject's performance and sense of orientation are considerably impaired if this information is denied to him. Provided that vision is retained, the absence of otolithic signals, muscle sense and pressure inputs can be compensated for, and learning takes place quickly; if, however, conflicting information is received, disorientation and deterioration of performance result. This seems to have been the cause of the "space sickness" experienced by the Russian cosmonaut Titov, in whom there was no apparent impairment of vestibular function as judged by simple clinical tests (Sissakian and Yasdovsky, 1962). The disturbances of central nervous function accompanying weightlessness have been attributed to the operation of the Weber-Fechner law, which, if it is indeed applicable, acts in two ways. In the first place, the law explains the apparent discrepancy between the effects of reducing the applied acceleration from a high value to 1 *g* and those accompanying the change from 1 *g* to weightlessness (von Beckh, 1954). Secondly, the Weber-Fechner law predicts that the sensory response to stimulation of the vestibular apparatus should be much greater under weightless conditions than when a normal gravitational environment is present (Gauer and Habner, 1950). However, experiments in which the weight was not abolished, though reduced to some intermediate value (Ballinger, 1952), cast doubt upon the validity of this attractive argument.

G. Long-term Adaptation

No studies have yet been made of the adaptation to long periods of increased acceleration. Experiments on centrifuges have lasted for seconds, minutes or occasionally an hour or two, with exposures of less than 1 min being by far the commonest. There has been more long-term experience of weightlessness. At the time of writing the most protracted spaceflight has had a duration of five days; the first (and shortest) orbital flight itself involved some 89 min of weightlessness and thus exceeded all but the longest of exposures to accelerations greater than 1 *g*. However, the number of subjects used and the amount of information gained have been understandably small; although the flights have shown that long periods of weightlessness are tolerable, they have given little or no indication of possible adaptive mechanisms.

In a sense, the erect posture represents the best evidence for long-term adaptation to increased gravitational forces, and the physiological differences that have evolved on this account between man and the lower animals indicate the means by which such readjustment has been made. It is well known that many species of domestic animal have a low tolerance for acceleration; the rabbit, for example, will rapidly faint under the influence of 1 *g* in the positive sense. It can be argued, however, that human adaptation to 1 *g* is far from complete, and the guardsman who faints on parade might well criticise the evolutionary quirk that relies upon the muscle pump for the maintenance of consciousness.

REFERENCES

Akesson, S. (1948). *Acta physiol. scand.* 15, 237.

Armstrong, H. C., and Heim, J. W. (1938). *J. Aviat. Med.* 9, 199.

Ballinger, E. R. (1952). *J. Aviat. Med.* 23, 319.

Barr, P. O. (1963). *Acta physiol. scand.* 58, Suppl. 207.

Barr, P. O., Bjurstedt, H., and Coleridge, J. C. G. (1958). *J. Physiol.* 143, 79.

Bayliss, W. M. (1902). *J. Physiol.* 28, 220.

Beckh, H. J., von (1954) *J. Aviat. Med.* 25, 235.

Beckh, H. J., von (1959). *Aerospace Med.* 30, 391.

Beckman, E. L., Duane, T. D., and Coburn, K. R. (1962). *In* "Visual Problems in Aviation Medicine" (A. Mercier, ed.), pp. 17–25. AGARDograph 61, Pergamon Press, Oxford.

Beckman, E. L., and Ratcliffe, H. L. (1956). *J. Aviat. Med.* 27, 117.

Bondurant, S., and Finney, W. A. (1958). *J. Aviat. Med.* 29, 758.

Browne, M. K., and Fitzsimons, J. T. (1957). Flying Personnel Research Committee Report No. 1009, Air Ministry, London.

Bührlen, L. (1937). *Luftfahrtmed.* 1, 307.

Clark, J. H., Hooker, D. R., and Weed, L. H. (1934). *Amer. J. Physiol.* 109, 166.

Clarke, N. P., and Bondurant, S. (1958). Wright Air Development Center Tech. Report 58-267, USAF Wright Patterson A.F.B., Ohio.

Dietrich, J. E., Whedon, G. D., and Shorr, E. (1948). *Amer. J. Med.* 4, 3.

Ernsting, J., and Parry, D. J. (1957). *J. Physiol.* **137**, 45P.
Folkow, B. (1953). *Acta. physiol. scand.* **27**, 99.
Franks, W. R. (1961). *In* "Bio-assay Techniques for Human Centrifuges and Physiological Effects of Acceleration" (P. Bergeret, ed.), pp. 14–22. AGARDograph 48, Pergamon Press, Oxford.
Gauer, O. H. (1938). *Luftfahrtmed.* **2**, 291.
Gauer, O. H. (1950). *In* "German Aviation Medicine in World War II" Vol. 1, chapter VI-B, pp. 554–83. U.S. Government Printing Office, Washington D.C.
Gauer, O. H. (1961). *In* "Gravitational Stress in Aerospace Medicine" (O. H. Gauer and G. D. Zuidema, eds.). J. & A. Churchill Ltd, London.
Gauer, O. H., and Haber, H. (1950). *In* "German Aviation Medicine in World War II" Vol. L, chapter VI-G. U.S. Government Printing Office, Washington, D.C.
Gauer, O. H., and Henry, J. P. (1956). *Klin. Wschr.* **34**, 356.
Gauer, O. H., and Hull, W. E. (1954). *Fed. Proc.* **13**, 52.
Gerathewohl, S. J. (1958). *Air Univ. q. Rev.* (summer number).
Gerathewohl, S. J., and Stallings, H. D. (1958). *J. Aviat. Med.* **29**, 504.
Glaister, D. H. (1963). Paper to Annual Meeting of Aerospace Medical Association, Los Angeles.
Graveline, D. E. (1963). *In* "Medical and Biological Problems of Space Flight" (G. H. Bourne, ed.). Academic Press, New York and London.
Graveline, D. E. and Barnard, G. W. (1961). Wright Air Development Division Tech. Report 61-257, U.S.A.F. Wright Patterson A.F.B., Ohio.
Graybiel, A., and Clark, B. (1961). *Aerospace Med.* **32**, 181.
Green, D. I., and Burgess, B. F. (1962). Flying Personnel Research Committee Report 1182, Air Ministry, London.
Greenfield, A. D. M. (1945). *J. Physiol.* **104**, 5.
Haber, F., and Haber, H. (1950). *J. Aviat. Med.* **21**, 395.
Hawkins, W. R. (1963). *In* "Physiology of Man in Space" (J. H. U. Brown, ed.). Academic Press, New York and London.
Hawkins, W. R., Young, F., and Roman, J. (1961). Unpublished observations cited by Hawkins (1963), p. 301.
Henry, J. P. (1950). U.S.A.F. Technical Report No. 5953, Wright Patterson A.F.B., Ohio.
Henry, J. P., Ballinger, E. R., Maher, P. J., and Simons, D. G. (1952), *J. Aviat. Med.* **23**, 421.
Henry, J. P., Gauer, O. H., Kety, S. S., and Kramer, K. (1950). *J. clin. Invest.* **30**, 292.
Hershgold, E. J. (1960). *Aerospace Med.* **31**, 213.
Howard, P. (1959). *J. Physiol.* **147**, 49.
Howard, P. (1964). Ph.D. Thesis, University of London.
Howard, P., and Byford, G. H. (1956). Flying Personnel Research Committee Memorandum No. 75, Air Ministry, London.
Jasper, H. H., and Cipriani, A. J. (1943). Unpublished report to Associate Committee on Aviation Medical Research, N.R.C., Canada.
Jongbloed, J., and Noyons, A. K. (1934). *Pflüg. Arch. ges. Physiol.* **233**, 67.
Lambert, E. H. (1945). *Fed. Proc.* **4**, 43.
Lambert, E. H., and Wood, E. H. (1946). *Fed. Proc.* **5**, 59.
Lawton, R. W., Greene, L. C., Kydd, G. H., Peterson, L. H., and Crosbie, R. J. (1958). *J. Aviat. Med.* **29**, 97.
Lindberg, E. F., Sutterer, W. F., Marshall, H. W., Headley, R. N., and Wood, E. H. (1960). *Aerospace Med.* **31**, 817.
Lindberg, E. F., and Wood, E. H. (1963) *In* "Physiology of Man in Space" (J. H. U. Brown, ed.). Academic Press, New York and London.

Lindberg, E. F., Wood, E. H., Sutterer, W. F., Marshall, H. W., and Headley, R. N. (1961). Wright Air Development Division Tech. Report 60.
Matthews, B. H. C. (1956). *Proc. Internat. Physiol. Congr.*, Brussels, p. 1038.
Miller, H., Riley, M. B., Bondurant, S., and Hiatt, E. P. (1959). *Aerospace Med.* **30**, 360.
Pryor, W. W., Sieker, H. O., and MacWhorter, R. L. (1952). *J. Aviat. Med.* **23**, 550.
Roddie, I. C., and Shepherd, J. T. (1957). *J. Physiol.* **139**, 377.
Rushmer, R. F., Beckman, E. L., and Lee, D. (1947). *Amer. J. Physiol.* **151**, 355.
Salzman, E. W., and Leverett, S. D., Jr. (1956). *Circulation Res.* **4**, 540.
Sissakian, N. M., and Yasdovsky, V. I., eds. (1962). *In* "Pervye Kosmicheskie Polety Cheloveka", pp. 161–4. Printing House of Academy of Sciences, Moscow.
Slack, W. K., and Walther, W. W. (1963). *Lancet* i, 1082.
Wagner, E. (1886). *Pflüg. Arch. ges. Physiol.* **39**, 371.
West, J. B. (1963). *Brit. med. Bull.* **19**, 53.
Whiteside, T. C. D., Graybiel, A., and Niven, J. I. (1963). Joint report to Flying Personnel Research Committee, Air Ministry, London and National Aeronautics and Space Administration, U.S.A.
Wilkins, R. W., Bradley, S. E., and Friedland, C. K. (1950). *J. clin. Invest.* **29**, 940.
Wood, E. H., and Lambert, E. H. (1952). *J. Aviat. Med.* **23**, 218.
Yasdovsky, V. I. (1958). Paper to Rocket and Satellite Symposium of Special International Geophysical Year Committee, Moscow.
Young, M. W. (1946). *Anat. Rec.* **94**, 531.
Zuidema, G. D., Cohen, S. I., Silverman, A. J., and Riley, M. B. (1956). *J. Aviat. Med.* **27**, 469.

CHAPTER 8

Nutritional Changes

ROBERT A. MCCANCE AND ELSIE M. WIDDOWSON

A. Introductory

If one regards a desirable intake of all nutrients as an essential part of one's well-being at every age, then any lapses from this optimum intake can be regarded as a stimulus, or stress, since they normally provoke some response, some "adaptation" on the part of the organism. A state of perfect well-being, however, can be achieved on many different diets, and people—all apparently in a state of excellent health—live on foods in different parts of the world bearing little superficial resemblance to each other. Some are exceedingly bulky and fibrous, others are highly concentrated and almost without residue. There is some evidence, perhaps not very good evidence by modern criteria (Lamb, 1893; Bryant, 1924; Cogswell, 1948; Pilling and Cresson, 1957), that races living on coarse bulky diets have longer intestines than those subsisting on food with little residue; if this is so, it argues a long-term genetic adaptation similar to the variations in build and stature said to have been evolved in response to high and low environmental temperatures (Barnett, 1961). Be that as it may, the observations of travellers and experimental studies of many kinds, including those of rationing (McCance and Widdowson, 1946), have shown that, if they are prepared to do so, healthy hungry men, women and children can adapt to changes in their dietary pattern so rapidly that they can accept them without exhibiting any signs of discomfort. This is at least one reason why man has successfully colonised practically the whole of the world's land surfaces.

These remarks may not seem strange to those accustomed to think of adaptation to heat or cold or to life at high altitudes, but there are differences between the adaptations to food and those to temperature and pressure. In adaptations to the physical environment, for example, the conscious and unconscious responses nearly always reinforce each other. It is true that in many of the responses to heat and cold the highest centres may not participate at all, but, if severe, these stresses are often emotionally disturbing and the physical adaptations are often consciously reinforced—from experience or "know-how". Men move into the shade consciously, for instance, as animals do also, if it is too hot, and they build a fire or huddle together if it is too cold. In many people, however, a surfeit of calories may not be appreciated consciously as a stress until the unconscious physical response, initially beneficial, has created a pathological state of affairs that is emotionally more satisfying than the attempt to correct it. It is also not uncommon for a man to subject himself to a severe and long-lasting nutritional stress for purely emotional reasons. There is a parallel to this in the realms of temperature and pressure, for man's adventurous spirit has led him to climb high mountains, or to take up skin diving, and thus to expose himself to severe physical stresses for the sense of achievement he gets from doing it.

Most of the well-known adaptations to changes in the physical environment again are primarily and wholly advantageous; only rarely are they not ultimately so, but here is one example. The dilatation of the peripheral vessels (Keatinge and Evans, 1958; Le Blanc, 1962) that comes with constant exposure to cold enables the adapted individual to work much better with his hands in cold water, but it means that his body as a whole must lose more heat, and this might be a disadvantage if he were completely immersed. The primary response to a nutritional stress is also immediately beneficial as a rule, but secondary strains are often set up, and the final result is commonly not beneficial. It is for reasons such as these that the word "response" rather than "adaptation" has been generally used in this chapter.

B. Calories

The stability of an adult's body weight is one of the niceties of homoeostasis. It is brought about by the regulation of the calorie intake to meet the energy expenditure, regardless of the consistency and nature of the food. This is why adaptations to changes in the physical structure of the diet go so smoothly. This form of regulating mechanism, however, has its limitations. Alcohol, for instance, provides calories, but nothing else; it enters the metabolic pool and reduces the consumption of foods that would otherwise have been eaten to provide calories. The unsophisticated alcoholic drinks of primitive races, for instance native beer, may contain proteins, carbohydrates and vitamins and constitute a valuable food, but wines probably and spirits certainly have little in them of any nutritional value except alcohol, and their consumption to excess may so reduce the consumption of other sources of energy that deficiencies of protein or vitamins may appear. Alcohol, therefore, taken as it nearly always is to escape from an emotional strain, may set up a physical strain that the body does not recognise and to which it makes no useful response. Alcohol, however, can do more than this. In its capacity as a drug it can interfere with the liberation of the antidiuretic hormone from the pituitary and so lead to a considerable loss of water from the body. This is a physical strain to which the conscious person sooner or later makes the appropriate response, but an alcoholic subject in a high environmental temperature may easily reach a stage of dehydration from which there is no return.

1. *Overnutrition*

a. The obese state. If a person in perfect nutritional balance finds himself surrounded by food on an ocean liner or cut off from his usual exercise, he eats more food than he immediately requires and gains weight. The increase may be 500 g, it may be 10 kg; whatever it is, fat causes most of it. This response to a surfeit of food and the opportunity of eating it, reinforced by the pleasurable sensations of doing so, is an adaptation that was probably wholly beneficial to primitive man and enabled him to provide himself in periods of plenty with an internal store of food that would stand him in good stead in a time of scarcity. So far so good, but any increase of body weight behind the desirable one calls for further adaptations to deal with it. The blood volume and the cardiac output increase to supply the new adipose tissue, the blood pressure rises and the heart hypertrophies (Alexander, Dennis, Smith, Amad, Duncan and Austin, 1962–3), and other changes go on that can only be regarded as physiological in so far as they are temporary (Davidson and Passmore, 1963). If the desire to eat does not naturally become less, and the weight continues to rise, there is generally a tendency to curtail the amount of exercise voluntarily taken and the individual becomes more and more immobile and breathless; the state that follows is no longer a physical adaptation but a handicap, which must often have been crippling in a violent world. Eglon, for instance, might have escaped Ehud's dagger if he had not been so fat (Judges: *3*). Man has evolved with an appetite that is often too great for him in the environment of modern civilisation, where it is much easier nowadays to get too fat than too thin. Obesity has always been one of the risks of prosperity, but prosperity was not so general 100 and certainly not 1000 years ago. In Western society nowadays even the factory workman can develop an alderman's paunch, and those in professions in which exercise is wellnigh impossible find it exceedingly difficult to maintain a steady weight. Spread over a period of years the increase may be almost imperceptible, but, once established in childhood or middle age, obesity generally persists with minor fluctuations for the rest of the person's life or at any rate until old age (McCance, 1953a; Duncan, 1959; Davidson and Passmore, 1963). The physical strain of this can be removed at any time by making the calorie intake less than the energy expenditure, but to do this requires the conscious intervention of the higher centres. For reasons personal or environmental, this generally sets up an emotional strain that the average person is not prepared to face—and may make no effort to do—until the results of the physical stress, such as heart failure, provide him with an overriding motive. Others, perhaps the majority, are prepared to try—for a time—provided it is made easy for them, and these men and women are the prey of the plausible food manufacturers and the pabulum of the obesity clinics. Something new, something a little spectacular, they may be prepared to try for a time, but results come slowly; nothing dramatic happens whatever they do, and sooner or later they stop doing anything. Even with every encouragement the continuous effort required to reduce one's weight by 10 or 15 kg and hold it there is more than most people can stand. Personality, however, and motive are all-important. Jockeys, coxes in rowing crews, actresses and models in the fashion shops manage to keep their weight low so long as it is desirable for their jobs.

Many people who are worried by the cares and troubles of this life find relief in eating and in the mental rest that follows it (Babcock, 1948). As the Rev. Sydney Smith put it, although not quite in this context,

> "Serenely full, the epicure would say
> Fate cannot harm me, I have dined today."
> (Lady Holland, 1850.)

The overeating that may be engendered in this way is a response to an emotional strain. It may be considered an adaptation for a time, if it enables the individual to cope with the distressing situation; prolonged, it leads inevitably to the physical strain of obesity and all that follows from it.

It can be seen from this brief survey that the causes, results and treatment of overeating give rise to a complex assortment of stresses, to which the responses (or adaptations) may be at the same time physiologically desirable and emotionally undesirable or vice versa. There is nothing quite like this among the strains set up by heat or cold, in which the psychological and physiological adaptations usually work together.

b. The weight of the body is only a rough expression of its composition, which can be broken down into four major components – the cell mass, the extracellular fluids, the bones and the fat. The weights of these can now be arrived at with fair accuracy during life, and Fig. 1 shows the composition of

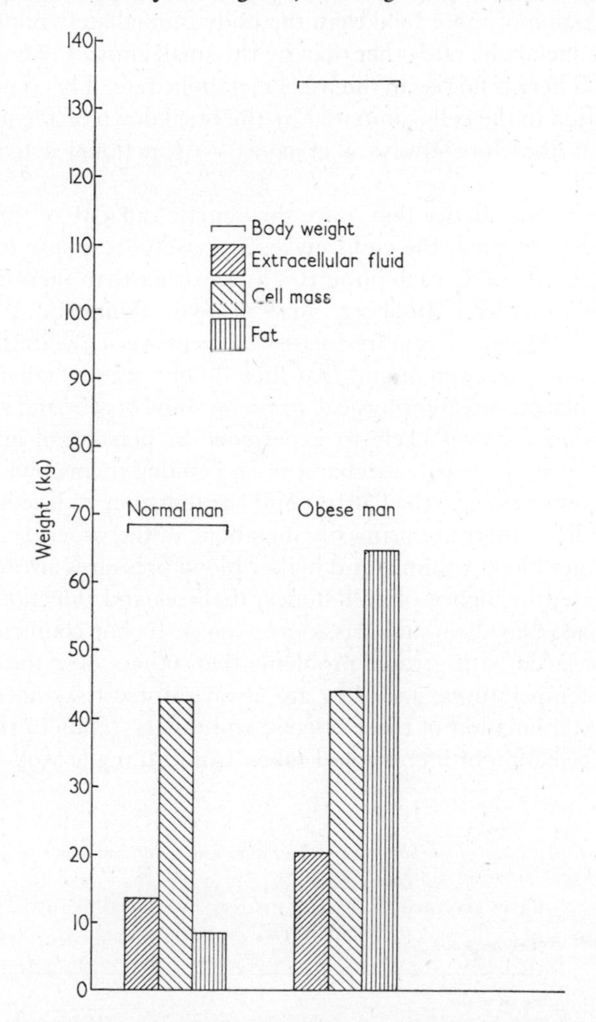

FIGURE 1. Amounts of extracellular fluid, cell mass and fat in the bodies of two men, one "average" and the other obese.

the body of two men, one normal and the other very fat. The excess of fat in the obese man is obvious, and this necessitates some increase in the blood volume and extracellular spaces to maintain it. One might suppose that some hypertrophy of the muscle mass would have taken place in response to the needs of mobility, and this may happen, particularly in young people, but often obese people become more sedentary, and the muscles, in fact, decrease from disuse. This is one of the bad physical results of the rise in body weight that so often accompanies taking up a profession in which exercise is wellnigh impossible.

The ingestion of more food than the body immediately requires does not increase the metabolic rate other than by the small amount of energy required to digest it. There is no rise in the basal metabolic rate. This is not surprising, since oxidation in the cell is initiated by the breakdown of adenosine triphosphate and is therefore always a response to functional activity at cell or somatic level.

There is little evidence that, once the genetic capacity of the organism to grow has been satisfied, the child makes any useful response to a surplus of food. Fat children may reach puberty a little earlier than their fellows and be just that much taller (Mossberg, 1948; Lloyd, Wolff and Whelan, 1961; McCance, 1962), but it requires statistics to prove it, whereas anyone can see that they are overweight and that they do not become taller adults.

Obesity places a greater physical strain on some organs and systems in the body than they are ever likely to experience in persons of normal weight. Some of these are primarily mechanical, and among them come the wear and tear of the joint surfaces, the flat feet and the tendency to lose balance and to fall awkwardly. Others are primarily metabolic in the widest sense and range from the larger blood volumes and higher blood pressures already mentioned to an increased incidence of gall-stones, diabetes and infections of the skin (Strang, 1959; Davidson and Passmore, 1963). In hot countries the "overweight" are faced with greater problems than others over the regulation of their body temperatures, and they are always worse risks on the operating table. The combination of these effects, and others, tends to shorten the fat person's expectation of life. Table I taken from Strang (1959) shows this.

Table I. *Mortality rates of persons regarded as overweight for insurance purposes*

Cases accepted from 1925 to 1934 and traced to 1950
Standard risk = 100

Age at issue of policy	*Men*	*Women*
20–29	180	134
30–39	169	152
40–49	152	150
50–64	131	138

2. *Undernutrition*

Shortage of food has been the fear of mankind since time immemorial, for it sets up physical and emotional strains of a most unpleasant character. Thus we read in "The Chronicle of Novgorod (1016–1471)" (Michell and Forbes, 1914): "This year it was cruel; the people ate lime tree leaves, birch bark, pounded wood pulp mixed with husks and straw; some ate buttercups, moss, horse flesh. Fathers and mothers would put their children into boats in gift to merchants or else put them to death. . . . Some of the common people killed the living and ate them. Brother had no sympathy with brother, nor father with son, nor mother with daughter, nor would neighbour break bread with neighbour. There was no kindness among us, but misery and unhappiness; in the streets unkindness one to another, at home anguish, seeing children crying for bread and others dying."

This loss of all moral sense and the breakdown of human relationships have been only too common in communities on the verge of starvation. They have been recorded in Samaria hundreds of years B.C. (II Kings: *6*, 28), in the accounts of the Middle Ages (Curschmann, 1900) and of the Thirty Years' War (Lammert, 1890) and elsewhere. Donovan (1848) gave an account of what he saw in Ireland in the middle of the last century. After remarking that the eyes acquired a most peculiar stare and that some people became very emotional, crying, whining, and appearing to be in a state of imbecility, he went on: "I have seen mothers snatch food from the hands of their starving children, known a son to engage in a fatal struggle with a father for a potato, and have seen parents look on the putrid bodies of their offspring without evincing a symptom of sorrow."

History is full of records like these and of changes in individual personality (Lusk, 1921; Morgulis, 1923; Keys, Brozek, Henschel, Mickelsen and Taylor, 1950; McCance, 1951; Davidson and Passmore, 1963). Undernutrition, however, does not always evoke such a response, as may be seen by these extracts from a famous diary.

January 2nd.

. . . my head is giving me trouble all the time . . .

January 4th.

We are reaching our limit, for we were so done up at noon with cold that the clinical thermometer failed to register the temperature of three of us at 94°.

January 26th. and 27th.

Tonight (the 27th.) we have had our first solid food since the morning of the 26th. We came to the end of all our provisions except a little cocoa and tea and from 7 a.m. on the 26th. till 2 p.m. on the 27th. we did 16 miles over the worst surfaces. . . . We fell into hidden crevasses time after time and were saved by each other and by our harness. . . . When we started at 7 a.m. yesterday, we immediately got into soft snow . . . The biscuit was all finished and with only one pannikin of hoosh, mostly pony maize, one pannikin of tea and once ounce of chocolate we marched till 4:45 p.m. We had one pannikin of tea. There was no more food. We marched till 10 p.m. then one small pannikin of cocoa. Marched till 2 a.m. when we were played out. We had one pannikin of cocoa and slept till 8 a.m. Then a pannikin of cocoa and marched till 1 p.m. and camped, about half a mile from the depot. Marshall went on for food and we got a meal at 2 p.m. We turned in and slept.

February 4th.

Cannot write more. All down with acute dysentery; terrible day. No march possible; outlook serious. Fine weather.

February 8th.

Did 12 miles . . . Adams and Marshall still dysentery. Wild and I all right. Feel starving for food. Talk of it all day.

February 16th.

We are down to about half a pannikin of half cooked horse-meat a meal and four biscuits a day.

February 17th.

We started at 6:40 a.m. and marched till 6 p.m. and today we had three pannikins of semi-cooked horse-meat and six biscuits on the strength of the day's march. We all have tragic dreams of getting enough to eat but rarely have the satisfaction of dreaming that we are actually eating. Last night I did taste bread and butter. We look at each other as we eat our scanty meals and feel a distinct grievance if one man manages to make his hoosh last longer than the rest of us.

February 22nd.

We found three small bits of chocolate and a little bit of biscuit at the camp after carefully searching the ground for such unconsidered trifles, and we "turned backs" for them. I was unlucky enough to get the bit of biscuit, and a curious unreasoning anger took possession of me for a moment at my bad luck. It shows how primitive we have become and how much the question of even a morsel of food affects our judgment. We are near the end of our food, but we have staked everything on the Bluff Depot; we had a good feed tonight. If we do not pick up the depot there will be absolutely no hope for us.

The party of four reached food on February 23rd, and the last man was on board the *Nimrod* by 1 a.m. on 4th March (Shackleton, 1909).

Many other parties and units, even within a famine area, have risen to the occasion and survived many weeks of undernutrition to emerge triumphant at the end. Several of them are quoted by Keys *et al.* (1950). Sometimes only a few of the party have survived, and in lifeboat voyages after shipwreck undernutrition has often not been the main cause of death (McCance, Ungley, Crosfill and Widdowson, 1956).

Fasting has been undertaken voluntarily on innumerable occasions, and it has been made an integral part of many religions, primitive and otherwise. The motive for this self-imposed physical stress is not always clear, even within the Christian religion (Frazer, 1927; Colloquium, 1959), but about some points there can be no dispute. Prayer is easy, but fasting is difficult – and unpleasant; the constant association of prayer and fasting suggests that the latter was often undertaken to demonstrate the sincerity of one's desire and also that one was prepared to make a considerable sacrifice for it. What did one achieve by fasting apart from this? God "giveth strength and power" (Psalms: *68*, 35; see also *18*, 32), and before God one fasted for the moral strength "to loose the bands of wickedness and to undo the heavy burdens" (Isaiah: *58*, 6). After Jesus had been led into the wilderness and fasted there for forty days, He had the strength of mind to resist all the temptations of the Devil (St. Matthew: *4*, 1–11; St. Luke: *4*, 1–14). The essence of much religious and ascetic fasting, therefore, was that it should be self-imposed, and that it should result in the uplift, strength of mind, sense of achievement, confidence and courage that we think of now as morale (Père X, 1959). At Novgorod, as in most of the great famines of history, there was widespread social degeneration and loss of morale. Yet some people retained theirs and lived to write the chronicle. In Shackleton's party morale was as high as it could be, and all came through. Shackleton never lost a man on any of his expeditions.

Naturally enough, fasting to mortify the flesh as a boost to individual or community morale is a two-edged weapon. It has its physical limitations and can be overdone—by patients with anorexia nervosa, for instance, although this is perhaps not an entirely appropriate example; certainly it was overdone in the past by members of the Greek church (Morgulis, 1923), but the principle remains, and a study of the records shows that when groups and communities are facing a long period of hardship and undernutrition, or thirst, morale is a man's best contribution to his survival. Without it, many have perished long before their powers of physical endurance were at an end.

"Hope, perseverance and subordination should form the seaman's great Creed and Duty as they tend to banish despair, encourage confidence and secure preservation" (Ancient Seaman's Guide, see Report, 1943).

a. The physical effects of undernutrition. The effects of undernutrition on the well-being of the body have been studied as thoroughly as those of over-nutrition, although it is less easy to obtain adult men and women for investigating those effects. It is, however, much easier to produce them experimentally. The best-known studies and experiments on man are those of

Benedict (1907), Benedict, Miles, Roth and Smith (1919) and Keys *et al.* (1950). Some classical work was done on the professional fasters (Lehmann, Mueller, Munk, Senator and Zuntz, 1893; Watanabe and Sassa, 1914; Benedict, 1915—see Morgulis, 1923; Lusk, 1931), and there have been investigations of civilians and prisoners of war (Lamy, Lamotte and Lamotte-Barillon, 1946a, b, c, d; Hottinger, Gsell, Uehlinger, Salzmann and Labhart, 1948; Helweg-Larsen, Hoffmeyer, Kieler, Thaysen, E. H., Thaysen, J. H., Thygesen and Wulff, 1952). When the calorie intake falls below the level of energy expenditure the weight begins to fall after a few days, and this loss of weight is almost entirely a loss of fat if the diet contains enough protein to balance the losses of N in the urine and other excreta. If it does not, there is a net loss of protein from the body, and the amount of this has been well investigated in both man and animals subjected to partial or complete starvation (Keys *et al.*, 1950; Medical Research Council, 1951). One of the responses to undernutrition noted over and over again is a fall in the basal metabolic rate below the figure predicted for a person of that weight and surface area (Zuntz and Loewy, 1918; Keys *et al.*, 1950). It is not immediately clear why this fall should take place. Possibly the prediction formula does not apply to the altered proportions of the undernourished body. If there is a fall in the rate of oxidation at cell level, it must originate from a reduction in functional activity. Associated with this fall in basal metabolic rate is a fall in skin temperature and later in deep body temperature. This reduction in metabolism and body temperature is an adaptation that must prolong life in a favourable environment, but it can also hasten the end if the environmental temperature is low (McCance and Mount, 1960; Klieber, 1961). The more important literature on undernutrition and famine was reviewed by McCance (1951) and McCance and Widdowson (1951). Most of the early work was done on adult animals, but recently the interest of physiologists and stockbreeders has turned to a study of children and growing animals, for they are potentially much more important than adults in a civilised community.

b. Growth. The child's response to undernutrition is to gain weight more slowly than its genetic capabilities would allow. All animals react in this way, and in some, notably in rats, raising or lowering the food intake during suckling can make a permanent difference to the size of the adult (McCance and Widdowson, 1962). This is not so, however, in all animals, even if the undernutrition is pushed back into foetal life, for twin lambs, which are always to some extent undernourished in utero and never reach the size of singletons, may ultimately attain the full stature to be expected in that breed (Potter; personal communication). The position has not yet been completely clarified in man (McCance, 1962). Over the last 100 years children have certainly been growing faster and faster. They are taller and heavier now than they were ten, twenty, fifty years ago (Tanner, 1955; Falkner, 1958; Cone, 1961;

Hagen and Paschlau, 1061), and adults may be taller also (Boyne and Leitch, 1954; Greulich, 1958; Boyne, 1960; Editorial 1961), but it would be premature to attribute the whole of this to better nutrition. The elimination of infections and of inbreeding may have had something to do with it (McCance, 1962). The older the animal, the greater its power of recovery from a given period of undernutrition. Rats undernourished from the sixth to ninth week of age may regain the weight to be expected in normal controls (Widdowson and McCance, 1963), and children (Prader, Tanner and von Harnack, 1963) may grow fast enough to get back "on course". Growth, however, is a much more complicated process than a mere increase in weight, and each part of the body should grow and mature in a characteristic way relative to all the others. At the appropriate age and stage of development some of these parts have a priority of growth, and their development will go forward even when there is not enough food available for the development of all. Undernutrition, therefore, accentuates the growth of some structures of the body relative to others. If there is not enough food to go round, bones develop at the expense of muscles and fat; when the time comes the teeth and sexual organs have even higher priorities than the bones. Undernutrition, therefore, not only prevents a child's normal increase in weight but also alters the whole pattern of its development, and the effect is particularly well seen if an undernourished child is compared in all respects with a well-nourished one of the same weight. Although the development of the teeth, bones and other organs with high priorities of growth goes forward to some extent even in undernutrition, the structures found are not normal. If full nutrition is restored, the structure of the bones and soft tissues may return to normal, but this can hardly be true of the teeth once they have erupted, certainly not of the enamel (McCance, Ford and Brown, 1961; Tonge and McCance, 1965).

Undernutrition, whether in a child or adult, alters the proportions of the three great compartments of the body. Figure 2 shows the response of an adult to a prolonged period of undernutrition. There are the inevitable losses of fat and cell mass and an increase in extracellular fluid. This may be much more spectacular than in the instance shown, and oedema is indeed one of the classical signs of undernutrition. McCance (1951) explored the aetiology of this without coming to any completely satisfying explanation, but since that time new evidence has come to light. It has been suggested that during undernutrition abnormal amounts of ferritin are released from the liver. This iron protein compound stimulates the hypothalamic posterior pituitary axis to produce antidiuretic hormone, which can be demonstrated in the urine and accounts for the retention of fluid (Gopalan, 1950; Baez, Mazor and Shorr, 1952; Srikantia, 1958; Srikantia and Gopalan, 1959). If true, this is a characteristically human, or at any rate primate, response to undernutrition, for lower animals do not get oedema in the same way, although they do show some increase in extracellular fluid.

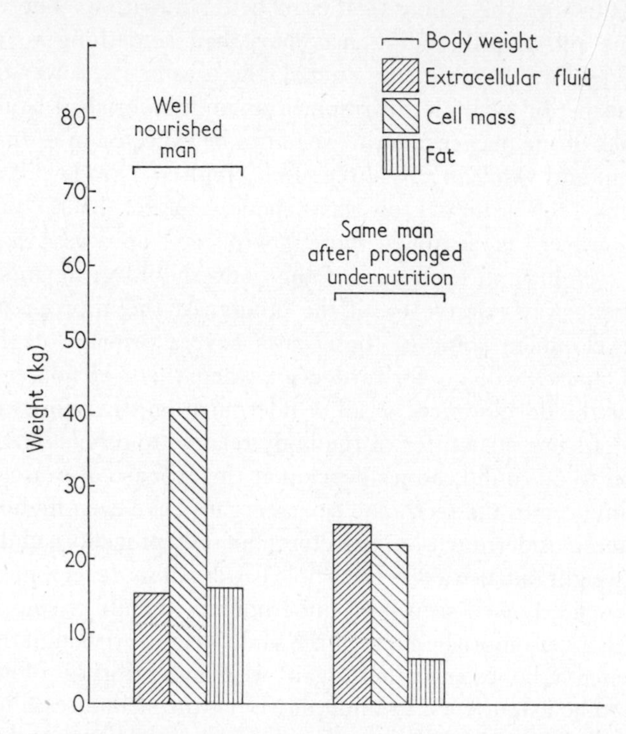

FIGURE 2. Amounts of extracellular fluid, cell mass and fat in the body of a well-nourished man and in the same man after a prolonged period of undernutrition.

C. Protein Deficiency

Adults react to a protein-free diet by a complete loss of appetite (Martin and Robison, 1922), but this is an experimental rather than a practical stress, and it is uncertain if any diets in use today have so little protein in them that the health of the adult population is likely to be affected thereby. Children, however, are different, and in many parts of the world – notably Central and South America, including the West Indies, Africa, parts of Turkey and the Middle East, India and South East Asia – children suffer from this nutritional deficiency. The disease has many names, but is usually now referred to as kwashiorkor. In its pure form this is the metabolic response of a weanling infant to a diet containing too little protein. There are several variants of the disease, for the diets given to the children differ from one locality to another, and the relation between the cause and the effect is still not always clear.

The children cease to grow and become miserable, but they may remain well covered with subcutaneous fat. The superficial layers of the skin peel off and leave large weeping areas that may resemble burns. Oedema fluid begins to accumulate and may become extensive, but effusions in the serous cavities are rare. The muscles shrink and the fibres become thin and atrophic (Montgomery, 1962). The liver may or may not be enlarged and fatty (Waterlow, 1948). The pancreas is usually pale and necrotic, the intestines transparent and distended (Trowell, Davies and Dean, 1954; Brock and Hansen, 1962; Valez, Ghitis, Pradilla and Vitale, 1963). There may be some anaemia, but the type varies (Kondi, MacDougal, Foy, Mehta and Mbaya, 1963; Sandozai, Haquani, Rajeshvari and Kaur, 1963; Valez, Ghitis, Pradilla and Vitale, 1963). The metabolism and distribution of protein within the body is changed (Waterlow, 1962). There are characteristic enzyme deficiencies in the duodenal and other body fluids, and hypoproteinemia is usual. Alanine, glycine, aspartic and glutamic acids, lysine and cystine are excreted in excessive amounts (Chaudhuri and Chaudhuri, 1962; Shendel and Hansen, 1962). Abnormal metabolic pathways develop in the breakdown of histidine and other amino acids, whose metabolites appear in the urine (Dean and Whitehead, 1963). In expert hands the response to treatment is good.

D. Multiple Stresses

Some recent Hungarian work (Andik, Donhoffer and Schmidt, 1963) has raised some problems of a general nature that will require further investigation before they can be properly integrated, but they are too important not to bring into the discussion now. If young rats at room temperature are offered a sufficiently low protein diet ad lib., they eat little, do not grow and die in a few weeks (Widdowson and McCance, 1957; Cabak, Dickerson and Widdowson, 1963). This is because their intake of food is regulated by both their calorie requirements and the percentage of protein in the diet, and consequently the food they eat does not provide them with enough protein to grow or enough calories with which to maintain life. The response is similar to that of the infant with kwashiorkor.

If now the same experiment is carried out at 5° instead of 20°, the rats not only survive but grow. The additional stress of a low temperature so increases the metabolic rate, and consequently the demand for calories, that the food eaten to meet this demand provides enough protein to remove the original strain and enables the young animal to survive. Forced exercise might have done the same. These multiple responses to simultaneous changes in the environment, having an overall survival value but all be reckoned deleterious individually, require further study. They may easily turn out to have a bearing on the relationship between food supplies and ecology.

E. Water †

About 72% of the fat-free components of the adult body is water, and the osmolar concentrations of the body fluids are of the order of 320m osmols/l (Black, 1960). In a healthy man these levels are maintained by the intake of water and salt, and deprivation of water is a much more compelling and imminent danger than the complete absence of food. Elijah, for example, was able to manage well enough on the bread and flesh brought him by the ravens, but as soon as the brook Cherith ran dry he had to move to Zarephath and take refuge with the widow woman (I Kings: *17*). Sarah Jacob, the Welsh fasting girl, who was alleged by her parents in 1869 to be able to live for weeks on end without food or drink, only lasted for eight days once a guard of nurses had been set to watch her day and night (Fowler, 1871), and she must be reckoned to have died of dehydration, not lack of food. Some loss of water through the lungs and skin is inseparable from life, and the amount varies with the external temperature, the altitude and the functional activity of the body. Over a period of 24 hrs the intake of water usually exceeds the requirement of the body in man, but this is not so often true of animals. The kidney excretes any excess, but once the electrolyte concentration of the body fluids has risen to the physiological limit, cells in the hypothalamus initiate the secretion of the antidiuretic hormone from the posterior lobe of the pituitary. In terms of the counter-current hypothesis, this makes the cells of the collecting ducts permeable to water, the osmotic pressure of the fluid in them rises to the level of that in the tip of the pelvis, and the output of water falls to the level corresponding with this. Little further direct renal adaptation can take place, although the volume of urine may fall fractionally farther if increased protein breakdown raises the concentration of urea in the body fluids while the kidney is still functioning as it should. As the volume of extracellular fluid falls, aldosterone and similar hormones are secreted, and sodium salts are almost completely reabsorbed even when the osmolar concentration of the body fluids begins to rise (Elkinton and Taffel, 1942; McCance and Young, 1944; Black, McCance and Young, 1944). This is a valuable response, in that it helps to maintain the volume of circulating blood and postpones the rise in haematocrit that must come later, but it can only take place at the expense of a rising concentration of electrolytes all over the body, and this is probably the ultimate cause of death from dehydration.

Dehydration sets up a desire to drink, which in man is referred to as thirst, and this soon becomes a conscious stress and later an overwhelming one. A book on the subject of thirst in all its aspects has been written by Wolf (1958) and should be consulted for points of detail, including references to the well-known books and papers of Adolph, Schmidt-Nielsen and many other authors, past and present. After deprivation of water for some time, and when

† See note, below, to Salt.

4 to 8% or more of the body weight has been lost, some animals, such as donkeys, camels and dogs, will drink enough to restore their weight to its normal level in a few minutes, but man's urge to drink may not be sufficiently acute to make him keep himself fully hydrated in a hot climate, particularly if he is working. This also applies with equal force to high altitudes, where hyperventilation and the difficulty of getting water may seriously dehydrate a man and undermine his efficiency before he even appreciates it. The reduction in efficiency, however, may not go on indefinitely at its initial level. Some degree of adaptation and functional recovery may take place if the dehydration persists as a steady state. However this may be, continuous loss of water from the body soon leads to physical and mental deterioration. The mouth becomes dry, and it is to this that man attributes his conscious and normal sensations of thirst, but if no fresh water can be obtained it soon becomes difficult to chew food and impossible to swallow it. Men may begin to talk unnecessarily, till the state of their mouths prevents it. Sooner or later judgement is impaired, and actions may become irrational. At this point adaptation finishes. Death may ensue in a matter of hours in a highly unfavourable environment, and life usually comes to an end in about fourteen days in a favourable one. Within the last 100 years there have been a number of records of men, lost in the desert for only a few days, struggling to safety in a state of extreme dehydration and collapse. The one most often quoted is that of Pablo Valencia (McGee, 1906; and see also Wolf, 1958). In some parts of the world prolonged thirst must always have been a nutritional risk to man (Adolph and associates, 1947). The references to it in the Old Testament are numerous, but then as now Palestine was "a thirsty land" (Psalms: *143*, 6) and water a highly prized commodity. With the introduction of seafaring and organised warfare, a satisfactory supply of water must often have meant survival in the face of an emergency. When castles were built and sieges envisaged in them, some provision for water had to be made inside the walls. If there were no wells, rainwater falling on the roofs was usually collected and conducted by most elaborate drainage systems to large underground tanks, often cut out of the rock on which the castle was built. The supply of water frequently set a limit to the length of time a sailing ship could stay at sea, and thirst has always been the dread of castaways (Nicholl, 1960). Many have lost all sense of morale, taken large quantities of sea water and committed suicide under the strain (Critchley, 1943; McCance *et al.*, 1956), but other castaways with far worse prospects have been saved under far worse conditions by their magnificent tenacity and their will to survive (Rickenbacker, 1943; Wolf, 1958).

Water requirements are not easy to define in absolute terms, for they vary so much with the environmental temperature and the nature of the work to be done. They also vary from one individual to another. They are probably of the order of 500 ml/24 hr for sedentary persons at sea in a covered raft and might be as high as 15,000 ml/24 hr for troops on a desert exercise. Too large an intake of water is not a nutritional risk of any practical importance to healthy men and women, but it has been produced experimentally (De Wardener and Herxheimer, 1957a, b). It is an uncomfortable experience and leads to temporary loss of renal concentrating power and to a desire for salt. It is a real clinical risk (Black, 1960), particularly to the newborn, and can easily be reproduced in animals (McCance and Widdowson, 1955).

F. Salt (sodium chloride) †

To be nutritionally satisfactory, a diet should contain enough salt to maintain an output in the urine of at least 3 and probably not more than 8 to 10 g/day. Even if large amounts of salt are being lost in sweat, an intake high enough to meet these losses and to provide a small surplus of, say, 3 g/day, which will find its way into the urine as a margin of safety, should not set up any nutritional strain. When men are first subjected to high environmental temperatures, and particularly to hard work in such temperatures, their losses of salt may be as high as 15 g/day—higher than they will be a week or so later, for the percentage of salt in the sweat falls with acclimatisation.

An intake of salt insufficient to fulfil the conditions laid down above constitutes a nutritional stress; if one assumes that the losses of salt continue to exceed the requirements, a sequence of events begins that may be regarded as a series of graded responses. As the days go by less and less salt appears in the urine, and sooner or later the amount becomes negligible. This is due to the action of aldosterone and other adrenal hormones. Coincidentally, the volume of the circulating blood and of the extracellular fluids begins to fall and so, to some extent, does the glomerular filtration rate. As the plasma volume falls, the concentration of serum proteins rises, and also the haematocrit, but there may be no fall in the blood pressure at this stage. As the drain of salt continues, the concentration of sodium chloride in the serum and extracellular fluids begins to fall and with it the osmolar concentration of the whole body. In spite of this, there is no diuresis, as there would be with a normal or raised plasma volume. There is probably continuous secretion of the antidiuretic hormone into the circulation, in response to the low blood volume. At all events, the urine volumes are small, and any diuresis set up by the ingestion of further water is limited and incomplete. Subjectively, the person feels tired and unwell. He may suffer from cramp in his hands, feet or limbs, and his efficiency as a working unit becomes less and less (McCance, 1936; 1956).

These responses are nevertheless all beneficial in that they enable the man to survive and even to remain tolerably active after losing 20 to 30% of his normal complement of sodium ions. A high intake of sodium salts will

† See Chapter 9 for further discussion of water and salt intakes.

restore normal function and produce a rapid amelioration of all the unpleasant symptoms. Taste, however, is the only sense that leads to an increased intake of salt, and it may be a most unreliable one. There may be no craving for salt, and it is possible for men to continue for some time in a state of chronic deficiency without appreciating the trouble or doing anything to remedy it. The classical work on this subject was that of Moss (1923–4) and Haldane (1928), and the subject was investigated experimentally by McCance (1936). See also Marriott (1950).

In rats and poultry a high enough intake of salt can lead to lesions in the kidney and a rise in the arterial blood pressure (Smirk, 1957; Youmans, 1957; Davidson and Passmore, 1963), but in a grown man, except under experimental conditions (Baker, Plough and Allen, 1963), the intake of sodium chloride by mouth can scarcely ever be large enough to set up a nutritional strain if water is available on demand. The response of the adult is to raise the excretion of salt in the urine to meet the expanding intake and to drink more water. The volume of the extracellular fluids may rise slightly, but there is little or no change in the osmolar concentration of the body fluids. This response is a satisfactory adaptation. Drinking sea water, however, as castaways have often done, sets up a physical strain to which man can make no useful response. In large amounts sea water generally leads to diarrhoea, and inevitably causes or accentuates dehydration, for the human kidney cannot excrete sodium chloride at the concentration found in sea water. If no fresh water is available, therefore, the salts in the sea water must either be excreted in association with water already present in the body or be retained, only to raise the osmolar concentration of the extracellular fluids.

The ingestion of an unnecessary amount of salt provides an interesting example of the effect of age on the way in which the body responds to this stress. Adults tend to excrete the salt even when hypertonic solutions are administered, and the volume of their extracellular fluids remains within normal limits or falls. Newborn infants have much less margin of safety and do not excrete much of the salt, even if it is taken as a relatively dilute solution; this results in hypertonic expansion of their extracellular fluids and massive oedema (McCance and Widdowson, 1957). This response can hardly be considered a useful adaptation. Age affects a number of the other responses to unfavourable environments, and among the best known are those to high and low temperatures.

Finberg, Kiley and Luttrell (1963) reported on a number of deaths in infants due to accidental administration of salt instead of sugar. The milk formula contained about 900 m eq of salt/l. Sea water contains a little over 600 (Wolf, 1958). The concentration of sodium in the serum rose to between 200 and 300 m eq/l before death, and one infant survived a serum sodium of 274 m eq/l. Multiple cerebral and meningeal haemorrhages were found after death, probably caused by osmotic dehydration of the cerebrospinal fluid and the brain cavity and a fall in extravascular cerebral pressure. The signs of hypernatremia in infancy are generally cerebral ones, whether the hypertonicity is brought about by too little water or too much salt (Finberg and Harrison, 1955; Weil and Wallace, 1956; Finberg, 1959; Finberg, Luttrell and Redd, 1959; Kerpel-Fronius, 1959; Sotos, Dodge, Meara and Talbot, 1960; Ussing, Kruhøffer, Hess-Thaysen and Thorn, 1960). Mental derangement is the ultimate response of the adult castaway to dehydration, whether or not this has been accelerated by the ingestion of sea water. The central nervous system appears to be the most vulnerable system at all ages to electrolyte and water imbalance, though in adults, certainly, the mental symptoms are reversible with rehydration.

G. Calcium and Iron

As far as we know, man has no "appetite" for calcium, iron or indeed for any electrolytes, except perhaps sodium chloride. The amounts he takes in his food depend on the kinds of food that are available and that he likes and on the total amount of food he eats (Widdowson, 1962). Consequently intakes of calcium and iron vary widely from one part of the world to another and from one individual to another, yet there are, except for some rather controversial issues such as senile osteoporosis, few records of an intake of calcium, however high or low, setting up a nutritional strain (Jackson, 1961a, b). This is even more surprising when it is remembered that the ability to absorb calcium, and possibly also iron, varies greatly from one individual to another and that there is no reason to suppose that the person who absorbs calcium or iron least readily is the one who takes the most in his food.

The amounts of both calcium and iron in the body are regulated by controlled absorption from the intestine. For iron this is really the only regulating mechanism, since the amount excreted in the urine is extremely small and the body has no known means of removing surplus iron once it has got in. Some individuals habitually excrete more calcium in the urine than others (Knapp, 1947; Nicolaysen, Eeg-Larsen and Malm, 1953); if such persons are to remain in calcium equilibrium they must absorb correspondingly more from the intestine. Indeed, the amount excreted by a normal individual varies largely with the amount absorbed (McCance and Widdowson, 1942b), but must depend ultimately on the level in the serum and the maintenance of homoeostasis there by the parathyroid glands and vitamin D.

1. *Calcium*

If an adult accustomed to a high calcium intake changes to a lower one, he absorbs less calcium, but urinary excretion does not fall as much as absorption, and he therefore loses calcium from the skeleton and continues to do so for some time. If the diet is particularly low in calcium and rich in some substance, such as phytate, that forms an insoluble salt with calcium, there may even be more calcium in the faeces than in the food (McCance and Widdowson, 1942a). Most people, however, mobilise calcium readily from their skeletons under the influence of the parthyroid hormone, homoeostasis is maintained, and the level of calcium in their serum remains normal. There is no immediate strain, although the amount of calcium in the body is falling. In a few people the parathyroid response may not be sufficiently great to maintain homoeostasis, the serum calcium may fall, and tetany may result (McCance, 1953b) without the subject's feeling any inclination to look for a source of dietary calcium.

Whether the person absorbs calcium freely or not, whether he excretes much or little in the urine and whether his serum calcium does or does not fall in response to the excretion of calcium from it, there is good evidence that in most people some change in the functional organisation of the intestinal epithelium takes place in the course of time if the person continues to eat the same low calcium diet, and more calcium is absorbed from it. Some, however, adapt more slowly, if at all (Malm, 1958, 1961). Nevertheless, in many parts of the world calcium intakes are known to be low, and the question has often been asked – how do the children obtain enough calcium for the growth and calcification of their skeletons? Between birth and adult life a person needs to lay down about 1200 g of calcium in the skeleton. If he does this over the course of eighteen years he must retain an average of 200 mg a day; in order to retain 200 mg he must take in considerably more, although the growing animal or child is able to absorb a greater proportion of the calcium in his food than an adult. Yet the calcium intake is said to be of this order among some populations. Walker (1954) estimated that the Bantu, for example, obtained only 200 to 450 mg of calcium a day from his food and the "poor" Bantu considerably less than 200 mg. In spite of this, the bones of the Bantu are reported not to differ from those of Europeans in their chemical and histological structure; the mean height of the Bantu is the same, and they have less dental caries. They are not particularly prone to fractures, nor do fractures heal less readily. No disorder arising from a calcium deficiency is known among the Bantu (Jackson, 1961a, b). The Bantu child cannot lay down in his skeleton more calcium than he absorbs from his food, and it is therefore possible that he has sources of calcium not usually taken into consideration in dietary surveys. Nevertheless, the Bantu must be able to utilise the calcium in his food more readily than the European who has been accustomed to a much higher intake. This may conceivably be due to genetic adaptation, but what enables the Bantu to do this we do not know. The level of calcium in the serum of the Bantu may be "set" rather low (Jackson, 1961b) by weak parathyroid activity; he may have a longer intestine than Europeans, and one highly responsive to vitamin D, or his intestinal secretions may contain less calcium, as a result of some enzymic adaptation that may have taken place. These are all matters of conjecture, but the fact remains – intakes of calcium that, judged by the results of balance experiments, should create a nutritional strain in many Europeans never appear to do so in the Bantu or in other races subsisting successfully on intakes equally low (Hegsted, Moscoso and Collazos, 1952).

2. *Iron*

The absorption and excretion of iron and the response to a low intake is in many ways less complicated than that of calcium, for only minute amounts of iron are excreted by the kidney, and the digestive juices of the adult contain little (Hahn, Bale, Hettig, Kamen and Whipple, 1939; Dubach, Moore and Callender, 1955; More, 1959–60). A normal adult's body contains 4 to 5 g of iron; urinary excretion amounts to about 0·5 to 1·5 mg a day, and this is balanced by the absorption of a similar amount. Even the poorest diets contain more than this, and iron deficiency in an adult male due to an inadequate intake is virtually non-existent. If, moreover, the need for iron increases, during growth or pregnancy, for example, or if rapid erythropoiesis is taking place, then more iron is absorbed, but how the signal for this is transmitted from the bone marrow to the intestine is still unknown (Brown, 1963; Charlton, Jacobs, Torrance, and Bothwell, 1965). Iron deficiencies may appear, however, if the demands for iron are for greater amounts than those the intestine can absorb from the amount the food contains—during periods of rapid growth, for instance, during pregnancy and in the non-pregnant woman to make up losses in blood during menstruation. The response of the body to a deficiency of iron in all these circumstances is the same. The blood volume is maintained, even when the production of haemoglobin and of red cells falls; consequently there is a reduction in the concentration of haemoglobin in the circulating blood. Wide variations in this are compatible with a reasonably normal life, and man's tolerance of such variations must be regarded as one of his responses to the stress of a diet containing too little iron for his physiological requirements.

Anaemia due to a dietary deficiency of iron is usual during the suckling period. The rate of growth, after the negative iron balance of the neonatal period (Cavell and Widdowson, 1964), is more rapid during the weeks and months after birth than it will ever be again, and the requirements for all the bodily constituents, including iron, are high. The food consists of milk alone, and this has little iron in it. It is likely that the intestine of the newborn baby absorbs more of the iron in its food than the intestine of an adult would do; even if the absorption were complete, however, the amount of iron available from breast milk would not be sufficient to maintain the circulating haemoglobin at its newborn level. Suckling anaemia is characteristic of all mammalian species when the young depend upon milk as their sole source of iron, and it appears to be almost physiological for the haemoglobin level to fall. Mackay (1931) demonstrated that there was a correlation between the fall in haemoglobin during the suckling period in man and the rate of growth—the more rapid the growth rate the greater the fall.

Until twenty years ago the possibility that the regulating mechanisms might break down in the face of an extremely high intake had never been seriously considered. A syndrome has now been described, however, largely confined to the Bantu, which is caused by overloading the body with iron. The sources of the metal are mainly the iron utensils used for cooking and for making native beer, and the absorption may be facilitated by the simultaneous consumption of the alcohol (Charlton, Jacobs, Seftel and Bothwell, 1964). The dried liver may contain 2% of iron and become cirrhotic. The pancreas, adrenals, thyroid, pituitary and heart may also be affected, but to a lesser extent. The patients are usually middle-aged males, for the beer is mostly consumed by this sex. They may present with diabetes, scurvy or osteoporosis, but the chemical pathology behind all this has not yet been fully explored (Bothwell, 1964).

H. Conclusions – *"It's human nature after all"* †

The study of disease has always been one of man's interests, and within the last 200 years it has become more and more intensive. Many diseases have a nutritional origin, but this has seldom been recognised unless the food produced evident and immediate signs of distress. The Eskimos described the poisonous effects of polar-bear liver for this reason and avoided eating it long before the trouble was known to be due to intoxication with excessive amounts of an essential nutrient, vitamin A (Rodahl and Moore, 1943). Once it was suspected, moreover, even if wrongly so, that a disease might be due to a particular food, man's response has usually been to avoid eating it, and this is the origin of many unfortunate dietary "taboos" (Campbell and Cuthbertson, 1963). The early history of scurvy illustrates man's groping response in the pre-experimental days to a disease with a nutritional remedy (Chick, 1953; Talbot, 1963).

Within the last 100 years, however, scientific man has been experimentally investigating the origin of his nutritional strains and by so doing has found the means of eliminating many of the causative stresses from his environment. Rickets is no longer the menace that it was in the northern countries seventy years ago, and, speaking generally, all the known protein, vitamin and mineral deficiency diseases are avoidable. They should no longer be problems in human survival. There are, however, other problems still to be solved.

Man is a conservative animal. He is, moreover, actuated by the feelings

† Old music-hall song.

of the moment and finds it hard to adopt a long-term point of view and act on it. "Carpe diem" will probably always be his motto (Horace, 30 B.C.) Consequently, within the last 1000 years man has created for himself nutritional strains that he now finds difficult to avoid. The people in Thailand, for example, live largely on rice because they have been brought up to do so and suffer from serious malnutrition in consequence, although their country will produce all the best food man requires, much of which is exported (Campbell and Cuthbertson, 1963). All those who have tried to improve the diets in backward countries and prevent malnutrition there have come up against the resistance to change engendered by ignorance, custom and conservatism (Burgess and Dean, 1962; McLaren, 1963). Western societies are no better. Although the vitamin diseases have largely disappeared, others have been allowed to take their place (McLaren, 1963). The availability of sugar and sweets is an irresistible attraction to most people and has led to widespread dental decay (Gordonoff and Minder, 1960; Holloway, James and Slack, 1963), which we have begun to take for granted. To avoid this would involve a revolutionary change in industrial organisation and in man's approach to his survival. It may safely be predicted that neither will take place. A water supply, however, containing too little fluorine may be regarded as an additional dental strain in that it increases the risk of caries in children's teeth. The scientific evidence in favour of adding minute amounts of fluorine to the drinking water would appear to be unassailable (Underwood, 1962; Symposium, 1963), but the proposal to do so has set up such psychological conflict in many minds that it has often had to be abandoned (Report, 1953; McNeil, 1957; Davidson and Passmore, 1963). An account of this emotional opposition, which can go to considerable lengths, has been given by Dalzell-Ward (1963).

Within the last forty years the incidence of ischaemic heart disease has increased to such an extent in the Western world that the expectation of life in middle age has scarcely gonc up at all, in spite of the remarkable progress that medical science has made possible in the treatment of most other diseases of this era. Man's response to this has been to initiate intensive research into the matter, and work carried out in the last twenty years has gone to show that a nutritional strain lies at the bottom of this threat to the survival of the individual into the old age that he has been led to believe he has deserved. Coronary thrombosis is a disease of the well-fed highly "Westernised" communities. Motor cars and a lack of exercise (Annotation 1963) may play a part, but a high consumption of fats, and particularly of certain fatty acids, is thought to be involved. Some evidence suggests that more of the so-called essential fatty acids in the diet would be beneficial and possibly also less butter, milk, eggs, animal fats and margarine (Report, 1961; Joliffe, Maslansky, Rudensey, Simon and Faulkner, 1961; Bronte-Stewart, 1962).

There is no complete solution to this problem yet, and meantime there has been practically no change in the food habits of the Western world, although the death-rate from ischaemic heart disease is increasing as fast as ever. When the cause is known any proposal to eliminate the disease is likely to set up intense emotional and political strains, for it looks as though its eradication may demand a reorganisation of ways of life that have come to be associated with success, prosperity and vested interests.

REFERENCES

Adolph, E. F., and associates (1947). "Physiology of Man in the Desert." Interscience, New York.

Alexander, J. K., Dennis, E. W., Smith, W. G., Amad, K. H., Duncan, C., and Austin, R. C. (1962–3). *Cardiovasc. Res. Center Bull.* 1, 39.

Andik, I., Sz. Donhoffer, M. F., and Schmidt, P. (1963). *Brit. J. Nutr.* 17, 257.

Annotation (1963). *Brit. med. J.* ii, 5.

Babcock, C. G. (1948). *J. Amer. diet. Ass.* 24, 390.

Baez, S., Mazur, A., and Shorr, E. (1952). *Amer. J. Physiol.* 169, 123.

Baker, E. M., Plough, I. C., and Allen, T. H. (1963). *Amer. J. clin. Nutr.* 12, 394.

Barnett, A. (1961). "The Human Species." Penguin Books, Harmondsworth, Middlesex.

Benedict, F. G. (1907). *Publ. Carneg. Instn.* No. 77.

Benedict, F. G. (1915). *Publ. Carneg. Instn.* No. 203.

Benedict, F. G., Miles, W. R., Roth, P., and Smith, H. M. (1919). *Publ. Carneg. Instn.* No. 280.

Black, D. A. K. (1960). "Essentials of Fluid Balance", (2nd Ed.) Blackwell, Oxford.

Black, D. A. K., McCance, R. A. and Young W. F. (1944). *J. Physiol.* 102, 406.

Bothwell, T. H. (1964). "Iron overload in the Bantu". *In* "Iron Metabolism" (F. Gross, ed.), p. 362. Springer Verlag, Berlin.

Boyne, A. W. (1960). *In* "Symposium of the Society for the Study of Human Biology", Vol. 3: Human Growth (J. M. Tanner, ed.) London.

Boyne, A. W., and Leitch, I. (1954). *Nutr. Abstr. Rev.* 24, 255.

Brock, J. F., and Hansen, J. D. L. (1962). "Protein deficiency." *In* "Clinical Nutrition" (Norman Joliffe, ed.). Hoeber, New York.

Bronte-Stewart, B. (1962). "Diet and Ischemic Heart Disease." *In* "Clinical Nutrition", 2nd Ed. (Norman Joliffe, ed.). Hoeber, New York.

Brown, E. B. (1963). *Amer. J. clin. Nutr.* 12, 205.

Bryant, J. (1924). *Amer. J. med. Sci.* 167, 499.

Burgess, A., and Dean, R. F. A. (1962). "Malnutrition and Food Habits." Tavistock Publ. Ltd., London.

Cabak, V., Dickerson, J. W. T., and Widdowson, E. M. (1963). *Brit. J. Nutr.* 17, 601

Campbell, R. M., and Cuthbertson, D. P. (1963). "Factors influencing man's selection of foods." *In* "Progress in Nutrition and Allied Sciences" (D. P. Cuthbertson, ed.), p. 395. Oliver and Boyd, Edinburgh and London.

Cavell, P. A., and Widdowson, E. M. (1964). *Arch. Dis. Childh.* 39, 496.

Charlton, R. W., Jacobs, P., Seftel, H., and Bothwell, T. H. (1964). *Brit. med. J.* ii, 1427.

Charlton, R. W., Jacobs, P., Torrance, J. D., and Bothwell, T. H. (1965). *J. clin. Invest.* 44, 543.

Chaudhuri, K. G., and Chaudhuri, A. (1962). *Turk. J. Pediat.* 4, 81.

Chick, H. (1953). *Proc. Nutr. Soc.* 12, 210.

Cogswell, H. D. (1948). *Ann. Surg.* 127, 377.

Colloquium (1959). "Redécouverte du Jeûne." Les Editions due Cerf, Paris.
Cone, T. E. (1961). *J. Pediat.* 59, 736.
Critchley, M. (1943). "Shipwreck-Survivors: A Medical Study." Churchill, London.
Curschmann, F. (1900). "Hungersnöte im Mittelalter: ein Beitrag zur deutschen Wirtschaftsgeschichte des 8. bis 13. Jahrhunderts." B. G. Teubner, Leipzig.
Dalzell-Ward, A. J. (1963). *Proc. Nutr. Soc.* **22**, 91.
Davidson, S., and Passmore, R. (1963). "Human Nutrition and Dietetics." Livingstone, Edinburgh and London.
Dean, R. F. A., and Whitehead, R. G. (1963). *Lancet*, i, 188.
De Wardener, H. E., and Herxheimer, A. (1957a). *J. Physiol.* 139, 42.
De Wardener, H. E., and Herxheimer, A. (1957b). *J. Physiol.* 139, 53.
Donovan, D. (1848). *Dublin med. Pr.* **19**, 67, 129, 275.
Dubach, R., Moore, C. V., and Callender, S. T. (1955). *J. Lab. clin. Med.* 45, 599.
Duncan, G. G. (1959). "Diseases of Metabolism", 4th Ed. Saunders, Philadelphia and London.
Editorial (1961). *Brit. med. J.* ii, 502.
Elkinton, J. R., and Taffel, M. (1942). *J. clin. Invest.* **21**, 787.
Falkner, F. (1958). *Arch. Dis. Childh.* 33, 1.
Finberg, L. (1959). *Pediatrics, Springfield*, **23**, 40.
Finberg, L., and Harrison, H. E. (1955). *Pediatrics, Springfield*, **16**, 1.
Finberg, L., Kiley, J., and Luttrell, C. N. (1963). *J. Amer. med. Ass.* **184**, 187.
Finberg, L., Luttrell, C., and Redd, H. (1959). *Pediatrics, Springfield*, **23**, 46.
Fowler, R. (1871). "A complete History of the Case of the Welsh Fasting girl (Sarah Jacob)." Henry Renshaw, London.
Frazer, J. G. (1927). "Golden Bough." Macmillan, London.
Gopalan, C. (1950). *Lancet*, i, 304.
Gordonoff, T., and Minder, W. (1960). *World Rev. Nutr. Dietetics* **2**, 209.
Greulich, W. W. (1958). *Science* **127**, 515.
Hagen, W., and Paschlau, G. v. R. (1961). "Wachstum und Gestalt." Thieme, Stuttgart.
Hahn, P. F., Bale, W. F., Hettig, R. A., Kamen, M. D., and Whipple, G. H. (1939). *J. exp. Med.* **70**, 443.
Haldane, J. S. (1928). *Brit. med. J.*, i, 609.
Hegsted, D. M., Moscoso, I., and Collazos, C. (1952). *J. Nutr.* 46, 181.
Helweg-Larsen, P., Hoffmeyer, H., Kieler, J., Thaysen, E. H., Thaysen, J. H., Thygesen, P., and Wulff, M. H. (1952). *Acta psychiat. scand. Supp.* 83.
Holland, Lady (1850). "A memoir of the Reverend Sydney Smith", chapter 11.
Holloway, P. J., James, P. M. C., and Slack, G. L. (1963). *Brit. dent. J.* **115**, 19.
Horace (*c.* 30 B.C.). "Odes." Book 1, 11, 7 and 8.
Hottinger, A., Gsell, O., Uehlinger, E., Salzmann, C., and Labhart, A. (1948). "Hungerkrankheit, Hungerödem, Hungertuberkulose." Benno Schwabe, Basel.
Jackson, W. P. U. (1961a). "Effects of altered nutrition on the skeletal system: the requirement of calcium in man." *In* "Recent Advances in Human Nutrition" (J. F. Brock, ed.), p. 293. Churchill, London.
Jackson, W. P. U. (1961b). *Voeding* **22**, 617.
Jolliffe, N., Maslansky, E., Rudensey, F., Simon, M., and Faulkner, A. (1961). *Circulation* **24**, 1415.
Keatinge, W. R., and Evans, M. (1958). *Lancet* ii, 1038.
Kerpel-Fronius, E. (1959). "Pathologie und Klinik des Salz - und Wasserhaushaltes." Akadémiai Kiadó, Budapest.
Keys, A., Brozek, J., Henschel, A., Mickelsen, O., and Taylor, H. L. (1950). "The Biology of Human Starvation", p. 934. University of Minnesota Press, Minneapolis.

Kleiber, M. (1961). "The Fire of Life." Wiley and Sons, New York and London.

Knapp, E. L. (1947). *J. clin. Invest.* **26**, 182.

Kondi, A., MacDougal, L., Foy, H., Mehta, S., and Mbaya, V. (1963). *Arch. Dis. Childh.* **38**, 267.

Lamb, D. S. (1893). *Amer. J. med. Sci.* **105**, 633.

Lammert, G. (1890). "Geschichte der Seuchen, Hungers-und Kriegsnoth zur Zeit des dreissigjahrigen Krieges." Von Bergmann, Wiesbaden.

Lamy, M., Lamotte, M., and Lamotte-Barillon, S. (1946a). *Pr. méd.* **54**, 431.

Lamy, M., Lamotte, M., and Lamotte-Barillon, S. (1946b). *Pr. méd.* **54**, 435.

Lamy, M., Lamotte, M., and Lamotte-Barillon, S. (1946c). *Pr. méd.* **54**, 510.

Lamy, M., Lamotte, M., and Lamotte-Barillon, S. (1946d). *Pr. méd.* **54**, 621.

Le Blanc, J. (1962). *J. appl. Physiol.* **17**, 950.

Lehmann, C., Mueller, F., Munk, I., Senator, H., and Zuntz, N. (1893). *Virchows Arch.* **131**, Suppl.

Lloyd, J. K., Wolff, O. H., and Whelan, W. S. (1961). *Brit. med. J.* ii, 145.

Lusk, G. (1921). *Physiol. Rev.* **1**, 523.

Lusk, G. (1931). "The Elements of the Science of Nutrition", 4th Ed. Saunders, Philadelphia and London.

McCance, R. A. (1936). *Lancet*, i, 643, 704, 765, 823.

McCance, R. A. (1951). *Spec. Rep Ser. med. Res. Coun. Lond.* No. 275. p. 21.

McCance, R. A. (1953a). *Lancet*, ii, 685 and 739.

McCance, R. A. (1953b). "The problem of adaptation to low calcium intakes." *In* "Metabolic Interrelations", p. 166. Trans. 5th Conference, Josiah Macy Jr. Foundation, New York.

McCance, R. A. (1956). *Canad. med. Ass. J.* **75**, 791.

McCance, R. A. (1962). *Lancet* ii, 621 and 627.

McCance, R. A., and Mount, L. E. (1960). *Brit. J. Nutr.* **14**, 509.

McCance, R. A., and Widdowson, E. M. (1942a). *J. Physiol.* **101**, 44.

McCance, R. A., and Widdowson, E. M. (1942b). *J. Physiol.* **101**, 350.

McCance, R. A., and Widdowson, E. M. (1946). *Spec. Rep. Ser. med. Res. Coun. Lond.* No. 254.

McCance, R. A., and Widdowson, E. M. (1951). *Post Grad. med. J.* **27**, 268.

McCance, R. A., and Widdowson, E. M. (1955). *J. Physiol.* **129**, 628.

McCance, R. A., and Widdowson, E. M. (1957). *Acta paediat., Stockh.* **46**, 337.

McCance, R. A., and Widdowson, E. M. (1962). *Proc. roy. Soc.* B. **156**, 326.

McCance, R. A., and Young, W. F. (1944). *Brit. med. Bull.* **2**, 219.

McCance, R. A., Ford, E. H. R., and Brown, W. A. B. (1961). *Brit. J. Nutr.* **15**, 213.

McCance, R. A., Ungley, C. C., Crosfill, J. W. L., and Widdowson, E. M. (1956). *Spec. Rep. Ser. med. Res. Coun. Lond.* No. 291.

McGee, W. J. (1906). *Interst. med. J.* **13**, 279.

Mackay, H. M. M. (1931). *Spec. Rep. Ser. med. Res. Coun. Lond.* No. 157.

McLaren, D. S. (1963). *Lancet*, ii. 86.

McNeil, D. R. (1957). "The Fight for Fluoridation." Oxford University Press, New York.

Malm, O. J. (1958). "Calcium requirement and adaptation in adult men." Oslo University Press, Oslo.

Malm, O. J. (1961). *Voeding* **22**, 567.

Marriott, H. L. (1950). "Water and salt depletion." American Lecture Series No. 32 (R. F. Pitts, ed.). Thomas, Springfield, Illinois.

Martin, C. J., and Robison, R. (1922). *Biochem. J.* **16**, 407.

Medical Research Council (1951). "Studies of Undernutrition, Wuppertal 1946–9" *Spec. Rep. Ser. med. Res. Coun. Lond.* No. 275.

Michell, R., and Forbes, N. (1914). Translation of "The Chronicle of Novgorod (1016–1471)". *In* "Publications of the Royal Historical Society," Camden 3rd Series, Vol. 25. Offices of the Society, London, and Arden Press, Letchworth.
Montgomery, R. D. (1962). *J. clin. Path.* **15**, 511.
Moore, C. V. (1959–60). *The Harvey Lectures*, **55**, 67.
Morgulis, S. (1923). "Fasting and Undernutrition." Dutton, New York.
Moss, K. N. (1923–4). *Proc. roy. Soc.* B. **95**, 181.
Mossberg, H. O. (1948). *Acta paediat. Stockh.* **35**, Suppl. 2.
Nicholl, G. W. R. (1960). "Survival at Sea." Sydenham, Bournemouth.
Nicolaysen, R., Eeg-Larsen, N., and Malm, O. J. (1953). *Physiol. Rev.* **33**, 424.
Père X (1959). "Expérience et Réflexions Postérieures d'un Religieux Prisonnier de Guerre." *In Colloruum* (1959).
Pilling, G. P., and Cresson, S. L. (1957). *Pediatrics, Springfield* **19**, 940.
Prader, A., Tanner, J. M., and von Harnack, G. A. (1963). *J. Pediat.* **62**, 646.
Report (1943). "A guide to the preservation of life at sea after shipwreck." *M.R.C. Memor.* No. 8.
Report (1953). "The Fluoridation of Domestic Water Supplies in North America as a Means of Controlling Dental Caries." Ministry of Health – Report of the United Kingdom Mission. H.M. Stationery Office, London.
Report (1961). *Circulation* **23**, 133.
Rickenbacker, E. V. (1943). *Life* **14**, 21, 78, 95.
Rodahl, K., and Moore, T. (1943). *Biochem. J.* **37**, 166.
Sandozai, M. K., Haquani, A. H., Rajeshvari, V., and Kaur, J. (1963). *Brit. med. J.* ii, 93.
Shackleton, E. H. (1909). "The Heart of the Antarctic." Heinemann, London.
Shendel, H. E., and Hansen, J. D. L. (1962). *J. Pediat.* **60**, 280.
Smirk, F. H. (1957). "High Arterial Pressure." Blackwell, Oxford.
Sotos, J. F., Dodge, P. R., Meara, P., and Talbot, N. B. (1960). *Pediatrics, Springfield* **26**, 925.
Srikantia, S. G. (1958). *Lancet* i, 667.
Srikantia, S. G., and Gopalan, C. (1959). *J. appl. Physiol.* **14**, 829.
Strang, J. M. (1959). "Obesity." *In* "Diseases of Metabolism" (G. G. Duncan, ed.), p. 529. Saunders, London and Philadelphia.
Symposium (1963). *Proc. Nutr. Soc.* **22**, 79.
Talbot, N. B. (1963). *Pediatrics, Springfield* **31**, 909.
Tanner, J. M. (1955), "Growth at Adolescence." Blackwell, Oxford.
Tonge, C. H., and McCance, R. A. (1965). *Brit. J. Nutr.* **19**, 361.
Trowell, H. C., Davies, J. N. P., and Dean, R. F. A. (1954). "Kwashiorkor." Arnold, London.
Underwood, E. J. (1962). "Trace Elements in Human and Animal Nutrition", 2nd Ed. Academic Press Inc., New York and London.
Ussing, H. H., Kruhøffer, P., Hess-Thaysen, J., and Thorn, N. A. (1960). "The alkali metal ions in biology." "Handbuch der expt. Pharmakol", Vol. 13 (A. Eichler and A. Farah, ed.). Springer, Berlin.
Valez, H., Ghitis, J., Pradilla, A., and Vitale, J. J. (1963). *Amer. J. clin. Nutr.* **12**, 54.
Walker, A. R. P. (1954). *Amer. J. clin. Nutr.* **2**, 265.
Watanabe, R., and Sassa, R. (1914). *Z. Biol.* **64**, 373.
Waterlow, J. C. (1948). *M.R.C. Memor. Spec. Rep. Ser.* No. 263.
Waterlow, J. C. (1962). "Factors influencing protein metabolism in the organism. Protein malnutrition and replenishment with protein in man and animals." *In* "Protein Metabolism": a Ciba symposium (F. Gross, ed.), p. 90. Springer, Berlin.
Weil, W. B., and Wallace, W. M. (1956). *Pediatrics, Springfield* **17**, 171.

Widdowson, E. M. (1962). *Proc. Nutr. Soc.* **21**, 121.
Widdowson, E. M., and McCance, R. A. (1957). *Brit. J. Nutr.* **11**, 198.
Widdowson, E. M., and McCance, R. A. (1963). *Proc. roy. Soc.* B. **158**, 329.
Wolf, A. V. (1958). "Thirst: Physiology of the urge to drink and problems of water lack." Thomas, Springfield, Illinois.
Youmans, J. B. (1957). *J. med. Ass. Ala.* **26**, 161.
Zuntz, N., and Loewy, A. (1918). *Biochem. Z.* **90**, 244.

CHAPTER 9

Water and Salt (sodium chloride) Intakes

PART I. BACKGROUND

WILLIAM S. S. LADELL

A. Introductory

It is only possible to define the "normal" water and salt content of the body if an allowance is first made for the amount of fat in it and if the age group and the state of nutrition are specified. It is impossible to define the "normal" water and salt intake, no matter what allowances are made. Even an average would mean very little; the water intake varies from the two or three cups of tea daily taken by the elderly spinster to the man who drinks five or six pints "of an evening", or farther afield to the African camel driver who wants eight or nine litres when he is in the desert, even in the cool season (Pales, 1950). Water intake is a matter of social custom, personal habits and the environmental demand. The salt intake varies just as widely, but less understandably (Bloch, 1963): in Indonesia it is reputed to be only 2 or 3 g per day; in West Africa (Nigeria) the intake of sodium chloride is certainly less than 7 g per day (Ladell, 1957). In other parts the intake is equally low and does not even always consist of sodium chloride (Porteres, 1950); yet in the climate of western Europe, which should not demand a high salt intake, it is at least twice as much, and in America it may be as high as 20 g per day (Marriott, 1950; Black, 1952).

The body can tolerate these wide variations in water intake because it behaves like a reservoir; the more water that flows in at one end the more flows out by the overflow; thus, no matter what the flow may be, the water level only varies between narrow limits, provided it is above a certain minimum. If the intake falls, the overflow diminishes; if, however, it falls too far and drops below the normal demands upon the "reservoir", the level begins to fall and then gradually "consumers" are cut off, the less important first. The body, like any other reservoir, has its leaks and losses, of which the greatest is the loss from evaporation. The first object of this chapter is to consider the nature and physiological significance both of the demands and of the "leaks". The control of these demands and the effects of failure to meet them will be discussed in detail; inevitably considerable reference will be made to work on animals, for much of what we know about water metabolism has been learned from experiments that could hever have been carried out on man.

To some extent the body also behaves like a reservoir for salt, but the factors involved are far more complex and the controls more subtle. For water there are some "consumers" that must always be satisfied, but except in special circumstances there need be no demand upon the body's stores of salt at all.

This reservoir of water and salt provides the essential environment, the *milieu intérieur* of Claude Bernard, within which all the vital functions of the body are carried on. Provided the overflows are working properly and are suitably adjusted, great variations in intakes may occur without any alteration in the total contents or any gross disturbance of function. As in a reservoir, the levels remain the same, but it is a dynamic equilibrium; the greater the inflow, the more rapid the turnover, both of water and of salt. Not only is there no such thing as "a normal salt and water intake", but, provided these are above certain minimums, which will be defined later, there is not even a physiologically desirable one.

B. Physiological Function of Water—Requirements

Of the four primitive "elements", air is the most essential, but water comes a close second. A man can muffle himself up against the cold or take refuge in the shade from the sun. A man can survive weeks without food, but only a few days without water. Water with its dissolved electrolytes is the main component of the *milieu intérieur*, and at one time it was suggested that the present composition of the body fluids represents the primeval *milieu extérieur* and reflects the composition of the ocean when man's coelenterate ancestor first closed off its coelom and included within its newly formed body space some of the sea water in which it was swimming. This attractive hypothesis, however, has now had to be discarded (Elkinton and Danowski, 1955). The composition of man's *milieu intérieur*, the extracellular fluid, is the result of evolutionary processes, with the development of the organs of active exchange and control on the one hand and adaptation of the evolution on the other of structures that discourage passive exchange or unwanted leaks, for example the change from the freely permeable frog skin to the relatively impermeable human skin, with its barrier layer deep in the stratum corneum (Tregear and Marzulli, 1961).

If the only function of the body water were to provide a stable internal environment, the body could operate on a closed water system like some atomic energy plants. But the body cannot consume its own smoke, and the excretory processes inevitably require water. In a temperate climate the supply of water for these excretory processes is the main demand upon the reservoir. The major losses are, like those in other reservoirs, by evaporation. The body needs a large reservoir, more than two-thirds of the fat-free mass of the body, because water also has a function of its own to perform. In addition to the relatively static water inside the cells, the intracellular fluid, there is also the freely moving, and in the plasma the circulating, fluid itself, which carries nutrients and oxygen to and the products of metabolism from the cells. This fluid provides a channel of communication between the brain and hypothalamus and the endocrine organs and also one from these to their target organs, a channel that can never fail during life, whatever happens to the nervous system. Water is also the medium for the body's central-heating system, distributing metabolic heat from muscles and liver to resting tissues elsewhere and keeping them functional. This working water is part of the extracellular fluid; unless this is maintained above a certain level, and the salt concentration in it is kept by the bodily controls within certain narrow limits, disaster can follow, beginning with circulatory failure secondary to diminished extracellular fluid volume and culminating in cellular death from desiccation as vital water is drawn out of the cells.

1. *Urinary losses*

Lizards and birds show a relatively small renal water loss, as they have a uricotelic protein metabolism; their final excretory product is semi-solid and hypertonic. Mammals never evolved this water-conserving mechansim, but nevertheless have highly efficient kidneys, whose main feature is their capacity to reabsorb water in considerable amounts. Some five-sixths of the water that filters through the glomerular membrances into the renal tubules is absorbed and with it the salts and metabolites that the body has need to conserve; in exchange go unwanted ions and metabolic end-products. It is these that set the lower limit to the volume of urine that must be excreted, depending upon the maximum concentrating power of the kidney. This maximum concentration varies widely among different species; the deer-mouse can concentrate its urine to a specific gravity of very nearly 1·100, whereas a man's highest is about 1·040. The urine of the little sand rat is more than three times as concentrated as sea water and fourteen times as concentrated as its plasma, whereas the most man can do is to produce urine only two-thirds the strength of sea water with the osmotic pressures having a ratio (U/P) of only 4·2 (Chew, 1951, 1961; Chew and Damman, 1961; Schmidt-Nielsen, 1962). In man the maximum renal concentrating power varies among ethnic groups; thus West Africans were unable to secrete a urine containing more than 1·0 osmole per litre, whereas some Europeans can achieve 1·4 osmoles (Kenney, 1955; Robinson, 1954). Thus if the sum of the day's metabolic end-products, mostly urea, and of the surplus electrolyte intake, amounts to 1·4 osmoles (a reasonable daily figure for a man on a mixed diet), a European need not excrete more than 1000 ml of urine, but the West African must excrete at least 1400 ml. This is the "obligatory urine volume" and is the first of the essential demands on the body water. This obligatory volume varies with the diet, being high on a high protein diet and low on a high carbohydrate one. Any volume of urine passed in excess of it represents overflow and, in a survival situation, waste.

2. *Evaporative losses*

Losses by evaporation are inevitable rather than essential and are mainly determined by the external environment. The evaporative loss by a non-sweating man is known as the insensible water loss (not to be confused with the insensible weight loss, which includes the cumulative weight difference between the inspired and expired air); it comprises the respiratory water loss and the insensible perspiration.

a. Respiratory water loss. The temperature and humidity of the expired air does not vary greatly. Osborne (1912) reported that normal expired air was saturated with water vapour at 33·9°; the dew point of alveolar air is higher, 34° (Christie and Loomis, 1933) or 35·5° (Seeley, 1940). Except under artificial conditions, inspired air never contains so much moisture. The tidal air takes up water first as it passes over the turbinates or through the mouth, and a little more is added as it flows down the pharynx and respiratory tract; by the time it reaches the lungs it is approaching the humidity of the alveolar air. On its way out through the nasopharnyx and the mouth or nose, moisture condenses on any surfaces below the dew points, so that the actual expired air may be less than 90% saturated (Burch, 1945; McCutcheon and Taylor, 1951). The increased water content of the expired relative to the inspired air is an inevitable loss; its extent depends on the atmospheric temperature and humidity, being less in hot humid conditions and more in dry cold ones. Burch and de Pasquale (1962) estimate that during the hot summer months in St Louis the respiratory water loss is about 5·0 g/m^2 body surface per hr. McCutcheon and Taylor (1951) give the loss in engineering units as 0·32 lb of water/lb of dry air breathed during the day in the desert and rather less at night. For men in a lifeboat the losses will depend upon the sea temperature, as this determines the dew point of the air in the first 5 m above the surface. Even in the hottest oceans, such as the Red Sea, this never exceeds 29° (U.K. Met. Office, personal communication, 1942), so that respiratory losses still occur.

As the saturation of the expired air increases with depth of respiration, moister alveolar air tending to be swept out, the rate of loss increases with work, and also with altitude, when the PO_2 is reduced. Brebbia, Goldman and Buskirk (1957) collected an average of 32 mg/l water from air orally expired by three men working in a sub-arctic environment; as the water content of cold air is inevitably low, most of this represented loss of body water. The greatest rate of respiratory water loss occurs with men working at altitude in cold air, especially if they breathe through their mouth, for very little condensation takes place in the mouth. In these circumstances as much as 1500 ml may be lost through the respiratory tract daily. A more conservative figure for men working in a moderately dry climate is 350 ml/day.

Under normal conditions the average respiratory water loss is about 200ml/day, and it may be calculated from the metabolic rate, as about 11% of the resting metabolic heat is lost by this route (Burch, 1945).

b. Insensible perspiration. There are considerable and unavoidable losses of water through the skin, usually known as the insensible perspiration. Water passes relatively freely through mammalian skin; even hippopotamus hide is not impermeable to water (Wright, 1964). Human skin provides relatively poor protection against water loss in a hostile environment. Victims of aircraft crashes in hot deserts rapidly become desiccated, so that their corpses become dry mummies within 48 hr if they are not promptly recovered and protected (Boulger, personal communications, 1943). Under laboratory conditions Berensen and Burch (1951) found that in air at 40°, r.h. 20%, water vapour diffused through dead skin at 1 mg per cm^2/hr; at this rate a dead man would dry out at the rate of 400 g/day. Hertzman, Randall, Peiss and Seckendorf (1952) reported insensible perspiration in live subjects from certain areas, for example, the face, at six times this rate. The passage of water through skin is mainly by simple diffusion (Brebner, Kerslake and Waddell, 1956), but according to Buettner and Odland (1957) there is also some active transport. Dirnhuber and Tregear (1960) were unable to confirm this, but the observation that the insensible perspiration may be sensitive to the antidiuretic hormone (ADH), if the effects of cutaneous vasoconstriction can be exluded, does suggest some active process (Mases, Falet, Joly and

Houdas, 1962). As most, if not all, water transport through human skin is by passive diffusion, the direction of flow must be determined by the vapour pressure or osmotic gradient across the skin, and in certain circumstances uptake rather than loss of water might be foreseen. This was first suspected by Whitehouse, Hancock and Haldane (1932) and has now been demonstated by Buettner (1959a, b). He finds that water may be taken up not only from air or free water but also from salt solutions. The neutral r.h., the point at which diffusion outwards becomes diffusion inwards, is 86%, and the skin will take up water from solutions up to 2-molar. Newling (1957) and Jacobi (1958) doubted whether water taken up in this way should ever be included in the water balance of the body, as they believed that the retained water was simply absorbed into the epidermis, where it remained. Even if this were so, it would still be useful for cooling purposes, but Pinson and Langham (1957) showed that radioactive water, HTO, taken up by the skin may subsequently be found in the bladder, and from the dynamics of the process they calculated that the diffusional resistance to HTO was the same as for H_2O; hence water coming through the skin enters the main reservoir of the body. As the insensible perspiration is a process of diffusion, it is not associated with any mineral loss; earlier observations to the contrary (Freyberg and Grant, 1937; McCance, 1938) were probably due to small amounts of epidermal detritus on the skin.

Mole (1948) calculated that insensible perspiration would not cease until the skin surface was at 100% r.h. The insensible loss might therefore continue even in the presence of relatively profuse sweating if the rate of evaporation was sufficient to prevent 100% wetted area (Winslow and Herrington, 1949). Webb, Garlington and Schwarz (1957) concluded that insensible perspiration would stop when the vapour pressure gradient fell to 23 mm Hg, compared with Buettner's (1953) figure of 21 mm. Buettner's demonstration that water diffuses in when the r.h. is more than 90% in normal individuals, or even only 80% in pregnant women and in certain diseases, suggests, however, that some water economy might occur during heavy sweating. The well-known rise in the electrolyte content of the sweat during profuse sweating could be due to this back diffusion as much as to the take-up of sweat water by the epidermis, as suggested by Weiner and van Heyningen (1952). As the rate of diffusion is dependent upon the vapour pressure gradient, the insensible perspiration is probably considerable in cold air, where the water vapour content is low even at high r.h. The "drying effects" of the wind have often been commented upon by Antarctic travellers; drying of the skin more rapidly than water can diffuse through it could be a factor in the development of the skin cracks often complained of by them and by climbers at high altitudes.

It is no more possible to give a firm figure for the insensible perspiration than it is for the respiratory water loss. For both the lower the atmospheric water-vapour pressure the greater the loss. Idachi and Ito, quoted by Newburgh and Johnston (1942), suggested that the sum of respiratory and skin losses is constant, a decrease in the one, whatever its cause, being counterbalanced by an increase in the other. However, this is intrinsically improbable and has never been confirmed. These two items are nevertheless sometimes treated as a single component in the water balance. The insensible water loss of a non-sweating man in a temperate climate is 900 to 1000 ml/day (Marriott, 1950; author, personal observations). Newburgh, Johnston, Lashmet and Sheldon (1937) concluded that 25% of the basal metabolic heat was lost by evaporation of the insensible water. Loss of insensible water increases with skin temperature (Pinson, 1942), and changes in the insensible weight loss with different environmental conditions have been studied by Hall and Klemm (1963). As the air temperature rose with constant aqueous vapour pressure from 22·0 to 26·7°, the insensible weight loss of clothed men only increased from 21·0 to 26·5 g/m^2 per hr; at this point the mean skin temperature reached 34·6°, there was then an abrupt rise in evaporation, and at 32·2° (air temp) the loss was 64 g/m^2 per hr. A similar inflection in the amount of water and hence of heat lost by evaporation was seen in subjects in the Russell Sage calorimeter (DuBois, 1948). Altitude also affects insensible weight loss; Hale, Westland and Taylor (1958) reported that, as the barometric pressure fell, the insensitive weight loss rose 12 mg/m^2 per hr per mm Hg.

3. *Gut losses*

a. Faecal losses. Formed stools on a normal diet account for about 100 ml/day (Elkinton and Danowski, 1955). When little water is drunk and there is a dry low bulking diet, few faeces are formed; only a few small "pills" totalling 10–20 g in all, are passed every two or three days; in extreme water deficiency the bowels cease to work altogether (McGee, 1906). On the other hand, water losses from the bowel may be great during diarrhoea. If this is infectious in origin, for example in cholera, the daily loss may be 7 l. Even from excessive purging, losses may reach 3 l/day (Marriott, 1950). Water is not the only loss in diarrhoea; there are also electrolytes, especially potassium, whose losses are reflected in alterations of plasma level (Elkinton and Danowski, 1955).

b. Vomiting. The potential losses by vomiting are considerable, but vomiting is not a normal body function, so that no predictions of its effect on loss either of water or of electrolytes can be made, but uncontrollable seasickness can be an important factor mitigating against the survival of men at sea in lifeboats.

4. *Eccrine sweat*

In hot conditions or during hard work in any climate there is one other call on the water of the body, namely the eccrine sweat. Here is another example of the use of water for a physiologically essential purpose. Water is actively secreted on to the skin surface, where it can evaporate and cool the organism. Although the sweat glands do not secrete pure water, the fluid is always hypotonic and in some men may contain less than 20 meq/l. The factors affecting sweat secretion are considered in detail in chapter 3 of this book; briefly, sweating only begins when the body is unable to lose all its metabolic heat or the irradiated heat gained from the environment by direct physical means. In a resting semi-nude individual sweating does not begin until the dry-bulb temperature is above 24° or 25°; with increasing work loads and heavier clothing, sweating occurs at lower and lower ambient temperatures. A cause contributory to the failure of the early Everest climbers may have been that they had not allowed for sweat as well as for respiratory water loss. Sweating does not begin until the skin temperature reaches a certain critical value, but this depends on many factors and may vary from only 31° to as high as 34·5°. Robinson (in Newburgh, 1949, p. 338). The amount of water lost as sweat cannot be predicted without knowledge of the environmental conditions, the work rate and the clothing and state of acclimatisation of the individual. Even under precisely the same conditions two similarly acclimatised men wearing the same clothing may have widely different sweat rates (Ladell, 1951a). The only limit to the speed with which sweat can be secreted is probably the rate at which blood can be supplied to the skin. In Iraq steady losses of 0·5 l/hr going on for many hours were observed (Ladell, Waterlow and Hudson, 1944). In Northern Nigeria I have recorded losses by marching men of 850 ml/hr and estimated that in certain circumstances this could rise to 1300 ml. Losses at night were only 70 ml/hr. Gosselin (in Adolph, 1947a) quotes sweat-rates for men in the desert varying from 250 ml/hr when sitting clothed at night to a maximum of 1200 ml/hr for men marching in the sun. I have myself sweated at 1 ml/sec, continuing for 30 min, but in general it is doubtful if rates more than 1200 or 1500 ml/hr could be sustained for long periods.

5. *Total water losses*

Enough has been said to show how misleading a guide can be the simple oft-quoted table

Urine	1500 ml
Respiratory loss	400 ml
Insensible perspiration	600 ml
Faeces	100 ml
Total per day	2600 ml

In practice the essential urine volume may be as low as 250 ml on a suitable diet, but a more usual figure is 700 ml. With quiet breathing, and provided the inspired air is not too dry, the respiratory loss may be cut to 200 ml, and the insensible perspiration may be as low as 500 ml in a moist climate. Thus, if the water lost in the stools is not considered, a man can manage with less than 1000 ml of water a day. Any increase in urine volume above 700 ml represents "overflow", but in cold climates and with hard work the respiratory loss may amount to 1·5 l/day, and in similar dry conditions the skin loss might also be doubled. If there is a little looseness of the bowels or profuse vomiting, a man could consume 4 or 5 l/day and still not be in water balance. The only way for a man to make sure he is in water balance is to drink sufficient to ensure that he is passing at least 750 or 1000 ml of urine a day; provided he is on a proper mixed diet (Sargent and Johnson, 1956) he then does not need to worry about his water intake.

C. Function of Salt—Salt Loss

Sodium chloride is the chief electrolyte of the extracellular fluid and blood plasma, but potassium salts are equally important, as potassium is the main intracellular inorganic cation. The osmotic pressure of the body fluids is governed mainly by the sodium salts outside the cell and the potassium salts within, the nature of the anion being less important; if necessary the body can provide bicarbonate ions, although at the expense of the acid-base balance. As neither sodium nor potassium salts enter directly into any metabolic processes, there is no obvious reason why the body's stocks of them should be depleted. In the process, however, of the autoregulation of the three factors controlling the circulating body fluids, the osmotic pressure, the electrolyte content and its volume, there are extensive electrolyte losses through the kidneys. These three factors are controlled by two, if not three, servo-mechanisms (Elkinton, 1960), which closely interact with each other.

1. *Osmotic pressure*

The osmotic pressure of the circulating body fluids is monitored by osmoreceptors in the supra-optic region of the hypothalamus. When these are stimulated by a rise in plasma osmotic pressure, there is a reflex release of antidiuretic hormone (ADH) (Verney, 1947). This hormone increases tubular reabsorption of water, but it does not affect solute excretion in man (O'Connor, 1962), who differs from other animals in this respect (Brooks and Pickford, 1956; Cross, Dicker, Kitchen, Lloyd and Pickford, 1960). In normal circumstances the plasma is never free from antidiuretic activity (Dicker and Tyler, 1952, 1953), but as the plasma osmotic pressure falls after drinking, so does ADH release. With a drop of 0·5%, which occurs on drinking 5 ml/kg, the rate of release of ADH is no longer high enough to keep up with the normal rate of destruction (O'Connor, 1962), the renal tubules are released from its restraining influence, and water diuresis begins. During a water diuresis, particularly after the infusion of hypotonic saline, there is increased sodium excretion; this is because there is both an increase in glomerular filtration rate and a decrease in tubular reabsorption in the proximal tubule (Burg, Papper and Rosenbaum, 1961). Changes in the extracellular osmotic pressure are succeeded by exchanges of both water and electrolytes across the cell membranes to bring the pressures within and without the cells into equilibrium. When the extracellular osmotic pressure rises, potassium comes out of the cells, possibly in exchange for sodium that passes into the cells with water; this potassium is promptly excreted by the kidneys and so is lost to the body (Elkinton and Winkler, 1944b; Elkinton and Danowski, 1948). This occurs in dehydration. Various explanations for this effect have been advanced.

2. *Circulatory and extracellular fluid volumes*

Although, as O'Connor (1962) points out, the idea of this control is still only a hypothesis, supported by what he considers insufficient and unsound evidence, teleologically it is most attractive, and the available evidence all points to the existence of such a control. Peters (1935) in his classic, "Body Water", first pointed out that "fullness of the blood stream may provoke the diuretic response on the part of the kidney". Attention was paid from 1948 onwards to the effect on renal function of iso-osmotic alterations in blood volume (Gauer and Henry, 1963); the corollary to Peters's dictum, that decrease in volume provoked anti-diuresis, was reported by Cort (1954). The correction of volume reduction, though not of expansion, comes before correction of osmotic pressure changes (Baratz and Ingraham, 1960). As Leaf and Frazier (1961) point out, this is teleologically desirable, "dilutional hyponatraemia is a lesser evil than circulatory collapse"; integrity of the circulation is more important for survival than constancy of osmotic pressure. Henry and Pearce (1956) showed that stretch receptors in the left atrium were probably involved in volume control. It was later found that afferent impulses pass up the vagus from these and control the release of ADH. There may also be receptors for the same reflex in the carotid sinus (Share and Levy, 1962). The left atrial reflex not only affects ADH release, but also influences the secretion of the salt-retaining hormone, aldosterone, changes being seen in the plasma level within 5 min of altering the circulating fluid volume in unanaesthetised dogs (Gann, Mills, Cruz, Casper and Bartter, 1960; Bojesen and Degn, 1961). Temporary constriction of the thoracic vena cava is also a potent and rapidly acting stimulus to aldosterone output in conscious sheep (Blair-West and Goding, 1962).

The evidence that this work on animals applies to man is mainly derived from observations on the effect of changes in posture (*Lancet*, 1961). From the changes in haemoglobin and haematocrit levels Widdowson and McCance

(1950) deduced that the plasma volume increased with a few hours recumbency, and antidiuresis occurs with quiet standing (Epstein, Goodyer, Laurason and Relman, 1951; Epstein, Kleeman, Lamdin and Rubini, 1956). Both sodium and water excretion decreased on standing up and increased on lying down (Goodyer and Seldin, 1953; Thomas, 1957). Diuretic responses, for example, to the injection of isotonic saline are also greater in recumbent than in erect subjects (Coxon, Dupré and Robinson, 1958). According to Camp, Tate, Fla, Lowrance, Wood and Va (1958), this effect of posture is the result of changes in ADH output brought about through the atrial stretch receptors, the volume being functionally less in the erect than in the recumbent posture. Posture effects may, however, also be seen in hydropenia, when ADH secretion is at its maximum (Hulet and Richardson, 1962), but Camp and Va (1958) found indications that vasoconstrictor substances released on standing might also be involved. For further details, see the review by Gauer and Henry (1963).

The control of volume changes involves variations in level, both of ADH, which has an indirect effect on electrolyte losses, and of aldosterone, which is directly concerned with sodium conservation and is as much responsible for alterations in the urinary electrolyte excretion as is the control of osmotic pressure. Inasmuch, therefore, as changes in posture that affect volume and pressure may vary from day to day, neither the postural changes nor electrolyte losses can be predicted (see also earlier review by Wrong, 1957).

3. *Electrolyte content of body fluid*

The electrolyte concentration of the extracellular fluid is determined by the osmotic pressure control, which ensures that water is retained when osmotic pressure gets too high, and by the volume control, which ensures retention of electrolytes. Both work through alterations in ADH and aldosterone release. In the simplest situation water is retained by ADH when the osmotic pressure is too high, and sodium is retained by aldosterone when the volume is too high. But sometimes circumstances may arise in which both hormones are acting at the same time, complementing each other's action. This is because of conflicting factors, each of which can elicit aldosterone release. The secretion of this hormone, which is the main hormone involved in sodium, potassium and chloride metabolism (Gaunt, Renzi and Chart, 1955), is under reflex control from the atrial stretch receptors, but in vitro experiments have also shown that a reduction in the Na:K ratio of the fluid perfusing the adrenal gland increases its aldosterone output (Rosenfeld, Rosenberg, Ungar and Dorfman, 1956), which suggests that there may also be some local mechanisms involved. Aldosterone decreases sodium excretion relative to potassium not only in the urine, but also in the sweat and saliva (Conn, 1963).

Rises in the aldosterone output of men on exposure to heat occur before there are any detectable changes in the plasma sodium concentration (Streeten, Conn, Louis, Fajans, Seltzer, Johnson, Gittler and Dube, 1960). This would appear to be an example of what one may call "biological pre-cognition". Kenney (1963) suggests that when a man enters a hot room his urine output diminishes before there is any detectable ADH output. In both events the body anticipates a loss, and appropriate action is taken in advance before the endocrine messenger arrives. Another example of biological pre-cognition is provided by Denton and Sabine (1961), whose parotid fistula sheep always managed to drink enough sodium bicarbonate solution to correct their balance, no matter whether the solution was presented for a short or for a long time, apparently adjusting their drinking rate in advance. Denton suggests that the sheep knows when to stop because it may monitor volume by its oesophagus and concentration by its tongue, but he still leaves unexplained how the sheep knows when to drink fast and when to drink slowly. If there really were such a phenomenon as biological precognition, it would be of immense importance in the whole field of survival, but what is more probable is that there are certain centres, probably in the hypothalamus, that receive warning signals from the cortex and then initiate protective auto-nomic reactions. In Conn's and Kenney's subjects the signals no doubt were the result of their knowing they were about to be exposed to heat; the sheep may have been warned from some small change in the pattern of the experiment that the drinking vessels were about to be removed. Water and salt metabolism is certainly sensitive to cerebral influences; Hulet, Shapiro, Schwa-rcz and Smith (1963) induced a water diuresis in hydropenic women by sugges-tion to them while they were under hypnosis that they were drinking water.

Aldosterone secretion is continuous; it can be detected in the urine of normal men, as can ADH (Leutscher, Johnson, Dowdy, Harvey, Lew and Poo, 1954). The output rises on changing from the recumbent to the erect posture (Gowerstock, Mills and Thomas, 1958), and it is also affected by changes in the sodium chloride intake (Rosnagle and Farrell, 1956). Man's capacity to maintain sodium equilibrium in a hot environment at widely different salt intakes and the changes in the electrolyte content of sweat during acclimatisation to heat are both probably due to alterations in aldos-terone output in response to changes in the salt intake (Ladell, 1945a, b; Streeten *et al.*, 1960). These changes may take place without a negative sodium balance being incurred; thus the chloride content of the sweat could be altered at will by raising or lowering the daily salt intake, although the subject remained in positive salt balance throughout, excreting chloride in the urine (Ladell, 1947b). A further factor is that aldosterone needs the supporting action (Ingle, 1954) of cortisone to maintain the diurnal rhythm of electrolyte secretion while the subject is up and active, but not when he is lying down (Muller, Manning and Riondel, 1958).

It is apparent from what has been written above that it is impossible to predict what the electrolyte needs may be. The controls are so good, however, that balance can be established at any level of intake over a wide range. Both the salt-starved but sweating Indonesian and the salt-replete sedentary American executive are taking exactly as much salt as will keep them in balance. Before this was realised high salt intakes, up to 40 g/day, were sometimes recommended for men in the tropics, probably contributing to rather than preventing their breakdown in the heat (Ladell, 1957). Regrettably this tendency to "oversalt" is not yet dead.

4. *Anion control*

The most important dietary anion is chloride. This is the main anion in the extracellular fluid, and it penetrates only sparingly into the cells. The osmolar chloride concentration is less than that of the sodium, but the difference is made up by other anions, including sulphate, phosphate, bicarbonate and lactate produced in the course of metabolism. The intracellular anions are mainly phosphate and sulphate, with metabolic acids and protein. As chloride is the electrometric filler, chloride excretion is usually less than that of sodium on a molar basis, the metabolic ions being preferentially excreted, according to the demands of the acid-base balance, along with any excess sulphate or phosphate.

D. Man's Normal Sources of Water and Salt

In addition to what he drinks, a man probably has a minimum daily intake of 1 l water in his food and also benefits from the water of oxidation: each 100 g metabolised carbohydrate gives over 55 g of water, protein 40 g and fat 107 g. Despite the high water equivalent of fat, it is not advantageous to metabolise fat rather than anything else in water deficiency; Mellanby (1942) pointed out that the water gain should be considered in terms of calories produced rather than grams metabolised. For each 1000 kcals metabolised when the source is carbohydrate, 130 ml of water are produced; if the calorifically equivalent amount of fat is metabolised, the amount is only 120 ml. Stored fat is therefore a food rather than a water store, yielding approximately 100% more calories and metabolic water than the same weight of carbohydrate. The yield from an ordinary mixed diet is between 120 and 130 g of water per 1000 kcals. An additional disadvantage of metabolising fat (pointed out by Mellanby) is that more oxygen is required for fat than for carbohydrate or protein; hence there is greater ventilation and the respiratory water loss may be increased.

Gamble (1944) suggested that a fasting man can also benefit from preformed water, that is, the water from tissues that are being metabolised. This, together with the water of oxidation, is enough to cover "a large part of total water outgo". This presupposes that water-containing tissue is metabolised, but during a fast the first to go is carbohydrate, stored within cells as glycogen granules and not dissolved in water, and then the fat stores are drawn upon; active tissue is initially spared. Even a thin man has 6 to 7% fat, and the normal person has 15 to 20% (McCance and Widdowson, 1951); the fat stores are therefore unlikely to be entirely exhausted from less than one week's fasting (see chapter 8). Fat tissue, moreover, contains little water. Hence in the early days of a fast, little preformed water is available; only when active cellular tissue begins to be metabolised will water be freed. Unless, therefore, a man survives at least longer than a week, he will not benefit from this source, but when he does the benefit is considerable; for every 100 g of fat-free tissue 73 g of water would be released, according to the body composition reported by McCance and Widdowson (1951). It is probable that, with prolonged periods on a calorie-deficient diet and little water, the apparent "adaptation to water loss" indicated by a decreased rate of weight loss (Ladell, 1943, and personal observations) is due to the switch from metabolising water-free fat-full tissue to water-rich lean tissue.

Man gets his salt partly as a natural component of his food and partly from added seasoning. The natural salts of food are preponderantly those of potassium; thus in 100 g of raw beef steak, there are only 69 mg of sodium, but 334 mg of postasium. There is an even greater discrepancy in vegetables; for example, in potatoes and peas the Na:K ratios are 11·4:1000 and 1·5:1000; the discrepancy is less gross in sea food, the ratio in the herring being 406:1000 Only in prepared foods, such as bread and biscuits, does the sodium content equal or surpass that of potassium (McCance and Widdowson, 1939). To a large extent, however, man's salt appetite is an acquired taste; the great trade routes of the world across Asia and through Africa were largely salt routes (Bloch, 1963; Pales, 1950). The salt of choice, the luxury article, has always been sodium chloride, but sometimes potassium takes the place of sodium, or the anion may be sulphate or carbonate (Porteres, 1950; Soula, 1950). This salt hunger may be part of the adaptation of the original tropical man to temperate condition (Macpherson, 1958; Ladell, 1963), but it has also been suggested that man added salt to the nourishing but unpalatable pap he turned to when he began to till the soil rather than hunt (Bloch, 1963, and personal communication). Salt balance may be maintained with a daily salt intake of less than 17 meq, not only under temperate conditons (Elkinton and Danowski, 1955), but even by a heavily sweating man provided he is properly adapted (Streeten *et al.*, 1960). A high salt intake is in theory physiologically unnecessary; it could therefore be considered as the result of habit or even addiction, possibly the first of mankind's major ones.

The daily intake of potassium, especially by vegetarians, is always more than enough to cover the accidental losses due to escape of fluid from inside the cells and its loss through the kidneys. Even in potassium deficiency some is lost in the urine, especially if there is an associated sodium deficiency (Peters, 1953). Such is the ubiquity of these two ions, however, that a simple deficiency in either potassium or sodium is a most unlikely danger. On the other hand, few men are as well adapted as was the subject of the experiments by Streeten *et al.* (1960), so that salt deficiency can and does occur in heavily sweating men.

PART II. THE SURVIVAL PROBLEM

A. Water Deficiency and Salt Excess

It is difficult to set a value on the normal let alone the physiologically correct amount of water in a man's body. If, for example, the weight on rising is taken as the "true weight" (and hence an index of water content), this value is immediately suspect, as respiratory and skin losses will have been taking place without replacement since the last drink the night before. The water content of the body is always changing. A man takes a drink and has a diuresis; he plays a hard game of tennis and sweats; he climbs a mountain and puffs an extra litre of water out in his breath. There are also seasonal changes, the total body water rising in summer (Yoshimura, 1958), and responses to extra stress, such as the fall in body weight, indicating water loss, during extremely hot weather (Ladell, Waterlow and Hudson, 1944). Even apart from the considerable amounts of water in the gut, which does not participate in the osmotic balance of the body, there are some 2 l of "free circulating water"; only when this has been lost are gross changes seen in bodily functions, such as sweat, salivation or urine flow (Ladell, 1955).

The seasonal changes can be associated with the increased ADH of the plasma in summer (Yoshimura, 1958); this may represent long-term over-hydration, with a corresponding increase in the volume of free circulating water. The falls seen in men living in severe heat are due to readjustment of the body water distribution after some electrolyte loss. Whether the extra water retained in the summer swells the extracellular fluid as a whole, or only the circulating plasma volume, is not certain. Although Dill's (1938) view that the blood volume increases during acclimatisation to heat has been questioned by Bass, Buskirk, Iampietro and Magger (1958), there is little doubt that some increase occurs (Glickman, Hick, Keeton and Montgomery, 1941; Doupe, Ferguson and Hildes, 1957); this could, however, be the result of redistribution of blood previously sequestered in the liver or lungs (Glaser, 1949; Glaser, Berridge and Prior, 1950).

The absence of gross physiological effects in a man from a temperate climate when he refrains from drinking until he has lost 2 l of body water (Ladell, 1947a, 1955) suggests either that most men are habitually over-hydrated or that each man has a fairly wide range of normal hydration. Failure by the kidneys to conserve water may result in more fluid being lost in the urine than by the insensible routes (Elkinton and Danowski, 1955); hence, it might appear at first sight that a good physiological indication of the beginning of water deficiency could be the appearance of antidiuretic activity in the blood, as the body's first step towards survival in the face of water shortage. However, vasopressin, the specific ADH, is found in the urine of men under normal conditions, although it cannot be detected in their blood. Moreover, ADH release is stimulated not by less water in the body as a whole but by a reduced mole fraction of water in the plasma reaching the osmoreceptors (Reeve, Allen and Robert, 1960). The close interaction between ADH, concerned with osmotic pressure of the plasma, and the adrenocorticoids, concerned with the electrolyte content of the extracellular fluid, means that normal water content and normal electrolyte content of the body are closely interdependent (Leaf, 1960; Ganong and Forsham, 1960). It is not possible to define one without the other, and the probability is that each man finds his own norms according to his habits and opportunity.

1. *Types of dehydration*

a. Deficiency dehydration: water. Despite the difficulty of defining the norm, degrees of water deficiency are usually expressed as percentages of body weight; probably this accounts for some of the discrepancies between the values of limiting water losses for various physiological effects, as published by different workers. Three distinct stages of water deficiency may be recognised: in the first the body is still drawing on its free circulating water; only when this is exhausted are various functions disturbed and the urine flow is at its minimum, about 0·5 ml/min; this usually occurs with a water loss of between 2 and 3 l (personal observations), but Adolph (1947a, b, d) sets the limit at less than 1 l, only 1% of body weight. The stage of real water deficiency now begins; there is steady though not clinically significant deterioration as the water loss increases, but a man remains active and fully capable of physical and mental activity until he has lost about 8 to 10% of his body weight. The third stage now begins, and deterioration becomes more rapid; death occurs when between 15 to 25% of the total body water has been lost.

The symptoms that develop with water deficiency depend on whether the main water loss has been from the extracellular fluid or from within the cells. This in turn depends on accompanying salt losses, which themselves depend on the route and manner of the original water loss. If a man loses water exclusively from his lungs and by insensible perspiration, the only electrolyte loss is through the kidneys. A true "water deficiency dehydration" develops, and the water loss is eventually borne evenly by both intracellular and extracellular fluid spaces; the loss is originally extracellular, but as the osmotic pressure of the depleted fluid rises there is an osmotic flow of water out of the cells which maintains osmotic equilibrium between the two compartments; the circulatory fluid volume is only affected when the dehydration is extreme.

b. Deficiency dehydration: salt. Dehydration can also occur as the result not of water but of salt depletion. The tendency for the osmotic pressure of the extracellular fluid to fall is then counterbalanced by the urinary excretion of the functionally excess water containing little or no electrolytes; the intracellular fluid is not affected (McCance and Widdowson, 1937). This steady drain on the extracellular fluid ends in the circulatory collapse typical of "salt deficiency dehydration". In practice both types of dehydration usually exist together; for example, a sweating man is losing both water and salt. The dehydration developing in these circumstances is a "mixed dehydration" (Nadal, Pederson and Maddock, 1941). The symptoms that predominate will depend on the relative degrees of water and salt loss. In general, the greater the salt loss the more severe the symptoms, but there is nearly always circulatory inadequacy or even circulatory shock (Leithead, Leithead and Lee, 1958).

c. Voluntary dehydration. Voluntary dehydration (Adolph, 1947a) develops in men in hot climates who fail to drink enough during the day to replace their sweat losses; they utilise their free circulating water and by the end of the day may be as much as 2 l in negative balance. Even though the opportunity is there, a man may fail to replace his losses fully when the evening comes and he stops work; he therefore begins the next day with a little less than his previous day's total body water. This may continue day after day until frank dehydration develops, leading to water deficiency exhaustion (Ladell, 1957). Voluntary dehydration occurs because thirst is an inadequate stimulus to drinking. To keep in water balance, a heavily sweating man may have to drink 10 or 11 l per day. This is a task that tends to be ignored unless a man is made aware of its importance; then, unwittingly, a man may sacrifice his free circulating water. The degree to which he does this can be monitored by watching his body weight; a steady slow fall suggests that he is not drinking enough. He may be adopting a new equilibrium, but is doing so at the expense of his reserves.

2. *Symptoms of water loss*

One of the best-known descriptions of water lack is McGee's (1906) account of the Mexican, Pablo, who was lost in the Arizona desert for eight days; his body weight decreased by 25%, but he made a complete recovery. Another classic account is by King (1878), of a detachment of U.S. cavalry who were lost in the Staked Plains of Texas for 86 hr. The two accounts differ only in minor detail. Mouths became dry, vision and hearing began to fail, they became feeble, and hands and fingers shrivelled. The cavalrymen's feet swelled, and they were unable to sleep. Symptoms developed more slowly in Pablo, and he was able to struggle over 100 miles in temperatures sometimes above 38° and r.h. often below 20%, mostly under nearly cloudless conditions. When he was picked up "his joints and bones stood out . . . the skin clung to them . . . suggesting shrunken rawhide used in repairing a broken wheel . . . his tongue was shrunken to a mere batch of black integument". Both Pablo and the rescued cavalrymen made full recoveries when given water, but there have been no recorded experiments on man in which such severe dehydration has been incurred. In the long desert hikes described by Brown (see Adolph 1947a), losses of body weight up to 7% occurred within a few hours, but gross changes, such as drying of the mouth and shrinking of the skin and superficial tissues, were not seen. Pablo's skin and surface tissues were so shrivelled that cuts would not bleed. Similar appearances have been reported in survivors in open boats after many days in tropical seas (Critchley, 1943). It takes not merely water-loss but time, and probably sun, to achieve the real Ancient Mariner desiccated look. In trials in which I lived without water on a completely dry low-calorie diet for three or four days, I sometimes lost 10% of my body weight, but I never approached the state described in Pablo or the cavalrymen, nor did Black, McCance and Young's (1944) subjects, who lost 3·5 l in 3 to 4 days.

It has been suggested that the different clinical pictures of dehydration shown by the man in a laboratory experiment and the man lost in the desert could be the difference between a nearly pure water deficiency and a mixed dehydration. This is not necessarily so. Even when salt loss is deliberately incurred, as by exercising in a hot room or by abstaining from water in a hot climate, there are still differences between the experimental and field effects. Men who had worked to impending exhaustion at 35·5° dry-bulb and 33·8° wet-bulb, without drinking, lost up to 4 l of body water and 10 g of sodium chloride in less than 2 hrs and looked drawn but not parched (Ladell, 1949a, 1955); a possible cause for this could be the effect of unevaporated sweat on the skin. Men in a lifeboat are in a humid environment; if it is hot they may sweat, but the sun dries their skins just as effectively as if they were in the desert. Sweating begins to fall off when the free circulating water is exhausted, but whether it ever ceases altogether in an otherwise normal man whose body temperature is not unduly raised is not known; if sweating did cease, this would hasten the drying out of the skin. In the desert the skin is probably always dry. Stott (1936) relates the experience of men who had to make a forced landing in the Arabian desert, when the temperature rose to 54·5° after the sun was up. Nine hours later their skins were dry with no elasticity, their eyes shrunken into the sockets and their cheeks hollow. Here, too, there was certainly salt loss, and the beginning of a "Pablo syndrome". Stott's experience may be compared with the results of an unreported test I carried out in Iraq; after 24 hrs on a low (930) calorie diet containing only 1 g NaCl I stopped drinking. The temperature at the beginning of the "thirst" was 34·4° dry-bulb and 19·7° wet-bulb; it got steadily hotter until in the early afternoon when the dry-bulb temperature was 43°3°. Temperatures then began to fall, but when the test was stopped after 12 hrs it was still 41·7° dry-bulb and 22·8° wet-bulb. In the 12 hrs without water I lost 3860 g, with an estimated salt loss of 7 g, and then looked "extremely dehydrated, with sunken eyes, scaphoid abdomen and a hot dry markedly inelastic skin" and showed signs of circulatory failure. Clinically I was in worse condition than Brown's desert hikers, probably because of the self-imposed salt deficiency. My skin did not dry out, but the test was conducted inside a laboratory; however, possibly if I had been exposed to the sun, it would have dried out. The implication is that three factors are concerned in the development of the "field" picture: a high salt loss, a water loss of about 4 l or 6% of the body weight and complete evaporation of the sweat.

Men in lifeboats become salt deficient even if they do not sweat. Until the sodium and chloride concentrations in the plasma fall below certain values, there is no stimulus to the secretion of aldosterone and renal tubular reabsorption of sodium. Sodium chloride losses in the urine therefore continue, and a negative sodium balance develops; this can amount to between 2 g and 3 g per day and may still continue for ten or eleven days of low water intake (Hervey and McCance, 1952). A non-sweating man loses nearly 1·3 l of water daily, through skin, urine and lungs; hence he is losing, effectively, about 0·25% saline. The sodium chloride content of sweat is between 0·2 and 0·3%

(Robinson and Robinson, 1954); thus the relative losses are the same in the non-sweating as in the sweating man, for the urinary salt losses fall to low values in sweating men who are not given salt (Ladell, 1949a). The man who has lost 10% of his body weight slowly, lying in the bottom of a lifeboat, will therefore be just as salt-deficient as the Mexican who has lost the same amount in the desert; hence the same syndrome occurs in both. There are two factors that may prevent the laboratory subject going the same way. First there is the small intake of salt in the rations; 1 g per day may be just sufficient to keep the cumulative negative salt balance from developing too rapidly (Wolf, 1945a, b): second, in the laboratory there is no mental anxiety, so that effects secondary to emotion are slight, if not absent. Of the two factors the second may be the more important; the lifeboat diet contains as much, or as little, salt as the laboratory one.

There is certainly increased aldosterone secretion during water restriction in normal men (Crabbe, 1961). This is the standard response to changes in the extracellular volume. Neither Black, McCance and Young (1944) nor Hervey and McCance (1952) report evidence of adrenal cortical activity in their experiments. Grande, Anderson and Taylor (1957) found that nitrogen excretion increased during dehydration, and the lower the water intake the greater this was; there is also an increased level of 17-hydroxycorticosteroids in the plasma (Huseby, Reed and Smith, 1959); this presumably is what stimulates protein metabolism, the extra urea giving rise to osmotic diuresis and so increasing the water loss. This adrenal response, according to these authors, "does not help to conserve water." If this occurs in the mild excitement of a laboratory investigation, how much greater would be effects induced by emotion in a real thirst situation? No laboratory trial, no matter how stringently conducted and to what lengths, could ever produce the same personal anxiety and despair as does being lost in the desert without water and no hope of rescue; the "guinea pig" knows that he can always walk to the tap and take a drink or can withdraw from the experiment. Part at least, therefore, of the Pablo picture may be due to excess adrenal activity. Only in trials in the desert have the subjects been in circumstances even approaching true "desert thirst" (McGee, 1906), but their clinical condition has never deteriorated so far, despite heavy water losses. The subjects in Brown's desert trials had no real cause for worry; they were carefully supervised and allowed to drop out and be restored as soon as they wished. Lemaire (1952-3) carried out trials in the French West African desert, but it was not so severe a physiological test, since each man was allowed 2·5 l water for three days. They lost from 2·8 to 5·7 l each, but the only comment on their appearance was that they "presented the clinical signs of dehydration".

Pablo must have suffered both intracellular and extracellular dehydration, though his circulation remained good enough for him to be able to "struggle forward". In my experience, there is an urge to keep moving when severely dehydrated; one feels better. The ability to keep going implies a reasonable circulating fluid volume; this is consistent with a high mineralo-corticoid activity, ensuring retention of sodium and an increased extracellular osmotic pressure. The extent of this retention can be calculated; thus the relatively unstressed subjects of Black *et al.* (1944) retained 702 m osmoles in three days. In my Iraq trial the functional retention was proportionately greater: the plasma chloride and sodium concentrations rose from 100·6 to 106·2 meq/l and from 133·7 to 157·5 meq/l, respectively, and the extracellular fluid volume (taken as equal to the thiocyanate space) fell from 14·05 l to 12·4 l. There was thus a drop in the extracellular sodium and chloride content from 3291·9 to 3269·9 m osmoles; there was also an intake of about 50 m osmoles NaCl. I lost 3·31 of sweat, of mean NaCl concentration 0·25% and 0·34 g of NaCl in the urine, a total loss of 294 m osmoles; hence, ignoring potassium, my extracellular electrolyte loss was 222 m osmoles. But there was available 1235 m osmoles from the 3·861 of body water lost (Black *et al.*, 1944), so that 1113 m osmoles must have been retained elsewhere. The minor extracellular functional osmotic retention ensures that the extracellular fluid volume is maintained, but only at the expense of water drawn out of the cells, and it is this that kills; as Winkler, Elkinton, Hopper and Hoff (1944) pointed out, "there is a lower limit beyond which intracellular dehydration cannot be tolerated; metabolism is not possible in desiccated cells."

Another factor may be the effect of stress on ADH activity. ADH is detectable in the urine of even mildly dehydrated men, especially after exposure to heat (Hellman and Weiner, 1953–4; Karvonen, Friberg and Anttila, 1955; MacFarlane and Robinson, 1957). J. Lee (1963) found up to 12 mu/hr of vasopressin in the urine of severely dehydrated men. Recently Moses (1963) has reported that methyl prednisolone and deoxycorticosterone increase the urine flow in dehydrated rats by, he believes, depressing vasopressin release in the absence of any plasma volume changes. Adrenocorticoids or other psychosomatically released hormones may do the same. Finally there is the effect of time; prolonged water deprivation in rats exhausts the antidiuretic activity of the hypophysis (Hare, Hickey and Hare, 1941); time and mental anxiety together may do the same to a man without water.

Sufficient has been said to indicate why the man under experiment in the laboratory never shows such striking clinical changes as the man in the desert or lifeboat with the same water loss. Besides water loss Pablo and the Ancient Mariner had also to face Time, Sun and Despair in their struggle for survival. The fundamental physiological changes, however, are the same as those seen in laboratory trials, described in the next section.

3. *Laboratory dehydration*

After a day or so without water a man feels lazy and listless, but he does not experience a conventional "thirst". The mouth feels dry, but there is a considerably less parched feeling than after a hard cross-country run or a long walk on a warm afternoon. Time passes slowly, but even with body-weight losses as much as 10% he experiences no more than some malaise. There is an increased awareness of cold; as this passes immediately on rehydrating, it could be due to increased circulating ADH having a vasoconstricting action in the skin. Sleep is not noticeably affected, but voluntary activity is reduced; after a week on lifeboat rations and restricted water, sailors became noticeably quieter and subdued, and they also complained of "leg weakness", of feeling "washed out" and of "black-outs". Hunger disappeared, but some subjects complained of indigestion and nausea. Sluggish tendon reflexes were reported in 9 of the 19 sailors after seven days or more of the trial, and after 10 days 12 out of 15 smokers reported they were smoking less (Ladell, 1943, 1947a, and unpublished observations). Searching exercise tolerance tests were not carried out on the sailors, but Beetham and Buskirk (1958) reported that trained men continued to retain their superiority over untrained individuals during water restriction.

a. Weight loss. The body weight falls steadily during water restriction (Fig. 1) at a rate depending on the degree of restriction (Fig. 2). In a temperate climate the weight loss of men restricted to 720 ml/day on the 400kcal/day lifeboat rations was less than 1% per day, most of which could be attributed to body wasting on the low-calorie diet. Grande, Taylor, Anderson, Buskirk and Keys (1958) required 1800 ml/day to keep their subjects in water balance, but the temperature was 25·6° and the men were exercised. With complete water deprivation the weight loss on the first day may be as much as 150 g/hr, including urine loss, but after two or three days the rate of fall may be only 60 to 75 g/hr. In my own experiments there was usually a short period of weight gain shown by all subjects; the greater the food intake the longer was this deferred, and it was associated with the appearance of ketone bodies in the urine. This suggests that the weight gain signalled the change-over from catabolism of stored carbohydrate to catabolism of fat; why this should have been accompanied by a transient weight gain was not determined. The change in pH consequent on the mild ketosis might lead to the release of protein-bound water, but the metabolism of fatty tissue releases little preformed water; the effect of ketosis and catabolism would be to slow up the weight loss, not reverse it. The most likely explanation is the retention of oxygen in a fat-carbohydrate conversion.

Weight loss is much more rapid in a hot climate and may on occasions be as much as 2 l/hr, the maximum recorded sustained sweat rate (Robinson, 1952; Newburgh, 1949; Adolph, 1947a; Ladell and Kenney, 1955, and personal observations).

b. Urine flow. The initial urine flow depends upon the subject's immediate drinking pattern; when a man stops drinking his "normal" urine flow falls rapidly at first, then slowly, until his free circulating water has been exhausted.

This takes 12 to 24 hrs, depending on circumstances; the urine flow is then at its minimum obligatory level, determined by the diet, which is only 4 to 5 ml/hr on a carbohydrate diet, 20 to 25 ml/hr on a protein diet (such as pemmican) or if fasting (Gamble, 1944 and personal observations). Even small intakes of

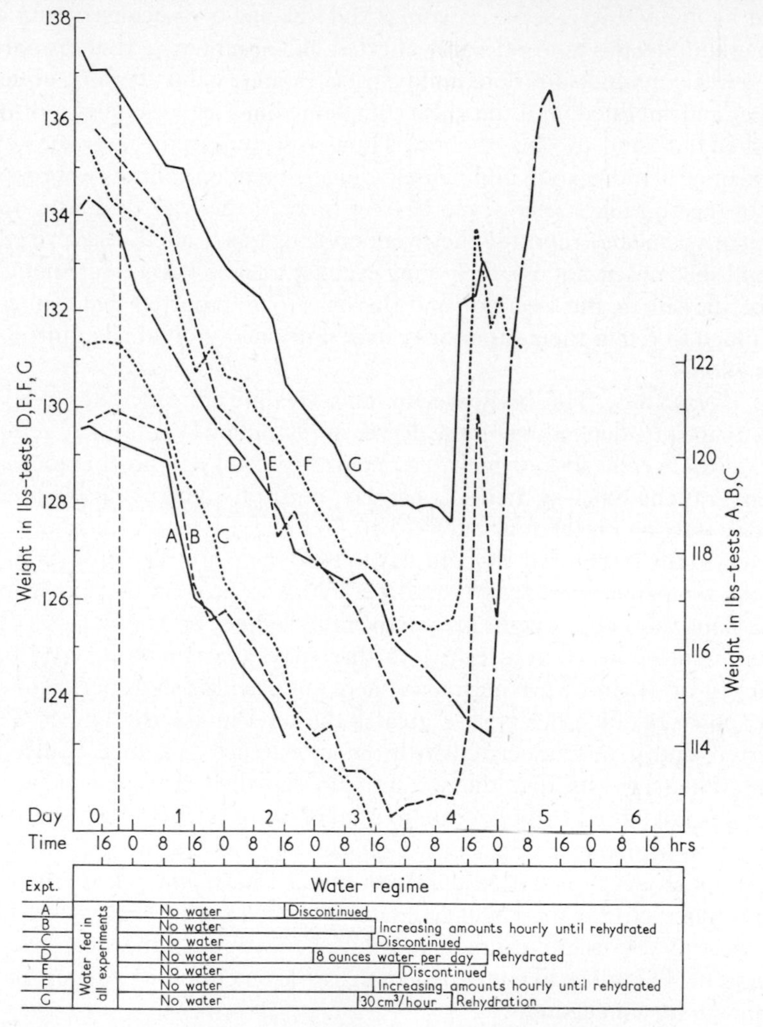

FIGURE 1. Rate of weight loss with different water intakes.

food when a man is water-deficient will increase his urine flow, especially if it contains protein. Table I gives the results of one unpublished experiment.

The increased urine flow after protein foods was associated with an increased urea output; in the 4 hrs before taking the gelatine and olive oil mixture the mean urea output was 0·72 g/hr, and in the four subsequent hours it was

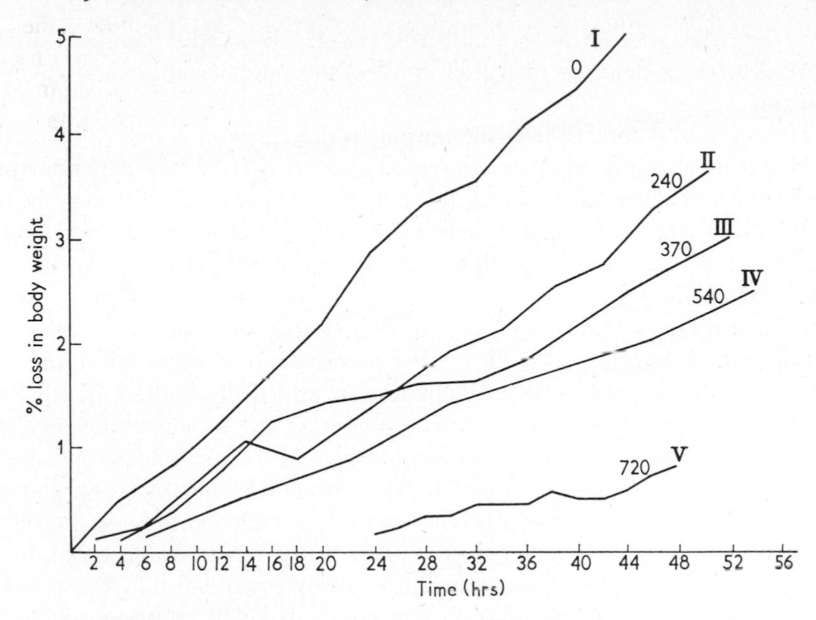

FIGURE 2. Body-weight curves during water deprivation.

Table I. *Diuretic Effect of Food during Dehydration*

Dry food	*Intake* g	*Mean urine flow ml/hr* *Subject R* *Before food*	*After food*	*Subject L* *Before food*	*After food*
Fat-rich biscuits	20	No change		No change	
	30	No change		24	27
	30			16	20
Chocolate	30	No change		No change	
Malted-milk tablets	10	No change		No change	
	20	15	21	21	26
	40	17	24	25	28
Pemmican	10	18	27	14	26
	20	11	21	16	22
* Gelatine / Olive oil	10 / 9			16	22

* Represents "synthetic pemmican".

1·2 g/hr. McCance, Young and Black (1944) observed a similar osmotic diuresis in water-deficient men after intake of urea and of sodium or potassium chloride.

How much of the fall in urine output in dehydration is due to increased reabsorption and how much to decreased glomerular filtration depends upon the degrees of dehydration. McCance, *et al.* (1944) found no change in the glomerular filtration rate until there was a 7% loss of body weight. Kenney (1949) found reductions in both glomerular filtration and renal plasma flow with body-weight losses of 3 to 6%. Smith, Robinson and Pearcy (1952) confirmed this, but noted even greater reductions when men were exercised in the heat. Robinson (1954) allows that there may be a slight fall in the glomerular filtration rate in severe dehydration, especially if there is salt loss. Much of the reduction in urine flow, however, is due to increased water and electrolyte reabsorption under the stimulus of ADH and aldosterone. Jones and de Wardener (1956) and Epstein, Kleeman and Hendrikx (1957) showed that vasopressin will increase the maximum urinary concentration power of human kidneys more in dehydrated than in water-replete subjects; in dehydrated rats, on the other hand, exogenous vasopressin had no effect (Dicker and Nunn, 1957). Epstein *et al.* (1957) concluded that in prolonged water deprivation up to 50% of the saving in urinary water loss is due to increased tubular reabsorption under the influence of ADH. There is some indication from these and others' experiments (Miles, Paton and de Wardener, 1954) that this effect becomes more marked the longer the water deprivation continues.

During water deficiency the urine may contain up to 2·3% of NaCl and 6·2% of urea (personal observations). The potassium content varies among individuals, being anything from one-third to two-thirds of the sodium excretion (McCance *et al.* 1944). In my experiments the s.g. rose to 1035, and Epstein *et al.* (1957) reported a maximal osmolar concentration of from 1200 to 1300 m osmols/l after three waterless days. There may be some orthostatic albuminuria, and the degree of ketonuria will depend on the level and content of the diet. On the wartime standard lifeboat diet of 1 oz each of malted milk tablets, chocolate, pemmican and biscuits, the urine usually became strongly positive to Rothera's test after two days.

c. Salivation. In subjects who salivate freely the flow falls off after about 24 hr of water deprivation, when the free circulating water has been expended. In naturally dry-mouthed individuals no change in the wetness of the inside of the mouth is detectable even after three waterless days, although there is subjective feeling of dryness (Ladell, 1947a, and unpublished observations). The same discrepancy between subjective sensation and objective measurement of mouth dryness was also noted by Epstein *et al.* (1957). Adolph (1947a, d) suggests that salivation stops when there is 8% loss of body weight; but I never had a completely dry mouth even after 10% body-weight loss.

d. Extrarenal water losses. Suggestions that the extrarenal water losses decrease during water deficiency have not been fully confirmed. The same 33% fall in the insensible weight loss of a 36 kg girl was seen after two days without water as when she was water fed on a ketogenic diet (Manchester, Husted and McQuarrie, 1931). Grande *et al.* (1958) found a decrease in insensible perspiration during water deficiency. There was sometimes a greater discrepancy between the dew points of nasally and orally expired air, the latter being the higher, apparent even on the first waterless day, when the weight loss was about 3%. This difference did not increase as the dehydration continued (personal observations). It suggests a drying up of the turbinate mucosa, possibly owing to vasoconstriction in the mucosa, the result of an increased vasopressin level in the blood. Theoretically the rise in the body's osmotic pressure should reduce evaporation from any exposed surfaces, but this would not provide any physiologically significant saving. Sargent, Sargent, Johnson and Stolpe (1954) found that the insensible perspiration was no more affected by 72 hr without water than by 72 hr starvation.

If there is sweating, then water deficiency does have an effect. Once there has been a loss of 2 to 3 l of body water, the sweat rate begins to fall off, irrespective of other factors such as temperature (Ladell, 1955). Hertzman and Ferguson (1959) and Leithead and Pallister (1960) consider that sweating diminishes with any degree of water loss. As the last-named workers say "reports . . . have often been contradictory"; this question has been reviewed elsewhere (Ladell, 1963). All would agree, however, that once 3% of the body weight has been lost, sweating gets progressively less; Adolph (1947a, d) considers that sweating falls off at an even smaller body-weight loss. Whether this diminished sweating assists survival is doubtful; in a desert environment less sweat means less cooling and a higher body temperature.

e. Body temperature. In a temperate climate there were no changes in the body temperature of men even after ten days of severe water restriction (personal observations); in the warm American summer, Sargent *et al.* (1954) found that mouth temperatures never rose above 37·6° during water deficiency. The thermal response to exercise, however, is greater in men on a restricted water intake (Johnson and Sargent, 1956). In hot or warm conditions the situation is different. Grande, Monagle, Buskirk and Taylor (1959) reported that the rectal temperature rise in men exercised at 25·6° was 0·8° greater in subjects after five days water restriction than in control subjects; they attributed this partly to inadequate cooling because of diminished sweating during dehydration and partly to the threshold for sweating being set higher and thus effecting a saving in body water—another example of hypothetical biological precognition. Hertzman and Ferguson (1958) found that the skin and oral temperatures of nude men resting at 43·4°, 38% r.h., remained constant if water was replaced, but if it was not replaced there was a 1·5° rise in both for each 0·5% body-weight loss. Adolph (1947a, b) reported a 2° rise in rectal temperature in men in the desert with a 6% body weight loss, and my own rectal temperature rose to 39·1° after 10 hr without drinking in Iraq and with only mild activity.

With more acute water deficiency the effects on body temperature are marked. The rectal temperatures of men working in hot wet conditions rose 0·5° more in 110 min when they were not allowed to drink than when their salt and water losses were replaced (Ladell, 1955). Pitts, Johnson and Consolazio (1944) found that the body temperatures and pulse rates of men working in either dry or wet heat rose to uncomfortably high levels if they did not drink.

f. Cardiovascular system. In my own trials heart rates tended to be above normal during the day and below normal during the night; rates as low as 50 beats/min were noted. According to Kenney (1949), his subjects' pulse rates slowed by 10 beats/min. Sargent *et al.* (1954), working in a warmer climate, considered the changes in pulse rate found in water deficiency no greater than would be expected from the increase in body temperature. Buskirk, Iampietro and Bass (1958), however, reported unmistakable increases in heart rate and body temperature with 5% loss of body weight.

Changes in the arterial blood pressure do not occur during slowly developing dehydration in temperate conditions, as compensatory changes have time to take place; in rapid desert dehydration marked alterations may be seen. In my own Iraq trial the diastolic pressure at first rose and the pulse pressure fell below 20 mm Hg; there then occurred a disastrous fall in the last 2 hr, ending in collapse. The heart rate rose 50% in the first 10 hr, partly on account of the increased body temperature; it then fell slightly in the next 2 hr. Hertzman and Ferguson (1959) found no changes in the arterial blood pressure of men resting at 50° and losing 0·94% of body weight per hour, despite a progressive decrease in the circulating plasma volume of about 2%/hr. There were, however, postural hypotension and a decreased cardiac output. Beetham and Buskirk (1958) also noticed an increased tendency for subjects to faint during passive tilting as water deficiency increased. Adolph (1947a) noted both diminished cardiac output and an increased heart rate in men with a 2 to 4% body-weight loss.

g. Respiration. Respiratory rate increases slightly during dehydration, even in temperate conditions. In hot conditions there may even be overbreathing, leading to tetany; much of this, however, is a direct effect of heat rather than of dehydration. Adolph (1947a) reports overbreathing in desert conditions when there is 2 to 4% loss of body weight, and Stott (1936) remarked on the "rapid shallow breathing" of men stranded in the desert for a few hours.

h. Changes in the blood. Qualitatively the blood becomes "thick and dark", indicative of rises in the haematocrit and in haemoglobin and plasma protein concentrations. Significant changes in the blood electrolyte concentrations do not occur unless there has been heavy sweating. When men alternated between long periods of work in the heat and shorter periods of rest in cooler conditions and refrained from drinking, their plasma chloride concentrations rose by about 2 meq/l only in 7 hr during two complete cycles of work and rest (Pearcy, Robinson, Miller, Thomas and Debrota, 1956). The rises in the Na and Cl content of my own plasma during 12 hr desert deprivation have already been described; at the same time my blood haemoglobin rose 1·8% and plasma protein 1·33 g/100 ml. The decreased urine flow leads to nitrogen retention. After two days water deprivation the blood urea may rise to 60 mg/100 ml even in a temperate climate; under desert conditions it may rise by nearly 2 mg/100 ml every hour. With rehydration this retained nitrogen is rapidly lost.

i. Intestinal changes. McKim and Marriott (1923) cite work showing decreased intestinal motility during anhydraemia, but Müller (1920) reported increased oesophageal movements during thirst, comparable with the stomach contractions of hunger. The existence of spontaneous contractions has not been confirmed, but there is increased irritability of the oesophagus during water deprivation (Ladell, 1945c). Fat is poorly absorbed (Dennig and Gurlor, quoted by McKim and Marriott, 1923), this possibly causing the marked distaste for fatty foods, such as pemmican or margarine, expressed by many water-deficient subjects. Constipation is marked; after the first few days only small stools the size and consistency of small nuts are passed. This can be a source of considerable discomfort to survivors; in the words of one old sea captain on being assured that surely this must have been the least of his worries in the lifeboat: "Nothing to worry about? How would you like to pass a three-inch wire hawser that's sprung?" (Platt, personal communication, 1943).

k. Sensory changes. Hearing itself is not affected at the degrees of dehydration permissible in laboratory trials, but most subjects complain of their ears "bunging up", because of Eustachian tube blockage. This may be correlated with the drying up of the oro-nasal mucosa.

After only two days without water in a temperate climate there may be difficulties in accommodation. After ten days on a water-restricted regimen, visual acuity declined somewhat, for example from 9/12 to 9/6. In four subjects in whom intraocular tension was measured the mean value fell from 15·2 mm Hg to 12·4 mm Hg in ten days, during which the mean body-weight loss was 6·4%. Some of these changes could have been due to the lifeboat diet; they could not be associated with the intake of small quantities of sea water (Ladell, 1943, and unpublished observations).

l. Mental changes. In the thirsting sailors there was the usual bravado of volunteers participating in a seemingly hazardous trial, exemplified by excessive lightheartedness and wisecracking, but there appears also to be a real diminution of judgement and a tendency to be rash. Performance in simple pencil and paper tests is not affected. Adolph (1947a) noted deterioration in morale in some of his desert tests, possibly associated with the fear that what started as an exercise was becoming a real emergency through lack of experience. Sargent *et al.* (1954) subjected their volunteers to extensive psychological and psychomotor tests, but the only unequivocal change for which no other cause could be found was an overestimate of the passage of time; this got worse with increasing water deficiency. This may be the objective indication of the subjective boredom so often complained of in these experiments.

4. *Accidental water deprivation; survival incidents*

In accidental water deprivation the only limit is the point at which death occurs when the loss in body weight is 20% or more; recent examples of death from thirst are quoted by Adolph (1947a). The dehydration is always of the mixed type; this becomes obvious sooner in a hot climate when the main salt losses are in the sweat than it does under temperate conditions when the loss is in the urine. The nearest to a pure water deficiency syndrome occurs when the victim is "dried out", either because the air he breathes is dry, as with climbers at high altitudes, or because the air blowing over his skin has a high saturation deficit, as with Stott (1936) and his party in the Iraq desert.

Most accidental water deficiencies occur in hot dry climates, apart from episodes in lifeboats. No one in the desert drinks enough during the day to maintain his water balance from hour to hour. Rothstein, Adolph and Wills (in Adolph, 1947a, p. 254) call "voluntary dehydration". It is not true dehydration; the subject is drawing upon his free circulating water; until these 2 or 3 l have been lost, he pays no physiological penalty. The survival value of the free circulating water is considerable; it allows men to stay fully active in the hottest desert for several hours without drinking. No Moslem could survive Ramadan without it, as it permits him safely to refrain from drinking from sunrise to sunset; at night the rules of the fast allow him to drink his fill and restore himself. In a hot desert a man may readily lose 11 or 12 l of water a day (Adolph, 1947a; Lemaire, 1952; Ducros, 1956; Ladell, Waterlow and Hudson, 1944). Even if he could find time to drink this amount while working, he would be hard put to absorb it all. He would have severe gastrointestinal discomfort and nausea and would probably vomit (Adolph, 1947a, d). Drinking at the same rate as one sweats is almost impossible; inevitably, therefore, a man ends the day with a water deficit. Unless this is made good each night before beginning work the next morning, a man will gradually develop a chronic water deficiency, and, depending upon the amount of salt he is taking, some degree of salt deficiency. This is the background of one of the commonest, and most successfully treated, of the heat disorders, although a sudden extra strain on the water and salt metabolism may precipitate the breakdown (Ladell, 1957; see also chapter 3). The better acclimatised a man is, however, the less will be his salt loss for a given water loss (Conn, 1963; Ladell, 1963).

The real survival problem occurs when a man under hot conditions has an insufficient water supply to draw upon. Sweat contains less electrolyte than plasma, and the kidney corrects this functional excess, maintaining as far as possible normal extracellular osmolarity; despite this, there is a gradual increase in osmolarity of both extracellular and intracellular fluid, combined with a fall in the volumes of both.

Thirst is an inadequate indication of water lack, and the associated salt and water deficiency may therefore develop insidiously. Disaster can come from niggardliness or straight misjudgement with the water bottle; men carrying nearly full water bottles have dropped from thirst. Lethal water deficiency may strike suddenly: Brown (in Adolph, 1947a) relates how two soldiers became separated from their unit in desert manoeuvres and were found dead 10 hrs later; here one suspects previously uncorrected voluntary dehydration. The need for water transcends all else; seasoned Afrika Corps men were sometimes, Brown says, forced to surrender in the 1942 North Africa campaign because of water lack.

If he is initially fully hydrated, a man can survive without water, even in the desert, for at least two or three days. Cannon (1917–18) puts the limit at seven days in the desert and eighteen days in a temperate climate. A Buddhist priest in Japan survived and was able to carry out his normal religious duties for eight days without a drop of water; his body weight fell by 23%. Longer periods of survival without drinking have been reported, but in these some chicanery cannot be excluded (Widdowson, 1957).

Deterioration may be a function of the speed of dehydration; Brown reports one man "at the end of his tether" after a 4 hr walk in the desert. His weight loss was only 5·5% of his initial weight that day, but he could neither stand nor march. Breathing was laboured, and he could hardly speak. This man may well have begun with an uncorrected voluntary dehydration. With weight losses of even less than 5% there may be slight decrements in performance, which get worse with increasing deficit, but complete collapse does not usually occur until there is at least a 10% loss. It is at this point that the determination to keep going and will to live become important. The collapse, when it comes, is a general weakness superimposed on a circulatory failure. Even with a severe water loss, the circulatory collapse may be deferred for a time by keeping gently on the move; it is precipitated by standing still (personal observation).

If a man is to survive in the desert and get back to safety, he must walk only at night and take advantage of what shade he can find during the day. Brown (in Adolph, 1947a) has calculated for various mean daily temperatures how far a man without water can expect to get before he collapses; an extract from his results is given below (Table II).

Table II. *Attainable Distance before Occurrence of Limiting Water Deficit, after Brown, in Adolph* (1947a)

Daily mean temp °F	*Number of miles if water supply is in U.S. quarts*†				
	0	1	4	10	20
80 (26.7°C)	45	50	70	110	170
90 (32.2°C)	20	35	35	50	80
100 (37.8°C)	15	18	20	30	50
110 (43.3°C)	9	10	15	20	30
120 (49°C)	7	8	10	15	25

† U.S. quart = 947 ml.

At 49° a man could only expect to travel seven miles without water before collapsing, and Brown doubts if a man could go more than forty-five miles even at 26·7° without water and last more than two days. These limits may be unduly pessimistic, however; Brown gives the marching capacity of a man, with 20 U.S. quarts of water, as 25 miles at 49°. In climatic chambers men can certainly march, or do the metabolic equivalent, at that temperature for 4 hrs if they have unlimited water and remain in reasonable condition. Laboratory experience would not support some of Brown's figures, but he based his table on field observations, and the difference between laboratory findings and desert experience could be due to the added burden of the sun.

Reference has already been made to McGee's (1906) account of Pablo, the Mexican, who survived for eight days in the desert on 2 gal of water. The mean daily maximum temperature was 35° and the mean daily minimum 28.0°. Pablo's limit, according to Brown, should have been fifty miles; in fact, he travelled thirty-five miles in the saddle and at least 100 miles by foot. When Pablo was finally picked up his estimated weight loss was 25%. His flesh was black and dry, and his lips, nose and eyelids had shrunk to nothing; his skin was covered with cuts and scratches, none of which bled. He was deaf to all but loud sounds and blind except to light and dark, but he had retained sufficient of his faculties to find his way back almost to the camp for which he had set out. He had a slow "roaring" respiration and a slow pulse, which could not be detected below the knees and elbows. Bad though his condition seemed, he made a complete recovery, "soaking up water like a dry sponge through his skin", then, after regurgitating it at first, drinking and retaining vast quantities of water. It was three days before his vision and hearing returned to normal. Four days later he was still "deliberately and methodically devouring water melons". One week later, a fortnight after he had been picked up, he was well, cheerful and back to his old weight, but his hair had turned grey. Pablo had been determined to live, and he succeeded.

5. *Salt excess*

Superficially salt excess and water deficiency present the same survival problems. Both initiate a drive to drink, and both upset the water distribution within the body; however, in water deficiency the kidney can conserve water by reducing the urine flow to the minimum obligatory volume, whereas salt increases the urine output. The osmotic pressure of the extracellular fluid rises more with salt excess than it does with water deprivation. This is corrected by a flow of water out of the cells with, inevitably, some potassium; there is also a diuresis as the result of a specific failure by the tubules to reabsorb sodium (McCance, 1945–6), irrespective of whether a man is already water depleted or not. This failure of reabsorption is due to the suppression of aldosterone secretion by the high salt intake (Garrod, 1958). Sodium chloride excess thus acts as a pump, sucking water from the cells and sending it out into the urine; in the process there is cell damage, indicated by the increased plasma potassium (Seldin and Tarail, 1949). It is not always appreciated how dangerous it is to give salt to a hydropenic man without giving him water at the same time; it is still sometimes said that: "A man can do without extra water in the tropics provided you give him plenty of salt." It is undeniable, however, that salt given to a hydropenic man may help to maintain his extracellular fluid volume and give temporary improvement in, or a prolongation of, his ability to work. Men given 10% saline just sufficient to replace their salt losses by sweat during hard work in a hot room, but no other water, did very well for a time; they felt "fine", but a few minutes later they collapsed (Ladell, 1955).

a. Drinking sea water. In trials in a temperate climate men drinking 200 ml of sea water daily remained in excellent condition, and some subjects believed that the sea-water supplement kept up their stamina (Ladell, 1943; Whillans and Smith, 1948). This is a real effect, due to the maintenance of adequate circulating fluid, but it is paid for by earlier eventual collapse when the cells reach their limiting osmotic pressure. If it is absolutely essential to keep a man on his feet a little longer, salt can be given; similarly, if survivors in a lifeboat wish to keep off any deterioration as long as possible for the first week or ten days, not only should they drink most of their fresh water in this period, but they may also supplement it with some sea water. The longer they continue to do this, however, the smaller their eventual chances of survival, as urinary water losses are increased almost by the amount of the sea water supplement. McCance *et al.* (1944) reported that the urinary minute volume of a dehydrated man was increased from 0·36 ml/min to 1·56 ml/min after drinking 500 ml of 3·6% saline (equivalent to sea water). McCance and Morrison (1956) studied the effect of excess salt during water deficiency on "model men", rats with five-sixths of their kidney tissues removed. When 3% saline was given daily in quantities equivalent to 2% of their body weight to the starved waterless animals, there was the predicted increase in water and salt loss and also a rise in the potassium excretion, indicating tissue breakdown. Foy, Altman and Kondi (1942) gave sea water enemas to men deprived of water for 80 hr. The mean weight loss of the subjects given 2900 ml of sea water rectally was 4377 g, compared with only 2873 g, by the controls, and their urine outputs were 2000 ml and 1400 ml, respectively; this high control figure suggests that the water restriction was by no means absolute. The effective weight loss of the experimental subjects was three times that of the controls and the effective water loss from the body six times as great, indicating nearly twice as much tissue wastage. The maximum sodium chloride concentration possible in the urine is 2·3% and in some people may be only 1·9%, about two-thirds the concentration of sea water. If Foy's subjects retained their enemas (admittedly improbable), they will have had an inevitable salt retention of 40 g, enough to raise the osmotic pressure of the extracellular fluid by nearly 0·8 m osmol/l. This manoeuvre does not therefore contribute to the chance of survival.

It is now accepted that "dry storage" of salt in the skin does not occur (Manery, 1954). The best any tissue can do is to hold fluid isotonic with the plasma, and this would not help to dispose of any salt retained as a result of drinking sea water. All who have worked on this problem (Rubini, Wolf and Meroney, 1956; Gamble, 1944; Elkinton and Danowski, 1955, and those already quoted) agree that small seawater increments may help to maintain the circulation, but only at the cost of an increased body osmolarity. An excess of solute in the body is a greater strain on the water excretion than is an excess of water on the solute excretion; whereas it is difficult to produce water excess in healthy men, it is only too easy to make them retain sodium (Elkinton and Danowski, 1955). Robinson and McCance (1952) give the warning that "modern castaways fare no better on sea water than the Ancient Mariner". After a brief "honeymoon period" of sea-water drinking, the predominant disturbances (from such clinical accounts as have appeared) are, especially in the later stages, in the central nervous system. It is true that sea water drives men mad (Elkinton and Winkler, 1944a). The oral lethal dose of NaCl to rats is 3·75 g/kg, with death from respiratory failure and convulsions (Boyd and Shanas, 1963). Men are probably more susceptible than rats, as human kidneys can only concentrate sodium chloride half as well;

hence it is possible that Foy's subjects may have retained as much as one-third of a lethal dose, and it is not surprising that wartime records show the mortality of castaways who drank sea water to have been ten times that of those who kept strictly to their limited supply of fresh water (McCance, Ungley, Crossfield and Widdowson, 1956).

Some of the optimism about man's ability to drink sea water, despite the wealth of human experience to the contrary, came from false analogies with animals. The human kidneys are among those with the lowest concentrating powers; a dog can do 20% better, and a kangaroo rat can produce urine twice as strong as sea water (Schmidt-Nielsen, 1962). Small mammals are the commonest laboratory test subjects, but small mammals are the least water-dependent vertebrates (Chew, 1961). Rats can survive water loss for seventy days (Barker and Adolph, 1953) and dogs for more than thirty days (Wolf and Eddy, 1957). Many physiologists, myself included, have nevertheless still not entirely discarded the idea that just a little sea water, "limited mariposia", may assist in survival and save some small fraction of body water (Wolf, Prentiss, Douglas and Smith, 1959).

b. Salt intake and hypertension. In rats with experimentally damaged kidneys a high salt intake predisposes to hypertensive vascular disease (Koletsky; 1957); a high salt diet in rabbits renders them more susceptible to circulating pressor substances (Gordon, Drury and Schapiro, 1953). Crane and Ingle (1959) produced hypertension in mononephrectomised rats, and occasionally in un-operated animals, by giving them up to 20% of their body weight of salt; after adrenalectomy, however, "full-blown hypertensive disease" rarely occurred. This hypertensive action of a high salt intake depends on circulating corticosterone (Hyde, Grefer and Skelton, 1962); adrenalectomised, mononephrectomised animals will not develop hypertension unless their plasma contains at least 0·25 μg corticosterone/100 ml, but in the presence of this the greater the NaCl intake the higher the systemic blood pressure. Rats drinking 2% saline normally develop a rise in the systolic blood pressure, but fail to do this if they are treated with the antithyroid drug propylthiouracil (Fregly, 1959). The thyroid gland may thus also be involved in the development of this syndrome.

Dahl, Heine and Tassinari (1962a) believe that such observations on animals are relevant to man and that dietary salt plays an aetiological part in human essential hypertension. Not all men on a high salt diet, however, develop hypertension, but only those of the right genetic make-up; just as two different strains of rats may be developed, one salt sensitive and the other not, so Dahl, Heine and Tassinari (1962b) believe there may be different strains of man. Certainly the fact that, in countries such as Nigeria where the salt intake is low, hypertension is not common (Brown, personal communication) would support Dahl's hypothesis that a high salt intake is a primary rather than a secondary factor in the development of hypertension;

nutritional and climatic effects would first have to be excluded. Although the hypertensive damage to rats produced by a high salt intake is permanent (Dahl, 1961), salt restriction in man is a well-tried therapeutic measure; the two conditions may not therefore be comparable. Crisman (1960) has marshalled some of the other arguments against Dahl's hypothesis, and Brown, Davies, Lever and Robertson (1963) report that plasma renin levels in men fall during salt loading.

Another argument against Dahl's hypothesis is the association between aldosteronism and hypertension (Conn, 1961, 1963), as a high salt intake reduces aldosterone release. The kidneys of a patient with aldosteronism, moreover, resemble those of hypertensives in that they no longer respond to aldosterone by salt retention (Rovner, Conn, Knoff and Hsueh, 1962).

c. Salt intake and the skin. One complaint that is sometimes associated with excess salt intake is prickly heat; this is a mild erythematous maculopapular skin eruption, sometimes with small bullae, which may weep. It is intensely irritating. Healed areas may show a local diminution of sweating. There is some evidence that it is either the first stage of or a contributing factor to tropical anhidrotic asthenia (Ladell, 1951b). In this condition a man is unable to exert himself once the temperature rises, though he is comparatively unaffected until then, a state scarcely compatible with survival. Horne and Mole (1949, 1951) concluded from clinical observations that a high salt intake was a contributory factor to the development of prickly heat, which they associated with a high chloride concentration in the sweat. Loewenthal (1961) produced skin lesions clinically and histologically resembling those of prickly heat in normal subjects by skin compresses of hypertonic saline; the clinical effects of prickly heat are, he suggested, secondary to the mechanical effects of the skin being bathed with partially evaporated and so hypertonic sweat. Although a high salt intake predisposes to a high chloride content of the sweat, in no circumstances is sweat ever hypertonic; moreover, prickly heat usually appears where evaporation is poor, whether in a tropical or a desert climate, for example, under belts or round the neck under the collar. There is thus no question of the skin being soaked in a hypertonic saline; moreover, prickly heat may occur in subjects who are slightly salt deficient. The rise in the sweat chloride-content that may occur in late or healing prickly heat could be a premonitory sign of a breakdown in adaptation (Ladell, 1963). Loewenthal's (1961) production of prickly heat by bathing the skin with hypertonic saline contrasts with Dobson and Lobitz's (1957) demonstration of apparently the same lesion after iontophoresis of distilled water into the skin from the anode. It does not appear to matter with what the skin is soaked; so far from salt loading adversely affecting the sweat glands, Dobson, Abele and Hale (1961) found that, when men are exposed for 6 hrs daily to 40·6°, there are marked cytological changes in the glands if the men are salt deficient, but only minor ones in salt loading.

6. *The nature of thirst*

a. The six stages of thirst. McGee (1906) described six stages of thirst in the field; he also distinguished between various types of thirst, according to when they occurred. We now know that all types of natural thirst are the same and differ only in degree; nevertheless, no account of thirst would be complete without the description of McGee's six stages.

The first stage is that of "normal dryness". There is an instinctive craving for fluid, although there is no real sense of deprivation. It is irritating and causes general discomfort and is hence called the "clamorous phase". The second phase (which could be associated with the exhaustion of the free circulating water) is marked by dryness of the mouth, lumps in the throat and blocking of the Eustachian tubes. The individual may become quarrelsome and do unreasonable things; he may even have mild mental changes with minor hallucinations. Body temperature and pulse rate rise. This phase was well known in Arizona and called the "cotton-mouth phase". Next the tongue begins to shrink and the mucosa of the mouth and lips dry out. There is a fullness of the head with tinnitus. The heart begins to slow; breathing becomes laboured and there is further fever. Hearing and vision deteriorate. This is the "shrivelled tongue stage". The fourth stage corresponds to Pablo's "structural degeneration". Further drying of the tissues brings cracks and fissures. There is a slight ooze of blood from the surface; hence this fifth stage is called "the blood sweat". Next the mind goes, and there is permanent cerebral damage, giving the final stage, corresponding perhaps to the last of Shakespeare's Seven Ages of Man, ". . . mere oblivion sans everything", McGee called it "the living death"; the victim is indeed oblivious of everything, tearing at cactus thorns, digging dry pools and "drinking" gravel; finally, with the senses gone, "dying from within out."

b. Modern theories of thirst. McGee's dramatic description of thirst does not, however, tell us what thirst is. Thirst is primarily a sensation, which often serves as a drive to drink, but the drive and the sensation are not necessarily identical. To equate the "receptors for thirst" with Verney's hypothalamic osmoreceptors is to confuse the issue. Thirst, wherever it may be sensed, is always referred to the periphery and tells a man it would be pleasant to drink. The hypothetical thirst receptors, on the other hand, not only tell a man he needs to drink, but also how much. Adolph (1945) claims that a dog, given free access to drink, will maintain his body water to within 1% of normal. Wolf (1952) suggests that there is in man a similar close control though it is slower in action, but it is difficult to accept this in the face of "voluntary dehydration". It may be, however, that man has evolved drinking habits by which he maintains himself in a perpetually overhydrated state; in contrast to all other terrestrial mammals, he rarely produces a highly concentrated urine and to do that he has severely to control his fluid intake.

In young adults the free circulating water amounts to at least 2 l, but whether it is as much as this at all ages and in both sexes it is not possible to say. All Adolph's work (1945, 1947b, c) would suggest that any overhydration is not normal among other mammals. Possibly the sensation of thirst as man knows it in his everyday life is geared to keeping the body overhydrated, the "drive" to drink only coming into play when the free circulating water has been expended. This hypothesis would allow thirst to be mediated peripherally, but not enough is known of the changes that occur in the mouth and nasopharynx during varying degrees of overhydration and underhydration to suggest the exact mechanism. Animals, on this hypothesis, would not experience thirst in the human sense, but only the drive, which is almost certainly mediated centrally. The thirst sensation may be mitigated or even induced artificially, but the drive, once initiated, cannot be modified in the intact animal except by drinking.

This hypothesis links with Cannon's (1917–18) distinction between "true thirst" of dehydration and false thirst, in which there is only a dry pharynx. According to Wittendorff, quoted by Cannon, the sensation of thirst may precede an increase in the serum osmotic pressure; his final hypothesis, based on experiment and still accepted by some, is that the sensations of thirst are due to a diminution in salivary secretion. Cannon's results were never completely confirmed; thus Critchley and Allison (1941) were unable to control thirst in patients with diabetes insipidus by inducing a flow of saliva. Dill (1938) accepted Cannon's hypothesis supplemented by Müller's (1920) suggestion that thirst was associated with spontaneous oesophageal contractions, comparable to gastric hunger pains; however, the existence of these "thirst pains" could not be confirmed (Ladell, 1945c). Müller associated thirst with a rise in the osmotic pressure of the plasma, which stimulated a thirst centre in the hypothalamus; this referred the sensation of thirst, the signal to drink, to behind the larynx. There must be other peripheral factors concerned, otherwise why, asks Wolf (1952), does a man feel "thirsty" after eating a salted anchovy. Elkinton and Danowski (1955), however, regard thirst as entirely central in origin; they call it "the cardinal symptom of dehydration", correlated with antidiuresis, hypertonicity and intracellular water deficit. They can find no evidence dissociating thirst from the state of hydration of the body. But Birchard, Rosenbaum and Strauss (1953) point out that thirsting men and hypertonic men are not necessarily in the same physiological state; naturally thirsting men can drink without having a diuresis, but men made hypertonic by intravenous injections of saline will always get a diuresis on drinking. Confusion and controversy will continue in this field until the clear distinction is recognised between thirst the sensation and "thirst" the drive.

This drive has so far been investigated mainly on animals. One of the difficulties has been differences in species; thus Pack (1923) reported that the water intake of rabbits deprived of water for seven days was reduced by the injection of pilocarpine; but Montgomery (1931) found that the same drug increased the water intake of dogs, irrespective of whether their salivary glands were intact or not. Pilocarpine reduces the water intake of rats deprived of water, but has no effect on the thirst of men in the desert (Adolph, 1947a, 1948). It is now generally accepted that there is a hypothalamic "thirst" centre, which controls the drive and is cholinergically controlled (Andersson, 1962; Stein and Siefter, 1962; Stein, 1963). The question of the "thirst" receptors is still under discussion. The process of taking in water, according to Adolph, Barker and Hoy (1954), has four phases: seeking water, drinking water, ceasing to drink and absorption and distribution. They suggest that there are separate hypothalamic receptors behind the drive to seek (which they suggest is true thirst) and the drive to drink, actuated by changes in the osmotic pressure and fluid volume; they also postulate some receptor mechanism that meters the intake and brings drinking to a close when the necessary quantity has been drunk. The identification of osmoreceptors in the hypothalamus (Verney, 1947), and the relationship between extracellular hypertonicity and the drive to drink in animals, led Andersson and McCann (1955) to experiments demonstrating that local hypertonicity in the hypothalamus or even electrical stimulation provoked drinking. They also showed that lesions in this "drinking area" would abolish drinking in dogs. Rats in whom this centre is destroyed die of dehydration within a few days (Gilbert, 1957).

Although the receptors for ADH stimulation and the "thirst" receptors are both stimulated by hypertonicity, the drive to drink is not mediated by ADH itself (Adolph, 1948; Zuidema and Clarke, 1957; Nicholas, Di Salvo and Corner, 1963); indeed, the existence of diabetes insipidus in man shows that compulsive drinking can occur in the absence of ADH. In rats, a drive to drink may be induced by depletion of body fluid without any change in osmotic pressure (Fitzsimons, 1961). Thirst is seen in men during depletion of either sodium (McCance, 1936b) or potassium (Fourman and Leeson, 1959). The existence of two different triggers setting off the drive to drink was also indicated by the observation of Barker, Adolph and Keller (1953) that dogs can be driven to drink through dehydration or through injection with hypertonic saline. The "thirst" centre therefore appears to be sensitive not only to tonicity but also to changes in circulating fluid volume, as are the receptors controlling ADH release (Cort, 1954, 1962). Andersson (1962) also suggests that the "thirst" receptors are sensitive to a third stimulus, temperature. Local warming of the preoptic region will elicit drinking in the water-fed goat, whereas preoptic cooling will inhibit it. The inhibition of drinking by preoptic cooling is only transient; there is an escape from the inhibition after 50 hr cooling, and drinking begins again; presumably there is a more powerful drive from the osmotic pressure and volume changes. Drinking may also be halted in animals, driven to drink through electrical stimulation of the centre, by giving them ice-cold water. If the same mechanism holds for man, this accounts for the extra thirst experienced after hard exercise in the heat, when the blood temperature is raised, and for the special thirst-quenching qualities of cold water.

The thirst receptors may be inhibited as well as stimulated (Adolph, 1950; Adolph *et al.*, 1954; Barker *et al.*, 1953). Drinking is inhibited in dogs by distending the stomach either with fresh or salt water or with an inflated balloon; the same phenomenon in man would explain the spurious satisfaction the thirst-maddened castaway gets from drinking sea water. There may also be some metering mechanism in the pharynx, as a dog with an oesophageal fistula drinks twice its water deficit before it stops drinking or has to have twice its deficit placed in its stomach before it loses its drive to drink. The sheep also seems to meter its intake as water passes down its pharynx (Denton and Sabine, 1961). The signal to stop drinking, however, probably comes from gastric distension (Stevenson, 1962); rats can keep themselves in water balance by gavage controlled by bar pressing (Epstein, 1960). The results of Denton and Sabine (1961) on sheep suggest concentration as well as volume monitoring; further evidence of this is the observation by Holmes and Montgomery (1960) that dogs given 20% saline intravenously drank two and a half times as much when they were given 0·9% saline as when they were given water. They also found that unless fluid got into the stomach of these animals, their drive to drink was not satisfied; fluid given intravenously did not satisfy their "thirst".

The presence of concentration monitors in the human oesophageal tract or mouth would account for the greater thirst-quenching quality of water over most flavoured beverages (Bagnold, personal communication). Men allowed to drink what they like in hot room experiments may, in my experience, try fruit squash or beer initially, but soon elect for neat water.

B. Salt Deficiency and Water Excess

1. *Salt deficiency*

The changes seen in water lack are related to hypertonicity and those in salt deficiency to hypotonicity. The two could therefore be considered as diametrically opposite conditions; both, however, give rise to thirst, and both result in dehydration. In water deficiency the water loss is evenly distributed throughout the different compartments of the body, whereas in salt deficiency the loss falls on the extracellular compartment, giving a reduced plasma volume, haemoconcentration and early peripheral circulatory failure (Nadal *et al.*, 1941). Pure salt deficiency like pure water deficiency, is almost a laboratory curiosity; in practice deficiencies are usually mixed, with the subject having lost both salt and water during severe sweating and failed to replace either adequately. The same circulatory failure develops in mixed deficiency; there is a low output of urine, practically sodium and chloride free, and, depending upon the amount of water drunk, severe muscular cramps, especially in well-used and fatigued muscles (Ladell, 1949b, 1957). Treatment consists of prompt and full replacement of both the water and salt, leading to recovery; if, however, the water is replaced more rapidly than the salt, the salt deficiency is exacerbated, and the cramps become worse.

An increased aldosterone output is pathognomonic of salt deficiency, but this may not be detectable until there is some extra strain placed on the sodium metabolism; thus unacclimatised men in the heat do not excrete aldosterone in their urine until they are made to work hard and sweat (Fletcher, Leithead, Deegan, Pallister, Lind and Maegraith, 1961), suggesting that there is a state of pre-deficiency, when the body is just in salt balance but has no reserve. This condition is a hazard of climate rather than of diet (Leithead *et al.*, 1958; Stenning, 1945).

Pure salt deficiency in a temperate climate has been studied extensively by McCance (1936a, b), but natural man is unlikely to meet salt deficiency on its own as a problem of survival. Man has had to learn to eat salt (Kaunitz, 1956a; Bloch, 1963) and has become addicted to it; Western man in particular has become adapted to a physiologically unnecessary high salt intake, but without it withdrawal symptoms develop.

McCance's experimental subjects lived on a salt-poor diet for eleven days. Their average daily intake was about 40 mg of sodium and nearly 200 mg of chloride; the potassium intake amounted to nearly 3 g/day, as they continued to eat vegetables. An additional strain was put on the salt metabolism by making the subjects sweat from time to time. Representative deficiencies in one subject after eleven days were: Na, 765 meq; Cl, 590 meq; K, 93 meq. In other subjects the losses were slightly more. The subjects looked ill; despite free access to water, they lost weight, and their cheeks fell in. They became irritable, fretful and apathetic; similar mental changes are seen in water deprivation or sea-water drinking. The sense of taste became dulled, the edge being taken off the flavour of everything. The subjects became physically fatigued, with muscles tiring easily and cramps developing. Recovery on taking salt was dramatic; half an hour after swallowing 15 g the sense of taste was restored, and within a few hours the subjects felt themselves again. During the period of salt deprivation there was a 25 to 30% fall in the extracellular fluid volume, greater, in one subject, than the loss in body weight (McCance, 1938). The average drop in the plasma chloride concentration was 16·5 meq/l and in the sodium 14 meq/l (McCance, 1937b). Both red-cell count and plasma protein content rose, the blood "thickening", so that it "would not flow satisfactorily through a large bore needle". At no time was there a massive diuresis; even with extra drinking, the maximum urine flow achieved was less than 6 ml/min (McCance and Widdowson, 1937); there was a slight fall in the glomerular filtration rate. Thus, although there was no change in the renal concentrating power, there was a reduced urea clearance and a rise in the blood urea concentration. When this was taken into account, considerable negative nitrogen balances were apparent, amounting in one subject to 33·6 g. The negative nitrogen and potassium balances indicated cellular breakdown. Salt deficiency in animals tends to reduce adrenal cortical activity, except for the release of aldosterone; thus, in rats the level of corticosterone in the blood falls to a low level (Eisenstein, 1959). In man the psychological effects of an experiment such as this cannot be discounted; hence, part of the nitrogen loss could have been due to general adrenal stimulation, with glucocorticoids stimulating nitrogen metabolism.

The NaCl content of the sweat of McCance's subjects fell; this has been described many times subsequently, especially when the salt restriction was combined with exposure to heat (Ladell, 1947b; Robinson, Nicholas, Smith, Daly and Pearcy, 1956). This is now known to be due to adrenal cortical activity (Ladell, 1945b, 1963) and specifically to that of aldosterone (Streeten *et al.*, 1960), which is released in chronic just as much as in acute sodium deficiency (Dahl, Silver, Christie and Genest, 1960). Clinically there may be an association between a high aldosterone level and hypertension (Conn, 1961), but in McCance's subjects, despite a presumptive aldosterone release, the arterial blood pressure was unchanged.

With more marked salt deficiency, however, such as that produced by jejunal suction in men on a salt-poor diet, there may be considerable falls in the arterial blood pressure (Nadal *et al.*, 1941). This is a peripheral circulatory collapse, due to the fall in the circulating plasma volume seen in both man and animals (Hopper, Elkinton and Winkler, 1944; Elkinton and Danowski, 1955). It rapidly responds to treatment with isotonic saline, both in men suffering from mixed deficiency and in animals (Danowski, Winkler and Elkinton, 1946; Ladell, Waterlow and Hudson, 1944). In severe deficiency the renal changes may also be more marked. McCance found a reduced urine flow attributable to a diminution in the glomerular filtration rate, but complete anuria may develop if the blood pressure falls too far (Nadal *et al.*, 1941). The low electrolyte excretion results from increased tubular reabsorption, partly caused by aldosterone and partly the direct result of lowered plasma osmotic pressure (Darragh, Welt, Goodyer and Abele, 1953). In extreme deficiency in rats there is also a loss of concentrating ability (Baker, Levitin and Epstein, 1961), but not in man (McCance, 1945). In salt deficiency, however, the human kidney is especially sensitive to alkalosis, and overbreathing does not result in a diuresis with an alkaline urine and increased sodium and potassium excretion (McCance and Widdowson, 1936).

In extreme salt deficiency, sodium is drawn from the bones as well as from the fluid phase. Nichols and Nichols (1956) removed 25% of the body sodium from dogs by vividialysis; the cells contributed 4% of this, the extracellular fluid 70% and the bones 26%, resulting in a 12% drop in the sodium concentration of the bones. Other changes in salt deficiency are alterations in glucose tolerance (McCance, 1937a) and, according to Castex and Septeingast (quoted by Keys, 1943), a lowered basal metabolic rate.

Salt deficiency fatigue, like fatigue in general, remains a mystery. Krnjevic, Kilpatrick and Aungle (1955) were unable to detect any significant changes either in the neuromuscular system peripherally or in the brain centrally. The changes in sensitivity of the taste buds is also unexplained; de Wardener and Herxheimer (1957a, b) noted a similar change during forced overdrinking, when they developed a small negative salt balance. They had a craving for salt and found all food tasteless without it. Their salt-taste threshold fell at the beginning of the overdrinking period, but rose to normal later. One possible cause is a change in the permeability to the chloride ion of the taste buds concerned, consequent upon alterations in the intracellular electrolyte. The observation of Herxheimer and Woodbury (1960) that the salt, but not the sucrose, preference threshold in rats is lowered by the administration of deoxycorticosterone acetate (DOCA) suggests that the adrenal cortex is concerned in these taste changes.

There is one survival situation in which salt depletion may occur even in a temperate climate, and that is in seasickness. Vomiting can cause acute and grave losses of both water and salt; Welt (1952) points out that severe vomiting, especially if there is also diarrhoea, can lead to acute hyponatraemia, with anorexia, lassitude, clouding of consciousness and coma. Motion sickness itself stimulates the release of ADH (Taylor, Hunter and Johnson, 1957), so the unfortunate traveller who drinks to restore his water loss after vomiting hastens his own downfall and develops hyponatraemia without the aid of diarrhoea. Welt's clinical sketch of acute salt deficiency resembles, as any bad sailor knows only too well, that of severe seasickness. The condition is rapidly relieved by infusing saline, and one is reminded of the old wives' tale that the best remedy for seasickness is to drink sea water; like many folk remedies, it may not be so senseless after all.

2. *Water excess*

It is difficult to visualise a survival incident in which the main hazard is an excess of water to drink. Though one may sympathise with Chesterton's "I don't care where the water goes if it doesn't get into the wine", this is rarely a matter of life and death. Even the term "water poisoning", introduced by Haldane (1923) for the syndrome described by Moss (1922) in miners working at high temperatures, was not brought about by excess water, but rather by loss of salt. The syndrome was, in fact, salt-deficiency heat cramp, associated with hard work and excessive drinking at a high temperature (Ladell, 1949a, b). Haldane appreciated that there was a salt loss, but attributed the condition to water excess; the "experienced mine workers" who suffered also blamed the water, and when thirsty they only rinsed their mouths out with water and never swallowed it. In one respect they were clearly right, as an excess of water results in a relative salt deficiency and will exacerbate an absolute one. Haldane blamed a "temporary paralysis of renal excretion"; we now know that this temporary paralysis is the normal physiological response to ADH, which is secreted in response to heat.

In a temperate climate heavy water-drinking in man, as in animals, will suppress the release of ADH (Welt and Nelson, 1952a). The delay between the end of drinking and the onset of diuresis represents the time taken for any previously circulating ADH to be inactivated. A man with a water diuresis is therefore in a state of "physiological diabetes insipidus" (Kleeman, Epstein and White, 1956). At the same time the expansion of the blood volume (Rioch, 1930) reflexly decreases the secretion of aldosterone (Vesin, 1958; Bartter, Biglieri, Pronove and Delea, 1958); the concomitant stimulus of electrolyte dilution to reflex aldosterone release is weaker and so inhibited. Only in special circumstances, such as the infusion of 5% saline and simultaneous salt loading, may tonicity in man be made to overrule the demands of volume changes, so that sodium, chloride and potassium losses decrease, when the opposite would be more desirable (Darragh *et al.*, 1953).

The maximum urine flow in a water diuresis is directly correlated with the glomerular filtration rate and the solute load (Kleeman *et al.*, 1956); in the absence of ADH, as in overdrinking, all or nearly all the urine that reaches the distal tubules is excreted (Welt and Nelson, 1952a). Two at least of the endocrine substances that might affect arterial blood pressure, ADH and aldosterone, are not circulating; hence, once the full diuretic flow is established, this represents the limiting rate of water loss by the kidneys in overdrinking, no matter how much water is drunk. In the subjects of Kleeman *et al.* (1956), this limit was 23·7 ml/min per man. Therefore with heavy drinking—more, that is, than 1·5 l/hr—water will accumulate in the body, and eventually symptoms of true water intoxication may develop,—headache, dizziness, nausea, sometimes cramps and, in animals, convulsions (Smyth, Callahan and Phatak, 1933). If the absorption is slow and the stomach becomes distended, both the distension and the nausea may stimulate the release of ADH and so increase water retention and cause further deterioration of the drinker's condition. Convulsions from overhydration are not seen in any rats until the body weight has been increased by at least 14%, but if the overhydration exceeds 22% all the animals have convulsions (Adolph, 1947d; Adolph and Northrop, 1952) which are preceded by changes in the EEG, but overdrinking does not produce any such changes in the EEG of normal men (de Wardener and Herxheimer, 1957a; personal observations).

In epileptics, however, and in those subject to petit mal, overdrinking may trigger off seizures.

Overdrinking increases both the insensible water loss (Manchester *et al.*, 1931) and true sweating (Lee, Murray, Simmonds and Atherton, 1941). Urinary electrolyte losses rise to as much as, for example, 1 mg Cl/ml of urine after several hours of diuresis (Wolf, 1945a, b). Trials by Darragh, *et al.* (1953), in which electrolyte excretion fell during hypotonicity from continuous infusion of 5% glucose, were complicated because ADH was also given. The urinary losses are increased by overbreathing, from, for instance, fear or heat; as with an alkalosis, tubular reabsorption of electrolyte is decreased (Ullman, 1958). There is also increased electrolyte diffusion into the stomach and lower end of the gut (Welt and Nelson, 1952b), further increasing the risk of salt deficiency. In rats, if not in man, there is an increased excretion of 17-oxosteroids, indicating, in view of the concomitant reduction of the adrenal ascorbic acid, extra adrenocortical activity (Nagareda and Gaunt, 1951). This could be considered a physiological survival response rather than a simple "stress reaction" manifestation of Selye's "general adaptation syndrome" (1949), as adrenal steroids help to maintain an adequate glomerular filtration rate (Renzi, Renzi, Chart and Gaunt, 1956).

The effects of overhydration were subjected to experiment by de Wardener and Herxheimer (1957a, b) on themselves. While on normal diet they drank an extra 10 l/day, in addition to their accustomed intake of tea and coffee, for eleven days. It was "not an enjoyable experience"; their skins were vasoconstricted and they felt cold and clammy. Their intellects dulled; food was tasteless; despite a daily salt intake of 1 meq/kg, they still had a salt hunger, a negative salt balance developing in one of them. They both gained weight, with an increased extracellular fluid volume. Renal concentrating power fell; the maximum attainable osmolarity dropped from a mean of about 1000 m osmol/l to only 350 m osmol/l. This inability to concentrate continued for some time after the experiment ended. Ladd (1952) noted a similar effect in men who overhydrated for 8 hrs before being infused with 0·9% saline; in comparison with controls they showed increased sodium excretion and a higher filtration rate. Rats, unlike men, rapidly adapt to water diuresis when the water is given by stomach tube; if it is given by mouth, the diuresis develops sooner and the chloride excretion is less (Adolph and Parmington, 1948; Adolph and Northrop, 1952). In overhydrated men, however, although there is a fall in the maximum osmolar concentration, there is an increased total solute excretion (Schoun, 1957). Possibly a man who wishes to overhydrate should do it by stomach tube. Another effect of overhydration is that post-exercise ketosis, the Courtice-Douglas effect, develops more quickly; in dehydration it is not seen at all (Passmore and Johnson, 1958). It is theoretically possible to drink oneself to death as easily with water as with alcohol, but, judging by de Wardener's and Herxheimer's experience, far less pleasant.

3. *Potassium gains and losses*

Potassium is as important as or even more important than sodium. It is the main intracellular cation and plays a vital part in the chemistry of muscular contractions; potassium deficiency therefore results in numerous disorders, ranging from drowsiness, irritability, confusion and languor to chronic ileus, abdominal distension and anorexia, and finally to depression of reflexes and flaccidity and paralysis of muscles (Womersley, 1953). Potassium deficiency is a particular hazard of starvation, as potassium, once it leaks out of a cell, is inevitably excreted. Apart from this, potassium deficiency will not develop unless a man is living on a peculiar diet; thus Elkinton (1952) describes how a man tried to slim by living on nothing but ginger ale for four weeks. He was admitted to hospital in potassium deficiency, weak, confused and with absent knee jerks (see also Womersley and Darragh, 1955).

The normal diet is potassium rich, and within limits the potassium content of the body is not affected by variations in intake; it does, however, vary with age, sex, race and body dimensions (Meneely, Heyssel, Ferguson, Weiland, Ball, Lorimer and Meneeley, 1962). Renal conservation of potassium is poor in comparison with that of sodium (Peters, 1953) and even worse if there is a high sodium intake (Squires and Huth, 1959). Nevertheless, it takes some time to develop severe deficiency, unless the potassium intake is virtually nil, as the body's potassium is almost entirely intracellular. Even after sixteen days' restriction to 1 meq of K/day, involving a negative balance of more than 500 meq and a fall in the potassium concentration in the serum from 4 meq/l to less than 3 meq/l (Huth, Squires and Elkinton, 1959), men were still symptomless. Faecal losses of potassium are maintained or even increased during potassium restriction and may be double the urine loss (Squires and Huth, 1959). There is also a metabolic acidosis, as more potassium diffuses out of the cells than sodium passes in, the electrometric discrepancy being made up by the retention of H ions (Huth *et al.*, 1959). The kidneys in potassium deficiency cannot produce a concentrated urine; Manitius, Levitin, Beck and Epstein (1960) showed that in rats this is due to impaired reabsorption of sodium in the collecting tubules, and Conn and Johnson (1956) have described reversible lesions in the renal tubules of potassium-deficient men. Provided there is no sodium deficit, potassium depletion increases the extracellular fluid volume. It may therefore have some survival value. This could be advanced as an argument for "limited mariposia", as the retained sodium would prevent salt deficiency and the amount of potassium in sea water, 10 m moles/l, would not be significant in the amount drunk.

Excess potassium retention is as undesirable as potassium loss. A raised serum potassium leads to paralysis, cardiac irregularities and eventually cardiac arrest (Womersley, 1953), but this only occurs when there is interference with its normal excretion. Normally potassium is not retained; hence the potash eaters of Africa are unaffected so long as they remain healthy (Porteres, 1950). Africans nevertheless may have a high concentration of potassium in their serum; Leschi (1948) quotes a figure of 296 mg for one tribesman, more like the level in infants (5·9 ± 1·4 meq/l) than the normal adult value (4·4 ± 0·3meq/l). Similar high figures have been reported from Brazil (Pales, 1950); the Brazililian negro has low sodium and high potassium contents compared with the local Europeans:

	Na	K
	mg per 100 ml plasma	
European	350 to 380	12 to 20
Brazilian negro	260 to 320	15 to 22

Pales attributed this to climate rather than to diet. The Mauretanians are less tolerant of a high potassium load than the local European "colons". An oral dose of 20 mg/kg of either potassium citrate or potassium chloride did not affect the plasma potassium concentration of the Europeans, but in the Africans caused rises ranging from 15% to 32% of the control levels within 30 min. There was a return to normal within the hour, but not before some subjects complained of nausea and headache (Leschi, 1947). Potash eaters seem to have such a high plasma potassium content that any extra potassium load brings them into the danger zone; a 32% increase on the 296 mg/l already quoted would bring the figure nearly up to 10 meq/l, which can be lethal (Elkinton and Danowski, 1955). There is neither need of nor virtue in a high potassium intake, even in a hot climate; sweat contains only about 4·5 meq K/l (Robinson and Robinson, 1954), and Fregly and Iampietro (1958) reported that a daily intake of 70 to 75 meq of K/day was sufficient for maintaining the performance of men working in the desert. With extremely heavy sweating, on the other hand, the potassium losses are similar to those that may occur in chronic diarrhoea; unless the diet contains at least the usual amount of potassium, a deficiency could consequently occur. This, according to Leithead (1963), could be a contributory cause to the weakness and lethargy seen in certain heat disorders. It is unlikely, however, that significant potassium deficiency would develop as an acute condition.

The rise in serum potassium seen in Africans after oral potassium was partially suppressed by deoxycorticosterone acetate (DOCA), the maximum rise after 10 mg being only 90%. When DOCA was given to subjects who had not had oral potassium there was a 10% fall according to Leschi (1948), who attributed this to inadequate adrenal activity in "melanoderms". Henrotte (1963) reported similar low potassium tolerances in the population of Madras, but he attributed it to climatic rather than racial factors. This reduced potassium tolerance occurs not only in the indigenous Southern Indians but also in resident Europeans (Henrotte, Krishnamurthi and Ranganathan, 1960; Henrotte and Krishnaraj, 1962). Henrotte suggests that adrenal activity varies with the ambient temperature. The significance of this work in relation to the heat tolerance of indigenous tropical men has been discussed elsewhere (Ladell, 1963).

C. Special Factors in Water and Salt Metabolism

1. *Infants*

In normal circumstances there are for infants no special survival problems related to salt and water intakes. With a larger surface area to body-weight ratio than the adult, an infant might be expected to have a proportionately greater insensible loss, but no observations seem to have been made on this point. The barrier layer in rodent skin appears in the last few days of foetal life (Tregear and Hunt, unpublished observations); hence, if it is the same in man, the skin of the infant will be no more permeable than that of the adult. The newborn pig responds to heat by increasing the respiratory evaporative loss, but there is little change in the evaporative skin loss (Mount, 1962), indicating that the piglet is already losing a maximum amount of water, and so of heat, through this route. Such limited evidence as there is, therefore, suggests that insensible perspiration is the same in the infant as in the adult.

Thermoregulation in infants is not as precise in adults. Their body temperatures rise more readily (McCance, 1959); consequently they might be expected to sweat more. Controlled observations on this matter, however, do not seem to have been made, but any difference might be marginal for water balance. A child is, however, as susceptible to the effects of heavy sweating as an adult, and watch must be kept for any who may have a high concentration of sodium chloride in the sweat, as in congenital cystic fibrosis (di Sant'Agnese, 1960). Greater threats to the water and salt metabolism of an infant are diarrhoea and vomiting.

Infants are more susceptible than adults to dehydration, because their kidneys cannot concentrate so well. Reabsorption is poor; thus, bicarbonate is not easily reabsorbed by the newborn piglet, and water containing bicarbonate is excreted as readily as pure water, whereas water containing chloride is retained. In certain circumstances this could lead to dehydration (McCance and Widdowson, 1963). There is also a lower ADH activity in infants; although Dicker and Tyler (1953) could detect vasopressin and oxytocin activity in the pituitaries of human foetuses from seventy days onwards, there is no detectable ADH in human circulating plasma until the fifth month of life. ADH secretion can, however, be elicited by hypertonic injections at only 2½ months (Martinek, Janovsky and Stanincova, 1962), showing that the infant can already respond to an emergency. Whether this is a process of natural development, or a response to the change in posture as the child begins to sit up, is not known; it would be interesting to have the results of investigations on the infants of those peoples whose mothers carry their babies strapped upright to their backs from the day they are born. Although the younger the infant the lower the osmolarity of the urine, the response of all infants to a diminished fluid intake is a hyperoncotic urine; this is mainly due to urea and reflects an increased urea content of the plasma (Martinek *et al.*, 1962).

In general, infants and children have the same survival problems as do adults in coping with salt and water restriction. It must be remembered, however, than an infant's salt and water metabolism is less finely adjusted and that it can neither signal its thirst nor take steps on its own to satisfy it. The only safe rule is that, if the mother is thirsty, so probably is the child. The water-drinking habits established in infancy continue until the teens; according to Walker, Margolis, Teate, Weil and Wilson (1963), the average intake of tap water by American children (from Michigan, Georgia, Florida and New Mexico) was about 500 ml/day for all children from a few months old until the age of 12 years; their total fluid intake rose meanwhile from an average of 707 ml/day at 2 months to 1247 ml/day at 12 years—from one bottle of "cola" per day to three.

2. *Interaction of water and salt metabolism with diet*

a. Foodstuffs. The most important dietary factors in water economy are the protein and electrolytes. If these are kept to a minimum, the obligatory urine volume may be reduced to less than 200 ml/day (Winkler, Danowski, Elkinton and Peters, 1944). Any increase in the protein content of the diet above that required for nitrogen equilibrium increases the urine output in proportion to the extra amount of urea to be excreted (Danowski, Elkinton and Winkler, 1944). Traditional survival foods, such as pemmican, are not good for this reason; thus, the ingestion of 30 g of "Board of Trade pemmican" by a water-deficient man led to sacrifice of 2 extra oz of water (personal observations). Bombard succeeded in his survival expedition, not because of his fish juice, but despite it. He may also have been lucky in his choice of fish; the seal, for example, can get all the water it requires from herring, but would not be able to do so from haddock (Prentiss, Wolf and Eddy, 1959). The castaway should therefore choose his fish carefully; even if he does so, the protein content of the juice would make the exercise of doubtful value. The cynic might say that he would be better occupied in praying for rain; at least the metabolic effort would be less.

It is important that a man should eat as well as drink. For effective water retention there must be a minimum food intake. Sargent and Johnson (1956) claim that even a dehydrated subject will not retain much of the water he drinks unless there is an osmotic excretion from his diet of at least 675 m osmoles. The body so adjusts its osmolar content that water cannot be held unless the missing osmoles are made good. They concluded that "diets of very low osmotic effect such as pure carbohydrate under conditions of enforced deprivation do not conserve total body water as effectively as diets of intermediate osmotic effect." Not everyone would agree with them. Hervey and McCance (1952) found that men restricted to 250 ml water and 100 g of carbohydrate per day lost 230 ml/day less body water than when drinking 100 ml/day and eating nothing, and their body tonicity rose only by 0·41% instead of by 0·64%. Sargent and Johnson's (1956) results may have been complicated by electrolyte losses in the sweat. Their experiments were carried out under comparatively warm conditions. Carbohydrate not only provides calories, but also saves its own weight in body water by sparing body protein and by reducing ketosis (Gamble, 1944). One-third of the water-sparing action of glucose was due to this reduction in starvation ketosis. Though for every gram of body protein spared from metabolism, between 2 g and 3 g less preformed water is released, this is counterbalanced by the absence of osmolar load from the metabolised protein. Gamble (1944) was the first to call attention to this metabolic legerdemain later observed by McCance's team; he recommends a high glucose intake by the castaway, not merely because it saves water but also because it prevents "the unhappy lassitude of fasting" and so "contributes to the courage of the castaway". Wolf (1956) endorses this. Sodium excretion is inhibited by carbohydrate (Bloom, 1962); at the beginning of a fast, therefore, in which small amounts of carbohydrate are given, this means sodium and hence water retention, indicated by a slight weight gain. Radford (1959) has queried the value of a protein-free diet in water conservation by men. He finds that the urine volume of hydropenic rats increases, although their osmolar excretion decreases when the animals are placed on a protein-free diet, suggesting that urea is essential to enable the kidneys to concentrate maximally. He also suggests that strong electrolytes are needed in the diet to evoke endogenous ADH release. His work should not be taken as giving *carte blanche* to the castaway to live on fish juice and drink sea water; men are not rats nor are the respective osmotic burdens of urea and electrolytes completely interchangeable.

b. Alcohol. Alcohol surpasses even fat as a source of metabolic water; 100 ml produces 117·4 ml. For every two molecules of alcohol metabolised, however, three molecules of oxygen are required; as with fat, therefore, the metabolic gains may be lost in the increased ventilation required (Mellanby, 1942). The diuretic effect of alcohol, moreover, is not merely the inconvenient social result of too large a fluid intake, but a real pharmacological effect (Eggleton, 1942). It blocks the release of ADH (Kleeman, Rubini, Lamdin and Epstein, 1955) in response to an increase in tonicity; after 120 ml of Kentucky Bourbon there were falls in the urinary clearances of sodium, potassium and chloride (Rubini, Kleeman, and Lamdin, 1955). Alcohol should therefore be avoided in water deficiency; it will increase the water loss and may cause electrolyte accumulation, apart from any effects it may have on physical ability (Hebbelinck, 1963). If Pablo had had his canteen filled with alcohol he would never have survived.

D. Man's Appetites and Stratagems for Survival

1. *Salt and water appetites*

a. Salt. Despite the failure of thirst as an adequate guide under extreme conditions, if a man relies on his appetite he will never be in trouble on a dry land in a temperate climate. If he finds himself adrift in the ocean, however, or strays into desert or tropical lands, or has to work hard under hot industrial conditions, then the two twin needs, of water and of salt, become imperative. Miners, steelworkers and stokers have long appreciated this need and traditionally have drunk pints of salted beer or barley water each night. The fluid is certainly a necessity, but Kaunitz (1956a) believes the salt is satisfying mainly an artificial appetite. The salt intake varies independently of what used to be considered normal physiological requirements; for example, the British high-temperature worker consumes as much as 24 g salt daily (Weiner and van Heyningen, 1952), whereas in Northern Italy, where carbohydrates form a larger proportion of the diet than in this country, the daily intake of the high-temperature worker may, if not supplemented, be as low as 4·1 g and never rises much above 10 g (Parmeggiani, 1959). Richter (1956) points out that the level of salt intake is a matter of cultural practice and that this may be so compelling that there is a danger, if there is not a good water supply, of a man getting "into trouble in using salt because of an inadequate ingestion of water". Other primates, such as the chimpanzee (Hill, 1962) also have a taste for salt; here Richter's second posited cause of salt hunger, an endocrine-based craving, may be operating, unless the chimpanzee, like man, just enjoys a salty tang to his food and "recognises the role of salt as a provocative agent in consummatory activity" (Kaunitz, Geller, Slanetz and Johnson, 1960).

b. Water. Although the salt appetites of different races and different cultures differ widely, there are no such variations with water requirements. Pales (1950) comments that "la soif d'eau des Noirs est classique" and says that the negroes of the Tchad drink 6 to 8 l, when Europeans would only take 4 to 5 l under the same conditions, and that in convoys lost in the desert "les Noirs" are the first victims; no doubt "les Noirs" are those who have to do the most work and sweat more. Ducros (1956) found little difference between the sweat losses of Mauretanians and Europeans under the same conditions, but Senegalese sweated 20% more. This difference was probably due to acclimatisation, but they drank 40% more than the Europeans, more than enough to cover the extra sweat loss, which suggests that "la soif des Noirs" is a matter of habit rather than of necessity As the concentrating power of the kidney in West Africans is lower than that in Europeans (Kenney, 1953), their minimum water requirements must be greater, but Ducros's subjects were passing volumes of urine well above the obligatory figure. In Singapore the mean fluid intake by Malays was found to be only 1368 ml/day, against a daily urine output of 937/ml. Clearly these men must have in their food a large fluid intake that was not accounted for (Whittow, 1956). Adam, Ellis and Lee (1953) found differences in water turnover between Asians and Europeans; after taking all factors into consideration, however, they concluded that these were more probably due to differences in acclimatisation than to racial differences. West Africans also sweat slightly more than unacclimatised Europeans and so require more water for the same reason (Ladell, 1950, 1963). Caucasians in the Arizona desert lost 10 l of sweat per day (Dill, Jones, Edwards and Oberg, 1933), but Welch, Buskirk and Iampietro (1957) found no increased water intake in U.S. soldiers until the mean ambient temperature reached 15·7°. It then rose steeply, and at 37·8° their mean intake was 8 l/day; this may be compared with the water turnover of West African soldiers under approximately the same climatic conditions in Northern Nigeria, which was 9 l/day if shade was available and an extra 2 l if they were out in the sun (Kenney and Ladell, 1953). The mean water loss, by all routes, of European oil drillers in the Sahara with temperatures up to 44·4° was 5334 ml/day, against a water intake, including what was in the food, of 4829 ml/day, the balance being made up, as always, by the water of oxidation (Metz, Lambert, Hasselman and van Hove, 1959). The ability of the Saharan oilmen to maintain health and efficiency on so much less water than others in desert conditions demonstrates the value of the modern way of avoiding water deficiency—the use of air conditioning.

As such large quantities of water are required by men under desert conditions, the first problem to be met in avoiding disastrous water deficiency is one of logistics. The official British Army allowance was, until recently, 9 l (2 gal) of water per man-day for all purposes; in the Nigerian manoeuvres the utilisation was found, in fact, to be 22 l (5 gal); of this 7 l (1·6 gal) were used in the cookhouse, 6 l (1·4 gal) for ablutions, vehicles and other purposes, leaving the remaining 2 gal for personal consumption, the 9 l already mentioned. Recent unpublished results from America paint an even more difficult logistic picture: the drinking requirements alone of troops working 8 hr/day in the sun at 43·30 is 17 l (4·5 U.S. gal). If a man is to remain effective, it is absolutely essential that these personal water needs be met. Sohar, Tennenbaum and Yaski (1962) have provided the logisticians with a nomogram, based on field observation in Israel; by using it the water requirements of marching men may be predicted from the wet- and dry-bulb temperatures and either the distance or the hours marched. Allowance should also be made for off-duty activities of the men, such as the British soldier's football or the African's propensity to go "hunting" in the bush, which may lead to considerable water losses.

2. *Fulfilling requirements*

The water supplies of an army in the desert are the responsibility of the all-powerful and ubiquitous "they", but "they" are not there to look after Jack adrift in a lifeboat or Pablo stumbling along in the desert. The African labourer toiling away under the sun in an opencast mine may wonder how he is going to get the salt with which to add savour to the handful of dry millet that will make up his main meal.

a. Salt. The African's problem is simpler than Jack's or Pablo's. There are some seventy different plants he can burn for their ash, ranging from the lily, *Gloriosa superba*, to certain palms. He can sample, if he is in the right areas, one of the numerous beds of saline soil just a few feet underground or partake of one of the saline springs, which are reputed to "stimulate the appetite, the generic and reproductive functions; regenerate the organism, cure diarrhoea", in fact "tout en étant dépurative" (Adriaens and Waegemans, 1943). Saline earths may contain anything from 2% to 94% of sodium chloride, but occasionally the main anion may be sulphate, which is particularly prized (Soula, 1950); calcium, magnesium and potassium carbonates are also found. Primitive man in Africa, Asia and elsewhere, however, may consume twice as much plant salt, which is almost exclusively potassium carbonate, as sodium chloride (Porteres, 1950), but there is no evidence that this does him any harm. Kaunitz (1956b) suggests that not enough attention is paid to the anion; supplements of sodium chloride increase the kidney and adrenal weights in the rat, whereas neither bicarbonate (sodium) nor potassium (chloride) has this effect. But the primitive man seeking salt doesn't have survival problems; with good reactive adrenals he will never become salt deficient, whatever the anion or cation with which he flavours his millet. A completely salt-free diet is a dietetic curiosity, and modern man is unlikely to lack salt.

b. Water. The man without water is an infinitely worse risk than the man without salt. Nothing can replace water, and without water a man will die as surely as he would if he lacked air. The first thing a man must do if he is going into the desert is to make sure that his water bottles are both full and large; the best canteen, McGee (1906) says, should hold at least 2 gal, "but three would be better." What water there is must be used to the best advantage; he should refrain from drinking until his "free circulating water" has been exhausted; hence the MRC's (1943) advice not to drink during the first 24 hrs adrift in a lifeboat. This advice should be extended to men lost in the desert, not with a time limit, but with the instruction that drinking should begin when a heavily pigmented concentrated urine indicates that minimum obligatory flow is being approached. Judicious drinking may then be begun without fear of a wasteful diuresis; the "biological precognition" seen in hot rooms is not seen under field conditions. When water is drunk, either in normal circumstances or in conditions of shortage, it is most economical to take it "little and often"; Kenney (1954) showed that up to 400 ml of water could be saved in this way over a 90 min period of work in the heat. The practice of slightly salting water is bad. Vaguely salty water tastes insipid, so not enough is drunk, and this leads to voluntary dehydration; if the water is clearly salty, man behaves like other animals and requires more to slake his thirst; this leads to waste, as extra urine is passed to excrete the excess salt (Holmes and Montgomery, 1960; Edmonds, 1960).

3. *Ensuring survival*

A man should drink his fill before setting out into a thirsty world. His stomach should be full of water; this is the only way in which a man can store water. Absorption is not instantaneous; a man may be sweating hard before it is well established. When water absorption begins, therefore, there are existing sweat losses to be replaced, and correspondingly less is lost in the urine; the greater the sweat rate, the greater the value of this manoeuvre. A man should "tank up", however, not with 0·9% saline, as suggested by Adolph (1947a), but with water; if a man consumed 1 l of 0·9% saline and saved 750 ml of his own body water, he would end up, owing to hypotonicity of the sweat, with a functional salt excess of 12 or 14 g, which would eventually have to be excreted in the urine. "Tanking up" could also be of value to a non-sweating man. In an emergency the dilute urine passed could be collected and drunk, either after treatment or as it was. It is only when the urine becomes concentrated that the kidney is being made to do the same job twice when urine is drunk. Urine passed during the first 12 hrs would certainly help with the water balance, and it should be possible to work out a simple test for physiological potability, based perhaps on the behaviour of a drop of blood dropped into the fluid.

Voluntary dehydration must be studiously avoided. A man in the desert never knows when he may not have need of his free circulating water. D. H. K. Lee (1963) shares my view that voluntary dehydration is not a true dehydration, but the veering towards the physiological norm of a man who is habitually superhydrated. Adolph (1947a, b, c, d) showed that overhydration does not improve performance in animals, whereas dehydration diminishes it. The fact, therefore, that a man can lose about 2·5 l of body water before his performance is affected (Ladell, 1955) shows that this "free circulating water" is in excess of requirements. This excess represents a valuable reserve, which a man in the desert should retain at all costs. It is not easy to do this; it is an unpalatable task to settle down each evening to drink several litres of insipid water, and to drink as one sweats during the day is not practicable, even if water is on the spot, as absorption is slowed up in the heat and the stomach becomes uncomfortably full of water. There would appear to be sound teleological reason for the development in many tropical and desert cultures of cheap, relatively or completely non-alcoholic beverages, for example, palm toddy in Africa or Asia and the delicate soft drinks of the Middle East.

E. Conclusions

The belief that man can be acclimatised to a low water intake has probably been the indirect cause of many deaths. Whittow (1956) claimed that there was evidence of acclimatisation to a reduced water intake from results he collected in Singapore; however, judging by the volumes of urine passed, his subjects were merely showing habituation to a lower but still adequate water intake. There are also suggestions, from unpublished observations on men flown out to a desert climate, that the body readjusts to a new water balance, probably losing both water and salt in the first few days in a hot climate, but there was no indication that water requirements lessened; total sweat losses, in fact, increased with acclimatisation to heat. D. H. K. Lee (1963) suggests that training of the posterior pituitary could occur; this might conserve water in the early stages of deprivation, but not once the urine was being maximally concentrated. Weiner (1954) states that there is no evidence that men can be hardened to a low water level, confirming the earlier conclusions of Adolph *et al.* (1947a) from their work in the Arizona desert. There is still no reason to revise this conclusion. A man must learn to conserve his water by keeping in the shade and not "going out in the midday sun."

Gamble (1944) found that the extra-renal losses of men in life-rafts was quadrupled when they sat in the sun; and Brown (in Adolph, 1947) advised men lost in the desert to walk only at night. In hot dry climates loose light clothing should be worn, both to keep the sun off and to prevent the desiccating effect of a desert wind blowing directly on the body or, if it is a hot moist wind blowing in from the sea, from warming the skin and stimulating greater sweating. Whenever the air temperature is below the skin temperature, however, as at sea or in the wet tropics, wind is an advantage, as it keeps the body cool and may prevent sweating.

Desert mammals make their own moist climate in their burrows and so reduce their cutaneous and respiratory water losses (Chew, 1961; Schmidt-Nielsen, 1962). Man cannot emulate the desert rat, but by breathing into a relatively confined space or into a blanket he might be able to reduce his respiratory losses by 100 ml or more per day. Dehydration and human heat killed the victims of the Black Hole of Calcutta, not lack of air; a party lost in the desert should not shelter from the sun in a cave unless it is large and well ventilated. "Black Hole" effects may also occur in temperate climates, as trials of underground air-raid shelters have shown, fortunately without fatalities; this may be a survival problem of the future.

A man who lacks water must be careful of his diet. The Bedouin's dates are an ideal desert or lifeboat ration; dried and stoned they contain 64% carbohydrate and 0·3% of nitrogen; their sodium content is negligible and they provide 14·6% of water (McCance and Widdowson, 1939). Nothing could be worse than the solid protein of "biltong" or "jerky", especially if salted. Occasionally a water bonus may be picked up in the desert – a cactus or a buried fungus or a water-rich root. Desert dwellers, such as the Kalahari bushmen or the Australian aborigines, know where to find these (Weiner, personal communications). There is also the desert snail, the haemolymph of which is potable, but a man would require 2,500 of these per day to keep himself alive (Billingham, 1961).

The man adrift in his lifeboat can adopt the same subterfuges. Shade from the sun and proper clothing are most important; unless a man is convinced he is going to reach safety within a few days, he should never drink sea water. On the principle, however, that the longer he remains reasonably fit the better chance there is of his making safety, he should not hoard his water; it is better to drink a cupful of water for ten days and still be relatively fit when all the water is gone, than ration it to a couple of teaspoonfuls a day and die of dehydration after a week, with water breakers still more than half full. Tragically there have been examples of this happening.

Body water can be conserved, not merely by avoiding protein foods, but by using sea water to cool the body and prevent sweating if it is hot. Sea water soaked clothes will cool as well as sweat, the sea itself offers some refuge from the sun, and if the skin takes up water into its stratum corneum this again will evaporate and cool.

The most important manoeuvre of all, however, is to make sure that no water is lost – not the miserable little quantity in the water breakers, 110 oz/man, but the water that falls as rain. Provision must be made for catching and storing this. Bombard survived, not because he drank fish juice, nor even despite it, but because he ran through heavy rain storms and was able to collect enough rain to fill his air bed; most of his voyage he was on ample water rations. Not even extreme fortitude nor confidence in one's own theories can ever compensate for lack of water. Water is the stuff of life.

REFERENCES

Adam, J. M., Ellis, F. P., and Lee, T. S. (1953). *M.R.C. Memor.* R.N.P. Rep. No. 53/749, Series No. 298: H.M.S.O.
Adolph, E. F. (1933). *Physiol. Rev.* **13**, 336.
Adolph, E. F. (1945). *Sci. Mon. N.Y.* **61**, 57.
Adolph, E. F. (1947a) (in association with Brown, A. H., Goddard, D. R., Gosselin, R. E., Kelly, J. J., Molnar, G. W., Rahn, H., Rothstein, A., Towbin, E. J., Wills, J. H., and Wolf, A. V.). *In* "Physiology of Man in the Desert". Interscience, New York.
Adolph, E. F. (1947b). *Amer. J. Physiol.* **151**, 564.
Adolph, E. F. (1947c). *Amer. J. Physiol.* **151**, 110.
Adolph, E. F. (1947d). *Annu. Rev. Physiol.* **9**, 381.
Adolph, E. F. (1948). *Amer. J. Physiol.* **155**, 309.
Adolph, E. F. (1950). *Amer. J. Physiol.* **161**, 374.
Adolph, E. F., Barker, J. P., and Hoy, P. A. (1954). *Amer. J. Physiol.* **178**, 538.
Adolph, E. F., and Northrop, J. P. (1952). *Amer. J. Physiol.* **168**, 320.
Adolph, E. F., and Parmington, S. L. (1948). *Amer. J. Physiol.* **155**, 317.
Adriaens, L., and Waegemans, G. (1943). *Mem. Inst. colon. belge* Tome XII, Fasc. 3. Van Campenhout, Brussels.
Andersson, B. (1962). *Proceedings of the XXII International Congress Physiological Sciences, Leiden.* **1**, 661.
Andersson, B., and McCann, S. M. (1955). *J. Physiol.* **129**, 33P.
Baker, G., Levitin, H., and Epstein, F. H. (1961). *J. clin. Invest.* **40**, 867.
Baratz, R. A., and Ingraham, R. C. (1960). *Amer. J. Physiol.* **198**, 565.
Barker, J. P., and Adolph, E. F. (1953). *Amer. J. Physiol.* **173**, 495.
Barker, J. P., Adolph, E. F., and Keller, A. J. R. (1953). *Amer. J. Physiol.* **173**, 233.
Bartter, F. C., Biglieri, E. G., Pronove, P., and Delea, C. S. (1958). *In* "Aldosterone", p. 110 (A. F. Muller and C. M. O'Connor, eds.). Churchill, London.
Bass, D. E., Buskirk, E. R., Iampietro, P. F., and Magger, M. (1958). *J. appl. Physiol.* **12**, 186.
Beetham, W. P., and Buskirk, E. R. (1958). *J. appl. Physiol.* **13**, 465.
Berenson, C. S., and Burch, C. G. (1951). *Amer. J. trop. Med.* **31**, 842.
Billingham, J. (1961). Flying Personnel Research Committee Report No. 1155.
Birchard, W. H., Rosenbaum, J. D., and Strauss, M. B. (1953). *J. appl. Physiol.* **6**, 22.
Black, D. A. K. (1952). "Sodium metabolism in health and disease." Blackwell Scientific Publications, Oxford.
Black, D. A. K., McCance, R. A., and Young, W. F. (1944). *J. Physiol.* **102**, 406.
Blair-West, J. R., and Goding, J. R. (1962). *Endocrinology* **70**, 822.
Bloch, M. R. (1963). *Sci. Amer.* **209**, 88.
Bloom, W. L. (1962). *Arch. intern. Med.* **109**, 26.
Bojesen, E., and Degn, H. (1961). *Nature (Lond.)* **190**, 352.
Boyd, E. M., and Shanas, M. N. (1963). *Arch int. Pharmacodyn.* **144**, 86.
Brebbia, D. R., Goldman, R. F., and Buskirk, E. R. (1957). Tech. Rep. No. EP-57. Quartermaster Res. and Dev. Center, Natick, Massachusetts.
Brebner, D. F., Kerslake, D. McK., and Waddell, J. L. (1956). *J. Physiol.* **132**, 225.
Brooks, F. P., and Pickford, M. (1956). *Proceedings of the 8th Symposium of the Colston Research Society, Bristol*, 9 to 12 April 1956, Vol. VIII. Butterworth, London.
Brown, J. J., Davies, D. L., Lever, A. F., and Robertson, J. I. S. (1963). *Lancet* ii, 278.
Buettner, K. J. K. (1953). *J. appl. Physiol.* **6**, 229.
Buettner, K. J. K. (1959a). *J. appl. Physiol.* **14**, 261.
Buettner, K. J. K. (1959b). *J. appl. Physiol.* **14**, 269.

Buettner, K. J. K., and Odland, G. F. (1957). *Fed. Proc.* **16**, 18.
Burch, G. E. (1945). *Arch. intern. Med.* **77**, 308; **77**, 315.
Burch, G. E., and de Pasquale, N. P. (1962). "Hot Climates, Man and his Heart", p. 112. Thomas, Springfield, Illinois.
Burg, M. B., Papper, S., and Rosenbaum, J. D. (1961). *J. Lab. clin. Med.* **57**, 533.
Buskirk, E. R., Iampietro, P. F., and Bass, D. E. (1958). *J. appl. Physiol.* **13**, 189.
Camp, J. L., Tate, F., Fla, M., Lowrance, P. B., Wood, J. E., and Va, C. (1958). *J. Lab. clin. Med.* **52**, 193.
Camp, J. L., and Va, C. (1958). *J. Lab. clin. Med.* **52**, 202.
Cannon, W. R. (1917–18). *Proc. roy. Soc.* B.**90**, 283.
Chew, R. M. (1951). *Ecological Monographs* **21**, 215.
Chew, R. M. (1961). *Biol. Rev.* **36**, 1.
Chew, R. M. and Dammann, A. E. (1961). *Science* **133**, 384.
Christie, R. V., and Loomis, A. L. (1933). *J. Physiol.* **77**, 35.
Conn, J. W. (1961). *Arch. intern. Med.* **107**, 813.
Conn, J. W. (1963). *J. Amer. med. Ass.* **183**, 775; 871.
Conn, J. W., and Johnson, R. D. (1956). *Amer. J. clin. Nutr.* **4**, 523.
Cort, J. H. (1962). *Proceedings of the XII International Congress of Physiological Sciences, Leiden* **1**, 665.
Cort, J. H. (1954). *J. Physiol.* **121**, 41P.
Coxon, R. V., Dupré, J., and Robinson, R. J. (1958). *Quart. J. exp. Physiol.* **43**, 86.
Crabbe, J. (1961). *Nature Lond.* **191**, 817.
Crane, W. A. J., and Ingle, D. J. (1959) *Endocrinology* **65**, 693.
Crisman, J. M. (1960). *Annu. Rev. Physiol.* **22**, 317.
Critchley, M. (1943). *Brit. med. J.* i, 726. "A Medical Story." Churchill, London.
Critchley, M., and Allison, R. (1941). *M.R.C. Spec. Paper* 15.
Cross, R. B., Dicker, S. E., Kitchin, A. H., Lloyd, S. and Pickford, M. (1960). *J. Physiol.* **153**, 553.
Dahl, L. K. (1961). *J. exp. Med.* **114**, 231.
Dahl, L. K., Heine, M., and Tassinari, L. (1962a). *Nature, Lond.* **194**, 480.
Dahl, L. K., Heine, M., and Tassinari, L. (1962b). *J. Exp. Med.* **115**, 1173.
Dahl, L. K., Silver, L., Christie, W. C., and Genest, J. (1960). *Nature, Lond.* **185**, 110.
Danowski, T. S., Elkinton, J. R., and Winkler, A. W. (1944). *J. clin. Invest.* **23**, 816.
Danowski, T. S., Winkler, A. W., and Elkinton, J. R. (1946) *J. clin. Invest.* **25**, 130.
Darragh, J. M., Welt, L. G., Goodyer, A. V. N., and Abele, W. A. (1953). *J. appl. Physiol.* **5**, 658.
Denton, D. A., and Sabine, J. R. (1961). *J. Physiol.* **157**, 97.
de Wardener, H. E., and Herxheimer, A. (1957a). *J. Physiol.* **139**, 42.
de Wardener, H. E., and Herxheimer, A. (1957b). *J. Physiol.* **139**, 53.
Dicker, S. E., and Nunn, J. (1957). *J. Physiol.* **136**, 235.
Dicker, S. E., and Tyler, C. (1952). *J. Physiol.* **117**, 28P.
Dicker, S. E., and Tyler, C. (1953). *J. Physiol.* **121**, 206.
Dill, D. B. (1938). "Life, Heat and Altitude", p. 211. Harvard University Press, Cambridge.
Dill, D. B., Jones, B. F., Edwards, H. T., and Oberg, S. A. (1933). *J. biol. Chem.* **100**, 755.
Dirnhuber, P., and Tregear, R. T. (1960). *J. Physiol.* **152**, 58P.
di Sant'Agnese, P. A. (1960). *J. Amer. med. Ass.* **172**, 2014.
Dobson, R. L., Abele, D. C., and Hale, D. M. (1961). *J. invest. Dermat.* **36**, 327.
Dobson, R. L., and Lobitz, W. C. (1957). *Arch. Derm.* **75**, 653.
Doupe, J., Ferguson, M. H., and Hildes, J. A. (1957). *Canad. J. biochem. Physiol.* **35**, 203.

Du Bois, E. F. (1948). "Fever and the Regulation of Body Temperature", p. 29. C. C. Thomas, Springfield, Illinois.

Ducros, H. (1956). *Contribution a l'Etude de la Deshydratation en Milieu Tropical et Desertique*, pp. 60 *et seq*. Drouillard, Bordeaux.

Edmonds, C. J. (1960). *Quart. J. exp. Physiol.* **45**, 163.

Eggleton, M. G. (1942). *J. Physiol.* **101**, 172.

Eisenstein, A. B. (1959). *Proc. Soc. exp. Biol. N.Y.* **101**, 850.

Elkinton, J. R. (1952). *In* "Advances in Medicine and Surgery", pp. 77–94. W. B. Saunders & Co., Philadelphia.

Elkinton, J. R. (1960). *Circulation* **21**, 1184.

Elkinton, J. R., and Danowski, T. S. (1948). *J. clin. Invest.* **27**, 74.

Elkinton, J. R., and Danowski, T. S. (1955). "The Body Fluids." Williams and Wilkins, Baltimore.

Elkinton, J. R., and Winkler, A. W. (1944a). *War Medicine* **6**, 241.

Elkinton, J. R., and Winkler, A. W. (1944b). *J. clin. Invest.* **23**, 93.

Epstein, A. N. (1960). *Science* **131**, 497.

Epstein, F. H., Goodyer, A. V. N., Laurason, F. D., and Relman, A. S. (1951). *J. clin. Invest.* **30**, 63.

Epstein, F. H., Kleeman, C. R., and Hendrikx, A. (1957). *J. clin. Invest.* **36**, 629.

Epstein, F. H., Kleeman, C. R., Lamdin, E., and Rubini, M. E. (1956). *J. clin. Invest.* **35**, 308.

Fitzsimons, J. T. (1961). *J. Physiol.* **159**, 297.

Fletcher, K. A., Leithead, C. S., Deegan, T., Pallister, M. A., Lind, A. R., and Maegraith, B. G. (1961). *Ann. trop. Med. Parasit.* **55**, 4.

Fourman, P., and Leeson, P. M. (1959). *Lancet* i, 268.

Foy, H., Altman, A., and Kondi, M. (1952). *S. Afr. med. J.* **120**, 113.

Fregly, M. J. (1959). *Proc. Soc. exp. Biol., N.Y.* **102**, 299.

Fregly, M., and Iampietro, P. F. (1958). *Metabolism* **7**, 624.

Freyberg, R. H., and Grant, R. L. (1937). *J. clin. Invest.* **16**, 729.

Gamble, J. L. (1944). *Proc. Amer. phil. Soc.* 88, 151.

Gann, D. S., Mills, I. H., Cruz, J. F., Casper, A. G. T., and Bartter, F. C. (1960). *Proc. Soc. exp. Biol., N.Y.* **105**, 158.

Ganong, W. F., and Forsham, P. H. (1960). *Annu. Rev. Physiol.* **22**, 579.

Garrod, O. (1958). *In* "Aldosterone", pp. 127–9. (A. F. Muller and C. M. O'Connor, eds.). Churchill, London.

Gauer, O. H., and Henry, J. P. (1963). *Physiol. Rev.* **43**, 423.

Gaunt, R., Renzi, A. A., and Chart, J. J. (1955). *J. clin. Endocrin.* **15**, 621.

Gilbert, G. J. (1957). *Amer. J. Physiol.* **191**, 243.

Glaser, E. M. (1949). *J. Physiol.* **109**, 366.

Glaser, E. M., Berridge, F. R., and Prior, K. M. (1950). *Clin. Sci.* **9**, 181.

Glickman, N., Hick, F. K., Keeton, R. W., and Montgomery, M. M. (1941). *Amer. J. Physiol.* **134**, 165.

Goodyer, A. V. N., and Seldin, D. W. (1953). *J. clin. Invest.* **32**, 242.

Gordon, D. B., Drury, D. R., and Schapiro, S. (1953). *Amer. J. Physiol.* **175**, 123.

Gowerstock, A. H., Mills, J. N., and Thomas, S. (1958). *J. Physiol.* **143**, 9.

Grande, F., Anderson, J. T., and Taylor, H. L. (1957). *J. appl. Physiol.* **10**, 430.

Grande, F., Monagle, J. E., Buskirk, E. R., and Taylor, H. L. (1959). *J. appl. Physiol.* **14**, 194.

Grande, F., Taylor, H. L., Anderson, J. T., Buskirk, E., and Keys, A. (1958). *J. appl. Physiol.* **12**, 202.

Haldane, J. S. (1923). *Brit. med. J.* i, 986.

Hale, F. C., Westland, R. A., and Taylor, C. L. (1958). *J. appl. Physiol.* **12**, 20.
Hall, J. F., Jr., and Klemm, F. K. (1963). *Fed. Proc.* **22**, 284.
Haight, A. S., and Weller, J. M. (1962). *Amer. J. Physiol.* **203**, 1144.
Hare, K., Hickey, R. C., and Hare, R. S. (1941). *Amer. J .Physiol.* **133**, 316.
Hebbelinck, M. (1963). *Arch. int. Pharmacodyn*, **143**, 247.
Hellman, K., and Weiner, J. S. (1953–4). *J. appl. Physiol.* **6**, 194.
Henrotte, J. G. (1963). *Fed. Proc.* **22**, 907.
Henrotte, J. G., Krishnamurthi, G., and Ranganathan, G. (1960). *Nature, Lond.* **187**, 328.
Henrotte, J. G., and Krishnaraj, P. S. (1962). *Nature, Lond.* **195**, 184.
Henry, J. P., and Pearce, J. W. (1956). *J. Physiol.* **131**, 572.
Hertzman, A. B., and Ferguson, I. D. (1958). *Fed. Proc.* **17**, 69.
Hertzman, A. B., and Ferguson, I. D. (1959). *Wright Air Develop. Center Technical Report* 59–398.
Hertzman, A. B., Randall, W. C., Peiss, C. N., and Seckendorf, R. (1952). *J. appl. Physiol,* **5**, 153.
Hervey, G. R., and McCance, R. A. (1952). *Proc. roy. Soc.* B. **139**, 527.
Herxheimer, A., and Woodbury, D. M. (1960). *J. Physiol.* **151**, 253.
Hill, W. C. O. (1962). *Nature, Lond.* **194**, 731.
Holmes, J. H., and Montgomery, V. (1960), *Amer. J. Physiol.* **199**, 907.
Hopper, J., Elkinton, J. R., and Winkler, A. W. (1944). *J. clin. Invest.* **23**, 111.
Horne, G. O., and Mole, R. H. (1949). *Lancet* ii, 279.
Horne, G. O., and Mole, R. H. (1951). *Trans. R. soc. trop. Med. Hyg.* **45**, 79.
Hulet, W. H., and Richardson, J. R. (1962). *J. appl. Physiol.* **17**, 67.
Hulet, W. H., Shapiro, T., Schwarcz, B. E., and Smith, H. W. (1963). *J. appl. Physiol.* **18**, 186.
Huseby, R. A., Reed, F. C., and Smith, T. E. (1959). *J. appl. Physiol.* **14**, 31.
Huth, E. J., Squires, R. D., and Elkinton, J. R. (1959). *J. clin. Invest.* **38**, 1149.
Hyde, P. M., Grefer, C. C., and Skelton, F. R. (1962). *Endocrinology* **71**, 549.
Ingle, D. J. (1954). *Acta Endocrinol.* **17**, 172.
Jacobi, O. (1958). *J. appl Physiol.* **12**, 403.
Johnson, R. E., and Sargent, F. (1956). Progress Report No. 17. Contract No. AF 18-(600)-80, S9(55-1190), Air Res. and Dev. Command.
Jones, R. H. V., and de Wardener, H. E. (1956). *Brit. med. J.* i, 271.
Karvonen, M. J., Friberg, O., and Antilla, E. (1955). *Ann. Med. exp. Fenn.* **33**, 326.
Kauntiz, H. (1956a). *Nature, Lond.* **178**, 1141
Kaunitz, H. (1956b). *Lab. Invest.* **5**, 132.
Kaunitz, H., Geller, L. M., Slanetz, C. A., and Johnson, R. E. (1960). *Nature Lond.* **185**, 350.
Kenney, R. A. (1949). *Acta. med. scand.* **135**, 172.
Kenney, R. A. (1953). *Brit. med. J.* i, 600.
Kenney, R. A. (1954). *Brit. J. industr. Med.* **11**, 38.
Kenney, R. A. (1963). *Int. Rev. trop. Med.* **2**, 293.
Kenney, R. A., and Ladell, W. S. S. (1953). Unclassified report to *M.R.C.* MRC 53/577, APRC 53/3.
Keys, A. (1943). *Fed. Proc.* **2**, 164.
King, J. H. (1878). *Amer. J. med. Sci.* **75**, 404.
Kleeman, C. R., Epstein, F. H., and White, C. (1956). *J. clin. Invest.* **35**, 749.
Kleeman, C. R., Rubini, M. E., Lamdin, E., and Epstein, F. H. (1955). *J. clin. Invest.* **34**, 448.
Koletsky, S. (1957). *Arch. Path.* **63**, 405.
Krnjevic, K., Kilpatrick, R., and Aungle, P. G. (1955). *Quart. J. exp. Physiol.* **40**, 203.

Ladd, M. (1952). *J. appl. Physiol.* 4, 602.
Ladell, W. S. S. (1943). *Lancet* ii, 441.
Ladell, W. S. S. (1945a). *Proc. biochem. J.* 39, 47.
Ladell, W. S. S. (1945b). *J. Physiol.* **104**, 13P.
Ladell, W. S. S. (1945c). *J. Physiol.* **104**, 43P.
Ladell, W. S. S. (1947a). *Brit. med. Bull.* 5, 9.
Ladell, W. S. S. (1947b). *Brit. med. Bull.* 5, 5.
Ladell, W. S. S. (1949a). *J. Physiol.* **108**, 440.
Ladell, W. S. S. (1949b). *Lancet*, ii, 836.
Ladell, W. S. S. (1950). *J. Physiol.* **112**, 15P.
Ladell, W. S. S. (1951a). *J. Physiol.* **115**, 296.
Ladell, W. S. S. (1951b). *Brit. med. J.* i, 1358.
Ladell, W. S. S. (1951c). *Brit. med. J.* ii, 177.
Ladell, W. S. S. (1955). *J. Physiol.* **127**, 11.
Ladell, W. S. S. (1957). *Trans. roy. Soc. trop. Med. Hyg.* **51**, 189.
Ladell, W. S. S. (1963). *In* "Handbook of Physiology" ("Environment", Chap. 39), p. 625. Williams and Wilkins, Baltimore.
Ladell, W. S. S., and Kenney, R. A. (1955). *Quart. J. exp. Physiol.* **40**, 283.
Ladell, W. S. S., Waterlow, J. C., and Hudson, M. F. (1944). *Lancet* ii, 491.
Lancet i (1961). Annotation 1037.
Laragh, J. H., Angers, M., Kelly, W. G., and Lieberman, S. (1960). *J. Amer. med. Ass.* **174**, 234.
Leaf, A. (1960). *Annu. Rev. Physiol.* **22**, 111.
Leaf, A., and Frazier, H. S. (1961). *Progr. cardiovas. Dis.* 4, 47.
Lee, D. H. K. (1963). *In* "Handbook of Physiology". (Chap. 35, "Environment"), p. 551. Williams and Wilkins, Baltimore.
Lee, D. H. K., Murray, R. E., Simmonds, W. J., and Atherton, R. G. (1941). *Med. J. Aust.* **11**, 249.
Lee, J. (1963). *J. Physiol.* **167**, 256.
Leithead, C. (1963). *Fed. Proc.* **22**, 901.
Leithead, C. S., Leithead, L. A., and Lee, F. D. (1958). *Ann. trop. med. Parasit.* **52**, 456.
Leithead, C. S., and Pallister, M. A. (1960). *Lancet* ii, 13.
Lemaire, R. (1952–3). *Bull. Mem.* Ec. Prep. Med. Pharm. Dakar. Tome 1, 150.
Leschi, J. (1947). *C.R. Acad. Sci., Paris* **225**, 1023.
Leschi, J. (1948). *C.R. Acad. Sci., Paris* **227**, 1050.
Leutscher, J. A., Johnson, B., Dowdy, A., Harvey, J., Lew, W., and Poo, L. J. (1954). *J. clin. Invest.* **33**, 1441.
Loewenthal, L. J. A. (1961). *Arch. Derm.* **84**, 2.
McCance, R. A. (1936a). *Proc. roy. Soc.* B. **119**, 245.
McCance, R. A. (1936b). *Lancet* i, 643, 704, 765, 823.
McCance, R. A. (1937a). *Biochem. J.* **31**, 1276.
McCance, R. A. (1937b). *Biochem. J.* **31**, 1278.
McCance, R. A. (1938). *J. Physiol.* **92**, 208.
McCance, R. A. (1945–6). *J. Physiol.* **104**, 196.
McCance, R. A. (1959). *Arch. Dis. Childh.* **34**, 361, 459.
McCance, R. A., and Morrison, A. B. (1956). *Quart. J. exp. Physiol.* **41**, 365.
McCance, R. A., Ungley, C. C., Crosfill, J. W. L., and Widdowson, E. M. (1956). "The Hazards to Men in Ships Lost at Sea, 1940–1944", *M.R.C.* (*War*) *Memor.* Spec. Rep. No. 291. H.M.S.O., London.
McCance, R. A., and Widdowson, E. M. (1936). *Proc. roy. Soc.* B. **120**, 228.
McCance, R. A., and Widdowson, E. M. (1937). *J. Physiol.* **91**, 222.

McCance, R. A., and Widdowson, E. M. (1939). "Chemical Composition of Foods", pp. 26–29. *M.R.C.* (*War*) *Memor.* Spec. Rep. Series No. 235. H.M.S.O., London.
McCance, R. A., and Widdowson, E. M. (1951). *Brit. med. Bull.* 7, 297.
McCance, R. A., and Widdowson, E. M. (1963). *J. Physiol.* 165, 569.
McCance, R. A., Young, W. F., and Black, D. A. K. (1944). *J. Physiol.* **102**, 415.
McCutcheon, J. W. and Taylor, G. L. (1951). *J. appl. Physiol.* 4, 121.
MacFarlane, W. V., and Robinson, K. W. (1957). *J. Physiol.* **135**, 1.
McGee, W. J. (1906). *Interst. med. J.* 13, 279.
McKim, W., and Marriott, W. M. (1923). *Physiol. Rev.* 3, 275.
Macpherson, R. K. (1958). *Nature, Lond.* **182**, 1240.
Manchester, R. C., Husted, C., and McQuarrie, I. (1931). *J. Nutr.* 4, 39.
Manery, J. P. (1954). *Physiol. Rev.* 34, 334.
Manitius, A., Levitin, H., Beck, D., and Epstein, F. H. (1960). *J. clin. Invest.* 39, 684.
Marriott, H. L. (1950). "Water and Salt depletion." Blackwell Scientific Publications, Oxford.
Martinek, J., Janovsky, M., and Stanincova, V. (1962). *Proc. XXII Inter. Congr. Physiol. Sci., Leiden*, Abstract 263.
Mases, P., Falet, R., Joly, R., and Houdas, Y. (1962). *C.R. Soc. Biol.* **156**, 73.
Medical Research Council (1943). War Memorandum No. 8, also *Brit. med. J.* **1**, 168.
Mellanby, K. (1942). *Nature, Lond.* **150**, 21.
Meneely, G. R., Heyssel, R. M., Ferguson, J. L., Weiland, R. L., Ball, C. O. T., Lorimer, A. R., and Meneely, E. U. (1962). *Proceedings of the XXII International Congress of Physiological Sciences, Leiden* Abstract 647.
Metz, B., Lambert, G., Hasselmann, M., and van Hove, R. (1959). "Metabolisme hydromineral et bilan calorique du travailleur de force en zone Saharienne." Centre d'Etudes et d'Information des Problems Humains dans les Zones Arides, Paris.
Miles, B. E., Paton, A., and de Wardener, H. E. (1954). *Brit. med. J.* **2**, 901.
Mole, R. H. (1948). *J. Physiol.* **107**, 399.
Montgomery, M. F. (1931). *Amer. J. Physiol.* 98, 31.
Moses, A. M. (1963). *Endocrinology* 73, 230.
Moss, K. N. (1922). *Proc. roy. Soc.* B. 95, 181.
Mount, L. E. (1962). *J. Physiol.* 164, 274.
Muller, A. F. (1920). *Dtsch. Med. Wschr.* **1**, 113.
Muller, A. F., Manning, E. L., and Riondel, A. M. (1958). *In* "Aldosterone" (A. F. Muller and C. M. O'Connor, eds.). Churchill, London.
Nadal, J. W., Pedersen, S., and Maddock, W. O. (1941). *J. clin. Invest.* **20**, 691.
Nagareda, C. S., and Gaunt, R. (1951). *Endocrinology* 48, 560.
Newburgh, L. H. (1949). "Physiology of Heat regulation and clothing test methods." W. B. Saunders & Co, Philadelphia.
Newburgh, L. H., and Johnston, M. W. (1942). *Physiol. Rev.* **22**, 1.
Newburgh, L. H., Johnston, M. W., Lashinety, F. H., and Sheldon, J. M. (1937). *J. Nutrit.* 13, 203.
Newling, P. S. B. (1957). *J. Physiol.* 137, 44P.
Nicholas, A., Salvo, Di, and Corner, M. (1963). *Proc. Soc. exp. Biol., N.Y.* **112**, 21.
Nichols, G., Jr., and Nichols, Nancy. (1956). *Amer. J. Physiol.* 186, 383.
O'Connor, W. J. (1962). "Renal Function". Physiol. Soc. Monograph. Edward Arnold, London.
Osborne, W. A. (1912). *J. Physiol.* 47, 12.
Pack, C. T. (1923). *Amer. J. Physiol.* 65, 346.
Pales, L. (1950). "Les Sels Alimentaires: Sels Mineraux." Gouvern. Gen. de l'A.O.F., Dakar.

Parmeggiani, L. (1959). *Med. Na. Lav.* 50, 725 (abstracted in *Bull. Hyg.* 1960, 35, 541.)

Passmore, R., and Johnson, R. E. (1958). *Quart. J. exp. Physiol.* 43, 352.

Pearcy, M., Robinson, S., Miller, D. I., Thomas, J. T. and Debrota, J. (1956). *J. appl. Physiol.* 8, 621.

Peters, J. P. (1935). "Body Water." Charles C. Thomas, Springfield.

Peters, J. P. (1953). *Journal-Lancet*, 73, 180.

Pinson, E. A. (1942). *Amer. J. Physiol.* 137, 492.

Pinson, E. A., and Langham, W. H. (1957). *J. appl. Physiol.* 10, 108.

Pitts, G. C., Johnson, R. E., and Consolazio, F. C. (1944). *Amer. J. Physiol.* 142, 253.

Porteres, R. (1950). "Les Sels Alimentaires: Cendres D'Origine Vegetale." Gouvern. Gen. de l'A.O.F., Dakar.

Prentiss, P. G., Wolf, A. V., and Eddy, H. A. (1959). *Amer. J. Physiol.* 196, 625.

Pugh, L. G. C. E. (1962). *Brit. med. J.* 2, 621.

Radford, E. P. (1959). *Amer. J. Physiol.* 196, 1098.

Reeve, E. B., Allen, T. H., and Roberts, J. E. (1960). *Ann. Rev. Physiol.* 22, 349.

Renzi, A. A., Renzi, M., Chart, J. J., and Gaunt, R. (1956). *Acta. Endocrinol.* 21, 47.

Richter, C. P. (1956). "Salt Appetite of Mammals: its Dependence on Instinct and Metabolism." Extrait du volume: Fondation Singer-Polignac. Masson et C[ie], Paris.

Rioch, D. McK. (1930). *J. Physiol.* 70, 45.

Robinson, J. R. (1954). "Reflections on renal function", p. 55. Blackwell Scientific Publications, Oxford.

Robinson, J. R., and McCance, R. A. (1952). *Annu. Rev. Physiol.* 14, 115.

Robinson, S. (1952). *Annu. Rev. Physiol.* 14, 73.

Robinson, S., Nicholas, J. R., Smith, J. H., Daly, W. J., and Pearcy, M. (1956). *J. appl. Physiol.* 8, 159.

Robinson, S., and Robinson, A. (1954). *Physiol. Rev.* 34, 202.

Rosenfeld, G., Rosenberg, E., Ungar, F., and Dorfman, R. I. (1956). *Endocrinology* 58, 255.

Rosnagle, R. S., and Farrell, G. L. (1956). *Amer. J. Physiol.* 187, 7.

Rovner, D. R., Conn, J. W., Knoff, R. F. and Hsueh, M. T-Y. (1962). *J. Lab. clin. Med.* 60, 1013.

Rubini, M. E., Kleeman, C. R., and Lamdin, E. (1955). *J. clin. Invest.* 34, 439.

Rubini, M. E., Wolf, A. V., and Meroney, W. H. (1956). Walter Reed Army Instiute of Research, Rep. WRAIR-190-56.

Sargent, F., and Johnson, R. E. (1956). *Amer. J. clin. Nutr.* 4, 466.

Sargent, F., Sargent, V. W., Johnson, R. E., and Stolpe, S. G. (1954). Wright Air Development Center Tech. Rep. 53-484. University of Illinois.

Schoun, E. J. (1957). *J. appl. Physiol.* 10, 267.

Schmidt-Nielsen, K. (1962). "Comparative Physiology of Desert Mammals." Brody Memorial Lecture II. Agric. Exp. Sta. Spec. Rep. 21. University of Missouri.

Seeley, L. E. (1940). *Heat. Pip. Air Condit.* 12, 377.

Seldin, D. W., and Tarail, R. (1949), *Amer. J. Physiology.* 159, 160.

Selye, H. (1949) "Textbook of Endocrinology" (2nd Ed.), p. 837, Acta Endocrinologica, Montreal.

Share, L., and Levy, M. N. (1962). *Amer. J. Physiol.* 203, 425.

Smith, J. H., Robinson, S., and Pearcy, M. (1952). *J. appl. Physiol.* 4, 659.

Smyth, F. S., Callahan, W., and Phatak, N. M. (1933). *J. clin. Invest.* 12, 55.

Sohar, E., Tennenbaum, J., and Yaski, D. (1962). "Biometeorology." *Proceedings of the 2nd International Bioclimatological Congress* (S. W. Tromp, ed.). Pergamon Press, London.

Soula, L. (1950). "Les Sels Alimentaires. Sels du Manga (Niger): La Thenardite de Maïné-Soroa." Gouvern. Gen. de l'A.O.F., Dakar.

Squires, R. D., and Huth, E. J. (1959). *J. clin. Invest.* 38, 1134.
Stein, L. (1963). *Science* **139**, 46.
Stein, L., and Seifter, J. (1962). *Amer. J. Physiol.* **202**, 751.
Stenning, J. C. (1945). *J. roy. nav. med. Serv.* **31**, 129.
Stevenson, J. A. F. (1962). *Proceedings of the XXII International Congress of Physiological Sciences, Leiden* Abstract 359.
Stott, H. (1936). *Indian med. Gaz.* 71, 712.
Streeten, D. H. P., Conn, J. W., Louis, L. H., Fajans, S. S., Seltzer, H. S., Johnson, R. D., Gittler, R. D., and Dube, A. H. (1960). *Metabolism* 9, 1071.
Taylor, N. B. G., Hunter, J., and Johnson, W. H. (1957). *Canad. J. Biochem. Physiol.* 35, 1017.
Thomas, S. (1957). *J. Physiol.* **139**, 337.
Tregear, R. T., and Marzulli, F. N. (1961). *J. Physiol.* **157**, 52.
Ullmann, E. (1958). *J. Physiol.* **141**, 11.
Van Heyningen, R., and Weiner, J. S. (1952). *J. Physiol.* 116, 395.
Verney, E. B. (1947). *Proc. roy. Soc.* B. **135**, 25.
Verney, E. B. (1955). *Lancet* ii, 1237, 1295.
Vesin, P. (1958). *In* "Aldosterone", p. 130 (A. F. Muller and C. M. O'Connor, eds.). Churchill, London.
Walker, J. S., Margolis, F. J., Teate, H. L., Weil, M. L., and Wilson, H. L. (1963). *Science* **140**, 890.
Webb, P., Garlington, L. N., and Schwarcz, R. M. (1957). *J. appl. Physiol.* **11**, 40.
Weiner, J. S. (1954). *In* "Biology of Deserts", p. 193 (J. L. Cloudsley-Thompson, ed.). Institute of Biology, London.
Weiner, J. S., and van Heyningen, R. (1952). *Brit. J. industr. Med.* 9, 56.
Welch, B. E., Buskirk, E. R., and Iampietro, P. F. (1957). *Metabolism* 7, 141.
Welt, L. G. (1952). *Arch. intern. Med.* 89, 931.
Welt, L. G., and Nelson, W. P. (1952a). *J. appl. Physiol.* 4, 709.
Welt, L. G. and Nelson, W. P. (1952b). *J. appl. Physiol.* 4, 713.
Whillans, M. G., and Smith, G. F. M. (1948). *Canad. J. Res.* 26, 250.
Whitehouse, A. C. R., Hancock, W., and Haldane, J. S. (1932). *Proc. roy. Soc.* B. 3, 412.
Whittow, G. C. (1956). *Med. J. Malaya* **11**, 126.
Widdowson, E. M. (1957). *Proc. Nutr. Soc.*, 16, 118.
Widdowson, E. M., and McCance, R. A. (1950). *Lancet*, 539.
Winkler, A. W., Danowski, T. S., Elkinton, J. R., and Peters, J. P. (1944). *J. clin. Invest* 23, 807.
Winkler, A. W., Elkinton, J. R., Hopper, J., and Hoff, H. E. (1944). *J. Chem. Invest.* 23, 103.
Winslow, C.-E. A., and Herrington, L. P. (1949). "Temperature and Human Life." Princeton University Press, New Jersey.
Wolf, A. V. (1945a). *Amer. J. Physiol.* **143**, 567.
Wolf, A. V. (1945b). *Amer. J. Physiol.* **143**, 571.
Wolf, A. V. (1952). 3rd Annual Edgar Ballenger Memorial Lecture, p. 1. Boca Raton, Florida.
Wolf, A. V. (1956). *Sci. Amer.* **194**, 70.
Wolf, A. V., and Eddy, H. A. (1957). Walter Reed Army Institute of Research Rep. 90–57.
Wolf, A. V., Prentiss, P. G., Douglas, L. G., and Smith, R. J. (1959). *Amer. J. Physiol.* **196**, 633.
Womersley, R. A. (1953). *Archives of Middlesex Hospital* 3, 26.

Womersley, R. A., and Darragh, J. H. (1955). *J. Clin. Invest.* 34, 456.
Wright, P. (1964). *Symp. zool. Soc., Lond.* No. 13, p. 17.
Wrong, O. (1957). *Brit. med. Bull.* 13, 10.
Yoshimura, H. (1958). *Jap. J. Physiol.* 8, 165.
Zuidema, G. D., and Clarke, N. P. (1957). *Amer. J. Physiol.* 188, 616.

CHAPTER 10

Exercise

J. V. G. A. Durnin

A. Introductory

Up to one hundred years ago it was unavoidable that almost everyone in the population frequently undertook exercise of a moderate degree. No public transport was available; to sit in a carriage and be conveyed anywhere, except for long journeys, was the prerogative of only a few. Walking was therefore a regular form of exercise for everyone, and considerable distances must have been covered daily by many people. Most of the population would also have been employed in work necessitating physical labour, sometimes of marked severity.

The development of public transport, first in cities and later over most of the country, together with the urbanisation of the community and, in the past forty years, the motor car, the most potent influence of all, have caused a major revolution in our way of life.

Instead of, as in the past, its being the prerogative of the few to sit in carriages, it is now the prerogative of the few to walk. Less and less physical activity is necessary during work, and few people seem to take part in exercise during their leisure time. None the less, even the small amounts of physical energy used up in leisure frequently lead to the position in which more energy is expended outside the work period than at work itself (Garry, Passmore, Warnock and Durnin, 1955; Durnin, Blake, Allan, Shaw, Wilson, Blair and Yuill, 1961). The general level of energy output reached by most people is a low one, particularly in housewives (Durnin, Blake and Brockway, 1957). It is conceivable that, at the present day, most men could now manage their daily round fairly well with primitive stumps for legs; backsides are becoming obviously better cushioned for their task of almost continuous sitting.

We are at the beginning of an era when exercise, if it occurs to any worthwhile extent at all, will take on an entirely different aspect. It will no longer occur at work; it will therefore no longer be obligatory for more than the few. It will be restricted to sports and games and, being voluntary, is likely to be the choice of a small, highly selected group, unless some positive action is taken to try to reverse or alter this tendency.

Large-scale adaptations are required by man if this state of circumstances develops, and the survival of many individuals may be severely impaired if the level of exercise continues to decrease. There is a mass of indirect evidence pointing to the disadvantages and even the dangers of minimal levels of physical activity.

In this chapter are discussed the chronic effects of continued muscular work, at high and low intensities, on some important physiological functions in man. An excellent brief summary has been given by Åstrand (1956).

B. Fitness and Health

1. *Intrinsic characteristics of communities*

There are still left in various parts of the world communities in which most of the men—although sometimes the role of the sexes is reversed, as in the ama or women divers of Japan—undertake prolonged hard physical work and have done so for generations. Sherpa porters are a good example. In such groups it is a relatively straightforward matter to investigate certain physiological attributes, such as general physical fitness, cardiac output and lung function (for example, Hong, Rahn, Kang, Song and Kang, 1963). Studies like these usually show that there are obvious differences between the groups engaged in frequent exercise and a control group. The results may be interesting, not in the qualitative sphere, since they usually follow an expected pattern, but in delineating the extent of the differences measured. Often this is something about which we may have only the vaguest notions. It would be an interesting study in the physiology of human survival as related to exercise to obtain information on inherited characteristics in such races or communities. Do these people inherit a superior respiratory, muscular or cardiovascular function, even after a few generations of changed circumstances of life, when no more than the average amount of physical exercise is taken? There seems no very positive evidence available. It would be fascinating to know, for example, whether the pure descendants of some of the North American Indian tribes, which must have contained phenomenally fit groups of men, were not just as unfit, with no better adaptation to exercise, as are apparently the remainder of the population of the U.S.A. and Canada. It seems probable that they and similar groups have no increased resistance to disease processes, such as atheroma, when they change their original environment to that of western Europe or the rest of the United States. However, even this is not exactly proven, since most of the groups studied in this context have come from large racial aggregations with no particular distinction in chronic levels of physical exercise.

2. *Is exercise beneficial?*

The obvious question asked by most people who are not basically enthusiastic about taking exercise themselves is, "Does exercise do you any good?" Some physiologists and clinicians find this a difficult question to answer. I have never been in any doubt myself and can only assume that these physiologists and others have seldom taken much exercise and therefore do not properly understand the outlook of people who routinely indulge in severe physical exertion. The question is a complex one with varying shades of meaning, but it can be answered indubitably and simply with an unqualified affirmative. There is not the slightest doubt that exercise "does you good", as anyone knows who has become accustomed to prolonged physical exertion and who ceases this for a considerable time and can also remember what his physical and mental states were before the period of exercise. There is not the slightest doubt that subjectively exercise is beneficial. But what, if any, are the measurable physical benefits of prolonged exercise? There is a common impression that it is not easy to assess these. The differences between the physically fit and trained individual and the untrained person are easily shown: but do these differences illustrate any benefits that have accrued to the trained person?

a. Exercise and liability to disease. This subject has been interestingly analysed by Davies, Drysdale and Passmore (1963). As they point out, the evidence can be made to appear indeterminate. The report of the Wolfenden committee (1960) in Britain, entitled "Sport and the Community", included a statement that no evidence had been produced showing an unequivocal connection between taking exercise and being healthy. The committee must have been looking for most peculiar evidence. Although it may not be a good criterion of "health", a diminished tendency to contract disease may provide some evidence for the beneficial aspects of exercise. This has been shown fairly conclusively by several authors. Morris, Heady, Raffle, Roberts and Parke (1953) have suggested that it is the extra exercise necessitated by the work of bus conductors, in running up and down stairs, that results in a much lower incidence of coronary heart disease than that suffered by the drivers. And again, the considerable amount of physical activity of postmen seems the most obvious difference between them and other Post Office workers, such as telephonists, clerks and administrative grades, who are much more prone to suffer from atheromatous heart disease. In a less severe form the same sort of tendency was found by Katsuki, Yamazaki and Hirao (1963), who record fewer changes in the ECG, indicative of degenerative changes in heart muscle, among those employees of a Japanese insurance company who had exercise habits than among those who had not. In another context Morris and Heady (1953), have stated that "physical activity itself may be a factor of health through promoting muscular fitness, cardiac and respiratory efficiency and through its promotion of leanness."

From a more negative approach, Morris and Crawford (1958) analysed the post-mortem reports on 5000 people who had died from various causes and whose occupations were known. On the matter of atheroma, there was a significantly lower incidence in occupations involving hard physical activity than in those involving light. They made the same point as previously: "Men in physically active jobs have less coronary heart disease in middle age, and what they have is less severe than men in physically inactive jobs."

True, this conclusion may be partly invalidated by the earlier statement in this section, to the effect that more physical activity may take place in leisure time than during work. But this is unlikely to have led to erroneous inferences in the groups studied by Morris and his colleagues, since the amount of physical exercise occasioned by the job was fairly well delineated. Rather is it that the energy expended on a job may be so low that more energy output occurs in leisure; it rarely means that so much energy is used up in the off-duty period that the total daily energy output approaches that required by people in strenuous occupations.

This argument may be used in reverse to dispute the findings of Spain and

Bradess (1960), whose post-mortem investigations on 207 men, aged 30 to 60, who had died suddenly from violence, suggested no differences in the incidence of atherosclerotic lesions between those having active occupations and those who were sedentary. The division, in this relatively small sample, between the exercise and non-exercise groups, is perhaps somewhat vague and indefinite; it depends partly at least on opinions of what relatives and friends thought of the physical activity of the deceased; these are notoriously open to gross error.

b. Plasma lipids. There is more difficulty in the interpretation of investigations on people still alive, when attempts are made to assess the degree of atheroma. This must be estimated indirectly, and the level of blood plasma lipids is supposed to be a helpful criterion. High levels for blood cholesterol and for other lipids seem to be associated with an increased tendency to severe atheromatous lesions and higher death rate (Kannel, Dawber, Kagan, Revotskie and Stokes, 1961). There is some confusion in the results on the relationship of blood lipid levels and exercise. However, most studies seem to show that exercise inhibits the rise in plasma lipids that normally occurs after a high fat meal. (Mann, Teel, Hayes, McNally and Bruno, 1955; Cohen and Goldberg, 1960; Nikkilä and Konttinen, 1962).

C. Exercise and Life Expectancy

1. *Athletes*

Many reports on the influence of exercise on life expectancy can be criticised on grounds of inadequate information. Frequently there are no reliable records of the general level of exercise throughout life – admittedly difficult to obtain accurately. Examples of such reports are those of Dublin (1932) and Rook (1954). Dublin studied the life expectancies of almost 40,000 graduates from ten eastern universities in the United States; their graduations covered the years 1870 to 1905. There seemed to be no difference between the so-called "athletes" (about 5000 of them) and the main group. Rook found a similar lack of difference between 700 Cambridge (England) athletes at the university between 1860 and 1900 and a random group.

When there is information about exercise habits throughout life, or when the group may be safely assumed to have continued to take hard exercise, the results are sometimes markedly different from those of the previous two studies. Karvonen, Kihlberg, Määtä and Virkajrävi (1956) showed that Finnish skiers had a life expectancy of several years longer than the general male population. In the United States, Pomeroy and White (1958) investigated health and age, and the causes of any deaths, in a group of 355 noted football players who were students at Harvard between 1900 and 1930. Although there had been less than one hundred deaths, and only twenty-five of these from coronary heart disease, the risk of this was reduced in those who continued to exercise after leaving the university, and no one who "maintained a heavy exercise program happened to develop coronary heart disease."

None of this evidence is indisputable.

2. *Mere mortals*

Retrospective studies, particularly when they involve subjective assessment of amounts of physical work—either by the person concerned or by relatives or friends—would not often be wholly accepted by impartial scientists. Most people grossly exaggerate, however unconsciously, the amount of exercise they take or have been in the habit of taking. In a study of the reliability of subjective assessments, Durnin (unpublished results) compared the estimated time spent in exercise, including walking, by some undergraduates in Glasgow with a more exactly recorded time study of the exact amount of their exercise during the subsequent week. The estimate was obtained by careful questioning, at individual interview, when the participant was asked how long he had spent walking and at sports or games during each day of the previous seven days. The time study entailed a direct recording on special diary cards (Durnin, 1955) of the amount of time passed in exercise during the seven days subsequent to the interview. The uniform finding was that there was a gross exaggeration of the estimated over the recorded time, of the order of three or four times. Almost all the subjects thought they spent much more time in exercise than they in fact did. Retrospective studies, therefore, seem liable to this potentially misleading factor.

The element of doubt about the benefits of exercise for health is surely minimal. The most sceptical might argue, with difficulty, that life expectancy was no greater for the person who partook of exercise than for the unfit man. He could not argue that the physical state of the two groups was similar in any other way, and it seems unphysiological, as well as illogical, to state that a series of systems in the human organism measurably superior as a result of exercise are nevertheless no better because of this.

D. Muscle and Body Build

There have been widespread impressions in the past that since exercise might not be a "natural" pursuit, especially for women, it might result in a change of body type. Thus Skerlj (1936) managed to discover, among a group of Polish athletes, that the female was more affected than the male and that strenuous exercise favoured the development of virile body types accompanied by menstrual irregularity. Contradictory evidence was supplied about the same time by Adams (1938), who found no evidence that prolonged and hard physical work affected the normal physical development of young American coloured women.

Both of these views, though held by many knowledgeable people, are perhaps only half-truths. There is probably a tendency for certain women athletes to be of a somewhat masculine type, but little evidence to show that the exercise has been the causative factor in making them masculine. Contrary evidence might be derived from an inspection of female ballet dancers, who spend a large part of their lives in strenuous exercise, with apparently little effect on their femininity. On the other hand, exercise undoubtedly may change body build by reducing fat depots and increasing muscle mass.

Increases in muscle mass, strength and endurance are among the most obvious and best-known lasting effects of physical exercise. Many early studies have reported an extension in the girths of skeletal muscles after repeated muscular work. These occurred over a wide range of ages, as demonstrated by Godin (1913) on children, Matthias (1916) and Herxheimer (1921) on youths, Kohlrausch (1924) on students and Herxheimer (1922) on adults. These studies showed that the increases were not always proportional to the work done; for example, in children and youths, when the body is growing longitudinally, there was less increase in muscle than when most of the long bone epiphyses had united. Similarly, muscle mass changed less in "asthenic" individuals. The increase in girths was more related to the rate of work rather than to a function of the duration or the total amount of work (Siebert, 1928).

It is generally agreed that muscle hypertrophy is due to a true hypertrophy of individual fibres and not to the appearance of new fibres (Morpurgo, 1897). More capillaries appear in the muscle with training (Petren, Sjöstrand and Sylven, 1936), and these are supposed to increase in number because of a special relationship with exercises requiring endurance, whereas strength exercises increase the fibre size. It is stated that the gain in muscle strength due to exercise is more marked than the hypertrophy.

The end result on the individual skeletal muscles varies with the length of the contraction, the amount of repetition, the force applied and the speed of contraction. An improvement in absolute strength does not necessarily imply an ability to perform more work over a long period, and weight-lifters, for example, may have a much lower endurance than middle-distance runners or swimmers. The exact mechanism within the muscle that produces these changes is, even now, little understood and probably involves complex changes in the energy pathways. However, the gross final result in the function of the muscle is of immense practical importance, not only physiologically, but also in rehabilitation after illness or disease. Much of the recent renewed interest in this field is due to the most interesting and potentially valuable work of Müller (1962), who has opened up new paths of investigation.

E. Cardiovascular System

1. *Cardiac output and stroke volume*

Within the past few years there has arisen some doubt and confusion of thought over the effect of exercise on the heart. Many of the accepted ideas about the extent of the increase in stroke volume during exercise have been disputed, and it is even suggested now that there may be practically no change in stroke volume during exercise. This is a difficult problem to investigate in man, since the techniques for measuring cardiac output are either insufficiently accurate or else entail enough risk to make their use unjustified in normal subjects during strenuous exercise. There is therefore an element of uncertainty about the values for cardiac output, especially in untrained subjects as distinct from athletes. Thus in a standard textbook of physiology (Bell, Davidson and Scarborough, 1961) values for cardiac output are given for "walking at 5 miles/hr—20 l/min" and "very severe exercise—34 l/min." No indication is given as to whether these results would apply to an exceptionally trained athlete, and one must assume that they are intended to represent the "normal subject". If this is so, then even at a cardiac output of 25 l/min there must be an appreciable rise in the stroke volume. There would be little argument that the cardiac output at rest is about 5 l/min; in the example cited above, then, there has been a fivefold increase in cardiac output. It is highly improbable that the proportionate rise in heart rate could be more than threefold or perhaps slightly greater—say, from a resting rate of 60 to 65 to an exercise level of 200. There must clearly have occurred a considerable rise in the stroke volume of the heart even at that level of cardiac output; if the volume of blood ejected from the heart can indeed reach levels of 30 l/min or more, the stroke volume may need almost to double its resting level. These high levels of cardiac output, which are frequently quoted in textbooks, often without any original reference, are possible but extremely hypothetical, for the reasons mentioned above. The values quoted in a recent paper by Åstrand, Cuddy, Saltin and Stenberg (1964) are much lower (about 26 l/min) and are perhaps more likely to be correct. In this respect, as in many other branches of medicine and physiology, the validity or otherwise of any absolute value—such as l/min of O_2 consumption or l/min of pulmonary ventilation—depends largely on the body size and build of the subject, and this must always be borne in mind when discussing such values (Durnin 1964).

The problem of cardiac output and the measurement of stroke volume is examined carefully and thoroughly in an excellent review by Rushmer and Smith (1959). Some of their findings may seem inconsistent with part of the above argument. They found that "the changes in various dimensions of the right and left ventricles during systolic ejection were unexpectedly small."

From a review of the literature, they concluded that the concept of heart rate and stroke volume increasing together during mild, moderate or severe exertion derived largely from studies on trained athletes conducted by indirect methods, such as the use of acetylene and similarly employed agents. From the available data on average normal human subjects, obtained by cardiac catheterisation and indicator dilution techniques, Rushmer and Smith thought that an increased stroke volume is neither necessary nor consistently observed during a wide range of exertion. "Indeed," they said, "the stroke volume in most subjects remains unchanged." However, they admitted that when an increased O_2 consumption was not accompanied by faster heart rates, then stroke volume was augmented, as in athletes. These are no new deductions and were precisely the conclusions reached by Eppinger, Kisch and Schwarz (1925), Ewig (1925), Henderson, Haggard and Dolley (1927) and Bansi and Groscurth (1930), all of whom found that the increased minute volume of the heart was brought about almost entirely by increases in heart rate in the untrained subject, whereas the trained person accomplished the same end result more efficiently and economically by a larger stroke volume. It may be difficult to agree completely with this thesis, unless the "average normal man" has much more restricted abilities in his cardiac system than has been supposed.

An important concept is emphasised by the uncertainty still prevailing in this section of cardiac physiology, namely, that it is often misleading to extrapolate from experiments, no matter how well controlled, on anaesthetised animals and apply the findings to intact animals and man.

2. *Cardiac hypertrophy*

With the above reservations in mind, it may none the less be reasonable to assume that, in animals and man as well, one of the adaptations made to long-continued and repeated exercise is an increase in stroke volume, both at rest and during exercise, accompanied by an enlargement of the heart. Many of the older studies reported in the literature appear to prove this beyond doubt and to show that the enlargement is both a dilatation and a hytertrophy of the muscle. That hypertrophy occurs has been shown indisputably by studies on intact animals; there is little doubt that man would not behave differently. Thus, Külbs (1906), Grober (1908) and Thörner (1930) showed that the ratios of heart weight to body weight of dogs exercised by running was markedly greater than those of their unexercised litter mates. Similar findings have been reported on rats by Hatai (1915), Secher (1921) and Borovansky (1930). Secher (1923) also showed that in rats the hypertrophy regressed rapidly when inactivity was imposed on previously exercised animals.

In man, Harvey (1628) stated positively that "the more muscular and powerful men are, the firmer their flesh; the stronger, thicker, denser, and more fibrous their hearts. . . ." Man, being a more awkward and less expendable experimental animal than dogs or rats, it is difficult to verify Harvey's statement. However, the probability remains, in support of Herxheimer (1927), Steinhaus (1933) and many others, that prolonged and frequent strenuous exercise, requiring endurance, will induce a growth of cardiac muscle. Hypertrophy of skeletal muscle, as opposed to heart muscle, is more likely to occur because of exercises needing strength, speed or effort. Exercise of a mild or moderate nature will apparently not have any measurable effect on heart size or stroke volume.

It should hardly be necessary to state that there is no convincing evidence that the hypertrophy and dilatation of the heart are anything other than beneficial.

3. *Heart rate*

a. Resting values. Since the heart may adapt itself to exercise by increasing semipermanently its stroke volume, there must be some concomitant alteration in heart rate. It is only by virtue of the greater ejection of blood by each heart beat of the athlete that it is possible for the heart rate to become appreciably reduced. There seems no doubt that slow heart rates are common among highly trained athletes. An early report of this finding among a large group of athletes was given by Hoogerwerf (1929): the resting heart rate of almost 200 Olympic contestants was 50. Such a finding seems significant, as also are the reported low heart rates found in individual athletes, especially long-distance runners. However, one must be careful about drawing conclusions from heart rates that appear low and have been found either in individuals or in small groups. The correct significance of these rates is sometimes misinterpreted because of a possibly erroneous impression as to the true resting heart rates of average normal healthy young adults. When control groups are carefully used, then this stricture does not apply. But when textbook values for resting heart rate are used for comparison, it is well to remember that there is a large scatter in these values and the mean value may be much lower than is generally thought. Thus Durnin (unpublished results) found in a group of forty-four men, not in training and of entirely average fitness, that the resting heart rate, measured in bed when the men were awake but before they had risen in the morning, was about 55 beats per minute.

There would be little dispute about the lower resting heart rates that result as an adaptation to a regimen of exercise, and it is almost impossible to visualise a physiological state in which stroke volume could be increased semipermanently without an accompanying reduction in heart rate. The conclusive studies are those in which individual resting heart rates have been followed carefully during a course of training. These have been reported by Lindhard (1915), Dawson (1920), Herxheimer (1924), Kaup and Grosse (1927), Schneider, Clarke and Ring (1927), Dill and Brouha (1937), Taylor (1941), Henry (1954) and many others.

b. Rates under exercise. The effects of adaptive reactions of the heart, caused by exercise, on the responses of heart rate during exertion are complex. Most studies demonstrate a progressive reduction in the heart rate during exercise as physical training progresses. This is not due to "practice", as is obvious if a proper control group is used. In a study (Durnin, Brockway and Whitcher, 1960) on four groups of men, one of which was a control group taking only minimal quantities of exercise, whereas the other three groups underwent different amounts of physical work as training (walking 10 km, 20 km or 30 km per day), there was no change in the heart rate of the control group during a standard exercise test on the treadmill, which was repeated at intervals during the training period. The "exercise" groups, on the other hand, showed significant reductions. However, this lowered exercise heart rate is more likely to be due to improved efficiency of muscles, neuromuscular co-ordination, respiratory and circulatory adaptations and other changes in the whole organism than to restricted alterations in the heart itself.

c. Maximum heart rate. There is still uncertainty about whether the maximum heart rate changes as an adaptation to exercise. There is insufficient evidence to show that athletes and other highly trained men have higher or lower maximum heart rates during exercise than the general population or that their training has altered their original highest possible rate. Superficially, there seems little difference between trained and untrained men and women in the highest rate reached by the heart during maximal work. This may not represent a true state of affairs: the attainment of maximum effort in the first place, with its combination of physical and psychological stress, is difficult, if not impossible, under laboratory conditions. The problems associated with measurements under severe exercise stress have been discussed almost *ad nauseam* and are excellently summarised by Taylor (1945), Åstrand (1956) and Dill (1963). The probability is that, even in the best of circumstances, there have been comparatively few recordings on athletes undergoing extreme physical strain in the laboratory. This is not likely to produce the same effect on heart rate as, say, an important athletic contest. Therefore, the recorded readings may be somewhat less than maximal, even in the best conditions. The techniques for measuring heart rate are also not always adequate at these levels of exertion, and often measurements are only made at the end of exercise, when the rate may have fallen slightly. With the improvements within the past year or two, resulting from cardiac electrodes that clip into the skin (Davies and Copland, 1964) and therefore give good contact even during marked muscular movements combined with better recording and transmitting devices, it should soon be possible to assemble enough information during truly maximum strain to satisfy our knowledge about maximum heart rates.

It is conceivable that the highest rate attainable by the normal adult human heart may approach 240 beats per min. In recordings on ballet dancers during a stage performance, Durnin (unpublished results) has found rates of upwards of 220 to 230 beats per min during the first 5 sec of recovery.

4. *Degree of strain required for adaptation*

A practical point of importance to heart rate is the level of exertion required to produce any adaptive changes in the heart. It has usually been accepted (Christensen, 1931; Steinhaus, 1933; Edwards, Brouha and Johnson, 1940) that for producing measurable changes in the heart the rate of work is of more importance than its amount. Indeed, Karvonen, Kentala and Mustala (1957) have stated that a heart rate of at least 150/min is necessary during the exercise of training before any lowering of heart rate will be produced. It is true that prolonged high exercise heart rates will exert the best effect on the heart. And it is equally almost certain that minimal exercise, such as walking at moderate speeds or playing a leisurely game of golf or tennis, will provide minimal stimuli to the heart and cause minimum results. On the other hand, exercise causing heart rates of 120 to 130/min carried on for several hours and repeated over days or weeks can undoubtedly improve overall muscular efficiency and result in a reduced heart rate during a standard exercise test (Durnin *et al.*, 1960).

Nevertheless this is an important problem for large numbers of the population in many countries, who will not normally have the time or the opportunity to spend many hours repeatedly on physical exercise, and it must be stressed that any physical benefit from exercise is only likely if the stress is reasonably severe. Mild exercise, except in convalescence from illness or in the elderly, is mostly of psychological rather than physical advantage.

F. Respiratory System

The significant differences in respiration caused by exercise are not apparently of much current interest. There is some discussion of these factors by Åstrand (1952, 1956), but a complete chapter in the "Science and Medicine of Exercise and Sports" (Johnson, 1960) devoted to "Pulmonary Function in Relation to Exercise" makes no mention at all of the effects of fitness or training. In the whole book the only reference to possible long-term effects on the respiratory system is a cursory half-page in the chapter on "Training". An even briefer account—one short paragraph—appears in the British Medical Bulletin on "Respiratory Physiology" (1963), and no obvious reference at all is contained in the full account of the Haldane Centenary

Symposium on "The Regulation of Human Respiration" (Cunningham and Lloyd, 1963).

1. *Breathing and oxygen consumption*

Most of the adaptation of respiration to exercise training appears obvious and has been known for a considerable time. As do many other skeletal muscles, the muscles involved in respiration become more efficient. The work of breathing therefore becomes lessened (Milic-Emili, Petit and Deroanne, 1962) and, partly because of this, the necessary pulmonary ventilation and oxygen consumption of a standard exercise also decreases (Gemmill, Booth, Detrick and Schiebel, 1931). Several volume dimensions become enlarged in the athlete (Shapiro, Johnston, Dameron and Patterson, 1964). For a given level of pulmonary ventilation the oxygen intake and carbon dioxide production increase, especially at the higher levels of work. Most marked of all, the maximum oxygen consumption and pulmonary ventilation attainable during exercise are much higher in the trained than in the untrained person; 3·5 l/min of oxygen consumption would be high for the average young man and low for the athlete. With the help of the newer methods of measuring oxygen saturation and pressure (P_{O_2}) as well as the measurement of P_{CO_2} and pH, it should soon be possible to have a much more complete and clearer picture of the exact differences between the fit and the unfit man during exercise. During rest, when most of the respiratory variables are easily and accurately measured, there seem to be no significant distinctions induced as an adaptation to frequent exercise. Though it is still possible that new work on the function of the muscles of respiration may resurrect the old controversy about abdominal and thoracic breathing, it seems unlikely that this will ever assume importance in the present context.

2. *Diffusing capacity*

Several authors have described higher levels of pulmonary diffusing capacity in athletes than in non-athletes (Bates, Boucot and Dormer, 1955; Bannister, Cotes, Jones and Meade, 1960; Newman, Smalley and Thomson, 1962). This seems a logical finding. Yet in an interesting paper Mostyn, Helle, Gee, Bentivoglio and Bates (1963), describing a similar study on groups of men and women, athletes and non-athletes, report that the only group to show significant superiority were champion (Olympic class) swimmers; no differences were seen between the control groups and swimmers of average ability, long-distance runners and older (26 to 45 years) ex-athletes still in training. This appears surprising; the problem clearly requires further investigation. That diffusing capacity does not always fit logical conceptions is again indicated by West (1962), who showed that there was still no change in this function after several months at 6000 m (19,000 ft).

G. Metabolism at Rest and at Work

There is fortunately little need nowadays, for present purposes, or indeed for almost any purpose, to discuss basal metabolism. The innumerable papers published on whether physical training caused an increase in the basal metabolic rate, had no effect on or possibly decreased it, were competently reviewed by Steinhaus and Jenkins (1930), and nothing has happened since to make the topic worth reviving. There are no good physiological reasons why training should affect resting metabolism (unless perhaps the athlete is also a disciple of yoga), and the combined results of published material support this contention.

The subject of metabolism during work is more interesting and important. It is not proposed to deal with the details of muscle metabolism in the trained and the untrained state. Our knowledge of this is limited in man and only fractionally more in animals. A comprehensive and most useful review on the whole subject of "exercise and metabolism" has been furnished by Taylor (1960). The purpose of this section is to consider the effects of high and low levels of exercise on the total energy metabolism of men and women and its implications.

1. *Daily energy expenditure*

Present-day levels of energy expenditure vary considerably, not only from individual to individual within any one area and from group to group, but also when areas at different stages of industrial development are considered. An illustration of the fine differences between countries can be given from a study on peasant farmers in the comparatively industrially developed country of Switzerland (Durnin, Taylor, Gsell and Verzar, unpublished results), where expenditure levels of over 4000 kcal/day were found among men of over 60 years of age; these levels would be difficult to duplicate for any typical population group in, for instance, Britain. It is therefore hazardous to generalise about average daily levels of energy expenditure, but perhaps representative information may be obtained from some interesting tables constructed by Harries, Hobson and Hollingsworth (1962). These tables summarise figures obtained from field studies in Britain; since Britain is possibly at a stage common to many countries, where technical and indusrial growth have reached a fairly advanced state but have not yet approached the level more uniformly present in the United States, an examination of their figures might be illuminating. The immediate impression is the limited range of values for energy in the groups of subjects. If one or two judicious exclusions are made, so that the "Army troops" and the "aged" and "elderly retired" are temporarily ignored, the range for all the male groups extends from 2600 to 4000 kcal/day, with the groups representing the great bulk of the whole population clearly being compressed around the 3000 $\pm$ 200 to

300 kcal level. For women, if Thomson's groups of pregnant women are omitted, the range is even more limited, with the overall mean about 2100 kcal/day. More recent unpublished findings provide results that fit into this pattern. The inescapable conclusion appears to be that a low, and even a minute, level of exercise is the common standard for adults in this and similar countries. Even in adolescence, when physical activity might be supposed to occupy a somewhat larger part of the daily routine, there is now evidence to suggest that the levels of energy expenditure are much lower than were supposed. Boys and girls at 14 years old seem to have an energy output several hundred kcal per day less than the internationally accepted standards, even though these children were fit and healthy and apparently typical of the range of social groups in the population (Durnin, McLees, Yuill, Busby, Gay and Blake, 1964).

2. *Intraspecific taxonomy*

From these pieces of evidence the implications for the amount of physical activity of the population have been the subject of a most interesting speculation by Passmore (1964). He suggests that the population can be divided into three hypothetical groups—*homo laborans*, *homo sedentarius* and *homo sportivus*. *Homo laborans* constituted the vast proportion of people in all countries up to recent times and is still the majority in large parts of the world today. Technical developments and a century of industrial legislation have resulted, however, in a considerable reduction in the hours of work. Often a 40 hr working week, and sometimes less, is taken as a temporary basis for wage structure, and the implict and occasionally explicit assumption is accepted that this period will inevitably soon be further reduced. In a matter of a few decades many industries may well be working a 20 hr week. Even at present much working time is spent in operating mechanical means of power without the necessity to use any appreciable muscular effort. For women who work in factories the situation is similar, and for housewives all of the available published data seem to imply a similar absence of physical work.

Homo laborans is thus well on the way to extinction. He is being replaced to a large extent by a breed who ought to be less happy and who probably is less healthy, *homo sedentarius*. Men and women of the new type like to spend much of their new-found leisure sitting—in motor cars, in front of their television sets, in social relaxation. Their presence is now most obvious in North America and western Europe.

The alternative development of *homo laborans* is, one must fear, likely to be less prolific, although one hopes healthier, happier and hardier than *homo sedentarius*. *Homo sportivus* spends some of his extra leisure in activities that use at least some of his muscles—football, swimming, walking, climbing mountains, dancing, bicycling and other pursuits. The extension of this third

Table I. *Variations in Energy Expenditure*

Source	*No. of subjects*	*Period of survey (days)*	*Information about subjects*	*Mean daily value for all subjects (kcal)*	*Minimum daily mean for any subject over the period (kcal)*	*Maximum daily mean for any subject over the period (kcal)*	*Standard deviation of subject means (kcal)*	*Coefficient of variation of subject means (%)*
Happold (1945)	29	—	Medical students ♂					
			November	2609	—	—	281	10·8
			January	2640	—	—	225	8·5
	20	—	Medical students ♀					
			November	2349	—	—	186	7·9
			January	2409	—	—	263	10·9
Edholm, Fletcher, Widdowson	6	14	Intermediate cadets	3488	2991	4099	327	9·4
and McCance (1955)	6	14	Junior cadets	3343	2972	3652	270	8·1
Garry, Passmore, Warnock and	10	7	Clerks	2800	2330	3290	353	12·6
Durnin (1955)	19	7	Miners	3660	2970	4560	459	12·6
			Army troops:					
Adam, Best, Edholm and Woolf	13	4	exercise I	3429	3124	3952	230	6·7
(1957)	13	5	exercise II	4923	4252	5554	370	7·5
	13	5	exercise III	3224	2897	3459	161	5·0
			or					
Booyens and McCance (1957)	4	7	Scientific workers: ♂	2797	1933	3897	597	21·3
	2	14	♀	2054	1428	[illegible]	[illegible]	[illegible]

Durnin, Blake and Brockway (1957)	12	7	Housewives	2090	1704	2320	164	[illegible]
	12	7	Their daughters	2255	1819	2848	282	12·5
Adam, Best, Edholm, Fletcher, Lewis and Woolf (1958)	29	21	Army recruits	3774	2868	4765	444	11·8
Adam, Best, Edholm, Goldsmith, Gordon, Lewis and Woolf (1959)	5	18	Army recruits	3764	3499	4105	251	6·7
Durnin, Blake, Brockway and Drury (1961)	15	7	Elderly women	1991	1492	2409	251	12·6
Durnin, Blake, Allan, Shaw and Blair (1961)	20	7	Elderly women	2290	1795	2962	299	13·1
Durnin, Blake, Allan, Shaw, Wilson, Blair and Yuill (1961)	33	7	Elderly industrial workers ♂	2957	2185	3958	486	16·4
Blake, Durnin, Aitken, Caves and Yuil (1962, in preparation)	9	7	Elderly retired ♂	2327	1754	2811	322	13·8

Table II. *Variations in Energy Intake: Adult Males*

Source	*No. of subjects*	*Period of survey (days)*	*Information about subjects*	*Mean daily value for all subjects (kcal)*	*Minimum daily mean for any subject over the period (kcal)*	*Maximum daily mean for any subject over the period (kcal)*	*Standard deviation of subject means (kcal)*	*Coefficient of variation of subject means (%)*
Widdowson (1936)	63	7	Various, middle class	3067	1772	4955	714	23·3
Happold (1945)	29		Medical students:					
			November	2674	—	—	695	26·0
			January	2713	—	—	526	19·4
Pyke, Harrison, Holmes and Chamberlain (1947)	12	7	Active aged (in small institutions)	2160	2050	2251	66	3·1
	12	7	Infirm aged (in large institutions)	2069	1903	2272	100	4·8
Bransby, Daubney and King (1948–9b)	15	3	(By calculation)	2306	1907	2796	243	10·5
			(By analysis)	2223	1849	2727	260	11·7
Kitchin, Passmore, Pyke and Warnock (1949)	61	7	Students: at home	3040	2140	4690	580	19·1
	47	7	in lodgings	3000	2150	3870	470	16·2
			in hostel	2960	2440	3590	290	9·8
Ministry of Food (1949, unpublished)	74	7	Industrial workers	3407	2108	4470	511	15·0
Bransby and Osborne (1953)	125	7	Elderly	2096	<1000	>3000	503	24·0
Bransby (1954)	152	7	Industrial workers	3540	—	—	620	17·5

Edholm, Fletcher, Widdowson and	6	14	Intermediate cadets	3524	2917	4222	407	11·5
McCance (1955)	6	14	Junior cadets	3340	2832	4085	348	10·4
Garry, Passmore, Warnock and	19	7	Miners	4030	3090	5410	557	13·8
Durnin (1955)	10	7	Clerks	3040	2500	3830	412	13·6
Adam, Best, Edholm and Woolf			Army troops:					
(1957)	13	4	exercise I	3502	2998	4130	370	10·6
	13	5	exercise III	5026	4374	6116	526	10·5
Booyens and McCance (1957)	4	7–14	Scientific workers	2868	2133	4412	676	23·6
Adam, Best, Edholm, Fletcher, Lewis and Woolf (1958)	58	21	Army recruits	3818	2027	5333	625	16·4
Adam, Best, Edholm, Goldsmith, Gordon, Lewis and Woolf (1959)	6	18	Army recruits	4077	3459	4563	415	10·1
Cook and Wilson (1960, personal communication)	561	7	Students	2984	—	—	507	10·1
Heady, Marr and Morris (1961 unpublished)	118	7	Bank officials	2851	1749	4055	430	15·1
Durnin, Blake, Allan, Shaw, Wilson, Blair and Yull (1961)	33	7	Elderly industrial workers	2993	2119	3963	376	12·6
Blake, Durnin, Aitken, Caves and Yuill (1962, in preparation)	9	7	Elderly retired	2054	1406	2606	398	19·4

Table III. *Variations in Energy Intake: Adult Females*

Source	*No. of subjects*	*Period of survey (days)*	*Information about subjects*	*Mean daily value for all subjects (kcal)*	*Minimum daily mean for any subject over the period (kcal)*	*Maximum daily mean for any subject over the period (kcal)*	*Standard deviation of subject means (kcal)*	*Coefficient of variation of subject means (%)*
Widdowson and McCance (1936)	63	7	Various, middle class	2187	1453	3110	388	17·7
Andross (1936)	109	3–5	Students	2035	—	—	371	18·2
Widdowson and Alington (1941)	57	7	Various, middle class	2137	—	—	420	19·7
Happold (1945)	20		Medical students:					
			November	2348	—	—	449	19·1
			January	2346	—	—	435	18·5
Pyke, Harrison, Holmes and	9	7	Aged: at home	1409	1034	2313	369	26·2
Chamberlain (1947)	18	7	in almshouses	1434	1079	1877	244	17·0
	12	7	in large institutions	1580	1252	1717	129	8·2
Bransby, Daubney and King	18	3	(By calculation)	1908	1360	2843	352	18·4
(1948–9b)			(By analysis)	1854	950	2506	395	21·3
Kitchin, Passmore, Pyke and	71	7	Students: at home	2180	1450	3160	370	17·0
Warnock (1949)	74	7	in lodgings	2280	1520	3220	400	17·5
	26	7	in hostel	2330	1750	3080	310	13·3
Yudkin (1951)	6	28	Students	2157	1100	2750	—	11·0
Bransby and Osborne (1953)	178	7	Elderly	1746	<1000	>3000	427	24·5
Booyens and McCance (1957)	2	7–14	Scientific workers	2007	1063	2471	—	—

Durnin, Blake and Brockway (1957)	12	7	Housewives	2100	1593	2435	245	11·7
	12	7	Their daughters	2220	1777	2720	333	15·0
Cook and Wilson (1960, personal communication)	205	7	Students	2429	—	—	462	19·0
Durnin, Blake, Brockway and Drury (1961)	17	7	Elderly, living alone	1894	1107	2283	299	15·8
Durnin, Blake, Allan, Shaw and Blair (1961)	21	7	Elderly, at home	1944	1243	2886	404	20·8
Copping (1962, unpublished)	82	7	Students	2157	1340	4014	646	21·5
			Pregnant women					
McCance, Widdowson and Verdon-Roe (1938)	120	7	Various	2347	1163	3522	497	21·2
Roscoe and McKay (1946)	35	7		2550	1660	3650	473	18·5
Hobson (1948)	111	7		2400	1600	3500	430	18·0
Thomson (1958)	101	7	Social class: I and II	2633	—	—	482	18·3
	109	7	class III	2521	—	—	487	19·3
	279	7	class IV and V	2354	—	—	513	21·8
Thomson (1958, unpublished)	489	7	All the above classes	2449	1126	4152	503	20·5

breed of *homo sapiens* will need encouragement and assistance, and it would seem only sensible for authorities at all levels, from government downwards, to provide education and facilities for suitable forms of active leisure.

H. Exercise and Obesity

A seemingly unavoidable result of the facts shown in the tables quoted is the widespread presence of obesity in the adult population of communities to whom these data pertain. Many theories have been proposed to account for the development of obesity in man—most of them effectively demolished by Meiklejohn (1953). An explanation that appears reasonable and logical might well be that most women and a large percentage of men are using up such small amounts of energy in their daily existence that the food required to replace that energy provides an unsatisfying quantity, and therefore that "appetite" is only gratified by an intake of food in excess of the absolute need. The thesis is difficult to prove in man, but there is supporting evidence from animal studies. Mayer, Marshall, Vitale, Christensen, Mashayeki and Stare (1954), found, for example, that an increased amount of physical activity beyond this minimum level resulted in no parallel increase of food intake.

However, not only is it unpopular (even scientifically) to state that obesity results most frequently because of a lack of exercise, but it is often denied that exercise can have any beneficial effect in treating or preventing obesity. The arguments are well summarised by Mayer (1960). The basis of the denial is not experimental work, nor even everyday experience (no athlete in training is fat), but is due to the plausible misconception that exercise requires comparatively small amounts of energy expenditure and is therefore unlikely to alter calorie balance. It is difficult to see how any knowledgeable person could maintain the first part of that argument; the tables quoted in this chapter, and any table of calorie requirements, will indicate unequivocally that occupations involving physical effort require considerably more energy than those involving little physical work. Tables showing details of the energy cost of individual physical activities, such as walking, cycling, playing golf, tennis and so on, will illustrate the fact even more clearly (cf. Passmore and Durnin, 1955). Whereas sitting may require about 1·5 kcal/min of energy expenditure, walking at 3 mph may expend three times that level, cycling perhaps 8 to 10 kcal/min, swimming anywhere from 4 to 12 kcal/min or even more, with large variations for other activities. Physical activity is therefore undoubtedly able to cause large increases in energy output.

The effect of the lack of physical exercise is also clear when examined. In both animals and in man studies have shown that obese groups may often eat less than non-obese, but that their exercise level is on a much lower plane. Mayer (1953) found that obese mice were 50 to 100 times less active than non-obese animals and that they exhibited these tendencies before they became obese, so that the inactivity was not the result of the obesity. Similarly observations have been made in man. Greene (1939) has reported on over 200 obese patients in whom the disability began with a sudden decrease in activity. Bruch (1940) and Johnson, Burke and Mayer (1956) in the United States have found inactivity to be the important factor in obesity in children and adolescents, and similar reports have been published in Denmark (Tolstrup, 1953; Juel-Nielsen, 1953). These authors, and also Swanson, Roberts, Willis, Pesek and Mairs (1955), Stefanik, Heald and Mayer (1959) and Chirico and Stunkard (1960), leave no doubt that the obese of all ages walk less, sit for longer and spend less time in active recreations than people of normal build.

If the prevention and cure for obesity—exercise—is difficult to attain for much of the population, it is as well to remember that the alternative of permanent mild or acute hunger (to restrict food intake) may be even more difficult. There is no suitable substitute except in remaining obese. The increased prevalence of mechanical disorders (rheumatism, arthritis), metabolic disorders (diabetes) and accidents among the obese are irrefutable. And insurance companies know well that death comes much earlier to the obese. The Metropolitan Life Insurance Company (1960) have provided mortality data for cases accepted between 1935 and 1953. When body weight was 20% over average the mortality increased 31% for men aged 15 to 39 years, 31% for men aged 40 to 69, 21% for women aged 15 to 39 and 32% for women aged 40 to 69. Much of this excess was due to coronary artery disease, vascular lesions in the central nervous system and diabetes mellitus. Overweight men who reduced their weight also reduced their likelihood of death to the level of the standard risk.

It would hardly be necessary to reiterate these comparatively well-known facts were it not so common to find respected medical authorities denying completely the influence of exercise on obesity and even any beneficial effect of exercise on the whole man. Perhaps this is the only part of this discussion on exercise in which all levels of exercise might be grouped together. Even low levels of exercise can prevent or reduce obesity, and there is no doubt from the evidence available that high levels of exercise are of direct benefit in this as in many other contexts.

REFERENCES

Adam, J. M., Best, T. W., Edholm, O. G., Fletcher, J. G., Lewis, H. E., and Woolf, H. S. (1958). M.R.C. 58/201, A.P.R.C. 58/3.

Adam, J. M., Best, T. W., Edholm, O. G., Goldsmith, R., Gordon, E. F., Lewis, H. E., and Woolf, H. S. (1959). *M.R.C.* 59/819, A.P.R.C. 59/9.

Adam, J. M., Best, T. W., Edholm, O. G., and Woolf, H. S. (1957). *M.R.C.* 57/93, A.P.R.C. 57/1.

Adams, E. A. (1938). *Res. Quart. Amer. Ass. Hlth. phys. Educ.* 9, 102.

Andross, M. (1936). Quoted by Cathcart, E. P., and Murray, A. M. T. (1936). *M.R.C. Memor.* Spec. Rep. Ser. no. 218, p. 47.

Åstrand, P. O. (1952). "Experimental studies of physical working capacity in relation to sex and age." Munksgaard, Copenhagen.

Åstrand, P. O. (1956). *Physiol. Rev.* 36, 307.

Åstrand, P. O., Cuddy, T. E., Saltin, B., and Stenberg, J. (1964). *J. appl. Physiol.* **19**, 268.

Bannister, R. G., Cotes, J. E., Jones, R. S., and Meade, F. (1960). *J. Physiol.* **152**, 66P.

Bansi, H. W., and Groscurth, G. (1930). *Klin. Wschr.* 9, 1902.

Bates, D. V., Boucot, N. G., and Dormer, A. E. (1955). *J. Physiol.* **129**, 237.

Bell, G. H., Davidson, J. N., and Scarborough, H. (1961). *In* "Textbook of Physiology and Biochemistry", p. 834. 5th Edition. Livingstone, Edinburgh and London.

Booyens, J., and McCance, R. A. (1957). *Lancet* i, 225.

Borovansky, L. (1930). *Bull. int. Acad. Sci. Boheme.*

Bransby, E. R. (1954). *Brit. J. Nutr.* 8, 100.

Bransby, E. R., Daubney, C. G. and King, J. (1948–9b). *Brit. J. Nutr.* **2**, 232.

Bransby, E. R., and Osborne, B. (1953). *Brit. J. Nutr.* 7, 160.

British Medical Bulletin (1963). "Respiratory Physiology." British Council, London.

Bruch, H. (1940). *Amer. J. Dis. Child.* **60**, 1982.

Chirico, A. M., and Stunkard, A. J. (1960). *New Engl. J. Med.* **263**, 935.

Christensen, E. H. (1931). *Arbeitsphysiologie* 4, 471.

Cohen, H., and Goldberg, C. (1960). *Brit. med. J.* ii, 509.

Cunningham, D. J. C., and Lloyd, B. B. (1963). "The Regulation of Human Respiration." Blackwell, Oxford.

Davies, C. T. M., and Copland, J. G. (1964). *J. appl. Physiol.* **19**, 325.

Davies, C. T. M., Drysdale, H. C., and Passmore, R. (1963). *Lancet*, ii, 930.

Dawson, P. M. (1920). *Amer. J. Physiol.* **50**, 443.

Dill, D. B. (1963). *Pediatrics* **32**, Supp. 653.

Dill, D. B., and Brouha, L. (1937). *Le Travail Humain*, 5, 3.

Dublin, L. I. (1932). *Statist. Bull. Metrop. Life Insce Co.* August.

Durnin, J. V. G. A. (1955) in Garry, R. C., Passmore, R., Warnock, G. M., and Durnin, J. V. G. A., *M.R.C.* Spec. Rep. Ser., no. 289.

Durnin, J. V. G. A. (1964). "Somatic standards of reference". *In* "Body Composition: Implications for Human Biology" (J. Brozek, ed.). (In press.)

Durnin, J. V. G. A., Blake, E. C., Allan, M. K., Shaw, E. J., and Blair, S. (1961). *J. Nutr.* **75**, 73.

Durnin, J. V. G. A., Blake, E. C., Allan, M. K., Shaw, E. J., Wilson, E. A., Blair, S., and Yuill, S. A. (1961). *Brit. J. Nutr.* **15**, 587.

Durnin, J. V. G. A., Blake, E. C., and Brockway, J. M. (1957). *Brit. J. Nutr.* **11**, 85.

Durnin, J. V. G. A., Blake, E. C., Brockway, J. M., and Drury, E. A. (1961). *Brit. J. Nutr.* **15**, 499.

Durnin, J. V. G. A., Brockway, J. M., and Whitcher, H. W. (1960). *J. appl. Physiol.* **15**, 161.

Durnin, J. V. G. A., McLees, W. J., Yuill, S. A., Busby, A. D., Gay, C. A., and Blake, E. C. (1964). *Proc. Nutr. Soc.* **23**, XIV.

Edholm, O. G., Fletcher, J. G., Widdowson, E. M., and McCance, R. A. (1955). *Brit. J. Nutr.* 9, 286.

Edwards, H. T., Brouha, L., and Johnson, R. E. (1940). *Le Travail Humain*, 8, 1.

Eppinger, H., Kisch, F., and Schwarz, H. (1925). *Klin. Wschr.* 4, 1101.

Ewig, W. (1925). *Münch. med. Wschr.* **72**, 1955.

Garry, R. C., Passmore, R., Warnock, G. M., and Durnin, J. V. G. A. (1955). M.R.C. Spec. Rep. Ser., no. 289.

Gemmill, C., Booth, W., Detrick, J., and Schiebel, H. (1931). *Amer. J. Physiol.* **96**, 265.
Godin, P. (1913). "La Croissance pedant l'age scolaire." Fischbacher, Paris.
Greene, J. A. (1939). *Ann. intern. Med.* **12**, 1797.
Grober, J. (1908). *Arch. exp. Path. Pharmak.* **59**, 424.
Happold, F. C. (1945). *Proc. Nutr. Soc.* **3**, 116.
Harries, J. M., Hobson, E. A., and Hollingsworth, D. F. (1962). *Proc. Nutr. Soc.* **21**, 157.
Harvey, W. (1628). "An anatomical dissertation upon the movement of the heart and blood in animals", p. 90. G. Moreton, Canterbury (1894).
Hatai, S. (1915). *Anat. Rec.* **9**, 647.
Henderson, Y., Haggard, H. W., and Dolley, F. (1927). *Amer. J. Physiol.* **82**, 512.
Henry, F. M. (1954). *Res. Quart.* **25**, 29.
Herxheimer, H. (1921). *Virchows Arch.* **233**, 484.
Herxheimer, H. (1922). *Klin. Wschr.* **1**, 480.
Herxheimer, H. (1924). *Zeitschr. f. klin. Med.* **98**, 484.
Herxheimer, H. (1927). *Klin. Wschr.* **6**, 21.
Hobson, W. (1948). *J. Hyg., Camb.* **46**, 198.
Hong, S. K., Rahn, H., Kang, D. H., Song, S. H., and Kang, B. S. (1963). *J. appl. Physiol.* **18**, 457.
Hoogerwerf, S. (1929). *Arbeitsphysiologie*, **2**, 61.
Johnson, M. L., Burke, B. S., and Mayer, J. (1956). *Amer. J. clin. Nutr.* **4**, 37.
Johnson, W. E. (1960). "Science and Medicine of Exercise and Sports." Harper, New York.
Juel-Nielsen, N. (1953). *Acta Paediat.* **42**, 130.
Kannel, W. B., Dawber, T. R., Kagan, A., Revotskie, N., and Stokes, J. (1961). *Ann. intern. Med.* **55**, 33.
Karvonen, M. J., Kentala, E., and Mustala, O. (1957). *Ann. med. exp. Fenn.* **35**, 307.
Karvonen, M. J., Kihlberg, J. Määttä, J., and Virkajrävi, J. (1956). *Duodecim* **72**, 893.
Katsuki, S., Yamazaki, T., and Hirao, Y. (1963). *Bull. phys. Fit. Inst. Tokyo* **1**, 14.
Kaup, J., and Grosse, A. (1927). *Münch. med. Wschr.* **74**, 1353.
Kitchin, A. H., Passmore, R., Pyke, M., and Warnock, G. M. (1949). *Brit. J. soc. Med.* **3**, 10.
Kohlrausch, W. (1924). *Z. Konstitutionslehre* **10**, 434.
Kulbs, F. (1906). *Arch. exp. Path. Pharmak.* **50**, 288.
Lindhard, J. (1915). *Pflüg. Arch. ges. Physiol.* **161**, 233.
McCance, R. A., Widdowson, E. M., and Verdon-Roe, C. M. (1938). *J. Hyg. Camb.* **38**, 596.
Mann, G. V., Teel, K., Hayes, O., McNally, A., and Bruno, D. (1955). *New Engl. J. Med.* **253**, 349.
Matthias, E. (1916). "Einfluss d. Leibesübungen auf. d. Korperwachstum." Rascher and Co., Zurich and Leipzig.
Mayer, J. (1953). *Science* **117**, 504.
Mayer, J., Marshall, N. B., Vitale, J. S., Christensen, J. H., Mashayeki, M. B., and Stare, F. J. (1954). *Amer. J. Physiol.* **177**, 544.
Mayer, J. (1960). Exercise and Weight Control. *In* "Science and Medicine of Exercise and Sports", p. 301 (W. E. Johnson, ed.). Harper, New York.
Meiklejohn, A. P. (1953). *Proc. Nutr. Soc.* **12**, 19.
Metropolitan Life Insurance Company (1960). *Statist. Bull.* **41**, Feb., p. 6; Mar., p. 1.
Milic-Emili, G., Petit, J. M., and Deroanne, R. (1962). *J. appl. Physiol.* **17**, 43.
Morpurgo, B. (1897). *Virchows Arch.* **150**, 522.
Morris, J. N., and Crawford, M. D. (1958). *Brit. med. J.* ii, 1485.
Morris, J. N., and Heady, J. A. (1953). *Brit. J. industr. Med.* **10**, 245.
Morris, J. N., Heady, J. A., Raffle, P. A. B., Roberts, G. G., and Parke, J. W. (1953). *Lancet* ii, 1053.

Mostyn, E. M., Helle, S., Gee, J. B. L., Bentivoglio, L. G., and Bates, D. V. (1963). *J. appl. Physiol.* **18**, 687.
Müller, E. A. (1962). *Rev. canad. Biol.* **21**, 303.
Newman, F., Smalley, B. F., and Thomson, M. L. (1962). *J. appl. Physiol.* **17**, 649.
Nikkilä, E. A., and Konttinen, A. (1962). *Lancet* i, 1151.
Passmore, R. (1964). An assessment of the Reports of the Second Committee on Calorie Requirements (FAO, 1957). Food and Agriculture Organisation of the United Nations, Rome.
Passmore, R., and Durnin, J. V. G. A. (1955). *Physiol. Rev.* **35**, 801.
Petren, T., Sjöstrand, T., and Sylven, B. (1936). *Arbeitsphysiologie* **9**, 376.
Pomeroy, W. C., and White, P. D. (1958). *J. Amer. med. Ass.* **167**, 711.
Pyke, M., Harrison, R., Holmes, S., and Chamberlain, K. (1947). *Lancet* ii, 461.
Rook, A. (1954). *Brit. med. J.* i, 773.
Roscoe, M. H., and McKay, H. S. (1946). *Edinb. med. J.* **53**, 565.
Rushmer, R. F., and Smith, O. A. (1959). *Physiol. Rev.* **39**, 41.
Schneider, E. C., Clarke, R. W., and Ring, G. C. (1927). *Amer. J. Physiol.* **81**, 255.
Secher, K. (1921). *Z. ges. exp. Med.* **14**, 113.
Secher, K. (1923). *Z. ges. exp. Med.* **32**, 290.
Shapiro, W., Johnston, C. E., Dameron, R. A., and Patterson, J. L. (1964). *J. appl. Physiol.* **19**, 199.
Siebert, W. W. (1928). *Z. klin. Med.* **109**, 350.
Skerlj, B. (1936). Prace Antropologiczne (Towarzystwa Naukowego Warszawskiego).
Spain, D. M., and Bradess, V. A. (1960). *Circulation* **22**, 239.
Stefanik, P. A., Heald, F. P., and Mayer, J. (1959). *Amer. J. clin. Nutr.* **7**, 55.
Steinhaus, A. H. (1933). *Physiol. Rev.* **13**, 103.
Steinhaus, A. H., and Jenkins, T. A. (1930). *Amer. J. Physiol.* **95**, 202.
Swanson, P., Roberts, H., Willis, E., Pesek, I., and Mairs, P. (1955). "Food intake and body weight of older women." *In* "Weight Control", p. 80 (Eppright, Swanson and Iverson, eds.), Iowa State College Press, Ames, Iowa.
Taylor, C. (1941). *Amer. J. Physiol.* **135**, 27.
Taylor, C. (1945). *Annu. Rev. Physiol.* **7**, 599.
Taylor, H. L. (1960). Exercise and Metabolism. *In* "Science and Medicine of Exercise and Sports", p. 123 (W. E. Johnson, ed.). Harper, New York.
Thomson, A. M. (1958). *Brit. J. Nutr.* **12**, 446.
Thorner, W. (1930). *Arbeitsphysiologie* **2**, 116.
Tolstrup, K. (1953). *Acta Paediat., Stockholm* **43**, 289.
West, J. B. (1962). *J. appl. Physiol.* **17**, 421.
Widdowson, E. M. (1936). *J. Hyg., Camb.* **36**, 269.
Widdowson, E. M., and Alington, B. K. (1941). *Lancet* ii, 361.
Widdowson, E. M., and McCance, R. A. (1936). *J. Hyg., Camb.* **36**, 293.
Wolfenden Report (1960). "Sport and the Community: Report of the Wolfenden Committee on Sport." Central Council of Physica Recreation, London.
Yudkin, J. (1951). *Brit. J. Nutr.* **5**, 177.

CHAPTER 11

Pregnancy, Childbirth and Lactation

F. E. HYTTEN AND A. M. THOMSON

A. Introductory

Other chapters in this book have described the physiological responses to various forms of stress caused by changes in the external environment. In pregnancy the changes are internal, and are often anticipatory rather than truly adaptive. Again, the adaptations of pregnancy do not aim at preserving the constancy of the *milieu intérieur*; on the contrary, the maternal *milieu intérieur* changes progressively, in a manner that is presumably appropriate to the successive stages of gestation. At its full development the composition of maternal tissues and body fluids may be so different from those during the normal non-pregnant state that the changes, though physiological, are difficult to distinguish from those usually associated with ill health. Much confusion has been caused by interpreting the characteristics of pregnancy in terms of ordinary non-pregnant standards; yet what may seem, at first sight, to be evidence of a pathological state may be not only normal, but necessary.

We shall deal with the adaptations of pregnancy mainly in the context of Western civilisation, that is to say of a comfortable external environment, an adequate supply of food and no need for hard physical exertion. Much less is known about the phenomena of pregnancy in an adverse environment and in conditions of food shortage and hard work.

We do not intend to discuss the demographic aspects of human reproduction, though in the opinion of many competent authorities the "population explosion" is the most important practical problem facing mankind today. Wynne-Edwards (1962) has pointed out that civilised man has lost the capacity of animals and primitive man to adjust population size in accordance with the means of subsistence. Yet even this is not wholly true. During the earlier years of this century, in face of intolerable industrial conditions, the population of Great Britain rapidly dropped its reproduction rate below replacement level, despite the crude nature of the available contraceptive methods and active or passive opposition to their use. Again, man has acquired, through modern technology, a remarkable capacity to adjust food production to the needs of population. Though it would be foolish to underestimate the dangers of the population explosion, it seems worth stressing that the difficulties to date have arisen more from the necessity for social adjustment in face of rapid change than from inability to produce enough food. Famine occurs, but probably no more commonly than it has done in past centuries. It may indeed become more common if populations fail to adapt quickly enough.

Nor do we intend to discuss the results of experiments on animals, except in so far as this is necessary to illuminate research on man. And, since we must be brief, it will be necessary to omit all but the more important references to authorities. A more detailed account has been prepared by Hytten and Leitch (1964).

B. Physiological Adaptations

1. *Volume and composition of blood*

During pregnancy, plasma volume and red-cell volume both increase, the former more than the latter. Plasma volume probably begins to rise towards the end of the third month, a maximum increase of about 1250 to 1500 ml being reached about the thirty-fourth week; during the last few weeks there is a slight fall (Paintin, 1962; Hytten and Paintin, 1963). Total red-cell volume probably increases linearly from the third month until term, the average rise being about 18% in the absence of iron medication. As a consequence of these changes, the haemoglobin concentration and packed cell volume fall, without any change in mean cell-haemoglobin concentration. The fall in haemoglobin level is often construed as anaemia. Other data are quoted in support of this view: reduction of serum iron level, increase in iron-binding capacity and in free protoporphyrin in red cells and, above all, the fact that administration of iron in therapeutic doses prevents the fall in haemoglobin concentration. Yet, for reasons discussed by Hytten and Duncan (1956), none of this is necessarily indicative of a pathological state. The changes in peripheral blood samples conceal an increase in the total oxygen-carrying capacity of the blood that more than matches the increased oxygen consumption. The "disproportionate" increase in plasma volume is probably designed to facilitate an increase in glomerular filtration rate and an increased blood flow in the skin for cooling purposes. Nevertheless, it needs to be said that in the opinion of many competent clinicians there is a genuinely increased liability to anaemia of the iron-deficiency type in pregnancy, and an increased liability to megaloblastic anaemia in pregnancy is beyond dispute. The mechanisms are not entirely clear, but seem unlikely to be those of simple dietetic deficiency states.

The increase of plasma volume is accompanied by changes in composition that are certainly not due to simple dilution. Total protein falls. A fall in albumin occurs early, before any noticeable increase of plasma volume is apparent. The α-globulins rise slightly, and β-globulin rises conspicuously, both in concentration and as a proportion of total circulating protein; γ-globulin probably changes little, if at all (MacGillivray and Tovey, 1957; Paaby, 1960; Alvarez, Afonso and Sherrard, 1961). The reason for these changes is obscure; they may be associated with the functions of circulating protein in the transport of hormones and nutrients. Fibrinogen concentration rises considerably and is accompanied by a decrease in fibrinolytic activity. Taken together with an increase in blood platelet concentration, these changes may indicate a protective mechanism against haemorrhage, but there is little evidence that intravascular clotting becomes more common in pregnancy, though it may be more common during the puerperium. The changing pattern of blood proteins raises the erythrocyte sedimentation rate to levels that would in other circumstances indicate pathology. The total lipid, neutral fat, phospholipid and cholesterol contents of blood increase (Alvarez, Gaiser, Simkins, Smith and Bratvold, 1959).

2. *Cardiovascular dynamics*

The resting pulse rate increases from about 70 to 78 during the first three months, then more slowly to about 85 near term. Cardiac output rises from about 4·5 to 5·5 l/min by three to four months and to 6·0 l/min in mid-pregnancy; it is maintained at that level until six to eight weeks from term and falls to 5·0 to 5·5 l/min at term (Hamilton, 1949; Palmer and Walker, 1949; Bader, Bader, Rose and Braunwald, 1955). These figures imply that stroke volume increases. The rise in cardiac output together with the increasing red-cell volume, exceeds the demand for oxygen carriage, especially in early pregnancy, so that the arteriovenous oxygen difference diminishes at first to about 33 ml/l in the third month, rising to the average non-pregnant value of 45 ml/l at term. Blood pressure probably falls at first, the diastolic pressure more than the systolic. Both pressures rise in the final weeks, when they may be further increased by the pathological but unexplained changes characteristic of pre-eclampsia. The reduced peripheral vascular resistance, which keeps the blood pressure constant as the cardiac output rises, is due to vasodilatation in the skin and a greater blood flow through low-pressure areas in the kidneys and placental site. This may be partly explained by the general relaxation of smooth muscle that is characteristic of pregnancy. Venous pressure probably does not change except in the leg veins, where it increases (McLennan, 1943), partly because of mechanical obstruction of the return blood flow by the growing uterus and its contents and partly because of hydrodynamic obstruction caused by a high-pressure outflow from the uterine vessels. Whatever the cause, any tendency to varicosity of leg, rectal and vulvar veins is accentuated during pregnancy.

The pregnant woman lying supine may experience an abrupt fall in blood-pressure, accompanied by faintness—the "supine hypotensive syndrome". Presumably this is due to obstruction of the inferior vena cava by the weight of the uterus and its contents (Quilligan and Tyler, 1959).

Many otherwise normal pregnant women develop oedema of the lower limbs and sometimes oedema of a more generalised character. Leg oedema may be due to obstruction of the venous return, but there is reason to suppose that the water content of skin generally is raised, possibly as a mechanism that assists the dissipation of heat. Certainly blood flow in the skin is considerably increased, with peripheral vasodilatation (Herbert, Banner and Wakim, 1958). The blood flow to the uterus is naturally increased enormously (Romney, Reid, Metcalfe and Burwell, 1955; Assali, Rauramo and Peltonen, 1960), so is that to the kidney (Sims and Krantz, 1958).

Though there is no evidence for any change in working efficiency, the pregnant woman usually has a considerable subjective impression of diminished capacity for work. This may be due mainly to the increased mechanical handicaps imposed by pregnancy.

3. *Respiration*

Vital capacity and respiration rate are unchanged, but the tidal volume is increased by about 40% at the end of pregnancy, with a proportionate increase in minute ventilation. The residual lung volume is reduced by about 20%. It follows that alveolar ventilation is increased more than total ventilation and that the mixing and distribution of gases in the lungs becomes more efficient during pregnancy (Cugell, Frank, Gaensler and Badger, 1953).

Values for increase in basal oxygen consumption vary widely; about 15% is likely. Since the minute volume increases much more, pregnancy gives rise to considerable "hyperventilation" and the P_{CO_2} in the alveoli, and therefore in arterial blood, falls. It is of interest that alveolar P_{CO_2} falls at about the time of ovulation in the menstrual cycle, so that this particular form of adaptation takes place even before the fertilised ovum imbeds (Bouterline-Young and Bouterline-Young, 1956). It may be an effect of progesterone, the injection of which has been shown to induce hyperventilation and to reduce alveolar P_{CO_2} in healthy males and in patients with emphysema and hypercapnia (Tyler, 1960). A reduced maternal P_{CO_2} (and slightly raised P_{O_2}) presumably assists the foetus to dispose of carbon dioxide.

It has for long been believed that in pregnancy the diaphragm is "splinted", but a careful X-ray study by McGinty (1938) indicates that, on the contrary, breathing in pregnancy is more diaphragmatic than costal. The level of the diaphragm rises and the transverse diameter of the chest increases. The work of breathing becomes proportionately greater with increasing rates of ventilation, and the effect is increased during pregnancy, but there is no satisfactory quantitative evidence on the matter.

Many pregnant women complain of dyspnoea. This is not necessarily related to exercise or to any of the changes in lung volumes. A possible explanation may be found in the theory of Campbell and Howell (1963) that dyspnoea occurs when the ventilatory response is "inappropriate" to the demand. As noted above, there is overbreathing in pregnancy, which results in unusually low alveolar P_{CO_2} levels, so that women are being compelled to breathe harder in biochemical circumstances that are ordinarily inappropriate.

4. *Respiratory exchange and supply of nutrients to the foetus*

Placental growth is initially ahead of that of the foetus. At its full development the placental surface area (excluding microvilli) has been assessed as anything from 6 to 14 m^2 (Clavero and Botella Llusiá, 1963); with this area, transfers to and from the foetal circulation take place across a two-layer membrane, cytotrophoblast and syncytiotrophoblast. Not much is known about the mechanisms of this transfer. Simple diffusion probably suffices for gases and water, with mechanisms of varying complexity for nutrients. Points to note are that foetal plasma proteins are not the same as those of maternal plasma and that the foetus probably manufactures its own large molecules from simpler precursors, with the important exception of γ-globulin, which it receives from the mother. A curious feature of placental physiology is that it fails to provoke an immune reaction against itself, although placental tissue appears to be capable of acting as an antigen. Possibly the placenta adsorbs circulating antibody to itself (Hulka, Hsu and Beiser, 1961).

Foetal and adult haemoglobins are not identical, and the bloods differ in such a way that transfer of oxygen to the foetus and release of carbon dioxide to the mother are both facilitated (Van Slyke, 1959). The quantitative aspects of oxygen transfer to the foetus are still confusing and obscure. There is no sound reason to believe that the basal metabolism of foetal tissue, per unit weight, is any higher than that of maternal tissue. It does not have to maintain body temperature against an adverse gradient, and foetal muscle tone is low. These facts (which have been obscured by the irrational habit of expressing even foetal metabolism in terms of surface area, and by the notion that a fast-growing tissue must have a higher metabolic rate than adult tissue, despite the higher water content of the former) are consistent with Barcroft's finding that in sheep foetal metabolism in utero is lower than that of the same foetus after extraction. Taking the maternal and foetal metabolic rates as equivalent, calculation indicates that the foetus at term requires about 12 ml oxygen/min and the uterus along with the whole product of conception about 18 ml/min. These requirements should as a rule be easily within the capacity of the mother to supply. There is, however, some evidence that near term and in post-mature gestations the capacity of the placenta to transfer oxygen diminishes. With lowered P_{O_2} levels, the haemoglobin concentration in cord blood may rise (Walker and Turnbull, 1953), and foetal mortality rates undoubtedly tend to rise when pregnancy proceeds beyond term (Baird, 1963).

The release of carbon dioxide from the foetal circulation is probably even easier than its acquisition of oxygen. Habitual cigarette smoking results in chronically raised levels of carbon monoxide in both foetal and maternal blood, which will impede foetal oxygenation (Haddon, Nesbitt, and Garcia, 1961).

There are few reliable quantitative facts about the transplacental transfer of nutrients. Studies with isotopes indicate a large increase during gestation in the transfer of sodium.

5. *Alimentary function*

There is a body of opinion that the mother has an enhanced ability to absorb and utilise presented nutrients, but the evidence for this in the human species is scanty. Hahn, Carothers, Darby, Martin, Sheppard, Cannon, Beam, Densen, Peterson and McClellan (1951) have produced results that suggest such enhanced abilities for iron. As will be shown later, the suggestion, based on the results of balance experiments, that pregnant women store enormous quantities of nitrogen in their own tissues is ill founded. Despite the lack of sound experimental evidence, the idea of improved alimentary function in pregnancy is supported by an interesting example described by Montgomery and Pincus (1955). A woman had only about 3 ft (1 m) of small intestine left, after extensive small bowel resection for stenosing ileitis, and was maintained in a precarious state of health with the aid of testosterone. During pregnancy she had an enormous appetite, gained 14 lb (6 kg), had a noticeable improvement in bowel movements and produced a normal baby weighing 6 lb (2·5 kg). Similar changes were noted in two subsequent, successful pregnancies; between pregnancies her condition reverted to its usual state.

Many pregnant women are plagued by nausea during early pregnancy and by heartburn and constipation later. There is evidence of reduced gastric tone, motility and secretion. Emptying time is probably increased with normal meals, but not after fluids that do not require digestion. The gall bladder shares in the general picture of sluggishness in muscular viscera, which may account for many of the gastric symptoms.

Again, many women report a marked increase in thirst and appetite before mid-pregnancy, when the foetus is still tiny and cannot be making large demands. Occasionally there are strange aberrations of appetite and taste, the reasons for which are not understood.

6. *Renal function*

As part of the general relaxation of smooth muscle, the ureters become dilated and contorted as early as the tenth week. In the later stages, these changes are more marked on the right side.

Renal plasma flow increases abruptly by about 45% in the first trimester, with a slight fall as pregnancy proceeds and a steeper fall almost to non-pregnant levels during the final month before parturition. The glomerural filtration rate is considerably increased throughout pregnancy, perhaps by about 60% (Sims and Krantz, 1958). Creatinine, urea and uric acid clearances increase, and the levels of these substances in blood are consequently diminished. On the other hand, there is no change in the level of sodium in tissue fluids. Although thirst and frequency of urination are frequently complained of, no evidence has been published that urinary volume changes during pregnancy. On the other hand, capacity to handle a water load shows dramatic alterations (Hytten and Klopper, 1963). During the second trimester urinary excretion after ingestion of 1 l water may rise to as high as 30 ml/-min. During the remainder of pregnancy the reaction diminishes steadily, until near term a similar load of water may cause little change in urine flow and most of the load is retained after 2 hr.

Benign glycosuria is common in pregnancy (Chesley, 1960), possibly as a result of the increased glomerular filtration rate without change of tubular reabsorption. Amino acids, too, are excreted by the kidney in unusually large amounts—too large to be explained by the rise in glomerular filtration rate, (Page, Glendening, Dignam and Harper, 1954; Christensen, Date, Schønheyder and Volqvartz, 1957). There is no direct evidence about whether reabsorption is impaired. Similarly, there are increased losses of inorganic iodine (Aboul-Khair, Crooks, Turnbull and Hytten, 1964) and probably of folic acid. The former may account in part for the common increase in thyroid gland size and the latter for an increased liability to megaloblastic anaemia.

One can only speculate at present on the reasons for these curious, wasteful and potentially dangerous changes in renal function. They are at their greatest when the need to get rid of foetal waste products is trivial. One has the impression of a change in many aspects of maternal metabolic turnover, in which nature tends to waste rather than to conserve, in the interests of some goals that cannot yet be recognised. It may be noted here that the trend towards increased elimination of some substances accompanies diminished uptake by maternal tissues. Thus, a glucose tolerance test during pregnancy frequently gives a "lag" type of curve, and similar effects have been described after the administration of a test dose of vitamin B_{12} (Hellegers, Okuda, Nesbitt, Smith and Chow, 1957). Though maternal metabolism is sometimes described as being anabolic during pregnancy, this should not be taken necessarily as signifying increased efficiency of absorption or utilisation of nutrients or of increased conservation by the kidney.

7. *Endocrine changes*

It seems reasonable to assume that many of the adaptations already discussed take place under endocrine influence, and certainly the main endocrine changes of pregnancy are extremely dramatic. Yet, despite an enormous literature, remarkably little is known about possible mechanisms.

The endocrine patterns of pregnancy are dominated by the production in the placenta of progesterone, oestrogens and chorionic gonadotrophin. Production and excretion of the last-named show a remarkable peak between about the fortieth and sixtieth days, falling to a much lower and fairly stable level during the remainder of pregnancy (Mishell, Wide and Gemzell, 1963); the significance of this is unknown, but it may be that it maintains the corpus luteum until the placenta can take over the production of progesterone. Progesterone metabolism has been mainly deduced from the excretion in urine of its principal metabolite, pregnanediol. This is produced, initially, by the persistent corpus luteum of the ovary, which is soon "overtaken" by placental production. Excretion of pregnanediol increases smoothly until the final weeks of pregnancy, when there is little further increase; yet the few available estimates of blood progesterone show continuously increasing levels during the final weeks; there may be a changing pattern of metabolism of progesterone (Shearman, 1959; Eton and Short, 1960). Oestrogen production in pregnancy, too, is almost wholly a function of the placenta. Levels of the principal metabolites in urine, especially oestriol, increase sharply for about the first six months, with an accelerating rate of increase in the final weeks (Klopper and Billewicz, 1963). It appears that an intact foetal circulation is necessary for both the normal production and the normal metabolism of oestrogen, but not of progesterone, during the later part of pregnancy (Cassmer, 1959).

Practically all the phenomena of pregnancy have been attributed at some time or another to progesterone or oestrogen, or to both, but there is little firm evidence. Progesterone may well be responsible for the general relaxation of smooth muscle, for the slight rise in body temperature after ovulation, and its persistence during the first half of pregnancy, and for the overbreathing of pregnancy. It probably has cerebral, perhaps hypothalamic, effects leading to reduced activity and increased appetite which may govern the tendency to fat storage. Oestrogens are believed to govern the development of the uterine decidua and to counteract the relaxant effect of progesterone on the uterus; they probably also cause development of mammary glandular tissue and preparation of the breasts and nipples for lactation (see below). More speculatively, oestrogens may be responssible for alterations in mucopolysaccharides leading to such phenomena as the softening and increasing elasticity of the uterine cervix, the increase of plasma volume and the tendency to store water in the skin. Yet, despite widespread belief, there is little evidence to confirm that high oestrogen levels promote water retention.

Even less is known of the pituitary hormones. There is a rise in adrenocortical hormone production (Martin and Mills, 1956; Cope and Black, 1959), which may be due to increased production of adrenocorticotrophin, and this may in turn be associated with increased melanophore-stimulating hormone. Many women experience darkening of certain areas of skin during pregnancy. There is no evidence of a change in thyrotrophic hormone production. Of the posterior pituitary hormones, oxytocin is probably responsible for the milk ejection reflex and, when injected, causes uterine contractions; vasopressin may account for the antidiuretic phenomena of late pregnancy and for vasospastic phenomena leading to raised blood pressure. It has to be borne in mind that normal pregnancy and labour have been recorded in hypophysectomised women and also in women with diabetes insipidus.

As noted above, corticosteroid production increases, and may cause amelioration of rheumatoid arthritis and Addison's disease. Aldosterone excretion increases, but probably not that of adrenaline and noradrenaline.

Although many symptoms and signs mimicking hyperthyroidism appear during pregnancy (raised basal metabolic rate, faster pulse rate, heat intolerance, raised serum protein-bound iodine), they can all be accounted for on other grounds. Increased losses of inorganic iodine in the urine lead to reduced serum inorganic iodine; thus, as in simple iodine deficiency, the thyroid gland has more work to do to maintain thyroxine output and may enlarge in the same way as in simple goitre (Aboul-Khair, Crooks, Turnbull and Hytten, 1964).

Pregnancy may unmask a latent diabetic or pre-diabetic trait or may even be "diabetogenic" (Fitzgerald, Malins and O'Sullivan, 1961). The mechanisms are unknown, as are those leading to increased foetal growth. There is no evidence that growth hormone is involved (Farquhar, 1962).

8. *Preparations for lactation*

One of the earliest signs of pregnancy may be a tingling and sensation of fullness in the breasts. These early signs are probably due to vascular engorgement, which rapidly passes off and is succeeded by growth of gland. The average increase in breast volume, as measured by Hytten (unpublished results), was about 200 ml, but the most remarkable feature was its variability, from little or no change to 880 ml. In civilisation, at all events, the need to lactate has probably never been essential to survival, and no doubt genetic selection for lactational ability has been attenuated. An interesting finding by Hytten was that, although increase in breast size was correlated with subsequent milk production, absolute breast size was not. It appears that large mammary glands have a purely symbolic significance in Western society.

Mammary glandular tissue may be the only source of galactose, which is required for the synthesis of chorionic gonadotrophin, of mucopolysaccharides and of cerebrosides in foetal growth.

In farm animals calcium is stored in the maternal skeleton during pregnancy, as a preparation for the great drain on calcium reserves during lactation (Benzie, Boyne, Dalgarno, Duckworth, Hill and Walker, 1955). Whether this occurs in women is unknown. On the other hand, as will be indicated below, earlier reports of large increases in maternal nitrogen stores are probably erroneous, and the drain of protein during lactation is small. But women do store fat during pregnancy, which may function as a reserve of energy during lactation, when milk production can involve the expenditure of 1000 kcal daily.

C. Variations and their Significance

What has been said above describes the changes believed to occur in the average normal pregnancy; we shall proceed below to discuss them in relation to weight change and energy metabolism. Here, it is necessary to stress that variations between individuals are strikingly large. The variations may to some extent have been exaggerated in the literature by failure to select for study women who may properly be regarded as representing the physiologically normal. Many subjects have been chosen from hospital series and are in one way or another manifestly abnormal. For example, the change in plasma volume during pregnancy, as reported in the literature, varies from zero or a slight decrease to more than a 100% increase. In a carefully selected "normal" series of primigravidae, Hytten and Paintin (1963) found the increase to range from 25 to 80%, most of their subjects showing increases from 40 to 60%. The conventional method of expressing increases in plasma volume, as a percentage of the non-pregnant volume, is misleading, since increase is not related to non-pregnant volume. The absolute increase averages about 1250 ml.

Further research is needed to define the true extent of the normal range, and to show the significance of deviations from it. If an increase in plasma volume of about 50% is regarded as normal, it seems unlikely that much smaller increases represent a fully satisfactory state of adaptation, and this should be capable of being assessed in clinical terms.

Meanwhile, we do know that all women are not equal in reproductive efficiency. Assessed in terms of stillbirth and neonatal death rates, a first pregnancy is less efficient than a second or third, and at higher parities efficiency falls off steadily. Age is also important. Baird, Hytten and Thomson (1958) have shown that in primigravidae efficiency deteriorates in many respects from a remarkably early age. It would appear that a woman is at her peak when she first reaches maturity and the completion of growth; thereafter, as with athletes, there is deterioration with age.

Physique and social status also bear upon efficiency. Thomson and Billewicz (1963) have summarised figures showing that tall women are better reproducers than short women, presumably because, on average, they are "better grown". Obesity appears to be associated with a number of reproductive disabilities.

D. Weight Gain and Energy Requirements

1. *Weight gain*

The weight gained by the average pregnant woman considerably exceeds the weight of the product of conception and of the enlarged organs of reproduction. The results given below, which were reported at the Sixth International Congress of Nutrition (Hytten, 1964), supersede some of those given in a preliminary communication (Thomson and Hytten, 1961).

The model used as a basis for estimates and calculations is the average clinically normal primigravida in Aberdeen who has eaten to appetite during pregnancy. She is aged 24 years, height 158 cm, and before pregnancy weighs 54 kg. During pregnancy she gains a total of 12·5 kg, and at the end of it produces a baby weighing 3·3 kg, with a placenta weighing 0·65 kg. When discharged from hospital about ten days after parturition she weighs about 4 kg more than she did at conception.

Table I gives total gain in weight and estimates of its components at different stages of pregnancy. The averages for foetal and placental weight are probably reliable, as are those for increase in maternal blood. Estimates for liquor amnii, uterus and mammary glands are less well founded, being based on fewer measurements. The final line of Table I shows the proportion of the total weight gained that remains to be accounted for. It must represent materials stored in the maternal tissues, excluding blood and reproductive organs; the amounts are large, and they remain large even if reasonable alternative figures are inserted in the upper part of the table.

Table I. *Analysis of Weight Gained in Pregnancy*

Tissues and fluids accounted for	*Increase in weight up to:* 10 *weeks*	20 *weeks*	30 *weeks*	40 *weeks*
	g	g	g	g
Foetus	5	300	1500	3300
Placenta	20	170	430	650
Liquor amnii	30	250	600	800
Uterus	135	585	810	900
Mammary gland	34	180	360	405
Blood	100	600	1300	1250
Extracellular extravascular water	0	0	0	1200
Total	324	2085	5000	8505
Total weight gained	650	4000	8500	12500
Weight not accounted for: ("Maternal stores")	326	1919	3500	3995

Though we do not know of any previous workers who have used total weight gain as the reference point for estimates, it would be concluded from a study of the literature that this "excess" maternal weight is probably mainly water and protein. Oedema is commonly noted during pregnancy, and measurements of extracellular water with thiocyanate and similar substances indicate a considerable increase in extracellular water. Again, several balance studies have indicated that astonishingly large amounts of nitrogen—much more than can be accounted for by the growth of the product of conception and enlargement of the organs of reproduction—are stored during pregnancy.

We believe that the conclusions drawn from such previous studies are completely misleading. In our hands, the thiocyanate method of estimating extracellular water has proved to be so erratic as to be useless; moreover, on theoretical grounds one cannot be sure that the factors governing thiocyanate distribution in tissues and fluids remain constant during pregnancy. Again, recent work on animals has indicated that nitrogen balance studies are liable to cumulative errors, which have the effect that more nitrogen appears to be stored than can be found from carcass analysis (Nehring, Lanbe, Schwerdtfeger, Schiemann, Haesler and Hoffman, 1957). Some of the amounts of stored nitrogen reported in studies of human pregnancy are impossibly large.

We approached the problem in a different way, by estimating change in total body water during pregnancy, using deuterium oxide as the tracer. This method gives consistent results on repeat estimations; even if the absolute values are assumed to involve some error, there is no reason to suppose that such errors will affect the changes that take place during pregnancy.

Table II gives the gains in total body water during pregnancy, together with estimates of the amounts of added water attributable to the growth of the product of conception, reproductive organs and maternal blood. Up to thirty weeks of pregnancy the two sets of figures agree fairly well, and there is certainly little or no excess water in the remaining maternal tissues. Up to this stage, therefore, it appears that there is no increase in maternal extracellular or intracellular water, apart from that already accounted for. By forty weeks, on the other hand, there is a slight excess of total body water, but apparently little more than about 1000 ml. This is alsmost certainly extracellular water.

Table II. *Water Component of Weight Gained in Pregnancy*

	Weeks of pregnancy		
	20	*30*	*40*
	g	g	g
Foetus	264	1185	2343
Placenta	153	366	540
Liquor amnii	247	594	792
Uterus	483	668	743
Mammary gland	135	270	304
Plasma	506	1058	920
Red cells	32	98	163
Total	1820	4239	5805
Measured increment (Hytten and Thomson, unpublished results)	1500	3750	7000

It appears, then, that up to about thirty weeks of pregnancy the excess weight shown in the last line of Table I is "dry weight" and that at forty weeks about four-fifths, or 4 kg, is "dry weight". It may be noted that the estimated gain in "dry weight" at forty weeks corresponds with the net gain of maternal weight, the difference between weight at conception and weight some days post-partum.

The capacity of the adult body to store protein, in the absence of severe malnutrition with wasting of muscle, is almost certainly small. Further, storage of protein is invariably accompanied by even greater storage of water. The weight of carbohydrate stores in the body is so small as to be negligible. The only alternative conclusion, therefore, is that the excess weight to be accounted for consists of fat. This is in agreement with the impression of most mothers that they "put on fat", especially round the lower trunk and hips, during pregnancy and with skinfold measurements (Taggart, 1961).

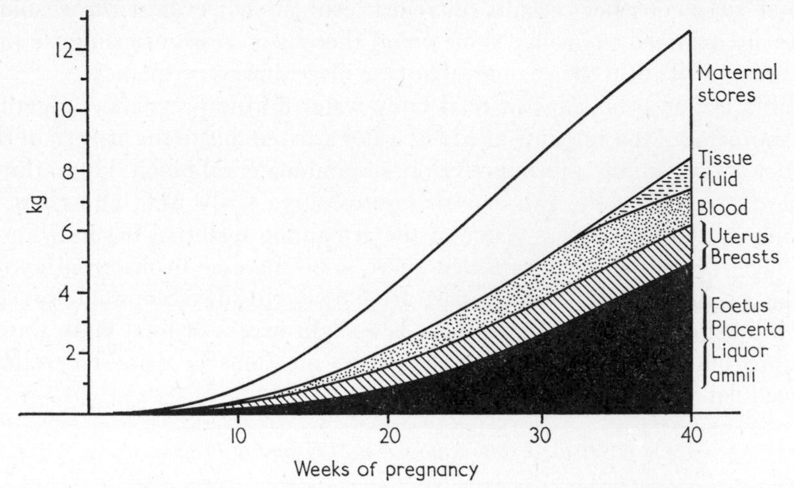

FIGURE 1. Components of weight gained in normal pregnancy, shown cumulatively.

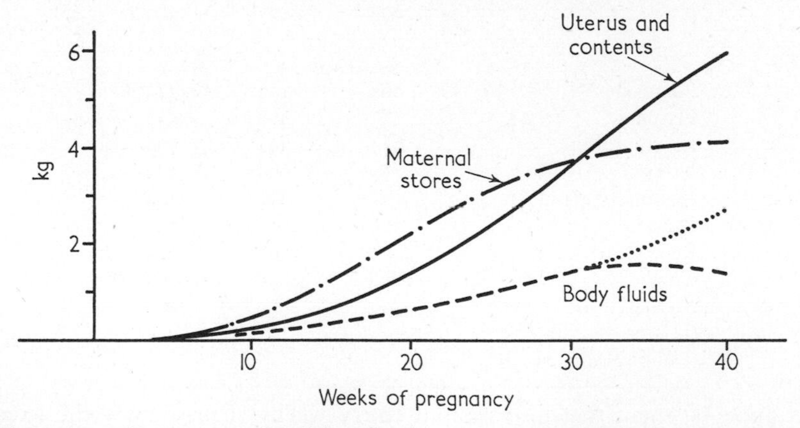

FIGURE 2. Components of weight gained in normal pregnancy, shown separately.

Figures 1 and 2 show the components of weight gain, cumulatively and separately. Gain due to growth of the product of conception occurs mainly during the second half of pregnancy. But "maternal stores", which we believe to consist entirely or mainly of fat, begin to be laid down much earlier; in our average normal subject they show no further increase during the final months.

This reserve of fat can only represent an energy store, and its early deposition is a remarkable instance of anticipatory adaptation. It is laid down before the growth of the foetus imposes any great metabolic strain and while the mother is still unimpeded by any severe physical handicap. Presumably, in primitive society, the energy reserve can be drawn upon during the later months, when capacity for work and food-gathering diminishes, and still later, when the mother is handicapped by the need to care for her young baby and to lactate.

2. *Storage of nutrients*

Table III gives estimates of the total amounts of protein and fat accounted for by the components of weight gain detailed in Table I. About 1 kg of new protein appears to be laid down during the course of pregnancy, of which more than half is in the foetus and placenta. Calculated as rate per day, the maximum deposition is at the end of pregnancy and amounts to only about 5 g daily. Of the fat laid down, by far the largest amount is in the maternal stores, probably mostly as subcutaneous fat. The daily rates of deposition are greatest around mid-pregnancy.

Table III. *Cumlative Storage of Protein and Fat in Pregnancy, Derived from the Components Detailed in Table I*

	Weeks of pregnancy			
	10	*20*	*30*	*40*
Protein g	35	210	535	910
Fat g	367	1930	3613	4464

The amounts of stored energy represented by these nutrients is easily calculated by multiplying the amounts by their heats of combustion: the total is about 42,000 kcal, much of which is in added maternal adipose tissue. It is not so easy to determine how much energy must be supplied in food to provide these stores, since little is known about efficiencies of absorption and utilisation. Calorific values in ordinary food tables give not heats of combustion but "metabolisable" energy, corrections having been made for losses due to incomplete absorption and for incomplete oxidation of protein. It may or may not be reasonable to assume that 10% of the metabolisable energy of food is lost during the process of conversion to new tissues in the product of conception and in the mother. On this basis, the stored energy given above must be increased to about 45,000 kcal, representing the amount of metabolisable energy that has to be supplied from diet. Here it should be noted that this figure indicates the total dietary energy required for the formation of new tissues during the whole of pregnancy. Further, the estimate takes no account of maintenance requirements, which are discussed below.

A little may be said at this point about storage of nutrients other than protein, fat and carbohydrate. About 30 g calcium has to be provided for the growth of the foetus, plus small additional amounts in other tissues. Total iron requirements probably amount to about 750 mg, two-thirds in the foetus and placenta and one-third in the enlarged volume of maternal blood. To this may be added a small amount of iron lost by haemorrhage during normal parturition, though the "saving" due to the cessation of menstruation during pregnancy is much larger. Practically nothing is known about additional vitamin requirements during pregnancy.

3. *"Running costs"*

We have discussed what might be termed the "accumulation of capital" during pregnancy. Having been laid down, the new tissues have to be maintained, involving additional energy requirements. The assessment of these requirements requires a great deal of approximation in the present state of our knowledge.

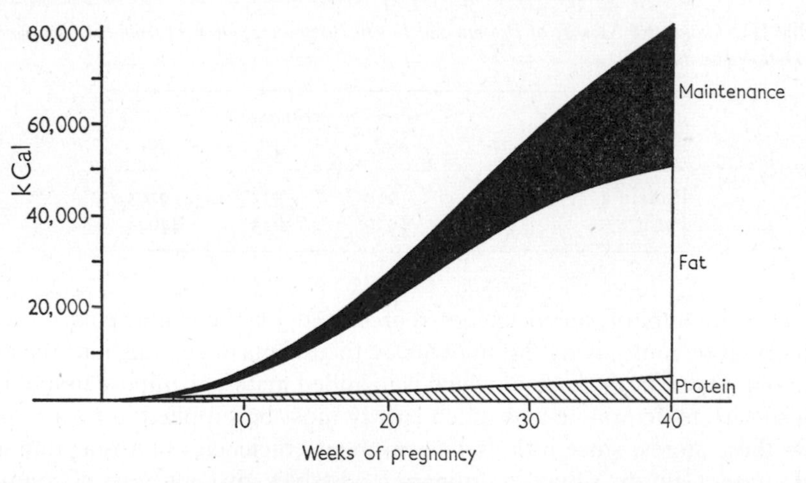

FIGURE 3. Cumulative energy costs of pregnancy.

Several investigators have measured basal metabolism during pregnancy. These indicate an increase in oxygen consumption amounting to about 50 ml O_2/min, possibly equivalent to an increased expenditure of 360 kcal/day between early and late pregnancy. Owing to the difficulty of achieving a truly basal state in pregnancy, especially in late pregnancy when full relaxation at rest cannot be achieved and hunger may produce additional discomfort, these figures may be too high. Calculations based on the increased work of the heart and lungs and the cost of maintaining the pregnant uterus and its contents give a lower total, about 200 kcal/day at term.

Fig. 3 gives a rough attempt to calculate the cumulative energy requirements of pregnancy, differentiating the cost of maintenance from those of building new tissue. The grand totals, for ten, twenty, thirty and forty weeks gestation are respectively about 3000, 25,000, 55,000, and 80,000 kcal.

4. *Activity*

The figures given so far take no account of the effect of altered activity. There is no doubt that activity does alter. In early pregnancy women often feel lethargic and easily nauseated; even when the need to work is undiminished, work output may decrease. In late pregnancy the mechanical disadvantages of pregnancy become obtrusive. On the one hand, the mother has more weight to move around and moves it under an increasing mechanical handicap. On the other hand, in civilisation at least, she may not only be able to "take it easy" but may be actively encouraged to do so.

We have at present no information from which the quantitative effects of these changes may be assessed. The Second Committee on Calorie Requirements of FAO (1957) thought that they might halve the "physiological" cost of pregnancy, from about 80,000 to about 40,000 kcal. The comment is made that the primigravida, by reducing her activity, may in practice require no additional energy supply, whereas the woman with several young children to look after may require to supply all or most of her physiological needs from the current food supply. However this may be, we have found, in a small-scale study of women in Aberdeen with several small children to look after, that the average daily calorie intake is less than that of primigravidae of the same social status.

There are few figures to show the pattern of calorie intake from the daily diet throughout the course of pregnancy. We have collected a limited amount of information that has not yet been adequately analysed. For what it is worth, it suggests an increased intake of about 250 kcal/day about mid-pregnancy, when appetite is often reported as being markedly increased, but the foetus is still very small. During the later part of pregnancy food intake appears to remain fairly steady or even to fall slightly.

E. Effects of Adverse Environment

What has been said so far refers, in the main, to pregnant women living in the relatively comfortable and healthy conditions of modern urban civilisation. What happens when food is short, or the diet ill balanced, or when pregnant women are obliged to undertake hard physical labour, especially in an environment where infections and infestations are widespread? Pregnant and lactating women have for long been considered as constituting a nutritionally "vulnerable" group, implying that adverse dietary and metabolic conditions have severe effects on them. Yet the scanty available evidence scarcely supports this view.

There is no doubt that animals placed on a severely restricted diet during the later stages of pregnancy have a greatly increased liability to produce undersized and weakly young (Thomson and Thomson, 1949). Similar experimental evidence for man is not available, but comparable situations occur in war and famine and when women are placed on restricted diets for therapeutic purposes.

1. *Therapeutic restriction of diet*

The idea of restricting the diets of pregnant women in order to secure easier childbirth first received general prominence from the writings of Prochownick (1889, 1901). He prescribed for women with a moderate degree of contracted pelvis a high-protein low-carbohydrate diet providing about 2000 kcal daily, but laid special emphasis on restriction of fluids. The intention was not so much to reduce the size of the foetus as to increase tissue mobility and to reduce "turgor". Prochownick's ideas were warmly debated, usually on the erroneous basis that they implied severe calorie restriction and reduction of foetal growth, and they directly or indirectly provoked some of the early studies of metabolism during pregnancy. The point at issue, whether foetal growth depends to any great extent on the mother's food supply, or whether within wide limits it will continue to grow normally, if necessary at the expense of maternal tissues, is still not completely resolved. It may be worth noting that many American obstetricians place rigorous limits on the amount of weight they permit pregnant women to gain, but there is no evidence that this reduces foetal vitality or has any obvious effect on foetal size.

2. *War and famine*

In Germany, during World War I, acute shortages of food were undoubtedly experienced by the civil population. Starling (1919), in an official report to the U.K. Government, said that during 1916 to 1917 those who had no priority for rations had just enough food to keep alive; the crude death rate in civilians increased fourfold, fulminating tuberculosis became a major cause of death, and famine oedema was reported in prisons and asylums. At the same time, the birth rate fell to about half of its former level, one reason being the widespread occurrence of amenorrhoea (Nilsson, 1920). Nevertheless, average birth weights showed little change; in some reports they are stated to have increased. Such rises in birth weight were attributed by some to a deficit of first births (in which birth weights tend to be lower than in subsequent parities), but the published figures are bedevilled by the arbitrary exclusion from many reports of "immature" babies. The same applies to figures from other central European countries affected by wartime food shortages. There was, too, a reluctance to present evidence that might damage morale.

During World War II conditions in Germany were altogether different, because preparations had been made to increase agricultural production and because conquered surrounding territories were made to yield additional supplies. The significant reports came from other countries. Antonov (1947) says that during the siege of Leningrad there was a steep fall in the birth rate associated with a high prevalance of amenorrhoea. In one clinic, during the second half of 1942, only 79 pregnancies were recorded, yielding 77 live births, 71 judged to be born at term; of the 79 women who became pregnant, 73 were women given priority rations. In other Leningrad clinics average birth weights fell by 400 to 500 g. Children were of low vitality, and women were able to lactate only poorly and for a short time. The siege of Rotterdam was of shorter duration, but also resulted in widespread amenorrhoea. Smith (1947) says without elaboration that most women gained no more than 2 kg during pregnancy and many lost weight, and the average birth weight fell by about 240 g. As in Leningrad, it seems that only the better-fed women were able to conceive and to sustain a pregnancy. From many other centres, which did not experience famine in such an acute form, there are reports that birth weights fell by 100 to 200 g.

Famines in peacetime have occurred in many areas of the world, but in none has the pattern of human reproduction been adequately documented.

3. *Poverty and disease*

In famine we are dealing with the effects of acute shortage of food of women who, under ordinary conditions, were presumably adequately nourished. But in many areas of the world malnutrition and undernutrition are chronic, affecting populations for generations. Pregnant women have been exposed all their lives to dietetic adversity, which continues during pregnancy itself. Dietetic adversity does not come alone, but is accompanied by exposure to infection, to parasitic diseases and to uncomfortable and unhygienic environmental conditions; women usually have to undertake hard physical work, even during pregnancy, and the standard of medical care is often low.

Adverse conditions are not limited to the "underdeveloped" countries. Despite much socio-economic "levelling" in recent decades, there is still a wide difference between the rich and the poor in Britain, and many of the present generation of British mothers in the poorer classes suffered from rickets and other florid deficiency states during childhood. Baird (1945) showed that the rates of stillbirth, neonatal death, prematurity and Caesarean section were considerably higher among the poor than among the wealthier women in Aberdeen, even though standards of medical care were similar for all. In the more affluent group the mothers were taller and were in a superior state of general health. Baird suggested that the nutritional background of the two groups was the primary cause of the difference in physique and health and in "reproductive efficiency". During World War II the national stillbirth rate fell dramatically, apparently as a result of an enlightened food policy (Duncan, Baird and Thomson, 1952). Nevertheless, Thomson (1958, 1959a, b) was unable to demonstrate any clear relationship between the food intake of Aberdeen primigravidae and many of the clinical features of their pregnancies. He thought that this negative result was not necessarily in conflict with the conclusion drawn from wartime experience: the fall in the stillbirth rate in England and Wales was from 38 per 1000 in 1940 to 28 per 1000 in 1945, a reduction of 1%. A change of this extent could scarcely be expected to be evident in a relatively small-scale intensive enquiry.

The investigations in Aberdeen have demonstrated beyond reasonable doubt the importance of the physique and health of mothers, which are almost certainly determined mainly by the conditions under which children grow (Thomson, 1959b). The ill-grown woman suffers a permanent handicap. not to be wholly overcome by excellent feeding during pregnancy itself. Conversely, the well-grown and generally healthy woman possesses reserves that enable her to overcome minor dietetic defects during pregnancy. Stature serves as a useful and simple index of nutritional status in population groups (Thomson and Billewicz, 1963).

The "poverty gradient" in stillbirth rates and birth weights that exists in the United Kingdom and similar countries exists also in less highly developed countries, from which many reports have come of, for example, differences in mean birth weight of the rich and the poor. And comparisons between different countries taken as a whole tend to reflect their general level of prosperity or otherwise. As already indicated, the interpretation of such trends is complex. They are not the result only of differing nutritional and metabolic patterns during pregnancy itself.

Despite the statistical trends referred to, the remarkable fact is that, even in countries where conditions of health and nutrition are undoubtedly and chronically bad, the course of the great majority of pregnancies is clinically satisfactory and the problem is one of increasing population rather than of attrition by wastage. Waterlow, Cravioto and Stephen (1960) go so far as to say that, "In many countries where protein malnutrition occurs the baby's weight and height at birth are normal, and growth and development are normal for the first 3 to 4 months." Even in South India, where adults are small and average birth weights are exceptionally low, Gopalan and Belavady (1961) have recorded their surprise that so many women of the poorest classes are not only capable of gestating but also of feeding from the breast babies whose initial vitality seems satisfactory. By Western standards, such women appear to be grossly underfed, but it is necessary to remember that they are much smaller than women in Western society. When their basal energy requirement, calculated from height and weight, is subtracted from energy intake, a surprisingly large surplus is left for activity and for the additional requirements of pregnancy.

4. *Energy metabolism under adverse conditions*

Little information is available as to the manner in which pregnant women living in such adverse conditions undergo gestation. Even weight gains during pregnancy have rarely been measured. The studies of Venkatachalam, Shankar and Gopalan (1960) among "poor class" women in South India enable a tentative pattern to be drawn up. Fig. 4, which may be compared

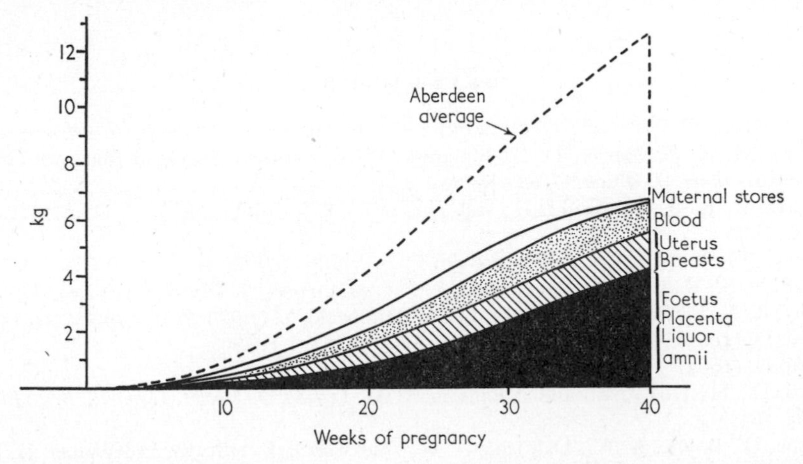

FIGURE 4. Possible components of weight gained in pregnancy by poor Indian women. Compare with Fig. 1.

with Fig. 1, for women in Aberdeen, is based upon measured weight gains and assumed values for the components of gain, of which only the birth weight of the baby is reasonably well established. Other named components have been scaled relative to foetal weight. The pattern suggests that the mothers put on a much smaller amount of fat during the first half of pregnancy than Aberdeen women and that they lose most of this energy reserve during the second half of pregnancy.

F. Conclusions

Pregnant and lactating women are generally considered to be important members of the nutritionally "vulnerable groups" in populations, and indeed there is no doubt that the necessity of providing for and growing a baby in utero, and of providing sufficient breast milk to feed the young child, constitutes a major additional metabolic load. Nevertheless, for nearly all women, and apparently even for those who live in adverse environmental and nutritional conditions, pregnancy at any rate represents not a period of loss but one in which energy reserves at least are laid down in the maternal body. The adaptations to pregnancy are extensive, involving most physiological systems, and the nature of some of them is such that in the absence of pregnancy they would be regarded as indicating a pathological state.

Pregnancy is the most common physiological stress to which women are exposed. The study of the adaptations that occur during pregnancy is not only of interest in itself, but also may help to increase understanding of the nature of responses to other forms of stress.

REFERENCES

Aboul-Khair, S. A., Crooks, J., Turnbull, A. C., and Hytten, F. E. (1964). *Clin. Sci.* **27**, 195.

Alvarez, R. R. de, Gaiser, D. F., Simkins, D. M., Smith, E. K., and Bratvold, G. E. (1959). *Amer. J. Obstet. Gynec.* **77**, 743.

Alvarez, R. R. de, Afonso, J. F., and Sherrard, D. J. (1961). *Amer. J. Obstet. Gynec.* **82**, 1096.

Antonov, A. N. (1947). *J. Pediat.* **30**, 250.

Assali, N. S., Rauramo, L., and Peltonen, T. (1960). *Amer. J. Obstet. Gynec.* **79**, 86.

Bader, R. A., Bader, M. E., Rose, D. J., and Braunwald, E. (1955). *J. clin. Invest.* **34**, 1524.

Baird, D. (1945). *J. Obstet. Gynaec., Brit. Emp.* **52**, 217; 339.

Baird, D. (1963). *J. Obstet. Gynaec. Br. Commonw.* **70**, 204.

Baird, D., Hytten, F. E., and Thomson, A. M. (1598). *J. Obstet. Gynaec., Brit. Emp.* **65**, 865.

Benzie, D., Boyne, A. W., Dalgarno, A. C., Duckworth, J., Hill, R., and Walker, D. M. (1955). *J. agric. Sci.* **46**, 425.

Bouterline-Young, H., and Bouterline-Young, E. (1956). *J. Obstet. Gynaec., Brit. Emp.* **63**, 509.

Campbell, E. J. M., and Howell, J. B. L. (1963). *Brit. med. Bull.* **19**, 36.

Cassmer, O. f1959). *Acta Endocrinol.* **32**, Suppl. 45.

Chesley, L. C. (1960). *In* "Clinical Obstetrics and Gynaecology", Vol. 3, No. 2. Paul B. Hoeber, Inc., New York.

Christensen, P. J., Date, J. W., Schønheyder, F., and Volqvartz, K. (1957). *Scand. J. Clin. Lab. Invest.* **9**, 54.

Clavero, J. A., and Botella Llusiá, J. (1963). *Amer. J. Obstet. Gynec.* **86**, 234.

Cope, C. L., and Black, E. (1959). *J. Obstet. Gynaec., Brit. Emp.* **66**, 404.

Cugell, D. W., Frank, N. R., Gaensler, E. A., and Badger, T. L. (1953). *Amer. Rev. Tuberc.* **67**, 568.

Duncan, E. H. L., Baird, D., and Thomson, A. M. (1952). *J. Obstet. Gynaec., Brit. Emp.* **59**, 183.

Eton, B., and Short, R. V. (1960). *J. Obstet. Gynaec., Brit. Emp.* **67**, 785.

FAO (1957). "Calorie Requirements." FAO Nutritional Studies No. 15. Food and Agriculture Organization of the United Nations, Rome.

Farquhar, J. W. (1962). *Postgrad. med. J.* 38, 612.
Fitzgerald, M. G., Malins, J. M., and O'Sullivan, D. J. (1961). *Lancet* i, 1260.
Gopalan, C., and Belavady, B. (1961). *Fed. Proc.* **20**, 177.
Haddon, W., Nesbitt, R. E. L., and Garcia, R. (1961). *Obstet. Gynec. N.Y.* 18, 262.
Hahn, P. F., Carothers, E. L., Darby, W. J., Martin, M., Sheppard, C. W., Cannon, R. O., Beam, A. S., Densen, P. M., Peterson, J. C., and McClellan, G. S. (1951). *Amer. J. Obstet. Gynec.* **61**, 477.
Hamilton, H. F. H. (1949). *J. Obstet. Gynaec., Brit. Emp.* **56**, 548.
Hellegers, A., Okuda, K., Nesbitt, R. E. L., Smith, D. W., and Chow, B. F. (1957). *Amer. J. clin. Nutr.* **5**, 327.
Herbert, C. M., Banner, E. A., and Wakim, K. G. (1958). *Amer. J. Obstet. Gynec.* **76**, 742.
Hulka, J. F., Hsu, K. C., and Beiser, S. M. (1961). *Nature, Lond.* **191**, 510.
Hytten, F. E. (1964). Proc. 6th Internat. Cong. Nutrit. E. and S. Livingstone, Edinburgh.
Hytten, F. E., and Duncan, D. L. (1956). *Nutr. Abstr. Rev.* **26**, 855.
Hytten, F. E., and Klopper, A. I. (1963). *J. Obstet. Gynaec. Br. Commonw.* **70**, 811.
Hytten, F. E., and Paintin, D. B. (1963). *J. Obstet. Gynaec. Br. Commonw.* **70**, 402.
Hytten, F. E., and Leitch, I. (1964). "The Physiology of Human Pregnancy." Blackwell Scientific Publications, Oxford.
Klopper, A. I., and Billewicz, W. Z. (1963). *J. Obstet. Gynaec. Br. Commonw.* **70**, 1024.
Macgillivray, I., and Tovey, J. E. (1957). *J. Obstet. Gynaec., Brit Emp.* **64**, 361.
McGinty, A. P. (1938). *Amer. J. Obstet. Gynec.* **35**, 237.
McLennan, C. E. (1943). *Amer. J. Obstet. Gynec.* **45**, 568.
Martin, J. D., and Mills, I. H. (1956). *Brit. med. J.* ii, 571.
Mishell, D. R., Jr., Wide, L., and Gemzell, C. A. (1963). *J. clin. Endocrin.* **23**, 125.
Montgomery, T. L., and Pincus, I. J. (1955). *Amer. J. Obstet. Gynec.* 69, 865.
Nehring, K., Lanbe, W., Schwerdtfeger, E., Schiemann, R., Haesler, R., and Hoffman, L. (1957). *Biochem. Z.* 328, 549.
Nilsson, A. (1920). *Zbl. Gynäk* 44, 876.
Paaby, P. (1960). *J. Obstet. Gynaec., Brit. Emp.* 67, 43.
Page, E. W., Glendening, M. B., Dignam, W., and Harper, H. A. (1954). *Amer. J. Obstet. Gynec.* 68, 110.
Paintin, D. B. (1962). *J. Obstet. Gynaec. Br. Commonw.* 69, 719.
Palmer, A. J., and Walker, A. H. C. (1949). *J. Obstet. Gynaec., Brit. Emp.* **56**, 537.
Prochownick, L. (1889). *Zbl. Gynäk* **30**, 577.
Prochownick, L. (1901). *Ther. Mh.* **15**, 387–446.
Quilligan, E. J., and Tyler, C. (1959). *Amer. J. Obstet. Gynec.* 78, 465.
Romney, S. L., Reid, D. E., Metcalfe, J., and Burwell, C. S. (1955). *Amer. J. Obstet. Gynec.* **70**, 791.
Shearman, R. P. (1959). *J. Obstet. Gynaec., Brit. Emp.* 66, 1.
Sims, E. A. H., and Krantz, K. E. (1958). *J. clin. Invest.* **37**, 1764.
Smith, C. A. (1947). *J. Pediat.* **30**, 229.
Starling, E. H. (1919). "Report on Food Conditions in Germany." Cmd. 280 (State Papers 1919, liii), H.M.S.O., London.
Taggart, N. (1961). *Proc. Nutr. Soc.* **20**, xxx.
Thomson, A. M. (1958). *Brit. J. Nutr.* **12**, 446.
Thomson, A. M. (1959a). *Brit. J. Nutr.* **13**, 190.
Thomson, A. M. (1959b). *Brit. J. Nutr.* **13**, 509.
Thomson, A. M., and Billewicz, W. Z. (1963). *Proc. Nutr. Soc.* **22**, 55.
Thomson, A. M., and Hytten, F. E. (1961). *Proc Nutr. Soc.* **20**, 76.
Thomson, A. M., and Thomson, W. (1949). *Brit. J. Nutr.* **2**, 290.
Tyler, J. M. (1960). *J. clin. Invest.* **39**, 34.

Van Slyke, D. D. (1959). *In* "Oxygen Supply to the Human Foetus" (J. Walker and A. C. Turnbull, eds.). Blackwell Scientific Publications, Oxford.
Venkatachalam, P. S., Shankar, K., and Goplan, C. (1960). *Indian J. med. Res.* 48, 511.
Walker, J., and Turnbull, E. P. N. (1953). *Lancet*, ii, 312.
Waterlow, J. C., Cravioto, J., and Stephen, J. M. L. (1960). *Adv. Protein Chem.* 15, 131.
Wynne-Edwards, V. C. (1962). "Animal Dispersion in Relation to Social Behaviour." Oliver and Boyd Ltd., London and Edinburgh.

CHAPTER 12

Time, Light and Diurnal Rhythms

MARY C. LOBBAN

A. Introductory

Man is essentially a creature of the daylight. He is apparently more dependent for his existence and continued well-being upon sight than upon his other senses. Since more than 90% of the human population of the world is distributed throughout those latitudes where day has always been associated with light and night with darkness, man has had a 24 hr periodicity imposed upon his activities from the time of his earliest history. Primitive man was abroad and active during the hours of daylight, when he could see his prey and his enemies and could use his superior powers of sight to pursue the one and to avoid the other; in the difficult and dangerous hours of darkness he retreated to the comparative safety of his cave and slept. This regular alternation of wakefulness and sleep has continued down the years to the more elaborate and civilised conditions of the present day. It is, in fact, the most widely recognised of the true physiological diurnal rhythms, in which man's activity pattern is accompanied by and reflected in similar regularly alternating fluctuations of physiological function.

B. Sleep and Activity

Sleep is the outward and universally recognisable sign of a specialised and complex physiological condition, which has attracted the interest of scientific observers from the early days of investigation into diurnal rhythms (Pieron, 1913; Kleitman, 1939a). Certain aspects of the study of sleep are dealt with elsewhere in this book (*v.* especially chapters 12 and 13). Recent studies of the human subject throughout the 24 hr of a normal day have shown that there are equally well-defined diurnal trends in many other physiological functions. The body temperature, heart rate and blood pressure (Kleitman and Kleitman, 1953), the blood eosinophil count (Conn, Kaine and Selber, 1955), the plasma iron content (Cartwright, Gubler, Hamilton and Wintrobe, 1950), the rate of urine production (Kleitman, 1939b) and the urinary excretion of electrolytes (Stanbury and Thomson, 1951) and of phosphates in particular (Mills, Thomas and Yates, 1954) all show well-marked diurnal fluctuations in man. In normal circumstances, when man's activity pattern is such that he is awake and active during the hours of daylight and is sleeping at night, all these physiological diurnal rhythms are approximately in phase, in that they exhibit their maxima during activity and their minima during sleep. The simplest explanation of this phenomenon would be that the observed physiological rhythms are a direct consequence of the activity pattern. The evidence against such an hypothesis is, however, overwhelming. The activity patterns of human subjects may be profoundly altered without giving rise to corresponding alterations in their physiological rhythms. For instance, experiments have been mounted in which the normal 24 hr routine has been replaced by 12 hr cycles of food and rest (Mills and Stanbury, 1952), by 21 or 27 hr "days" (Lewis and Lobban, 1957a) or in which there has been a complete reversal of activities within the normal 24 hr framework (Mills, 1951; Sharp, 1962); in all these abnormal circumstances certain of the physiological functions have continued to show a normal 24 hr periodicity. Such observations would point to the existence of an intrinsic "biological clock" in man or at least to some system far more sensitive to time or to the diurnal fluctuations in the environment than to changes in bodily activity.

C. Biological Clocks

The existence of biological clocks is well recognised in lower animals, and many excellent reviews of this subject have been written in recent years (Kleitman, 1949; Harker, 1958; Aschoff, 1963). The possibility that similar systems can be demonstrated in the human subject presents an intriguing subject for research, albeit a difficult one. Man is a complex experimental animal, particularly because many of the observations made on him are complicated by the intervention of social and emotional factors. With the advent of permanent housing, lighting, heating and modern methods of food production and distribution, man has become more independent of his natural environment and less aware of diurnal and seasonal changes. Indeed, at first sight it would appear that the individual is influenced more by the activities and demands of other members of the community in which he lives than by his external environment. Similarly, man's emotional—and, to some extent, his physical—make-up is such that relatively minor changes can produce large fluctuations in physiological function.

1. *Choice of rhythms for measurement*

Unfortunately, some of the physiological processes so affected are the very ones that have been used as indicators for the existence of diurnal rhythms: the pulse rate and blood pressure, for example, may vary much more as a result of emotional or physical stress than they do in diurnal observations upon the sedentary human subject. On the other hand, there are certain physiological rhythms that are extremely stable and can be used as reliable indicators of diurnal trends. Outstanding amongst these are the rhythms of urinary excretion, and it is for this reason that recent workers have concentrated upon them—and, to a lesser extent, upon the body temperature rhythm—in the investigation of physiological diurnal rhythms in man.

2. *Rhythms and work patterns*

The social aspects of the problem tend to complicate the interpretation of results, in that civilised conditions diminish the impact of the environment, as has already been stated: but there is one characteristic of life in a modern, elaborately organised community that provides an added stimulus for investigating human diurnal rhythms, in that it points to a practical application for such studies. This is the growing demand for workers, both in industry and in medical, transport and social services, to be alert and on watch for 24 hr of the day. Complicated machinery must be operated, heavy lorries driven over the roads, locomotives negotiated through the unstandardised and complex signal system and patients watched and treated, all just as efficiently during the hours of the night as during normal hours of activity. More dramatically still, the growing use of air transport and the increased speed of aircraft have compressed the lines of longitude and have made nonsense of the ordinary concept of "time of day". With certain aircraft on specific routes "take-off time" and "landing time" may, in fact, coincide—or even overtake each other—with resulting chaos in the personal times of the pilots and navigators responsible for the safety of those on board. This alone could lead to considerable fatigue; when it is realised that evidence is accumulating for the existence of fluctuations in man's efficiency and performance related to the time of day (Halberg, Hauty, Hawkins and Steinkamp, 1960; Colquhoun, 1960, 1962), the problem of relating human diurnal rhythms to the demands of modern work programmes becomes of great practical importance.

D. Methods of Study

The scientific study of physiological diurnal rhythms in man falls naturally into three phases. First, that of observational studies on human subjects in a normal 24 hr light day and dark night environment, to provide evidence for the existence of diurnal rhythms; secondly, the induction of changes into the activity patterns and the environment of the human subject, in order to reveal the endogenous or exogenous nature of the rhythms and the factors that influence them; thirdly, attempts to determine the site and nature of the biological clocks involved in the initiation and maintenance of the rhythms.

1. *Observational*

a. Renal excretion. Description and discussion of experimental work in this field will here be limited almost entirely to observations on the diurnal rhythms of renal excretion, since these may be most satisfactorily and accurately followed under field conditions, even in the untrained subject and over prolonged periods of time.

Typical normal diurnal fluctuations in the renal excretory rates of water, potassium, sodium and chloride are shown in Fig. 1. The plots are drawn from the averaged figures for a group of human subjects living under experimental conditions on a strictly defined routine, when times and types of food and fluid intake, rest and work were all rigidly controlled and bodily activity was restricted. It can be seen that the rates of excretion for all four urinary constituents are substantially greater during the hours of the normal working day (08.00 hr to 20.00 hr) than during the hours of rest. The conditions under which these observations were made were ideal for the emergence of diurnal rhythms, and it is pertinent to compare them with similar recordings obtained from human subjects under more ordinary conditions. Such recordings are shown in Fig. 2, which is plotted from the averaged figures for twenty-two laboratory workers who were carrying on with their normal everyday pursuits, interrupting them only to collect urine samples at frequent intervals. No attempt was made to standardise food or fluid intake or times of meals; hours of work varied widely from subject to subject, as did the amount of physical exercise taken in off-duty periods, yet the results obtained are closely similar to those yielded by the group submitted to a strictly controlled routine. This similarity is most marked in the rhythm of potassium excretion; water and salt excretion show some small differences in phase and a reduction in amplitude (difference between "peak" and "trough" values). An explanation for these differences will be put forward later. The rhythm of potassium excretion, however, is a remarkably stable one, whose study has proved to be rewarding in a number of different situations.

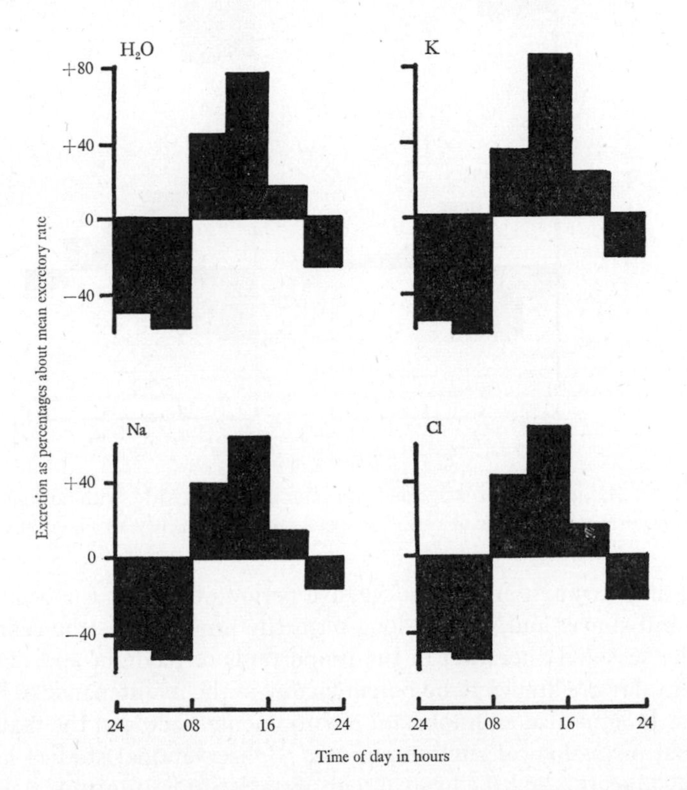

FIGURE 1. Typical normal diurnal fluctuations in renal excretion of water, potassium, sodium and chloride. Averaged values for eight human subjects living under experimental conditions. Four-hourly rates of excretion expressed as percentages about the mean daily rate of excretion for each urinary constituent: time of day in hours.

b. Temperature fluctuation. Fig 3(a) shows the normal diurnal fluctuations in body temperature for seven of the eight subjects whose renal rhythms are depicted in Fig. 1; the average oral temperature is higher during the normal waking hours than it is during sleep. Temperature recordings for the eighth subject are plotted separately in Fig. 3(b); this subject shows a reversed rhythm, sleep temperatures being higher than waking temperatures, a somewhat unusual finding, to which later reference will be made when the preservation of individual peculiarities of rhythm is being considered.

c. Excretion at high latitudes. In the lower animals the alternation of

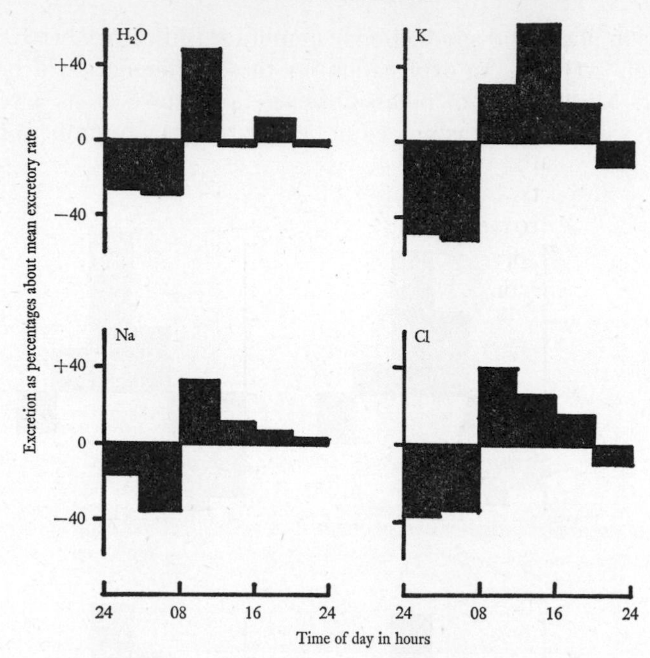

FIGURE 2. Diurnal patterns of renal excretion in everyday life, with no standardisation of diet, times of work, exercise or rest. Averaged values for twenty-two laboratory workers.

daylight and darkness in each successive period of 24 hrs, the occurrence of sunrise and sunset and, for the long-term rhythms such as the reproductive cycle, the seasonal alteration in the proportions of daylight and darkness in each 24 hr day are known to be potent factors in the maintenance of biological rhythms. Is light also an important environmental factor in the maintenance of diurnal physiological rhythms in man? Observational studies have been made on subjects who have lived at high arctic latitudes for prolonged periods

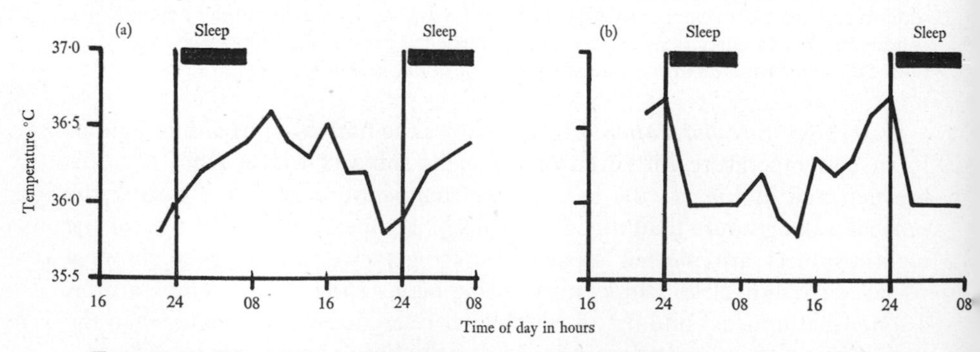

FIGURE 3. Diurnal fluctuations in body temperature: (a) averaged values for seven normal subjects; (b) values for a subject with a reversed temperature rhythm.

(Lobban, 1958). In these latitudes the normal daily rhythm of daylight and darkness is interrupted twice a year by periods when the natural stimulus of sunrise and sunset is withdrawn: about midsummer there is a period of continuous daylight and about midwinter a similar period of continuous darkness. The renal excretory patterns of water and of potassium for four typical arctic subjects, three from residents in an Indian village in the Yukon Territory and one from an Eskimo village on the north coast of Alaska, are shown in Fig. 4. Subject OCA, a 28 year old white man who had lived at 68°N for only two months of the perpetual daylight of the arctic summer,

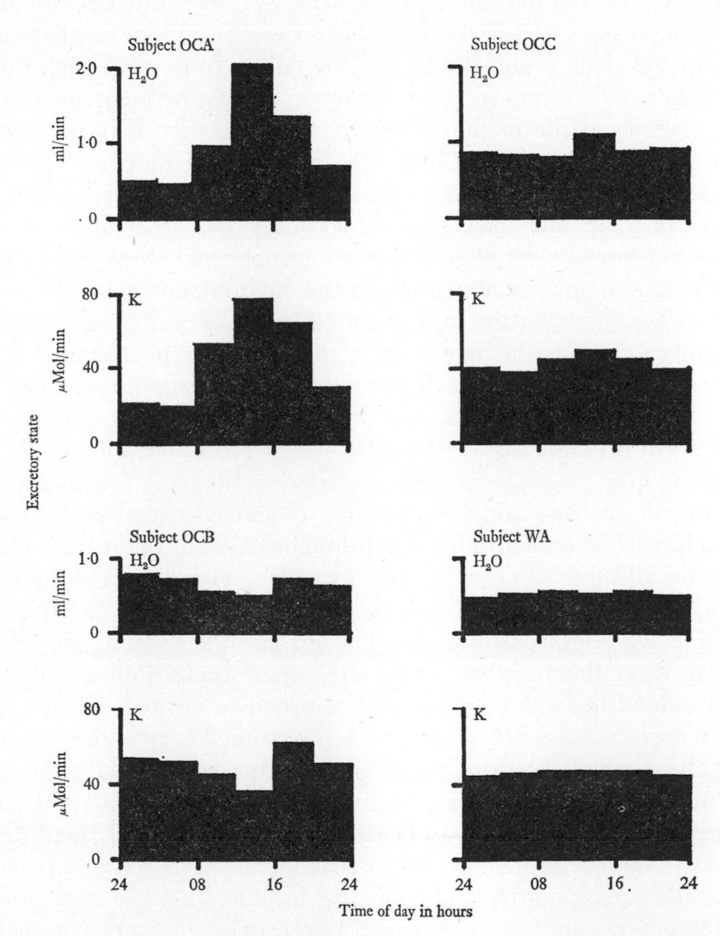

FIGURE 4. Diurnal patterns of renal excretion for four individual arctic-dwelling subjects. OCA, 28 year old white male after two months in the summer arctic; OCB, 33 year old white male after twenty-three months in the arctic; OCC, 34 year old North American Indian male, arctic resident for the whole of his life; WA, 36 year old Eskimo male, arctic resident for the whole of his life.

shows renal rhythms indistinguishable from those of temperate zone (light day and dark night) subjects; subject OCB, a 33 year old white man who had spent twenty-three months in the same latitude and who was subject to an exactly similar routine to that of subject OCA at the time when the measurements were made, shows disruption and disappearance of the normal renal rhythms. Subject OCC, a 34 year old Indian who had passed the whole of his life in the village at 68°N, shows renal rhythms that are severely reduced in amplitude, and subject WA, a 36 year old Eskimo who had spent thirty-two years at 71°N, shows completely arhythmic renal recordings. These observations may well indicate that alterations in the normal environmental pattern of light days and dark nights have caused the loss of physiological rhythms in the elder white subject and the failure to develop such rhythms by the Indian and Eskimo subjects; however, it must be mentioned that the irregular pattern of life in the arctic environment, especially noticeable in the Eskimo village in summer, may also have played its part. Two points remain clear: first, the excretory patterns of high arctic subjects do differ from those of temperate zone subjects; secondly, the observations made on the younger white subject show that the normal renal rhythms can persist for some weeks in spite of alterations in the environment.

Some observations on the immediate effects of exposing temperate zone subjects to continuous daylight were made in the arctic summer of 1960 (Lobban and Simpson, 1961). Three subjects, all with pronounced renal diurnal rhythms, were to take part in field work in Spitsbergen. They travelled north through Norway from Bergen to Tromsø by motor van, going easily, taking little physical exercise and stopping on every fourth day of the journey to collect urine samples according to a strict experimental routine. As they approached the latitudes of continuous daylight the pattern of water excretion for all three subjects became disturbed, with a marked diminution of amplitude about the latitude of the Arctic Circle (Fig. 5). Arriving at Tromsø (70°N), they remained at this high latitude, waiting for the ship that was to carry them farther north; after three weeks both the phase and the amplitude of the water rhythm had returned to normal. The pattern of potassium excretion was not so affected, but remained normal throughout the whole of the northward journey: thus there is a marked dissociation between the rhythms of excretion of water and potassium in the first eight days of exposure to continuous daylight. These observations not only lend support to the theory that a continuous light environment affects some physiological rhythms, albeit temporarily, but they also indicate that the control of the renal rhythms is not simple. There would appear to be at least two mechanisms involved, of which one is concerned with water excretion and is disturbed by the withdrawal of a period of darkness in each 24 hr day, the other, concerned with potassium excretion, remaining unaffected by the loss of this stimulus. Once the water rhythm has been re-established it can be maintained by

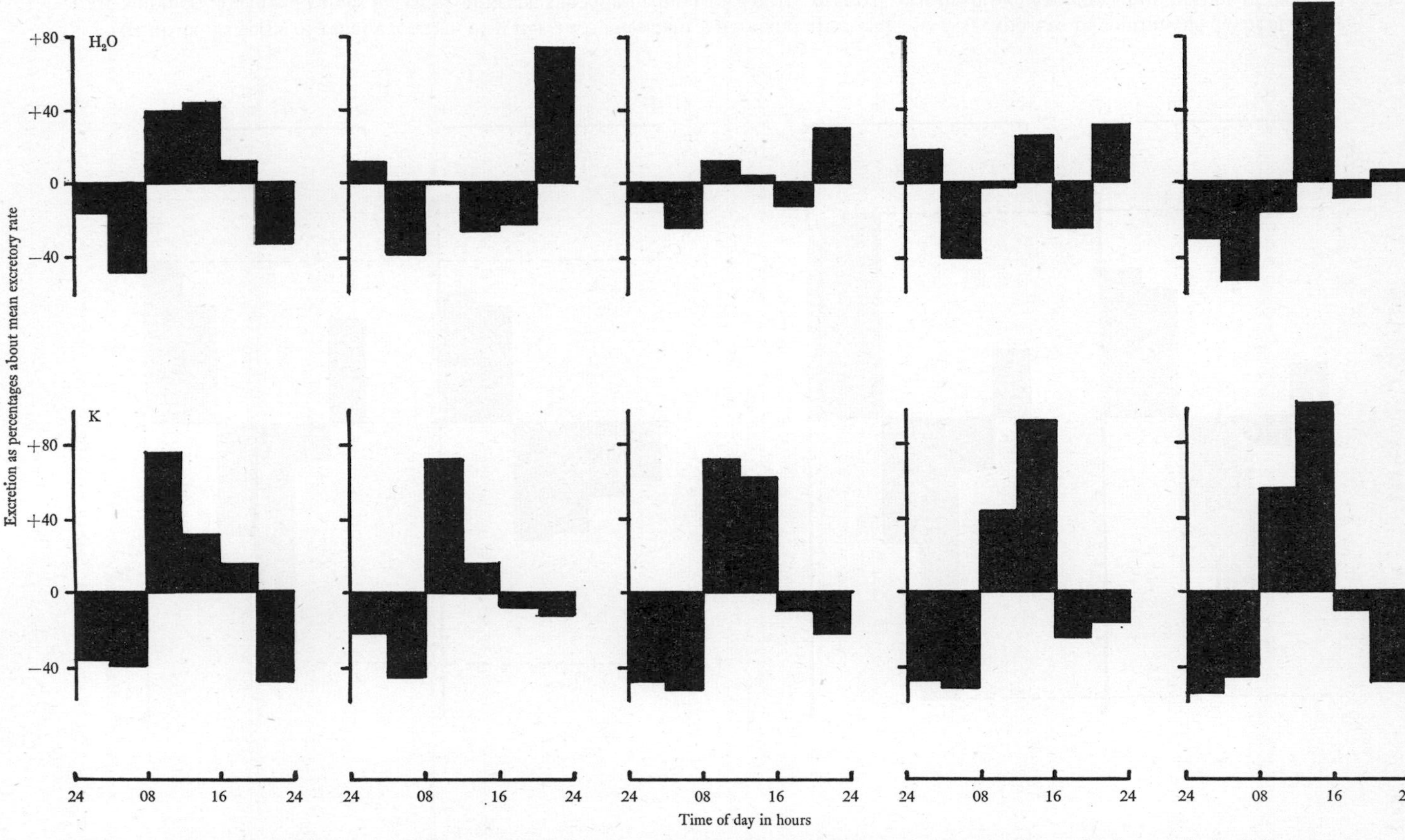

FIGURE 5. Renal excretion of water and potassium by subjects during a journey from temperate latitudes to a high arctic latitude in summer. Averaged values for three subjects, showing maintenance of potassium rhythm throughout, but temporary disruption of water rhythm on first exposure to continuous daylight.

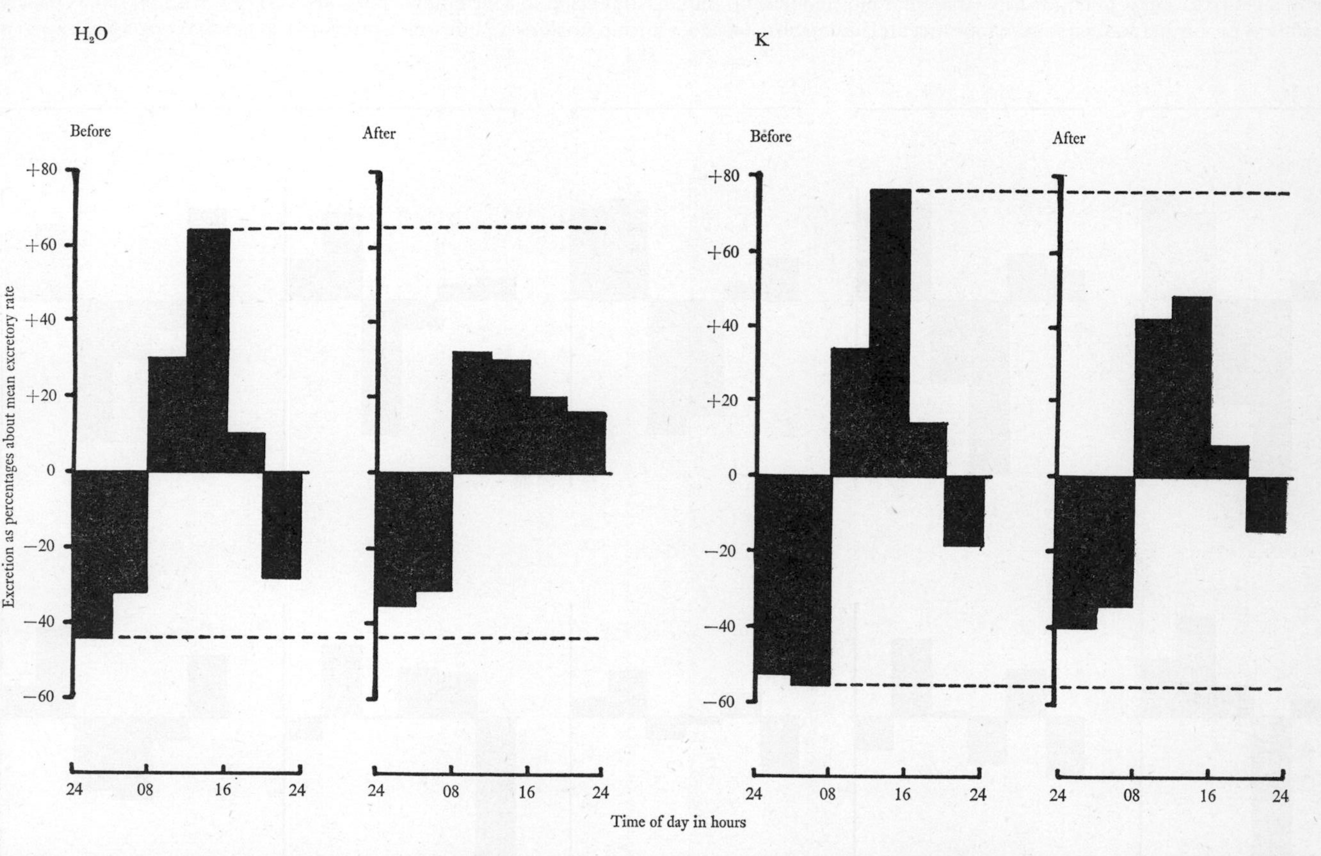

FIGURE 6. Patterns of renal excretion of water and potassium before and after eight weeks' exposure to continuous daylight in the arctic summer. Averaged values for eight subjects, showing maintenance of the original relationship of excretory rate to time of day, but diminution in the observed amplitude of the rhythm.

subjects from environments of light day and dark night over a period of several weeks' continuous daylight; this is an important consideration when interpreting experiments in which the time scale of human beings is altered in an arctic summer environment (Lewis and Lobban, 1957a, b). The phase and amplitude of the renal rhythms are satisfactorily maintained for at least eight weeks of continuous daylight, although a careful comparison between the recordings made on control days on 24 hr time at the beginning of such a period with similar days at the end of the exposure to summer arctic conditions shows some indications of a slight diminution in the amplitude of the original renal rhythms (Fig. 6).

d. Excretion by the blind. The most recent evidence in support of the hypothesis that the perception of diurnal alternation of light and darkness plays a part in the maintenance of human diurnal rhythms is emerging from the study of blind subjects (Lobban and Tredre, 1964). Only a few observations have as yet been made, but the results show a real difference between the normal renal rhythms of subjects who still retain the power of differentiation between light and darkness and the disrupted and diminished rhythms of totally blind subjects (Fig. 7).

2. *Experimental*

a. Changing activity patterns. Civilised man is more aware of the passage of time than he is of any other changing factor in his environment, and we have reason to believe that he is only at his most efficient when his physiological diurnal rhythms and his activity patterns are temporally synchronised. Although many people in this country now regularly work during the night and sleep by day, the effect of this time shift on their physiological processes is difficult to evaluate by straightforward methods. It is true that their activity pattern is reversed, but they may still be subjected to the normal light day and dark night fluctuation, and they are certainly well aware that other people around them are awake and active at a different time from themselves. It is indeed difficult to isolate man from his fellows and to insulate him completely from his environment except by confining him in an insulated soundproof room, artificially lighted and heated, and such conditions are tolerable only for a short period of time. But since we now know that the original normal 24 hr intrinsic renal rhythms persist for several weeks of exposure to continuous daylight, field experiments in the high arctic during the summer months may be designed to study the effects of alterations in time scale. Small groups of human subjects have been taken from Britain to Spitsbergen, where the differences between night and day in either light or temperature are extremely small during the summer months (Lewis, Lobban and Shaw, 1956; Lewis and Lobban, 1956, 1957a, b). In Spitsbergen, at 78°, these subjects lived as isolated communities in uninhabited country

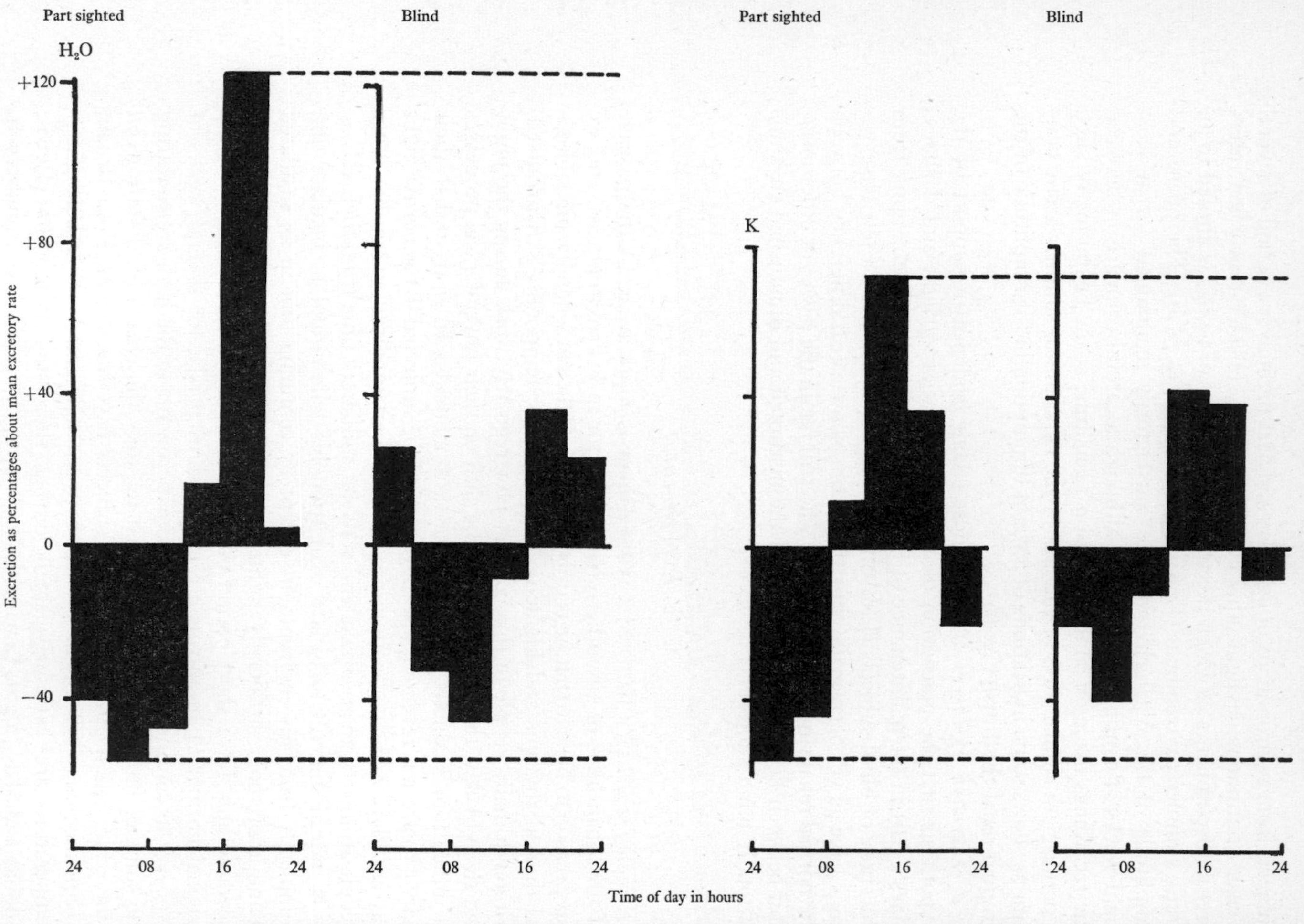

FIGURE 7. Renal excretion of water and potassium by subjects with impaired vision. Comparison of the averaged values for five subjects with partial vision (to the left of each plot) with those for five blind subjects. The pattern of water excretion is temporarily

and constant environmental conditions. Their ordinary wristwatches were replaced by watches that had been adjusted so that they recorded 12 hr in 11, $10\frac{1}{2}$ or $13\frac{1}{2}$ real hr, giving routines of a 22 hr, 21 hr or 27 hr "day". The field bases at which the subjects lived were reasonably comfortable, and the surrounding country and wildlife were new and interesting; under these conditions the subjects co-operated fully, living entirely by experimental time as shown by their specially adjusted watches and thinking not at all of true solar time. False clues to the experimental time routines were given not only by these watches, but also by the times of meals and by the heating of the base huts, where the fires were kept up during the experimental "day-time" and were allowed to die down at "night". The experimental time scales did not differ widely enough from those of the normal 24 hr day to cause any undue stress, and subjects lived on the new time routines for periods of about six weeks. On both 21 hr and 27 hr routines urine samples were collected, and oral temperature measurements were taken at frequent set intervals throughout each recording "day". During recording periods, all subjects lived on a strict experimental regimen and activity; times of meals and rest and types and quantity of food and fluids taken were rigidly controlled for each subject on each "day". Most of the subjects had no difficulty at all in sleeping, in spite of the absence of nocturnal darkness, and the pattern of wakefulness and sleep was satisfactorily maintained. In the intervals between recording periods, subjects were allowed more freedom, in that they could be away from base for short periods and could pursue a variety of field studies, such as physiography, geology and ornithology; but wherever they were and whatever their occupation, they lived always with reference to their special watches, and their activity patterns were related always to the experimental time.

On both 21 hr and 27 hr routines eight experimental "days" are lived through in seven and nine real days, respectively, and during this eight "day" period, which is referred to as the experimental cycle, the relationship of experimental time to real time goes through a regular sequence of change. At the beginning of each cycle the two time scales are in phase; they diverge by a progressively greater amount each "day" up to mid-cycle, when experimental time is reversed in relation to real time and subjects are then awake and active during solar night and are sleeping during solar day. After this period the two time scales converge again, to become perfectly in phase at the end of the cycle, the only difference being that in each cycle the subjects on the 21 hr routine have gained a "day" and those on the 27 hr routine have lost one. If measurements of urinary excretion and of oral temperature are made throughout the whole of an experimental cycle, every aspect of the relationship between real time and experimental time may be investigated, and the results lend themselves to mathematical treatment. This is particularly convenient with renal rhythms, for which consideration of the overall excretory rhythm for each urinary constituent for each "day" on the experimental time scale reveals significant differences in the way in which the intrinsic (24 hr) rhythms respond to the abnormal time scale. The ideal form of mathematical procedure for the examination of such findings is one that reduces the individual's values for each "day" to the minimum number of variables necessary for defining the overall excretory patterns, so eliminating possible random fluctuations. Such a method, devised by P. R. Lewis, was first fully described in 1956 (Lewis and Lobban, 1956) and will be outlined in the paragraphs below.

b. Normal patterns. It would appear that the normal pattern of renal excretion approximates to a sinusoidal oscillation with a 24 hr period, maximum rates of excretion occurring during the afternoon and minimum rates during the small hours of the night. Such a pattern can be completely defined by three variables: the mean rate of flow (M), the amplitude of the sinusoidal oscillation (A) and the value of some angle (α), defining the phase of the oscillation. The magnitude of the 24 hr component in each individual urinary rhythm can be calculated, and the standard errors of the variables defining the component can be obtained by a modified Fourier analysis in which a sine curve is fitted to twelve values equally spaced throughout one cycle of the oscillation (Whittaker and Robinson, 1944). For the urinary rhythm on a normal 24 hr routine, therefore, the values used are the excretory rates of the urinary constituents obtained from successive two-hourly urine samples.

To quote Lewis's original description of the mathematical procedure: "If the best fit to twelve consecutive values is given by $R_\theta = M + A \sin (\theta + \alpha)$, where M is the mean rate, A the amplitude, α the phase angle and R_θ the individual value for the rate at a time given by θ, θ having values of 0°, 30°, 60°, . . . for each successive 2 hr period measured from some arbitrary time, the values of the three constants and their standard errors may then be calculated from the expressions given by Whittaker and Robinson." When a series of thirteen control days for two subjects on 24 hr time was analysed by this procedure: "There appeared to be significant day-to-day fluctuation in A and M, but not in α or A/M, . . . so the fluctuations in A and M must be closely correlated. Parameters α and A/M are, therefore, the obvious ones to consider in any study of an abnormal routine."

In the Spitsbergen experiments urine samples were collected at intervals of 2 "hours" by experimental time, so that it was a comparatively simple matter, by using the mathematical procedure outlined above, to analyse the individual values for a 21 hr or 27 hr component. The best sine curve with a 21 hr or 27 hr period was fitted to the values for each urinary constituent in each recorded experimental "day". Then, to demonstrate the progressive changes in α during a complete experimental cycle, the time of origin by British Summer Time (B.S.T., in which the subjects had been living for some weeks before the beginning of the experiment; it is equivalent to solar time at the base longitude 15°E) of each fitted sine curve (when $\theta = \alpha$) was plotted against the difference between experimental time and B.S.T. This is equivalent to a plot of the phase of the observed sine curve against the experimental phase difference; on such a plot the points for a true adapter to experimental time would lie along the diagonal and points for a complete non-adapter along the horizontal axis. Two such plots are reproduced in Fig. 8, where differences in the adaptation to abnormal time routines

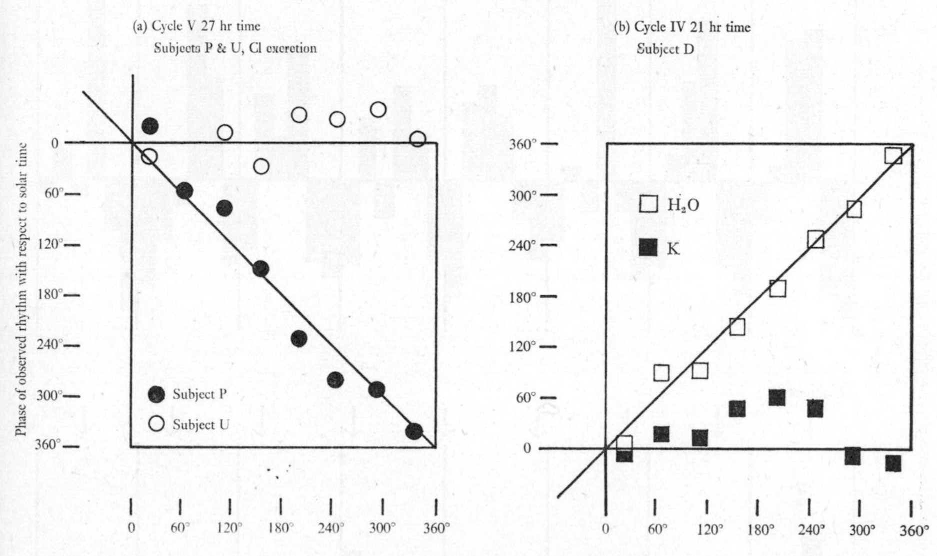

FIGURE 8. Progressive changes in the phase of the excretory rhythm relative to the control day rhythm in subjects living on abnormal time routines. (a) Excretion of chloride by two individuals during their fifth cycle on 27 hr time; for subject P the values follow the diagonal, indicating good adaptation to the experimental time routine, but for subject U the values lie on or near the horizontal axis, indicating that the excretory pattern is following 24 hr time. (b) Excretion of water and of potassium by a subject during the fourth cycle on 21 hr time; here the excretion of water follows the experimental time scale, whereas the excretion of potassium follows 24 hr time.

are clearly shown, both between the individual subjects and between the excretions of different urinary constituents by the same subject.

c. Extent of adaptation. Of the nineteen subjects exposed to abnormal time scales under the conditions described, only three showed rapid and complete adaptation to the new routine; in all the others the 24 hr component retained ascendancy over the 21 hr or 27 hr components throughout the whole of the experimental period in some or all of the renal rhythms. This maintenance of the intrinsic 24 hr renal rhythm was most strongly marked in the excretion of potassium. These findings show up well even when far less critical procedures than the Fourier analysis are used, and typical recordings for an adapter and a non-adapter throughout the first experimental cycle on 27 hr time are shown in Fig. 9. For the adapter, subject O, the

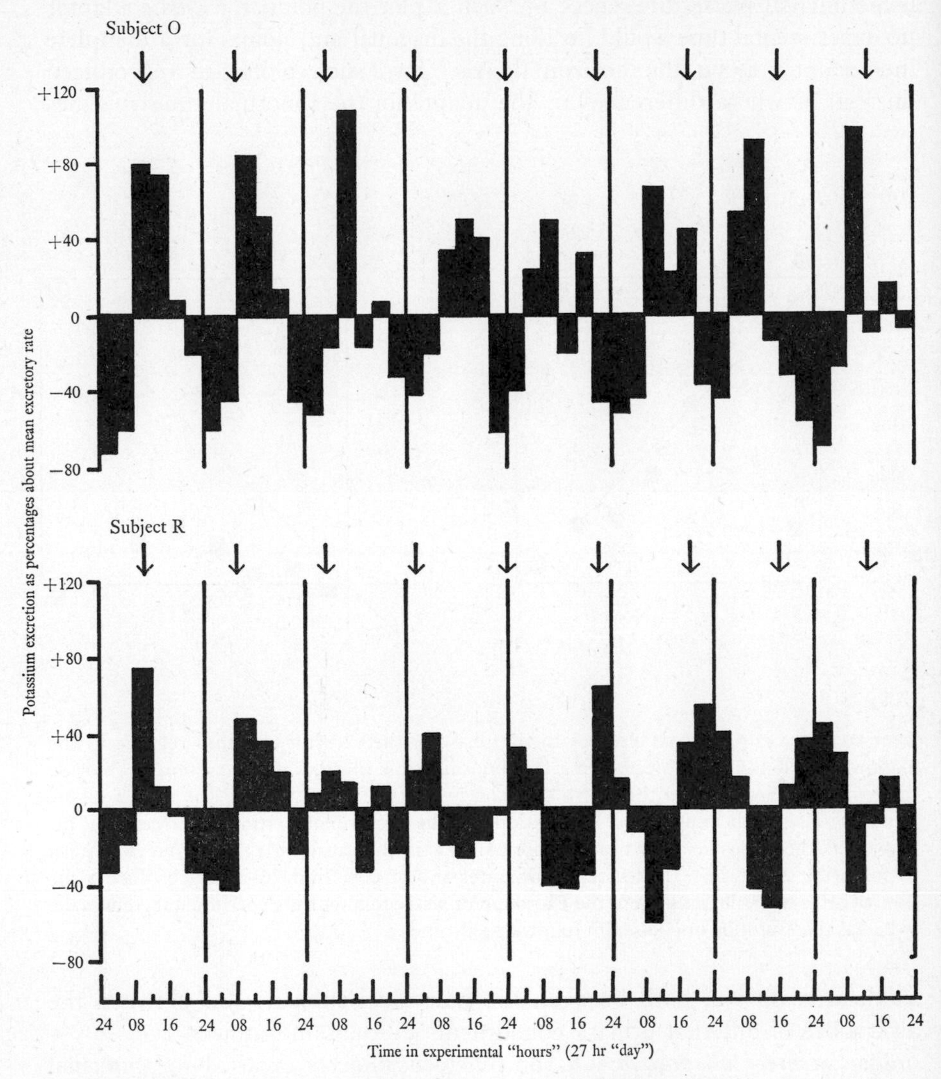

FIGURE 9. Excretion of potassium by two individuals throughout the whole of their first experimental cycle on 27 hr time. Arrows indicate midday by solar time. Subject O shows reasonable adaptation of the excretory pattern to the experimental time scale, while in subject R the peak periods of excretion are always related to midday by real time, irrespective of the subject's activity pattern.

(Redrawn from *Quart. J. exp. Physiol.*, 1957, 42, 361.)

excretion of potassium follows experimental time from the outset, the highest rates of excretion being always associated with activity and the lowest with sleep; eight excretory "peaks" are shown, in phase with the eight experimental "days". For the non-adapter, subject R, however, the excretion of potassium follows 24 hr time, with consequent divergence from the activity pattern of the subject. About mid-cycle the highest rates of excretion occur during sleep, and nine periods of "peak" excretion occur during the whole cycle, related to the occurrence of true midday by solar time, for the eight experimental "days" on a 27 hr routine occupy nine real days.

Adaptation of the rhythm of potassium excretion to an abnormal time routine is extremely unusual. The excretion of water, however, shows adaptation in most subjects, although this adaptation may require exposure of the subject to the abnormal time routine for some weeks before becoming complete. The excretory patterns for water and for potassium throughout the fifth cycle on 27 hr time for subject R (the non-adapter in Fig. 9) are shown in Fig 10. Here the pattern of water excretion shows eight main "peaks" throughout the cycle, corresponding to the eight experimental "days", although small subsidiary "peaks" bearing some relationship to 24 hr time may still be detected. The pattern of potassium excretion, however, is still related to solar time rather than to the activity pattern of the subject, with nine "peaks" corresponding to the nine real days occupied by the cycle and a complete reversal in relation to the activity pattern of the subject and to the rhythm of water excretion about mid-cycle. This is a good example both of the persistence of the intrinsic 24 hr rhythm in the excretion of potassium and of the marked dissociation to be observed between different physiological diurnal rhythms, even those in the same organ, under certain conditions. A further example of the dissociation of diurnal rhythms in subjects exposed to abnormal time routines is shown in Fig 11. Here the rhythms of water excretion for two subjects are compared with their oral temperature rhythms throughout the first experimental cycle on 21 hr time. Though the rhythm of water excretion becomes considerably disturbed about mid-cycle, the temperature rhythm is in phase with the activity pattern of the subjects from the outset. Subject D is the one, already represented in Fig. 3, who possesses a reversed temperature rhythm, and it can be seen that this individual peculiarity of rhythm shows perfect adaptation to the new time routine: on another occasion this subject's oral temperature rhythm was maintained in its complete specialised form throughout seven days of mild pyrexia. A similar maintenance of individual peculiarities in the form of diurnal rhythms, for example a double temperature peak during the hours of activity or a small but undoubted secondary peak in the excretory rates, has been seen in many subjects. Another observation of some significance in the consideration of the mechanisms that underlie physiological diurnal rhythms is also shown in Fig. 11; this is that the disruption of the water rhythm occurring about

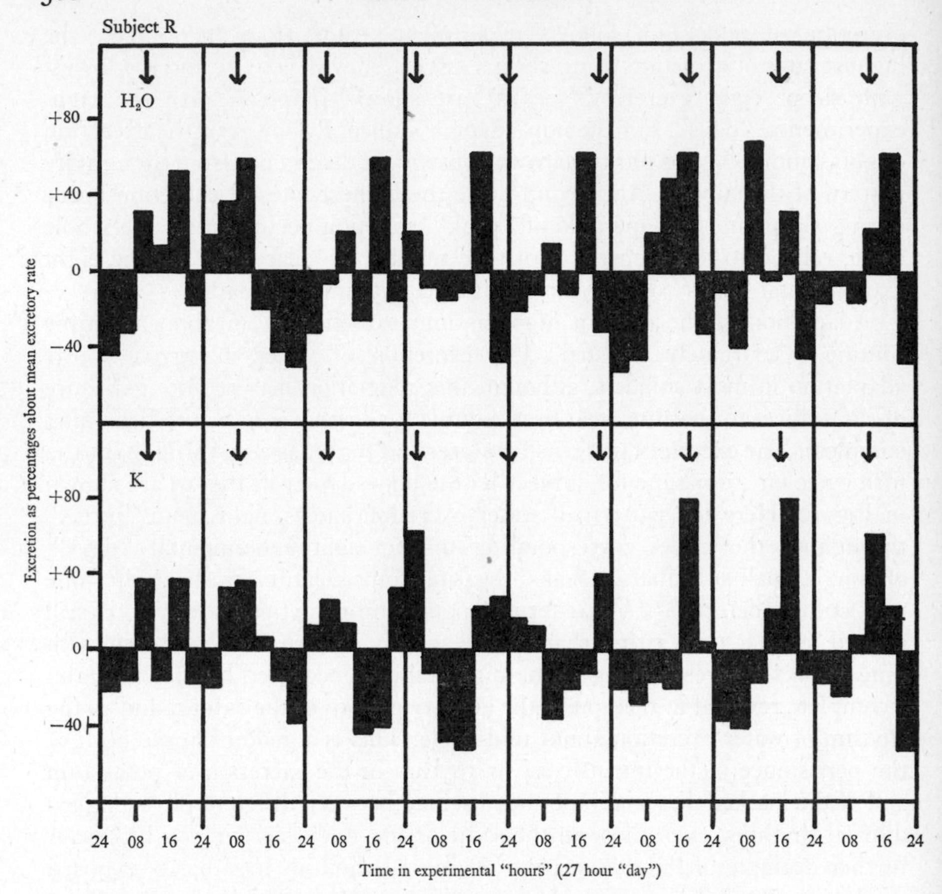

FIGURE 10. Excretion of water and of potassium throughout the fifth cycle on 27 hr time: arrows indicate midday by solar time. The pattern of water excretion shows reasonable adaptation to the experimental time scale, but potassium excretion still follows 24 hr time. Thus there is dissociation between the two excretory patterns about mid-cycle. (Redrawn from *Quart. J. exp. Physiol.*, 1957, 42, 381.)

mid-cycle in the unadapted subject is accompanied by a severe reduction in amplitude. It must be emphasised that this reduction in amplitude is not accompanied by a reduction in the total amount of water excreted, the mean excretory rate for water for each experimental "day" varying little throughout the cycle; it represents a true qualitative difference in the excretory rhythm.

d. Separate mechanisms. To summarise, the experiments carried out under carefully controlled experimental conditions on subjects in an environment of continuous daylight who are exposed to the false clues of an abnormal time routine show, first, that adaptation of some of the renal diurnal rhythms does occur to such an abnormal routine; secondly, that there are individual

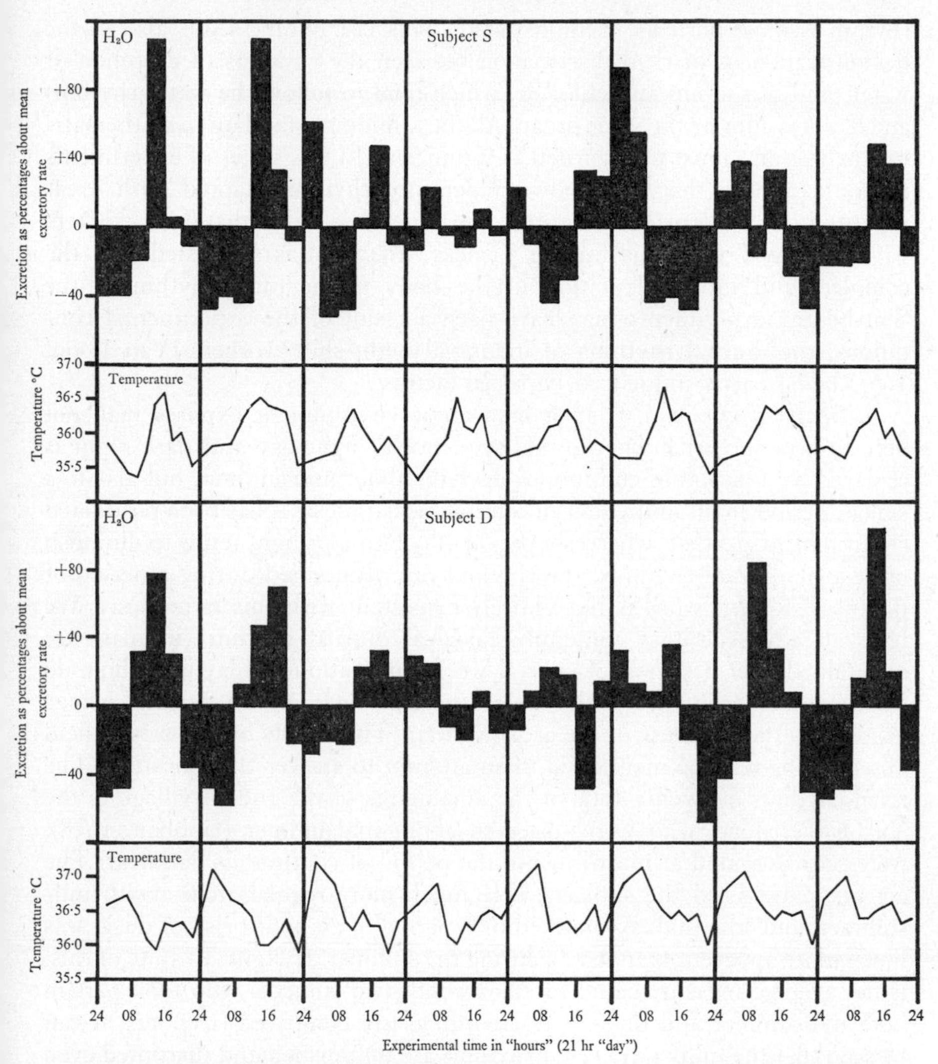

FIGURE 11. Comparison of the rhythms of water excretion and of body temperature for two subjects during their first cycle on 21 hr time. For both subjects the water rhythm is disorganised and does not follow the experimental time scale, but the temperature rhythm shows good adaptation to 21 hr time from the outset—even for subject D, who has a reversed temperature rhythm.

(Redrawn from *Quart. J. exp. Physiol.*, 1957, 42, 375.)

differences in adaptation from subject to subject, rapid adaptation being uncommon and slow adaptation the rule for the renal rhythms; thirdly, that the rhythm of potassium excretion is the most resistant to change and the one in which the maintenance of the intrinsic 24 hr physiological diurnal rhythm is most perfectly demonstrated. This last finding leads also to the demonstration of marked dissociation between the rhythms of excretion of water (and of sodium and chloride, which tend to follow the water rhythm) and of potassium by the same organ. All these findings stand up to mathematical analysis and must be regarded as significant. In this series of experiments it would appear that the body temperature rhythm adapted with great rapidity to an abnormal time routine, but it must be stated that other workers have disagreed with this finding (Teleky, 1943). It is suggested that the complete and rapid adaptation of the body temperature rhythm in the Spitsbergen experiments may have been a result of the experimental conditions, the diurnal rhythms of industrial night-shift workers (Van Loon, 1963) being partly influenced by social factors.

c. Seasonal effects. Life in a high arctic environment exposes man not only to a period when the normal sequence of light day and dark night is replaced by a period of continuous daylight about midsummer, but also to a similar period about midwinter of continuous darkness. It has been postulated (Kleitman, 1949) that, whereas a period of continuous light tends to diminish or destroy diurnal rhythms, the rhythms are maintained during a period of darkness. Recent work is not entirely consistent with this hypothesis. We have already seen that some physiological diurnal rhythms in man are maintained over a period of several weeks in continuous daylight; how do these findings compare with observations carried out on human beings exposed to the darkness of the arctic winter? The results of two sets of field observations may be mentioned in an attempt to answer this question. The renal rhythms of twenty-four of the inhabitants of the Indian village in the Yukon Territory, who were observed about midsummer (Lobban, 1958), were reinvestigated at midwinter in the period of continuous darkness. The activity patterns of the subjects were much more regular than about midsummer, and true midday, marked by a period of an hour or so of dusk, was much more apparent than it was during the summer daylight. In spite of this, if the average measurements for the twenty-two subjects who took part in both midsummer and midwinter recordings are compared (Fig. 12), it can be seen that the midwinter renal rhythms are diminished and disrupted even more than those observed at midsummer. Again, this is a true qualitative difference, for the mean excretory rates of the urinary constituents are closely similar for the two periods.

The indigenous arctic residents of Indian and Eskimo communities are highly specialised human subjects, in that for the whole of their lives they have been exposed to the arctic environment. It is probable that they have never had the opportunity to develop physiological diurnal rhythms to the extent that these are present in subjects from the temperate zones; in any event, their dietary intakes differ somewhat from those of men living under more "civilised" conditions. Reference has already been made to Spitsbergen as an ideal venue for experiments on man in the comparatively constant conditions of environmental light and temperature of the high arctic summer at a little above 78°N. Some fifty miles south of the bases where imported subjects lived in isolation and carried out the experiments on abnormal time routines already described, there is a Norwegian mining community of about 800 inhabitants. This is a true mixed community, comprising miners who

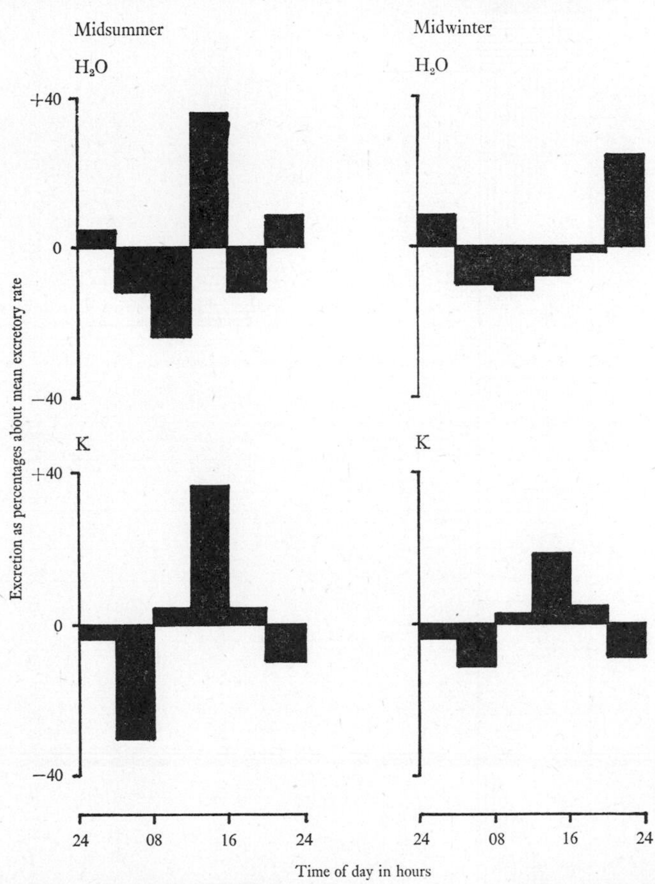

FIGURE 12. Excretory patterns for water and potassium in arctic-dwelling North American Indians at midsummer and midwinter. Averaged values for twenty-two subjects, showing that the midwinter patterns are more disorganised and more diminished in amplitude than are the midsummer patterns.

carry out heavy physical work underground, heavy manual workers above ground, sedentary workers, women and children. The latitude is such that in each year the inhabitants are exposed to four months of continuous daylight and four months when the sun never rises in winter; thus there are only two months about each equinox when the normal temperate-zone conditions of

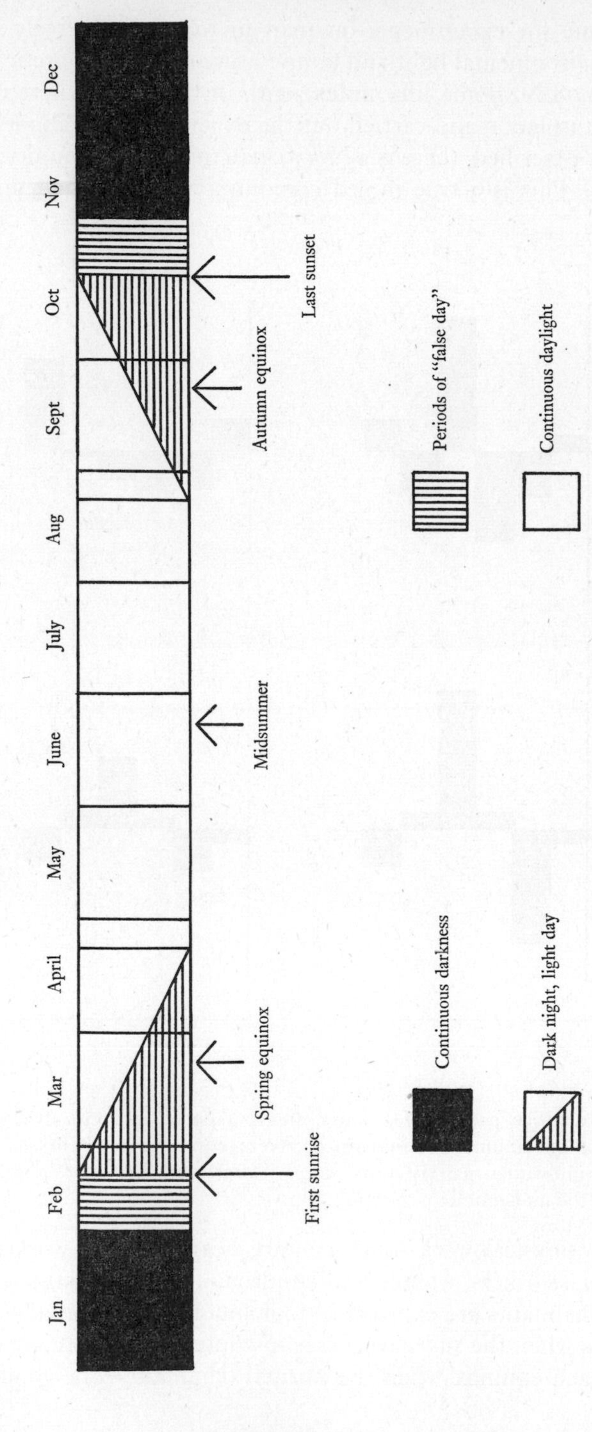

FIGURE 13. Diagrammatic representation of the annual cycle of daylight and darkness at 78°N in Spitsbergen.

light day and dark night are experienced. The community is completely isolated by sea-ice from mid-November to mid-May in a normal year, and the standard of living compares favourably with that of communities in temperate zones. I have had the unparalleled opportunity of living throughout two winters in Longyearbyen and of making observations on the renal diurnal rhythms of representative sections of its community.

The seasonal fluctuations of daylight and darkness at Longyearbyen, this Norwegian coal-mining community in Spitsbergen, are diagrammatically represented in Fig. 13. In the periods of "false day", which precede the first rising of the sun towards the end of February and follow the last sunset towards the end of October, solar day is marked by a lightening of the southern sky and is apparent to all members of the community who are out of doors in the hour or two about midday. The renal rhythms of sedentary workers, sixteen subjects, all of whom were working during normal daytime hours, are little different from those of temperate-zone subjects. The rhythm of water excretion shows some diminution of amplitude and a slight shift of phase in the months of January (after three months without the sun) and March (when there is a marked increase in physical activity during off-duty hours), but the original 24 hr rhythm is reasonably well maintained. The rhythm of potassium excretion is indistinguishable from that of temperate-zone subjects (Fig. 14). For dayshift workers on heavy manual work, however, the diurnal excretory patterns show some disorganisation and diminution of the renal rhythms during the winter darkness. This effect is least pronounced in the rhythm of potassium excretion and most marked in that of water excretion for underground workers. There is some indication that the improvement in the rhythms that occurs as the sun comes back begins a month earlier in those workers who are above ground (Fig. 15). In the mines work is going on for 24 hr a day, six days a week; thus there was an opportunity to examine the renal rhythms of night-shift workers. These are of particular interest, since the men do not rotate shifts; once a man is on the night shift, he normally remains on it for the whole of the winter; also the social conditions are such as to favour adaptation to unusual times of work. In these night-shift workers a totally unexpected finding emerged: the excretory pattern for water shows marked disruption during the months of darkness, but the potassium rhythm is well marked and is perfectly entrained to the activity pattern, being reversed in relation to normal working time, throughout the year (Fig. 16). About the spring equinox a diurnal rhythm that is in phase with normal solar daylight and not with the working day makes its appearance in the rhythm of water excretion, giving rise to complete dissociation between the rhythms of water and of potassium excretion at the time of the return of the sun (Lobban, 1963). This indicates that, given the right conditions, even the relatively stable potassium rhythm can become adapted to a new routine.

The observations on the permanent Norwegian population in Spitsbergen, then, would indicate some disorganisation of the renal rhythms that may be correlated with seasonal changes in the environment, but that these are only seen in subjects carrying out work involving the expenditure of considerable physical energy. The observed changes in the rhythm of water excretion could be due to the water loss caused by heavy sweating, but this would not explain the return of the 24 hr rhythm at about the time of the spring equinox. Also, if sweating were involved, one would expect the diminution

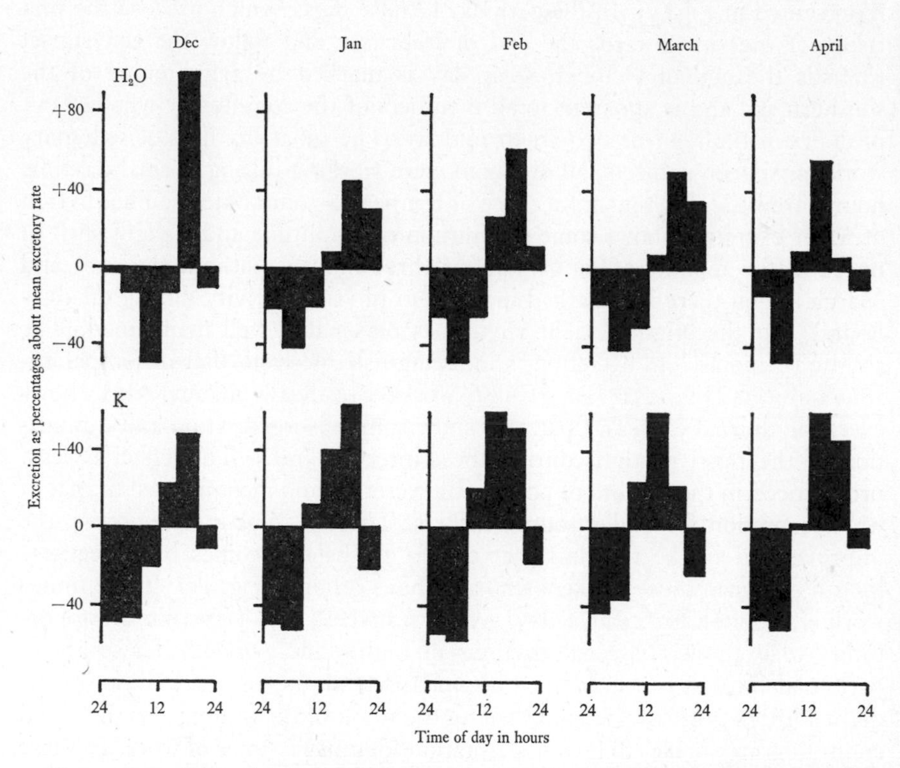

FIGURE 14. Diurnal patterns of renal excretion of water and potassium by sedentary day workers in Spitsbergen throughout the winter season. Averaged values for sixteen subjects, showing good maintenance of both temporal relationship and amplitude in environments of continuous darkness, light day and dark night or continuous daylight.

in amplitude of the rhythm to be accompanied by a diminution in the total amount of urine produced, and this does not happen; again, the mean excretory rate of water for each subject remains remarkably constant throughout the year. We are thus faced with the enigma of a qualitative, rather than a quantitative, difference in rhythm.

f. Effects of exercise. The effects of physical exercise upon the renal rhythms have been noted in other situations. In the spring of 1960 observations were made on a party of young men who were members of a glaciological team working above Zermatt, in April, under conditions of light day and dark night (sunrise 06.15 hrs, sunset 18.45 hrs, approximately). The rhythms of excretion of water and potassium for three of these subjects are shown

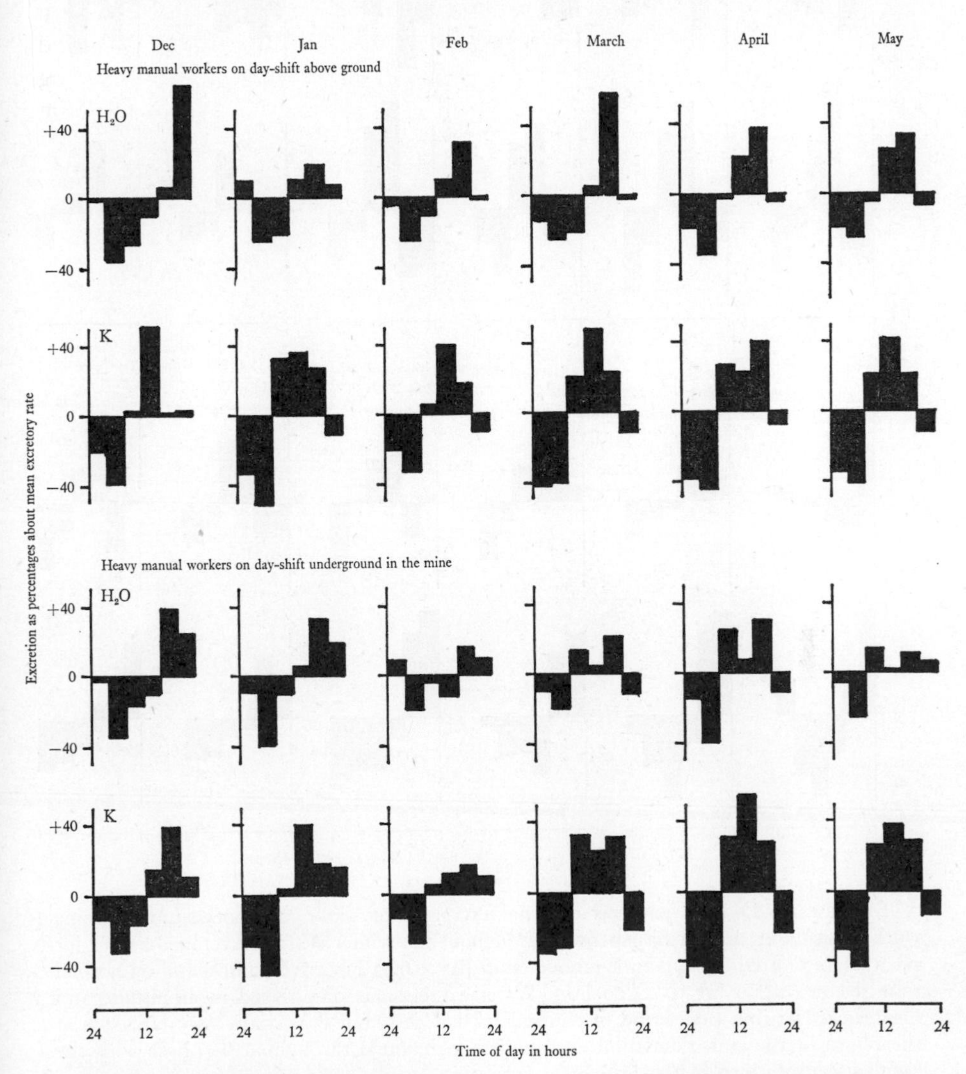

FIGURE 15. Diurnal patterns of renal excretion of water and potassium by heavy manual workers on day shift in Spitsbergen throughout the winter season. Averaged values for eight workers above ground (above) and eight workers in the mine (below). Disorganisation and diminution of amplitude occur during the period of continuous darkness, and re-establishment of the rhythms occurs one month earlier in the workers above ground than in the mine workers.

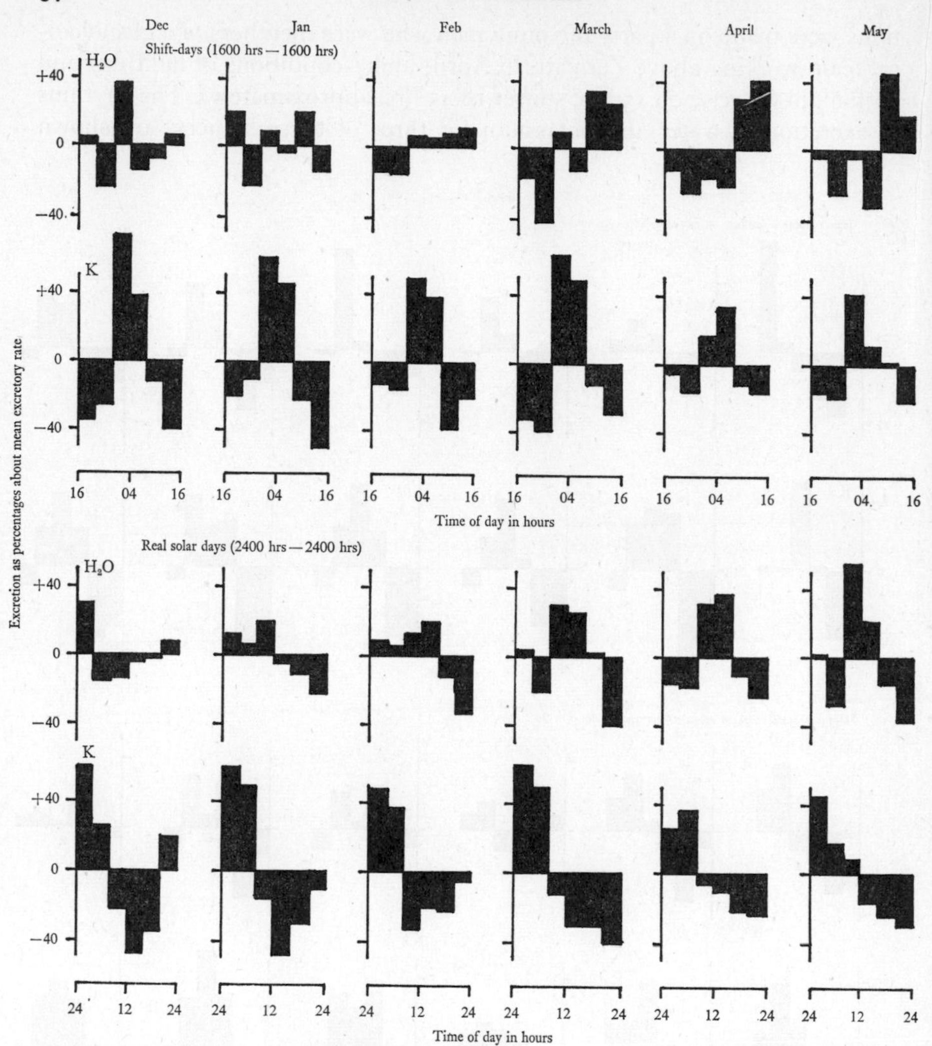

FIGURE 16. Diurnal patterns of renal excretion of water and potassium by mine workers on night shift in Spitsbergen throughout the winter season. Averaged values for six workers plotted on shift time (above; shift day: 16.00 hrs to 16.00 hrs) and on normal time (below; 24.00 hrs to 24.00 hrs). Water excretion is diminished in amplitude and disorganised during the period of continuous darkness, but is partially re-established at the return of the sun. Potassium excretion is maintained throughout in phase with the men's activity pattern on the shift day.

in Fig. 17. The three subjects were of similar age, had come up to the survey hut only five days previously (altitude, 2700 m) and were all unaccustomed to that particular type of strenuous field work. Subjects ZA and ZB worked on the glacier, subject ZD remaining at the base hut and spending a comparatively inactive day. Breakfast and supper were taken by all members of the party at the hut, and field lunches, accompanied by 250 ml of fluid, were eaten by all three subjects at 13.00 hrs. The subject who remained at base shows excellent diurnal rhythms of excretion of both water and potassium, whereas the rhythm of excretion of water for the two subjects on the glacier is so disorganised as to become completely reversed and dissociated from the rhythm of potassium excretion. The latter maintains a good phase relationship

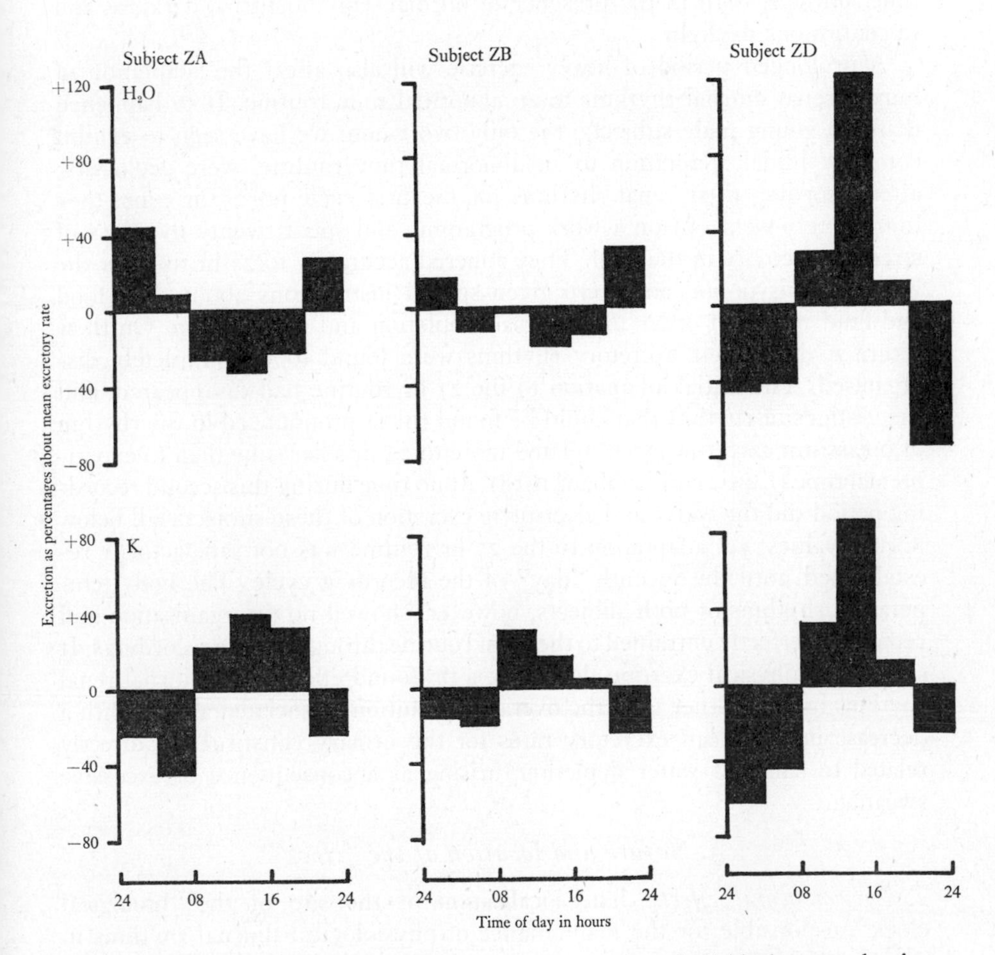

FIGURE 17. Excretion of water and potassium by three individuals on 24 hr time routines in conditions of normal light day and dark night in Switzerland. Subjects ZA and ZB undertook severe exercise in the field from 09.15 hrs to 19.00 hrs, and subject ZD was employed in a sedentary occupation at base. The excretory patterns for subjects ZA and ZB show marked diminution in amplitude (water and potassium) and temporal disorganisation (water).

(Redrawn from *Cold Spr. Harb. Symp. Quant. Biol.*, 1960, 25, 330.)

with solar time, but shows a qualitative change in that the amplitude of the excretory rhythm is much reduced, although the mean excretory rate for potassium for all three subjects was closely similar (Lobban, 1960). Similar observations have been made during an arduous sledging journey in Spitsbergen by subjects adhering rigidly to a true solar-time 24 hr routine during the summer daylight. Thus it would seem that strenuous physical exercise can give rise to dissociation of certain of the renal rhythms on 24 hr time routines, both in the presence of normal daily nocturnal darkness and in continuous daylight.

A prolonged period of heavy exercise will also affect the adaptation of human renal diurnal rhythms to an abnormal time routine. It so happened that two young male subjects, the only two whom we have seen to exhibit complete initial adaptation to an abnormal time routine, were geologists; after recording their renal rhythms for the first cycle on 27 hr time, they immediately went out on a work programme and spent twenty-five days of strenuous activity in the field. They adhered accurately to 27 hr time for the whole of this period and were given special instructions about their food and fluid intake in order to avoid salt depletion and dehydration. On their return to base their excretory rhythms were found to be completely disorganised. The initial adaptation to the 27 hr routine had disappeared, and the vestiges of rhythm that could be found (most pronounced in the rhythm of potassium excretion) were related more to 24 hr solar time than to experimental time (Lewis and Lobban, 1958). At no time during this second recording period did the water and electrolyte excretion of these subjects fall below normal values, yet adaptation to the 27 hr routine was not satisfactorily re-established until the seventh "day" of the recording cycle. The body temperature rhythms of both subjects, however, showed no disorganisation and remained perfectly entrained to the 27 hr routine throughout the recordings. It seems that physical exertion does have a profound effect upon diurnal renal rhythms in man, other than the overall diminution (associated always with a decrease in the mean excretory rates for the urinary constituents) directly related to salt and water depletion, arising as a consequence of excessive sweating.

3. *Nature and location of the "clock"*

a. Environmental effects. The localisation of the site of the "biological clock" responsible for the maintenance of physiological diurnal rhythms in man must be largely a matter for conjecture. Whereas Harker's classic experiments (1960) on the cockroach, *Periplaneta americana*, involving experimental parabiosis and transplantation of the suboesophageal ganglion, can give precise information as to the site and nature of the "biological clock" in insects, no such direct approach is possible for the human subject. The information at our disposal about diurnal rhythms in man is limited to what may be obtained from experiments on the whole healthy human subject and to a limited number of observations made upon some medical and surgical cases. The interpretation of such observations may sometimes be modified by subjective impressions, but such an effect for the most part lies outside the scope of this book. Observational and experimental studies carried out on the normal human subject show that true diurnal rhythms do exist in many of man's physiological processes and that these rhythms differ one from another in their response to environmental factors, the rhythm most resistant to change being that of the renal excretion of potassium. The main environmental factors involved appear to be the alternation of daylight and darkness in each 24 hr day and the regular sequence of solar time; subsidiary to these are severe physical exertion (not the mere alternation of night sleep and daylight wakefulness observed in sedentary workers or those engaged in work that involves only moderate bodily activity) and social factors. The whole picture in man is extremely complicated, as might be expected from the high degree of specialisation of function in the human body.

b. Changes with age. The frequency with which dissociation between certain of the physiological diurnal rhythms has been observed in man would suggest that no one biological clock, but at least two and possibly a series of them, is involved in the maintenance of these rhythms. Evidence of this occurs early in the development of the individual, in that different physiological rhythms are established at different stages in the life of the human infant.

Hellbrügge (1960), in the course of a lengthy and critical study of developing physiological diurnal rhythms in a large number of infants, made observations on the initiation of such rhythms of heart rate, body temperature, sleep and wakefulness, the electrical resistance of the skin of the forearm and the functioning of the kidney. These rhythms are not present at birth, the first indication of a regularly occurring diurnal trend being seen in the electrical resistance of the skin during the first week of life. Then comes the development of daily rhythms in the rate of urine flow, sleep and wakefulness, body temperature, heart rate and, finally, at the end of the second month of life, the urinary excretion of sodium and potassium. Hellbrügge (1960) emphasises that the human organism is dependent upon a fully functional neuro-hormonal system, organs of perception and effector organs for the establishment of physiological diurnal rhythms and their entraining to the 24 hr environment. He also points out that the difference between night and day levels of function is more often produced by a lowering of night values than by an increase in the daytime levels. In the two types of diurnal rhythms with which this discussion is particularly concerned, that of body temperature first arises by means of a lowering of night temperature values, whereas those of renal excretion are established by an increase in the daytime excretory rates. All the diurnal rhythms that Hellbrügge (1960) studied develop independently of physical activity and of feeding cycles

(Engelmann and Kleitman, 1953). Finally, all the typical monophasic diurnal rhythms of the older child and the adult appear to be developed from polyphasic rhythms (the most common form being a diphasic one, with two "peak" periods of function during the daytime hours) occurring in early childhood. This conversion of polyphasic rhythms, which seem to appear independently of external stimuli, into monophasic rhythms in phase with the sequence of events in a normal day would seem to be good evidence for the existence of true endogenous physiological diurnal rhythms in man, which may be entrained to a 24 hr light day and dark night environment.

Hellbrügge's (1960) findings for the human infant are not entirely in agreement with those on renal diurnal rhythms in the adult as shown in the experimental work described in this chapter. The developmental sequence would lead one to expect that the body temperature rhythm could be considered as one entity, the urinary excretion of water as another and the urinary excretion of sodium and potassium, taken together, as a third. In the adult, however, the separation of function (as shown by the dissociation of rhythms in human subjects living on abnormal time routines) would appear to lie between body temperature, the excretion of water and sodium, taken together and, finally, the excretion of potassium as a separate entity. This separation is supported further by the fact that although the overall sodium to potassium ratio in the total amount of urinary electrolytes for each recording period does not diverge from normal on an abnormal time routine, marked variations in the excretory sodium to potassium ratio can be demonstrated at different times of the experimental "day" in partly adapted subjects. Hellbrügge (1960) suggests that the sequence of events he has observed in early post-natal renal function in the human infant may be explained by the fact that water excretion is mainly a function of the glomerulus of the kidney and electrolyte excretion is a function of the kidney tubule, contending that glomerular development is farther advanced at birth than is tubular development; hence the diurnal rhythm of water excretion is manifested before similar rhythms can arise in the excretion of electrolytes. It is extremely doubtful whether such an explanation is correct. In the first place, his "division of labour" in the structures of the nephron is not in agreement with the usually accepted theories of kidney function; secondly, although it is true that the human kidney, in common with that of many other mammalian species, is not totally functional at birth, this incomplete development is quantitative rather than qualitative: thus in certain areas of the kidney, mainly those situated in the extreme outer part of the cortex, closed glomeruli can be found in association with tubule systems in which the lumina of the tubules are not yet patent. It is suggested that the separation in initiation and maintenance of diurnal functioning in the kidney is more likely to be caused by different controlling mechanisms than by developmental differences in the organ itself.

c. Physiological periodicals. Spontaneous rhythmicity is an inherent property of many mammalian cell groups. Cells from the foetal myocardium will exhibit spontaneous rhythmic activity when grown in complete isolation in tissue culture; various areas in the cerebral cortex show continuous rhythmic discharge, even during sleep; the cell groups that constitute the respiratory centre at a lower level in the brain maintain a regular basic rhythm throughout life. By definition, all living cells are capable of response to environmental changes. The responses to such changes, whether they be brought about by alteration in the internal or the external environment, must be favourable to the organism as a whole if the individual is to survive. The natural periods of spontaneous rhythms in nervous tissue are short, and their responses to environmental changes are as a rule immediate: but there are also rhythms in the human being—the hormonally controlled sexual cycle in the human female is an excellent example—in which the regular sequence of events making up the whole rhythmic cycle takes much longer. This being so, it is not difficult to accept the possibility of the existence in the human being of physiological rhythms with a 24 hr periodicity.

It has already been stated that man appears to exhibit daily rhythms of performance and efficiency in phase with his physiological diurnal rhythms. If this is accepted, it is of obvious advantage to the average human being, who lives out the whole of his life on a routine of daytime activity and nocturnal sleep and is not normally asked to participate in polar expeditions, that all his physiological processes should show a diurnal periodicity in phase with solar time. Such an individual will be best adjusted to his environment if he is most alert and active during the daytime hours and can enjoy unbroken nocturnal sleep without having to get up to empty his bladder several times during the night. There would seem to be little scope in such an organisation of physiological function for response to environmental changes: indeed, when the environment does change, as when the individual's time routine is altered, certain of his physiological diurnal rhythms would appear to adapt but poorly. The fact that human beings in increasing numbers are now being asked to undertake shift-work, and thus to alter their normal time routines, points to a wryly amusing paradox, for the demands of the complex and orderly organisation built up with such care are now beginning to make a nuisance of the established social and physiological order. In the average individual, however, there is some indication of response to environmental changes by the physiological rhythms. As a general rule, the periodicity of the longer term rhythms is less precisely defined than is that of the rapidly oscillating systems. This has been well shown by experimental methods for a number of lower organisms, in which, if the normal daily fluctuations in their environment are replaced by a constant environment, the natural functions continue to show a rhythmicity, though this does not have a period of precisely 24 hr. It is in order to describe such rhythms—those with a periodicity of the order of 24 hr—that the term "circadian rhythm" has come into use. As has been most beautifully demonstrated by Pittendrigh (1960), once the free-running rhythm has emerged in a constant (usually dark) environment, the re-introduction of regularly occurring environmental clues (usually a period of light) can cause the natural rhythm to be trained to a new routine, always provided that this new routine is not too widely divergent from that of a normal 24 hr day. Engelman and Kleitman (1953), following the development of the diurnal rhythm of sleep and wakefulness in a human infant, found that this particular child first showed clearly a 25 hr periodicity, which then became adjusted to the usual 24 hr periodicity only after the eighteenth week of life. It seems more than likely that this is an example of response to the environmental fluctuations of night and day. Hence it is suggested that true endogenous physiological circadian rhythms do exist in man and that they exhibit their sensitivity by responding to environmental clues or *Zeitgeber* (Aschoff, 1960) in early life. Thus the role of the environmental fluctuations is to bring about a synchronisation of the separate physiological rhythms and their entraining into the routine of the normal 24 hr day.

Until recently it has been difficult to find any evidence of a free-running circadian rhythm in the adult human subject. A series of ingenious and carefully documented experiments carried out by Aschoff and Wever (1962), however, has now provided some extremely valuable observations that point clearly to the existence of physiological circadian rhythms in man. In these experiments each subject was confined alone in an underground chamber, where he was completely cut off from the external environment. Subjects remained in this chamber for periods of up to eighteen days, during which time they controlled their own environment by switching the lighting on or off and sleeping, waking, working and eating as they felt inclined, without any reference to solar time. Each subject was under constant observation for the whole of his period in the chamber, and the courses of various physiological functions, including the body temperature and the renal excretion of water, electrolytes and steroids, were followed throughout the period. In all, nine subjects took part in the experiment, and eight of them clearly took up daily rhythms with a period greater than 24 hr. The periodicity varied among individual subjects from 24·5 hr to 26·0 hr, but was absolutely constant for each individual subject, a 26 hr man being always a 26 hr man and so on. The ninth subject showed a period of 24·2 hr; no subject showed a period of less than 24 hr; six out of the nine subjects showed a period of more than 25 hr. The circadian rhythm appeared after the third day in the chamber; as a general rule, the patterns of activity, sleep and physiological function appeared to be in phase. After release from the chamber, a similar three-day period of exposure to the normal *Zeitgeber* of the external environment was required before the subjects became trained again to a 24 hr rhythm.

It is of interest to compare the results of Aschoff and Wever's (1962) experiments with some of the observations made on subjects on 21 and 27 hr routines in the constant daylight of the arctic summer. The most striking difference lies in the fact that, when these subjects went back to a 24 hr routine in phase with solar time, the physiological rhythms always took on a 24 hr rhythm immediately, irrespective of whether the subject had become adapted to the abnormal time routine or not. Aschoff and Wever's (1962) subjects needed a period of days before the 24 hr periodicity was re-established. It may be that the slight *Zeitgeber* elements present in the environment in the arctic summer were, in fact, producing some effect throughout the experimental period, even in those Spitsbergen subjects who had apparently adapted to the abnormal time routines. This adaptation would then be the result of a critical balance between true and false environmental clues and would tend to disappear as soon as the false clues were removed: the large number of the subjects in whom an appreciable 24 hr component was detected in the renal rhythms throughout the experiment would support this view. In Aschoff and Wever's (1962) experiments the isolation of the subjects from their external environment was probably more nearly perfect: also, the free-running rhythms that emerged in their experiments can be regarded as the intrinsic circadian rhythms that, once allowed to emerge, were more resistant to subsequent training, even to a normal 24 hr routine. It might also be expected that, since the circadian rhythms in man appear to be of a periodicity greater than 24 hr, adaptation to a 27 hr routine would tend to occur with greater ease than it did to a 21 hr routine, but there is no evidence of this in the Spitsbergen results.

In the special environmental conditions found at high arctic latitudes, it may be that indigenous human subjects are not exposed to sufficient environmental clues in early life, especially those individuals who are born during the periods of summer daylight or winter darkness. This might explain the arythmia or markedly low amplitudes of renal rhythms observed in some, though not all, arctic-dwelling Eskimo and Indian subjects (Lobban, 1958). But it cannot account for the loss of renal diurnal rhythms observed in some subjects, originally from temperate light day and dark night zones, who have lived for a considerable time in the high arctic environment. It may be that the original rhythms tend to die out if the environmental clues are withdrawn for a sufficient period, although this is difficult to equate with the idea that true endogenous rhythms do exist in the human subject. Relevant to this problem is the observation that renal diurnal rhythms are slightly reduced in amplitude, albeit unaltered in phase, after six to eight weeks in an arctic summer environment and are disorganised and reduced in amplitude if the subjects are carrying out hard physical work in an arctic winter environment. The position is by no means clear; the problems of the interplay of endogenous rhythms and exogenous factors and the exact role of physical exercise, which appears to affect the whole balance of the system, are exceedingly complex. If these factors could be separated, some indication of the true nature of the human diurnal rhythms might be obtained. For example, it has been frequently observed, during the course of experiments involving exposure of human subjects to time routines other than the normal 24 hr time scale, that double "peaks" make their appearance in the rhythms of renal excretion. These can often be related one to the experimental time scale and the other to solar time. If this is a true effect, it would be good evidence for the existence of a multi-oscillator system in man, an idea supported further by the marked diminution in amplitude of the renal rhythms observed at mid-cycle on experimental time routines, when experimental time is reversed in relation to solar time. On the other hand, it must be remembered that the original rhythms tend to become disorganised, or even to disappear, if environmental clues are withdrawn for sufficiently long, so that the appearance of multiple "peaks" may merely represent a partial return to those polyphasic rhythms that Hellbrügge (1960) has found in the human infant during the first few weeks of post-natal life, before environmental synchronisation and training of the spontaneously occurring physiological rhythms has taken place.

E. Conclusions

The present state of knowledge is not sufficiently far advanced for understanding the nature of the "biological clock" in man. Also, although the evidence available is all in favour of there being more than one clock involved in the control of physiological diurnal rhythms in man, the precise sites of these several clocks is a matter of conjecture. The more familiar type of temperature regulation in the human being is usually thought to be controlled, at least in part, by a centre in the region of the hypothalamus. It is reasonable to suppose that diurnal fluctuations in body temperature may be regulated from the region of this centre; in any event, the centre would seem to be extremely sensitive to environmental changes of time and of activity pattern, for the rhythm of body temperature appears to be one of the more labile of the physiological diurnal rhythms. The experimental observations described in this chapter demand a double system for the control of kidney function, one part of which is primarily concerned with the excretion of water and of moderate lability in most subjects and the other with the excretion of potassium and resistant to environmental changes. Such a double mechanism does exist in the body. The hypothalamic-hypophyseal axis does control the excretion of water and is a satisfactory mechanism, in that it comprises the supraoptic nucleus, a nervous centre in which it is not difficult to envisage rhythmic activity, and an endocrine portion, the pars nervosa of the pituitary gland. It is also known that extirpation of the pituitary gland or the interruption of the supraoptico-hypophyseal tract by section of the pars tuberalis produces a marked diuretic response in man, accompanied by a complete disruption of the diurnal rhythm of water excretion. The mineralocorticoids of the adrenal cortex are vital for preserving the normal mineral balance of the body, but it is not known how the separation between the rhythms of excretion of sodium and potassium is brought about. It has been shown, however, that marked dissociation between the rhythm of water excretion (normal) and of potassium excretion (disruption of the diurnal rhythm) does occur in Addisonian patients whose mineral balance is being satisfactorily maintained by the administration of aldosterone (Prunty, McSwiney, Mills and Smith, 1954). A most relevant observation is that there is a diurnal rhythm in the plasma level of 17-hydroxycorticosteroids (Halberg, 1960) and in the urinary excretion of these substances (Aschoff and Wever, 1962; Simpson, unpublished observations). But in one patient subjected to extensive surgery over a period of years for the treatment of mammary carcinoma, I have seen a polyphasic (three regularly spaced peaks in 24 hr) excretory rhythm for potassium after bilateral adrenalectomy which was converted to a monophasic diurnal rhythm of excretion after hypophysectomy. The adrenal cortex obviously does play a part, but it would seem that the clock for the regulation of potassium excretion is the end result of an extremely complex hormonal balance.

It is unlikely that the maintenance of the physiological diurnal rhythms in man could have a direct effect upon the chances of the individual's survival under civilised conditions, but it is reasonable to suppose that if his physiological processes are well entrained to a normal 24 hr day, and he is then asked to be alert during the period of minimal physiological function, some strain will result. This will certainly be so if similarly entrained diurnal rhythms of performance and efficiency are also involved. If man is to work at unusual times of day, ideally he should be given a chance to adapt his diurnal rhythms to the new time routine. This would appear to present considerable difficulties, however, for few human subjects adapt rapidly and completely to such working conditions, and long periods of work on night shift, for example, are unpopular with most workers for social reasons. The arctic-dwelling Eskimo, in whom physiological diurnal rhythms are reduced to a minimum, is little affected by sudden calls for activity at any time of the day or night and is extraordinarily efficient and contented in his special environment. Perhaps the best that can be hoped for is that irregular and unusual hours of work in a dark night and light day environment can lead to a diminution of amplitude in the diurnal rhythms of the human subject, so that differences between day and night levels of physiological function are reduced.

REFERENCES

Aschoff, J. (1960). *Cold Spr. Harb. Symp. quant. Biol.* **25**, 11.
Aschoff, J. (1963). *Annu. Rev. Physiol.* **25**, 581.
Aschoff, J., and Wever, R. (1962). *Naturwissenschaften* **15**, 337.
Cartwright, G. E., Gubler, C. J., Hamilton, L. D., and Wintrobe, M. M. (1950). *Proc. Soc. exp. Biol. Med.* **75**, 65.
Colquhoun, W. P. (1960). *Ergonomics* **3**, 377.
Colquhoun, W. P. (1962). *Bull. Etud. Rech. psychol.* **11**, 27.
Conn, J. W., Kaine, H. D., and Selber, H. S. (1955). *J. Lab. clin. Med.* **45**, 247.
Engelmann, T. G., and Kleitman, N. (1953). *J. appl. Physiol.* **6**, 269.
Halberg, F. (1960). *Cold Spr. Harb. Symp. quant. Biol.* **25**, 289.
Halberg, F., Hauty, G. T., Hawkins, W. R., and Steinkamp, G. R. (1960). *Fed. Proc.* **19**, 54.
Harker, J. E. (1958). *Biol. Rev.* **33**, 1.
Harker, J. E. (1960). *Cold Spr. Harb. Symp. quant. Biol.* **25**, 279.
Hellbrügge, T. (1960). *Cold Spr. Harb. Symp. quant. Biol.* **25**, 311.
Kleitman, E., and Kleitman, N. (1953). *J. appl. Physiol.* **6**, 283.
Kleitman, N. (1939a). "Sleep and Wakefulness", p. 4. 1st Edition. University of Chicago Press, Chicago.
Kleitman, N. (1939b). "Sleep and Wakefulness", p. 242. 1st Edition. University of Chicago Press, Chicago.
Kleitman, N. (1949). *Physiol. Rev.* **29**, 1.
Lewis, P. R., and Lobban, M. C. (1956). *J. Physiol.* **133**, 670.
Lewis, P. R., and Lobban, M. C. (1957a). *Quart. J. exp. Physiol.* **42**, 356.
Lewis, P. R., and Lobban, M. C. (1957b). *Quart. J. exp. Physiol.* **42**, 371.
Lewis, P. R., and Lobban, M. C. (1958). *J. Physiol.* **143**, 8 P.
Lewis, P. R., Lobban, M. C., and Shaw, T. I. (1956). *J. Physiol.* **133**, 659.
Lobban, M. C. (1958). *J. Physiol.* **143**, 69 P.
Lobban, M. C. (1960). *Cold Spr. Harb. Symp. quant. Biol.* **25**, 325.
Lobban, M. C. (1963). *J. Physiol.* **165**, 75 P.
Lobban, M. C., and Simpson, H. W. (1961). *J. Physiol.* **155**, 64 P.
Lobban, M. C., and Tredre, B. (1964). *J. Physiol.* **170**, 29 P.
Mills, J. N. (1951). *J. Physiol.* **113**, 528.
Mills, J. N., and Stanbury, S. W. (1952). *J. Physiol.* **117**, 22.
Mills, J. N., Thomas, S., and Yates, P. A. (1954). *J. Physiol.* **125**, 466.
Pieron, M. (1913). "Le probleme physiologique du sommeil." Masson et Cie, Paris.
Pittendrigh, C. S. (1960). *Cold Spr. Harb. Symp. quant. Biol.* **25**, 159.
Prunty, F. T. G., McSwiney, R. R., Mills, I. H., and Smith, M. (1954). *Lancet*, ii, 620.
Sharp. G. W. G. (1962). *Nature, Lond.* **193**, 37.
Stanbury, S. W., and Thomson, A. E. (1951). *Clin. Sci.* **10**, 267.
Teleky, L. (1943). *Industr. Med.* **12**, 758.
Van Loon, J. H. (1963). *Ergonomics* **6**, 267.
Whittaker, E. T., and Robinson, G. (1944). "The Calculus of Observations", pp. 267–71. 4th Edition. Blackie and Son, London.

CHAPTER 13

Patterns of Sleep

J. P. MASTERTON

A. Present Knowledge

Although much information has accumulated on the physiology, psychology and psychiatry of sleep (Ciba Symposium, 1961; Oswald, 1962; Kleitman, 1963), its survival value, both short term and long term, is not known. Ability to do without sleep continuously over short periods has been the subject of numerous apocryphal stories usually related to some other feat of prowess ranging from ballroom dancing to continuous action in the face of an enemy. The objective evidence on acute sleep lack over short periods is reviewed in the next chapter of this book.

Long-term deprivation of sleep cannot be studied unless the normal sleep patterns for modern man are accurately known. Until appropriate and completely objective means of recording can be developed, resort must be made to the partly subjective technique of having the subject record his own sleep. The details of this method have been described by Lewis and Masterton (1957) and depend entirely on the co-operation of the subject, in that he is requested to complete a card (Figs. 1 and 2) on which each day is divided into 24 hrs and each hour is represented by a small square. Such a record is to some extent inexact, since it depends on the recall of the subject of his time of going to sleep. However, this defect need not be overstressed, because after a few days subjects acquire the habit of noting the time of going to bed or putting out their light, and thereafter are able to make a reasonably accurate estimate of when they fell asleep. It is unlikely that accuracy to 1 min is obtained, but errors of more than 15 min are probably rare, and this applies also to the time of waking. Periods of interruption of sleep during the night are recorded, as also are naps during the day. In both instances recordings of such interruptions depend on memory, but this increases in reliability as subjects become more trained.

Previous work (Gesell and Amatruda, 1945) has indicated that man's cycle of sleep and wakefulness is determined largely by the influences of the community, and this in turn derives its pattern from the diurnal solar cycle of light and darkness. Kleitman and Kleitman (1953) show that even in arctic Norway the community conforms to a diurnal pattern of sleep and wakefulness, as is customary in the rest of the world, in spite of the presence of the midnight sun. The habit of the world at large has influenced the diurnal pattern of the dwellers in the north. Results, obtained by the techniques

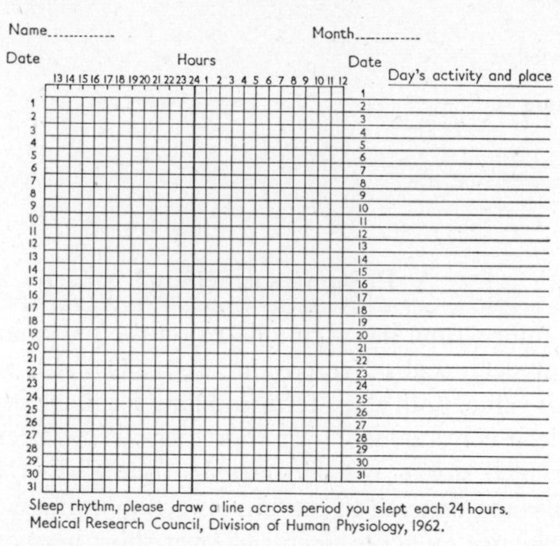
Name............ Month............
Date Hours Date
13 14 15 16 17 18 19 20 21 22 23 24 1 2 3 4 5 6 7 8 9 10 11 12
Day's activity and place
1 2 3 4 5 6 7 8 9 10 11 12 13 14 15 16 17 18 19 20 21 22 23 24 25 26 27 28 29 30 31
Sleep rhythm, please draw a line across period you slept each 24 hours.
Medical Research Council, Division of Human Physiology, 1962.

FIGURE 1. Blank sleep card. The most convenient size is 21 — 21 cm.

described, on the British North Greenland Expedition showed, first, that when individuals could sleep as much as they liked they did so to the extent of approximately 8 hrs in the 24; secondly, that when social factors were removed sleep was often broken throughout the same period. Eight hours can, therefore be taken as a satisfactory mean period against which the performance of men under stress or subject to community pressures can be judged. Many groups in the community are obliged to sleep at times and for periods dictated by the demands of work and also of leisure, but little information has so far been obtained on the degree to which these demands encroach on the amounts viewed as normal by long tradition and observation.

B. An Investigation

Recently we have undertaken investigations into the sleep patterns of individuals subjected to a wide variety of work disciplines—the medical, nursing

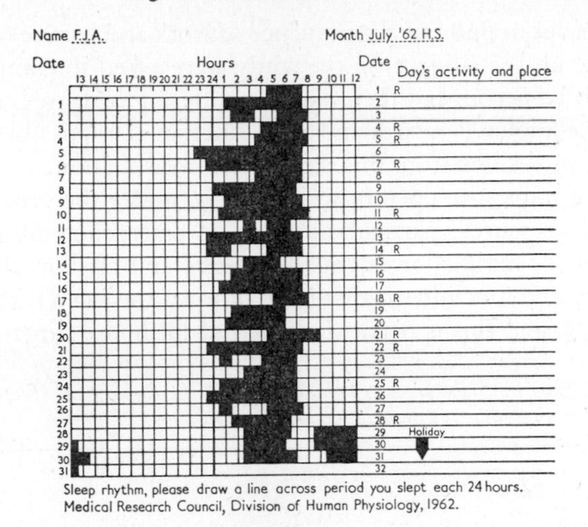

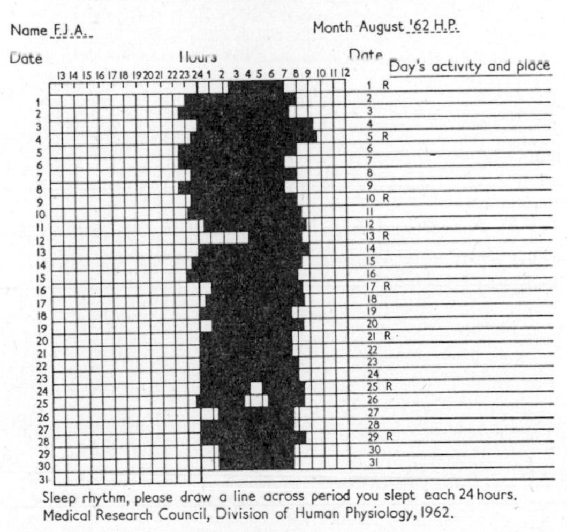

FIGURE 2. Sleep cards completed by the same individual while working as house surgeon and house physician. Note the great encroachment on sleep during July as a result of surgical receiving days.

and other staff of a busy general hospital (the Aberdeen Royal Infirmary) and secondary school children in Aberdeen. The groups studied included a number of different grades of doctors, some of whom changed jobs within the hospital during the course of the study; a class of student nurses on both day and night duty; medical students in their fourth year of studies during both the Christmas vacation and the University term; male and female clerical staff and male tradesmen in the hospital; a group of boys and girls from two secondary schools, all about 17 years old, studied during both term and vacation. In order to find out the influence of work and leisure on the amount of sleep taken, all the subjects in the study were asked to complete as far as they could a brief account of their daily activities. The degree to which these remarks were completed varied, and accordingly it is impossible to draw firm quantitative conclusions from this part of the investigation.

Of all the groups, the doctors seemed to be under the greatest influence from external pressures, particularly those of emergency admission at any hour of the day or night. For example, Tables I and II show the mean sleep of six surgeons and four physicians: in all but two the sleep taken is less than 7·5 hr per night, and this is particularly true of the house surgeons. Table III

Table I. *Sleep Taken by Six Surgeons Who Worked in the Same Unit (Note the progressive fall in mean sleep from Senior Lecturer to House Surgeon)*

	Days	*Hours*	*Mean*
Senior Lecturer A	114	856·3	7·5
Senior Lecturer B	90	655·9	7·3
Senior Registrar C	92	645·3	7·0
Senior Registrar D	92	590·0	6·4
House Surgeon E	92	532·1	5·8
House Surgeon F	78	370·0	4·7
Mean for all subjects			6·5

Table II. *Sleep Taken by Four Physicians Who Worked in the Same Unit (Compared with the surgeons, the residents sleep more than the two senior members of staff)*

	Days	*Hours*	*Mean*
Senior Lecturer	72	501·5	7·0
Senior Registrar	30	203·8	6·8
House Physician A	61	482·9	7·9
House Physician B	57	423·2	7·4
Mean for all subjects			7·3

Table III. *Comparison between Sleep Taken as House Surgeons and House Physicians (Three of these subjects did both jobs consecutively. The first two recorded sleep for a year)*

	Duty	*Days*	*Mean*
W.S.	H.S. Ward 8	184	5·3
W.S.	H.P. Ward 1	181	6·0
R.C.	H.P. Ward 4	184	7·5
R.C.	H.S. Ward 8	181	6·9
F.A.	H.S. Ward 10	87	6·1
F.A.	H.P. Ward 4	184	7·5
J.S.	H.S. Ward 8	184	6·2
G.C.	H.S. Ward 10	51	6·5
J.P.	H.S. Ward 9	52	6·8

emphasises this and records the amount of sleep taken by three men who held posts first as house surgeons and thereafter as house physicians. It is clear that there is considerable variation between subjects doing resident jobs as housemen in different units; further, there is a variation between the amounts of sleep taken by different men doing the same job. There is, however, a general overall lowering of the expected mean sleep throughout the whole period, below that found for the men in Greenland. Moreover, as Table IV shows, during holidays there was no conclusive evidence that anything but the smallest amount of lost sleep was recovered.

Table IV. *Sleep Taken by the Group of Surgeons* (Table I) *during Holidays* (*Senior Lecturer A was the only one to achieve a moderate surplus of sleep*)

	Days	*Hours*	*Mean*
Senior Lecturer A	22	210·5	9·6
Senior Lecturer B	20	160·4	8·0
Senior Registrar C	29	218·8	7·5
Senior Registrar D	20	153·8	7·7
House Surgeon E	17	137·3	8·1
House Surgeon F	7	56·0	8·0
Mean for all subjects			8·1

Senior doctors also show continued evidence of sleep lack. Some of this was necessary because of long operating lists and emergency admissions that required visits to the hospital at a late hour, but much of this encroachment was self-imposed by the desire to keep up with current literature and to write papers. This was particularly so with the physicians and is likely to continue for many years if not throughout the whole working life of the individual.

A different pattern is apparent for the nurses studied (Table V). Of twenty-two girls, all but five took an adequate amount of sleep even when on night duty, although the pattern of sleep was then disrupted. The five who slept less appeared to be girls who, judging from the study of their activities, had fairly busy social lives; for example, one girl managed to go to six dances in sixteen days. In such circumstances there is little wonder that they reduced their sleep ration. In contrast to the doctors, therefore, the nurses' sleep was more influenced by their off-duty pursuits than by their professional activities. There was wide variation in the amount of sleep taken by students. Eleven of the twenty-one slept less than 7·5 hr. However, there was little difference among individuals between sleep during holidays and University term, except in a small number (Tables VI and VII). Four students only showed much difference: in three this was still not great and varied between 0·9 and 1·1 hr, but the fourth student averaged 4·2 hr per night for forty-four days of term and 9·5 hr for twenty-six days on holidays. This is the

Table V. *Sleep Taken by Twenty-two Student Nurses (Days on night duty in parentheses)*

	Age	*Days*	*Mean* Night duty	Whole period
E.B.	22	102 (72)	6·7	7·1
M.B.	19	91 (10)	8·4	8·5
A.B.	19	101 (14)	8·5	9·4
E.C.	18	101 (0)	—	8·4
M.D.	20	102 (74)	6·6	7·1
N.E.	18	100 (20)	7·5	7·9
M.G.	19	42 (9)	7·5	7·1
J.G.	19	101 (11)	9·1	8·9
M.G.	20	101 (12)	7·4	8·0
E.H.	18	102 (65)	8·9	8·2
K.M.	19	89 (46)	8·1	7·9
T.M.	20	102 (0)	—	7·3
M.M.	22	103 (75)	7·8	7·8
D.P.	23	103 (73)	6·7	7·0
J.R.	18	102 (33)	8·4	8·9
R.S.	18	103 (0)	—	8·9
C.S.	20	103 (0)	—	8·7
C.S.	18	103 (25)	8·1	8·5
I.S.	18	104 (43)	7·8	7·9
J.T.	20	120 (0)	—	8·7
V.W.	19	101 (35)	8·2	8·1
G.W.	19	104 (36)	8·3	7·9
Mean for all subjects				8·1

Table VI. *Sleep Taken by Eleven Male Medical Students during Work, Holidays and for Whole Period (For all the subjects twenty-six days' holiday: number of working days in parentheses)*

	Days	*Work*	*Mean* Holiday	*Whole period*
D.A.	62 (36)	6·4	7·5	6·9
D.D.	71 (45)	8·2	8·7	8·4
G.D.	62 (36)	8·4	8·5	8.5
A.G.	62 (36)	7·5	7·5	7.5
A.H.	62 (36)	6·4	7·0	6·7
M.H.	70 (44)	6·7	7·0	6·8
A.I.	71 (45)	7·5	8·5	7·9
D.McL	62 (36)	7·9	8·4	8·1
G.R.	62 (36)	7·4	7·3	7·4
J.S.	71 (45)	6·7	7·1	6·8
F.W.	62 (36)	7·9	8·6	8·2
Mean for all subjects		7·4	7·8	7·6

Table VII. *Sleep Taken by Ten Female Medical Students during Work, Holidays and for Whole Period (For all the subjects twenty-six days' holiday: number of working days in parentheses)*

	Days	Mean Work	Mean Holiday	Mean Whole period
J.B.	70 (44)	7·2	7·0	7·2
S.B.	70 (44)	6·4	6·6	6·5
N.G.	62 (36)	7·4	7·8	7·5
P.G.	62 (36)	6·8	6·7	6·7
S.G.	61 (35)	7·6	7·2	7·4
J.J.	71 (45)	6·6	6.8	6·7
S.P.	62 (36)	7·3	8·3	7·6
D.R.	62 (36)	7·6	7·9	7·8
E.W.	70 (44)	4·2	9·5	6·2
U.W.	62 (36)	8·4	8·3	8·3
Mean for all subjects		6·9	7 6	7·2

only instance we have recorded of any substantial recovery of lost sleep, and even for this girl the holiday period increased the overall mean by only 2 hr. Clerks, male and female, recorded a satisfactory amount of sleep, 7·8 hr per 24 hr (Tables VIII and IX), although there was considerable variation among individuals, and five slept less than 7·5 hr per night, whereas one woman clerk averaged 9·5 hr over a period of forty-eight days. Neither age nor social activity appeared much to influence the amount of sleep in this small group of sixteen people. Tradesmen behaved similarly (Table X). There was wide variation among the ten studied, one 46-year-old man sleeping only 5·9 hr over the period of sixty days. However, the mean for all was 7·7 hr per night. Schoolboys and girls returned sleep cards for two periods (Tables XI and XII). One was for the last month of the school term, and the other was for a month of their summer vacation. The gross figures suggest that they took less sleep than might have been expected during both periods. During work it

Table VIII. *Sleep of a Group of Six Clerks (The mean for male and female clerks combined was* 7·8)

	Age	Days	Mean sleep
A.P.	30	61	6·9
B.P.	32	60	7·1
S.G.	37	61	8·1
C.K.	40	61	7·8
J.G.	50	62	7·3
M.S.	52	60	8·8
Mean for all subjects			7·7

Table IX. *Sleep of a Group of Ten Female Clerks*

	Age	*Days*	*Mean sleep*
A.C.	19	62	8·3
A.T.	19	48	9·5
M.A.	20	62	7·7
R.R.	22	52	7·9
J.W.	22	61	8·4
J.K.	24	60	7·9
U.M.	24	60	7·9
W.T.	28	61	7·8
F.R.	33	61	6·9
N.F.	36	62	6·9
Mean for all subjects			7·9

Table X. *Sleep of a Group of Ten Tradesmen*

	Age	*Days*	*Mean sleep*
J.K.	27	62	7·5
D.B.	32	23	8·3
J.B.	45	62	8·2
N.W.	47	62	7·7
E.G.	48	62	7·6
A.B.	49	16	7·4
J.C.	49	62	8·3
W.M.	49	60	5·9
J.G.	56	47	7·9
J.B.	58	60	8·2
Mean for all subjects			7·7

was not apparent their study that kept them out of bed, in spite of the fact that they were all engaged in a strenuous course aimed at fitting them for university and other higher education. Their range of extracurricular activities was wide, and during the holidays a number took casual employment as labourers or waitresses. Travel abroad was also popular. At this turbulent age many of them seemed to go to dances regularly, sometimes as often as once or twice a week, even during the school term. The mean sleep for the whole period for all schoolchildren was 8·0 hr per night (boys 7·9, girls 8·0), and the mean sleep for holidays was 8·0 hr per night (boys 8·0, girls 8·0).

C. Conclusions

These results, taken from various groups in the community, show that the doctors incurred a marked sleep deficit, which rarely seemed to be made up.

Medical students tended towards this state of affairs also, and many of them would be used to lack of sleep by the time they graduated. In others there was wide variation within the group; in the young, social factors may much modify the sleep pattern, but nevertheless the average values are such as to suggest that not much sleep is lost in the long run. It is both salutary and

Table XI. *Sleep of Twenty-eight Schoolboys during School and Holidays (Number of holiday days in parentheses. The mean sleeping times for boys and girls during school and holidays were* 8.0 *and* 8.0, *respectively)*

		Means	
	Days	*School*	*Holiday*
J.A.	31 (31)	7·1	8·0
J.B.	31 (31)	7·7	7·0
C.C.	31 (31)	7·3	7·3
J.D.	30 (31)	6·8	5·7
D.F.	19 (0)	9·1	—
R.F.	30 (0)	8·2	—
G.G.	30 (31)	8·0	8·0
M.G.	31 (0)	6·9	—
W.H.	30 (31)	8·2	8·8
A.K.	31 (28)	7·9	9·1
A.L.	30 (31)	8·1	8·4
J.L.	30 (31)	8·0	8·1
M.L.	31 (14)	8·4	9·3
J.McB.	31 (3)	8·2	8·7
G.McE.	30 (16)	8·6	9·4
D.McL.	30 (31)	8·4	9·1
H.M.	25 (0)	7·6	—
G.M.	30 (31)	8·4	8·7
E.M.	30 (0)	8·0	—
W.N.	30 (31)	7·8	7·4
G.P.	30 (31)	7·8	6·9
N.R.	31 (24)	7·3	7·9
D.S.	31 (32)	9·1	8·3
D.S.	30 (30)	7·5	8·1
J.S.	30 (15)	7·7	7·3
A.T.	31 (30)	8·5	8·3
G.T.	30 (30)	8·8	8·6
J.W.	30 (31)	8·2	7·9
Mean for all subjects		7·9	8·0

disturbing to find how little sleep some of the house surgeons in particular really get. Did they give as good service as they might have? Certainly, it was not uncommon during the course of this study to find these young men and women slumped at their desks, having completely succumbed to the demands of sleep. A phenomenon repeatedly noted by surgeons when they are operating is that, although they themselves may have been up all night, they fortunately never fall asleep, yet their equally tired assistants every now and again nod off while holding a retractor. The effect of boredom versus interest clearly plays a part, as is discussed more fully in the next chapter. If the jaded assistant half asleep or wholly somnambulistic at his post behind the retractor is asked to cross to the other side of the table and carry out the operation

Table XII. *Sleep of Twenty-seven Schoolgirls during School and Holidays (Number of holiday days in parentheses)*

		Mean	
	Days	*School*	*Holiday*
J.A.	30 (31)	8·0	8·5
P.B.	31 (31)	7·9	8·6
J.B.	19 (0)	8·6	—
A.C.	30 (31)	8·7	8·9
M.C.	31 (31)	8·4	8·7
I.C.	31 (24)	7·3	7·0
M.D.	31 (0)	6·9	—
M.D.	30 (31)	6·7	6·4
S.E.	30 (31)	8·1	8·3
H.F.	31 (31)	7·8	8·1
C.G.	29 (32)	8·4	8·2
R.G.	31 (31)	8·6	8·9
J.H.	31 (31)	8·3	8·4
P.H.	11 (0)	8·9	—
J.I.	30 (31)	8·5	8·7
P.McC.	30 (31)	7·6	7·9
A.McC.	30 (31)	7·2	7·4
W.McF.	30 (31)	8·2	8·4
M.McK.	30 (31)	8·5	8·4
S.McK.	30 (31)	7·4	7·0
J.McM.	29 (31)	9·0	9·0
K.McP.	30 (31)	6·6	7·6
A.M.	30 (31)	8·3	8·0
R.R.	30 (31)	8·3	7·8
S.S.	30 (31)	7·9	5·6
M.S.	30 (31)	8·1	7·6
P.W.	30 (31)	8·6	7·4
Mean for all subjects		8·0	8·0

himself, he immediately has a stimulus and is capable of performing the operation to the satisfaction of his supervisor. However, this ability to compensate should not be regarded as permission to continue a state of affairs that is clearly unsatisfactory and one that the medical profession must try to remedy.

Information is lacking on how much sleep the critically ill patient gets and how essential it is to his survival. Nowadays intensive patient care results in the acutely ill person being under close scrutiny all day for many days; there is thus an opportunity to study the sleep taken by patients with grossly disordered *milieu interieur*. A man recently under our care was struggling to survive with severe peritonitis and pulmonary insufficiency. He had a tracheostomy, and his respiration was being assisted with a mechanical respirator. A casual observer gained the impression that he could rarely be asleep, yet over eleven days his mean sleep as recorded by his special nurses at 15 min intervals was 7·2 hr per day. The pattern was grossly disrupted, but he did get more than was thought possible, though less than what may be necessary for smooth convalescence. However, we cannot be dogmatic about the latter, since the desiderata for speedy recovery from operation or illness are not known.

We are assailed in this age of pressure advertising with the dogma that sleeplessness is a hazard to health necessitating ingestion of vast numbers of different pills or potions to promote deep restful sleep. Modern man is coerced into believing that he must have 8 hr sleep or he will crack mentally and physically, and one wonders how many people have gone to their doctor because they have been persuaded into this state. There is no knowledge allowing us to say that sleep deficits have a deleterious effect on life expectancy or general well-being. The question remains unanswered, as does the other imponderable: how much benefit occurs from deep compared with that from light sleep for the same time.

REFERENCES

"Ciba Foundation Symposium on the Nature of Sleep" (1961). (G. E. W. Wolstenholme and M. O'Connor, eds.) J. and A. Churchill Ltd, London.

Gesell, A., and Amatruda, C. S. (1945). *In* "The Embryology of Behaviour", p. 156. Hamish Hamilton, New York.

Kleitman, N. (1963). *In* "Sleep and Wakefulness". University of Chicago Press, Chicago and London.

Kleitman, N., and Kleitman, H. (1953). *Science Monthly* New York, 76, 349.

Lewis, H. E., and Masterton, J. P. (1957). *Lancet* i, 1262.

Oswald, I. (1962). *In* "Sleep and Waking". Elsevier, Amsterdam.

CHAPTER 14

Sleep Deprivation

ROBERT T. WILKINSON

A. Introductory

This chapter will be confined to considering the effects of sleep deprivation, with particular reference to studies of the resulting biochemical, physiological and psychological changes. Occasionally more than one of these measures has been taken in the same study, but only rarely has any attempt been made to correlate them. This is unfortunate for the theorist, because the point that leaps from many of these pages is that such measures have a limited value when taken separately: the biochemical and physiological changes that occur in response to sleep deprivation may depend considerably on what the subject is doing when they are taken; correspondingly, performance may be difficult to interpret without knowledge of the bodily changes that accompany it. For present purposes, however, the situation is more fortunate, in that it permits presentation of most of the evidence under the three separate heads: first, the biochemical changes resulting from sleep deprivation; secondly, the physiological ones; and last, the changes in performance and behaviour.

B. Biochemical Changes

At the Ciba Foundation's symposium on the nature of sleep (Wilkinson, 1961a), Bremer asked the question, "Are there acceptable data concerning the biochemical aspects of sleep deprivation—I mean changes in the blood plasma and spinal fluid composition?" Wilkinson replied, "I know of no important biochemical changes which have been found as a result of loss of sleep." This chapter provides the opportunity to fill in the detail behind that answer and to show that two years later this statement still summarises the position, with but one possible exception. This is that changes in the specific activity of adenylic phosphate compounds in the blood have been found to accompany prolonged sleep deprivation (Luby, Frohman, Grisell, Lenzo and Gottlieb, 1960; Luby, Grisell, Frohman, Lees, Cohen and Gottlieb, 1962). The first of these papers was published at about the same time as the symposium was being held. The result is of particular interest, as it suggests that sleep deprivation may in some way be placing a strain on the energy transfer systems of the body.

1. *Blood constituents*

The main measure of agreement about changes in the constituents of the blood from sleep deprivation is among the negative findings, of which there are many. One promising positive result has been mentioned already, that of Luby and his associates on changes in the energy transfer systems. Of the remaining positive findings, most involve changes bearing upon the secretion of adrenocortical hormone. Unfortunately there is little agreement as to either the degree or even the direction of these changes.

A surprisingly large number of constituents of the blood appear to show no change even from prolonged sleep deprivation. These include blood sugar (Rakestraw and Whittier, 1923; Kleitman, 1923; Tyler, 1947), alkali reserve (Rakestraw and Whittier, 1923; Kleitman, 1923; Mangold, Sokoloff,

Connor, Kleinerman, Therman and Kety, 1955), haemoglobin (Kleitman, 1923; Mangold *et al.*, 1955) and creatinine (Kleitman, 1923; Tyler, 1947). Negative findings must always be interpreted cautiously, but as far as they go these results suggest that the body is well equipped to maintain homoeostasis when deprived of sleep, even for periods as long as 100 hrs.

Of the positive findings, those of Rakestraw and Whittier (1923) about a "possible increase" in lactic acid and phosphates in the blood after 30 hr without sleep are perhaps what might be expected if we regard a period of sleep deprivation as being one of prolonged muscular activity at the waking level. We might also expect sleep deprivation to cause an increase in the secretion of adrenocortical hormones, such as has been shown to occur in a number of other potentially threatening situations, for example, in aeroplane pilots carrying out test flights or instructional duties (Pincus and Hoagland, 1943). Five studies are relevant here. Two of them do indeed suggest such an increase. A fall in the level of reduced glutathione in the blood is an indicator of increased adrenocortical activity, and Susuki (1961) found this to occur in five out of five subjects during four to six days of curtailed sleep. A second indicator of such activity is the eosinophil count of the blood, and Koranyi and Lehman (1960) have reported a significant fall in this count over the first two days of sleep deprivation. When measures were continued into the third and fourth day of prolonged wakefulness in their study, however, the eosinophil count was found to return to normal levels. The implication is that adrenocortical activity does not necessarily increase monotonically with time spent awake; it appears rather to be a response that the body makes in the earlier stages of the vigil and abandons later. In three other studies adrenocortical activity has been examined by measuring the level of 17-hydroxycorticosteroids in the blood. This more specific measure gives less positive findings. Both Tyler, Marx and Goodman (1946) and Bliss, Clark and West (1959) found these levels to be unchanged during periods of sleep deprivation in excess of 70 hr. A third study by Murawski and Crabbe (1960) found the level of plasma 17-hydroxycorticosteroids to be lower than normal at 8 a.m. on the second day without sleep. This change is, it will be noted, in a direction opposite from that found by Susuki (1961) and by Koranyi and Lehman (1960) at the equivalent stage of moderate sleep deprivation.

It has often been reported that the process of staying awake for long periods is accompanied by considerable outward signs of effort, even if the subject is not being called upon to carry out any particular task. It is of special interest therefore that work has been undertaken recently on changes in the energy-transfer systems during the course of periods of prolonged wakefulness (Luby *et al.*, 1960, 1962). These may well be the most significant biochemical changes yet found to result from sleep deprivation. Ion exchange chromatography was used to isolate adenosine triphosphate

(ATP), diphosphate (ADP) and monophosphate (AMP) as well as fructose 6-phosphate (F1, 6P) from whole blood incubated for 1 hr with 0·05 mc of P 32. The levels and specific activity of these substances, which play a central part in the energy-transfer processes of the blood, were then determined. Analyses were performed on the blood withdrawn from subjects after normal sleep and on the fourth and seventh days of sleep deprivation (Luby *et al.*, 1960). On the fourth day awake "the specific activity of ATP rose markedly from 640,000 to 2,000,000 counts per milligram per minute. On the seventh day the specific activity of ATP had dropped to 670,000 counts/mg per minute. The specific activity of ADP and fructose 6-phosphate (F1, 6P) changed in the same direction. . . ." AMP (adenylic acid) is a by-product of an emergency procedure that provides additional quantities of ATP when presumably, normal supplies approach exhaustion. Normally AMP shows no specific activity, but by the fourth and seventh day of sleep deprivation its specific activity had risen to 200,000 and 120,000 counts/mg per min, respectively. The levels of these substances changed as shown in Fig. 1. The

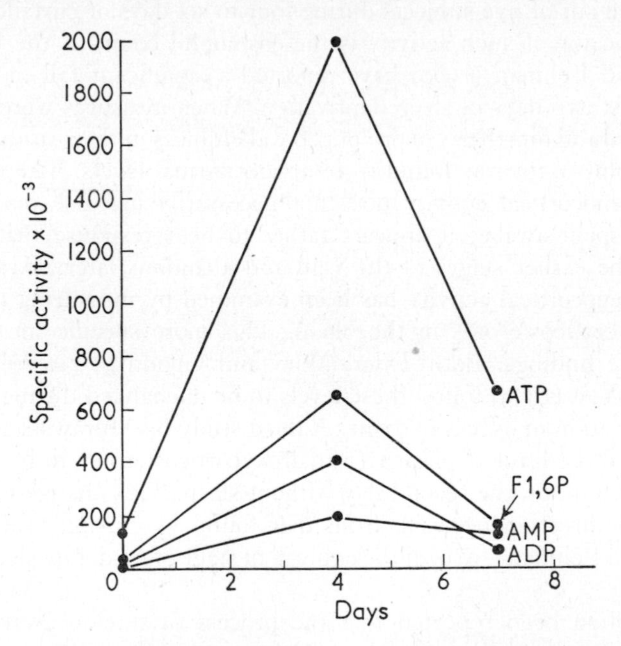

FIGURE 1. Specific activity of ATP, F1, 6P, AMP and ADP one hour after incubation with $^{32}PO_4$ during successive days of sleep deprivation. (From Luby *et al.*, 1960).

authors comment, "During the first half of this study" (that is, the first four days awake) "sleep deprivation apparently acted as a stressor on the chemical mechanisms producing energy. Energy production had increased by the fourth day, as evidenced by the change in specific activity of ATP and

F1, 6P during this period." The appearance of radioactive phosphorus in the AMP on the fourth day implies that neither the conventional Emden-Meyerhof or the tricarboxylic acid cycles were wholly responsible for the increase in specific activity of the ATP. Luby *et al.* (1960) suggest that the emergency methods involved synthesis of the AMP from adenine and either inorganic phosphate or the triphosphate. The fall in the rate of triphosphate synthesis and the return of the triphosphate : diphosphate level to normal in the seventh day of wakefulness "could well represent a failure of the mechanisms responsible for the increased synthesis of ATP." In the later study from the same laboratory (Luby *et al.*, 1962) these results were essentially reproduced in a group of twelve subjects who remained awake for 123 hr. Specific activity of the adenlyic phosphates (AMP, ADP, ATP) rose sharply and then fell back to normal levels. The only important difference was that this change, which took place at the 100 hr point of sleep deprivation in the earlier study, appeared at the 48 hr point in the later one. This result is reminiscent of the similar lack of monotonicity already noted in the results of Koranyi and Lehman (1960), in which the eosinophil count fell during the first half of the vigil and rose again during the last two days. The possibility that the two measures, specific activity of ATP and the level of blood eosinophils, show correlated changes under sleep deprivation may be worth investigating.

2. *Urine constituents*

Analysis of the urine during sleep deprivation has produced results no less contradictory and confusing than those derived from the analysis of the blood. The only real measure of agreement is on the levels of nitrogen, phosphate and phosphoric acid, all of which appear to be increased. Urinary catecholamine excretion which should reflect the degree of adrenal activity during sleeplessness is, in particular, a topic on which there is little agreement.

Considering the position in more detail, there are three studies that are unanimous, if not overconfident, in claiming an increase in urinary nitrogen, phosphates and phosphoric acid (Patrick and Gilbert, 1896; Kleitman, 1923; Rakestraw and Whittier, 1923). Although these are all early studies on few subjects and perhaps not based on the best of experimental techniques, the measure of agreement is impressive in an area where there is scant information.

Luby *et al.* (1960) report that "the urinary excretion of creatinine increased significantly during the period of sleep deprivation; uric acid decreased." These authors conclude that the fall in uric acid is what would be expected from the conservation of purines, including adenine, which is part of the body's emergency response in preserving the constant level of adenylic phosphates under adverse conditions. Rakestraw and Whittier (1923) found no such decrease in uric acid, but their 30 hr period of sleep deprivation would hardly constitute as severe a stress as the 200 hr wakefulness imposed by Luby *et al.* (1960). On the topic of creatinine excretion, however, the results of two studies disagree more seriously with the findings of Luby *et al.* (1960). Neither Kleitman (1923) nor Tyler (1947) observed any increase in urinary creatinine over periods of about 100 hrs wakefulness. About the level of chlorides and the total acidity of the urine under sleep deprivation there is further disagreement. Kleitman (1923) found a fall and an increase, respectively, in these constituents, whereas Rakestraw and Whittier (1923) reported no change, although their short period of deprivation may again have been the reason for this. To complete the list of negative findings, there appear to be no changes in the levels of ammonia and urea (Rakestraw and Whittier, 1923). Koranyi and Lehman (1960) kept six psychotic patients awake for 100 hr and yet found the urine to be, in general, normal, except for a possible increase in specific gravity.

Analysis of the urine can also provide evidence on adrenal activity to supplement that from the blood. Changes in the blood leaned slightly in favour of an increase in adrenal activity as a result of sleep deprivation. Urinary analysis, though still contradictory, leans about equally in favour of reduced adrenal activity. Murawski and Crabbe (1960) found a "trend toward lowered 17-hydroxycorticosteroids in the urine following sleep loss" (two days), which implies lowered adrenocortical activity. Tyler, Marx and Goodman (1946) analysed the urine for the same constituents and reported that "prolonged wakefulness" (100 hrs) "did not change the level of urinary ketosteroids." These results of Tyler and his associates appear, however, on closer examination to show a significant ($P < \cdot 02$) fall in the steroid output compared with control levels. This occurs over the second, third and fourth days of wakefulness. Again, reduced adrenocortical activity is implied. The third study on this question, by Bliss, Clark and West (1959), did not show any change in ketosteroid output; the subjects were seven medical students who remained awake for 72 hrs. Finally Hasselman, Schaff and Metz (1960) analysed the urine for adrenaline and noradrenaline, whose levels would reflect activity of the adrenal medulla. This study is particularly interesting in that it is the only one in which biochemical measures of this type have been taken under sleep deprivation and in different working situations. It was found that the excretion of these catecholamines was higher during work under sleep deprivation than during the same work after a normal night's sleep and that this trend was greater when the subjects' work was associated with physical activity and also when the work took place at a moderate rather than a high environmental temperature.

Perhaps the most important aspect of the work of Hasselman *et al.* (1960) is the emphasis it places on the situation of the subject in deciding the biochemical changes that result from sleep deprivation. The contradictory nature of many of the biochemical findings we have been considering may well have been due to this situational factor. In other words, the kind of biochemical changes occurring under sleep deprivation depend considerably upon what the subject is doing when the measures are taken and upon the environment in which he happens to be at the time.

C. Physiological Changes

At first sight the evidence about the physiological responses to sleep deprivation appears to be as contradictory as that already noted for the biochemical findings. The physiologists, however, provide rather more detail than the biochemists do about what may be called the behavioural content of these measurements. This turns out to be an important factor in helping us to understand the physiological responses, and this again makes it seem likely that the failure to control this factor may have been responsible for much of what is puzzling in the biochemical evidence. Briefly, it appears that not only the degree but even the direction of various physiological changes under sleep deprivation may vary according to what the subject is doing at the time of measurement, and in particular upon whether he is engaged in some activity, for example a performance test, or whether he is just resting. Most of the physiological measures taken during sleep deprivation fall into the class that may be loosely described as arousal indices. These will be considered first; a second miscellaneous section will be devoted to the other measurements made.

1. *Indices of bodily arousal or activation*

Malmo (1959) has described the concept of arousal as a new "neuropsychological dimension". It is thought to reflect the intensive as opposed to the directional aspect of behaviour and bodily function and, as such, to vary from a low point in the deepest stage of sleep to a maximum at the highest pitch of waking excitement. The concept is clearly highly relevant to our present concern with sleep deprivation, and it would be convenient indeed if we had at our disposal physiological "arousal measures" to show us how this state varies when loss of sleep occurs. Various physiological measures, some of them accepted indicators of autonomic activity, have indeed been proposed as arousal measures. These include heart rate, respiration rate, blood pressure, skin conductance, galvanic skin response (g.s.r.), muscle tension and the degree of activity in the α-range of the electroencephalogram. Unfortunately the very use of these so-called arousal measures in association with states of sleeplessness has helped to call into question their qualifications for the title. The experimental findings presented in detail below show that when a sleep-deprived subject is resting or performing a simple and not very demanding task the indices will reflect abnormally low arousal. But if he is in a situation that provokes a high level of anxiety or provides high incentive to react with normal efficiency the arousal measures may well indicate abnormally high arousal. Intuitively this seems an unlikely outcome of losing sleep, but this conclusion is unavoidable if these physiological indices are indeed reliable indicators of the level of arousal. An alternative hypothesis, however, is that in highly stimulating conditions, at least, these physiological indices are reflecting not abnormally high arousal but a high level of effort to maintain normal levels of arousal and therefore performance. The effort is perhaps necessary to counteract the influence of sleep deprivation, which on this hypothesis would be defined as being always in the direction of lowering the level of arousal. Bearing in mind this possibility, we can now consider in more detail how the various "indices of arousal" respond to sleep deprivation.

a. Heart rate. Of the seven studies in which heart rate has been recorded during sleep deprivation, four revealed no change, one showed a rise and two a fall. The best conclusion is that heart rate probably shows little variation with loss of sleep in most circumstances. The one study in which heart rate rose was that of Koranyi and Lehman (1960); it was found to be higher than normal after the second day of a 100 hr vigil. Patrick and Gilbert (1896) and Kleitman (1923) reported that heart rate fell over some 80 to 100 hr without sleep, the latter noting that no such change was observable if the subject was sitting instead of lying relaxed. To set against these positive but conflicting findings, there are four studies (Katz and Landis, 1935; Tyler, 1947; Malmo and Surwillo, 1960; Scholander, 1961), covering periods of 30 to 200 hr sleep deprivation, in which heart rate remained essentially normal.

b. Respiration rate. The evidence permits no firm conclusion on the effect of loss of sleep upon respiration rate, even if the conditions during measurement are taken into account. Two investigators report a decrease: Kleitman (1923), again with the subject lying prone but not with him sitting, and Scholander (1961) during the prestimulus interval of a g.s.r. trial. Two other groups have found an increase; Malmo and Surwillo (1960), in subjects who were performing a tracking task with pain as the punishment for errors, and Ax and Luby (1961) whose subjects, in contrast, were sitting quietly. These findings could not be much more equivocal, but it may well be that a simple rate measure may not be the best one to take. The variability of respiration rate is an obvious alternative that has not been examined. Another that has been considered is the ratio of sinus arhythmia to the long wave variations in heart rate. A fall in this ratio may represent merely the frequency with which the subject takes long breaths. All the same, this ratio has been found to fall markedly as wakefulness is continued up to 123 hr (Ax and Luby, 1961).

c. Blood pressure. It seems unlikely that there is any important alteration in blood pressure with sleep deprivation. Kleitman (1923), Katz and Landis (1935), Mangold *et al.* (1955) and Tyler (1947) are unanimous in reporting no change, except that Kleitman found, in the same way as he did with pulse rate, that a marked fall in blood pressure occurred during sleeplessness if the subject was lying prone instead of sitting.

d. Skin conductance. So far as they go, the two studies in which this measure has been taken suggest that the experimental situation may play an important part in deciding whether skin conductance rises or falls with sleep deprivation. In the active working situation of Malmo and Surwillo (1960) it rose; in the passive subjects of Ax and Luby (1961), on the other hand, it fell.

e. Galvanic skin response. The final autonomic measure included under this general heading of arousal indices is that of the galvanic skin response (g.s.r.) or electrodermal response, as some writers call it. As with all the other autonomic responses to sleep deprivation, there appears to be no simple quantitative change. Either the response is subject to too much moment to moment variation to show any consistent trend over a period of sleep deprivation, or the direction of its change may vary from study to study according to the conditions accompanying its measurement. Burch and Greiner (1958) distinguished between specific g.s.r. (responses in the level of skin conductance to a light electric shock to the foot) and the non-specific g.s.r. (conductance responses to no specific external stimuli). Figure 2 shows how the specific

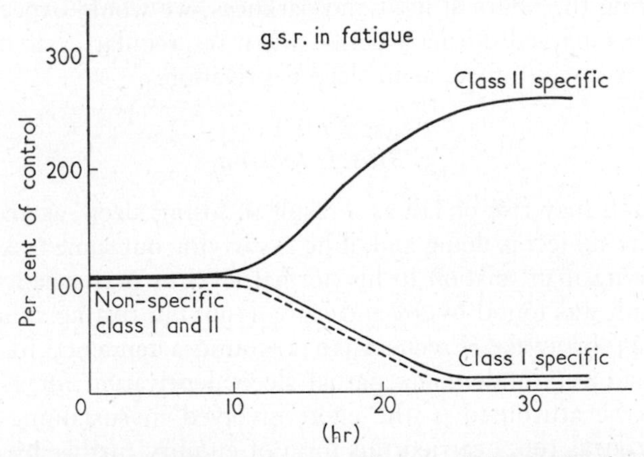

FIGURE 2. Changes in specific and non-specific g.s.r. during 30 hr sleep deprivation. (From Burch and Greiner, 1958.)

g.s.r. increased in magnitude during the course of 30 hr without sleep, whereas the non-specific g.s.r. decreased, its curve being almost the mirror image of the other. These results appear to show that the responsiveness of the sleep-deprived subject may be lower than normal when no external stimulation is being applied, but can be higher than normal when such stimulation is present. The results of Scholander (1961) appear to show the same thing. He presented bursts of "white noise" at the high sound intensity of 105 decibels and recorded the g.s.r. before and after these stimuli. The prestimulus g.s.r., which may correspond roughly with the nonspecific response of Burch and Greiner (1958), fell below normal levels towards the end of 30 hr without sleep. The amplitude of the g.s.r. to the stimulus (the equivalent, perhaps, of the specific response) was higher than normal under the same conditions.

2. *Body temperature*

Though it is agreed that the average level of body temperature falls during the course of a period of sleep deprivation, there is disagreement about whether the diurnal swings in temperature are accentuated or diminished. Kleitman (1923) reported a gradual damping of the diurnal swings over some 100 hr of sleeplessness; Murray, Williams and Lubin (1958) and Loveland and Williams (1963), on the other hand, found an increase in the amplitude of the swings. All three groups of investigators agree with Ax and Luby (1961) that a general fall in average body temperature occurs with loss of sleep. This fall in the average level is understandable as reflecting a gradually decreasing level of arousal, but the increase in amplitude of the diurnal cycle found in two of the later studies is surprising. If the diurnal rhythm is a learned response and conditioned by the alternation of activity and rest during the hours of light and darkness, we would expect this habit to become less marked during a period when the regular cycle of sleep and wakefulness is discontinued, as in sleep deprivation.

3. *Muscle tension*

Muscle tension may rise or fall as a result of losing sleep, again depending upon what the subject is doing and, if he is carrying out some task, upon how well he is doing it in relation to his normal level of performance. A fall in muscle tension was found by Ax and Luby (1961) in resting subjects over a period of 123 hr awake. Freeman (1932) found a tendency for tension to rise in two subjects undergoing partial sleep deprivation for a number of nights. This he attributed to the effort involved in sustaining the stress. Wilkinson (1961a, 1962) carried this form of enquiry further by correlating changes in muscle tension with those of performance under sleep deprivation. He found that tension rose in subjects who maintained normal levels of performance, but remained the same or fell in those whose performance declined. Malmo and Surwillo (1960) found different changes in tension in different muscles. All three of their subjects showed a fall in the pronator teres muscle, but in each subject there were muscles that showed a rise in tension. This result is difficult to interpret, especially in relation to the accompanying performance, in which two subjects showed a decline and one a rise in accuracy during the course of the 60 hr sleep deprivation.

4. *The electroencephalogram*

There is little doubt that the amount of activity at the α-range of frequencies (around 10 c/sec) in the electroencephalogram is reduced by sleep deprivation. From the point of view of the level of arousal under sleep deprivation, however, this is little help to us, because a reduction in α-activity (or α-depression) can indicate a change in the level of arousal either towards or away from sleep; unless a record of frank sleep appears subsequently, there is no way of knowing which has occurred. As a result, although five studies have revealed α-depression during sleeplessness (Blake and Gerard, 1937; Tyler, Goodman and Rothman, 1947; Bjerner, 1949; Armington and Mitnick, 1959; Rodin, Luby and Gottlieb, 1962), those authors who attempt an interpretation have to make it in the light of other observations accompanying the depression. These include the incidental observation of the subject, the evidence of other physiological indices of arousal (Malmo and Surwillo, 1960) and performance measures (Bjerner, 1949). Thus α-depression, unless it can be clearly shown to be leading towards a stage of sleep (which is difficult), adds nothing to the information that other observations can provide about the level of arousal when sleep has been lost. Nevertheless, the firm establishment of α-depression during sleep deprivation is of considerable interest, even if the interpretation of it is difficult at present. One possibility is that when sleep has been lost the state corresponding to α-activity in the electroencephalogram may be an unstable one that cannot persist for long. A situation conducive to the presence of α-activity may be one in which the sleep-deprived man cannot stay awake without some effort. If he succeeds in staying awake, the effect, as Tyler *et al.* (1947) have suggested, may be to replace α-activity by a fast activated record. If he fails, α-activity may give way to patterns more characteristic of sleep. Finally, on this topic of α-activity, Rodin *et al.* (1962) noted that in six of their sixteen subjects the effect of sleep deprivation was to cause α-activity in the waking state with eyes open, which is unusual in normal subjects. This may imply that a level of visual stimulation sufficient to maintain arousal above that corresponding to the presence of α-activity in the normal subject may not be sufficient to do so in the sleep-deprived one. The implication is that when sleep has been lost there is, at some point in the system, an attenuation of, perhaps, non-specific stimulation provided by this visual input.

The final point about the electroencephalogram under sleep deprivation does not involve the α-rhythm. Rodin *et al.* (1962) noted an interesting change in pattern in five out of their sixteen subjects. This consisted of the presence of high-voltage diffuse paroxysmal activity of about 1 sec duration and occurring four to ten times in the 15 min recording. Only one of these subjects had shown similar paroxysms in the pre-test records. Three others had slightly dysrhythmic but not paroxysmal resting electroencephalograms, and the fifth had a low-voltage normal resting record. The paroxysms were most pronounced in the 24 hr and 48 hr periods and decreased subsequently as sleeplessness progressed. "This activity was not related to the degree of drowsiness because it was absent at the 96 and 120 hr levels, when the subjects were markedly more drowsy. The recordings seemed similar to those seen in certain patients with centrencephalic disorders." The authors comment further, "This would suggest that sleep loss is initially associated with increased excitability of the brain which may manifest itself in certain predisposed individuals in epileptic-like phenomena. . . . One would expect therefore that epileptic patients might be more susceptible to loss of sleep than a normal individual."

5. *Body weight*

Body weight appears either to rise or to remain constant as a result of sleep deprivation. In four reported studies there was a rise (Patrick and Gilbert, 1896; Kleitman, 1923; Katz and Landis, 1935; Koranyi and Lehman, 1960), and in all of them the vigil was over 80 hr. There is one study, however, of a similar period in which no change occurred (Tyler, 1947); it differed from the rest in that more physical exercise was undertaken during the vigil.

6. *Pain threshold*

What evidence there is on this topic suggests that the threshold for pain tolerance is lowered by sleep deprivation. Kleitman, Cooperman and Mullin (1934) measured pain threshold by means of the sets of hairs of Von Frey (1926) and found it to be reduced in all of the six subjects they kept awake for 60 hrs. Malmo and Surwillo (1960) used a modified Hardy-Wolff pain threshold apparatus, which inflicts pain by focusing heat upon the forehead. In all three subjects "the mean pain threshold for the last day of vigil was lower than the mean pain threshold for determination made prior to the last day, and following sleep after the vigil." Schumacher, Goodell, Hardy and Wolff (1940) failed to find a similar lowering of the threshold, using basically the same methods as Malmo and Surwillo (1960), but this was probably because the period of wakefulness was only 24 hrs. Finally Ax and Luby (1961) inflicted heat pain but have not reported the tolerance levels as a function of loss of sleep. They did, however, measure the change in hand conductance and diastolic blood pressure to standard pain stimuli and found these changes to be reduced when sleep was lost. Whether this implies higher or lower tolerance is hard to say!

The finding of a lowered threshold for pain has been interpreted by Malmo and Surwillo (1960) as further evidence for their contention that sleep deprivation increases the level of activation or arousal. The argument rests upon Hebb's (1949) suggestion that pain intensity will increase with the quantity of high-level central nervous discharge. This is a possibility, but a simpler hypothesis would be in terms of the effect of sleep deprivation upon the level of motivation, a topic elaborated in section D below, on performance changes. It is simply that the sleep-deprived person may be less motivated to withstand a high intensity of pain than a normal one and therefore cries halt sooner. If the achievement of a high threshold were rewarded substantially, it might be that sleep deprivation would have little effect upon the tolerance level.

D. Changes in Performance and General Behaviour

We shall here consider the response of different performance measures to sleep deprivation and also how this response varies with other factors of the situation. The evidence will show what the biochemical and physiological evidence has only hinted at, namely, that the impact of sleeplessness is highly sensitive to such independent influences and that these have to be controlled carefully if any attempt is to be made at comparing the effects of loss of sleep upon different aspects of performance.

Important effects of sleep deprivation are to be observed in psychological fields outside those of formal performance tests, namely, those of social interaction, perceptual disorientation and psychopathological symptomatology at various degrees of severity. Although much of this evidence is based upon casual observation, it is still considered a necessary part of this survey.

1. *Effect on performance*

We now have considerable evidence, contributed to and crystallised by the Walter Reed group of workers (Williams, Lubin and Goodnow, 1959), that the way in which sleep deprivation affects performance is by imposing periodic lapses in efficiency interspersed with periods of normal or near normal functioning. The organism falters, recovers and runs normally for a time, falters, recovers and so on. Thus the effect on simple reaction time is to lengthen greatly the tail of the distribution of response times, while having little effect on most of the responses at the short end of the distribution. The same effect can be seen in a test of serial reaction in which the subject works at his own speed. An example of such a task is the five-choice test of serial reaction (Fig. 3) of Leonard (1959), which has been used extensively in studies of stress at the Applied Psychology Research Unit, Cambridge. The subject uses a stylus to tap one of five discs, corresponding to whichever of five lights is on. This extinguishes the light and brings another on; the appropriate disc is again tapped, and so the cycle continues with the subject responding at his own speed for about 30 min. On the day after one night

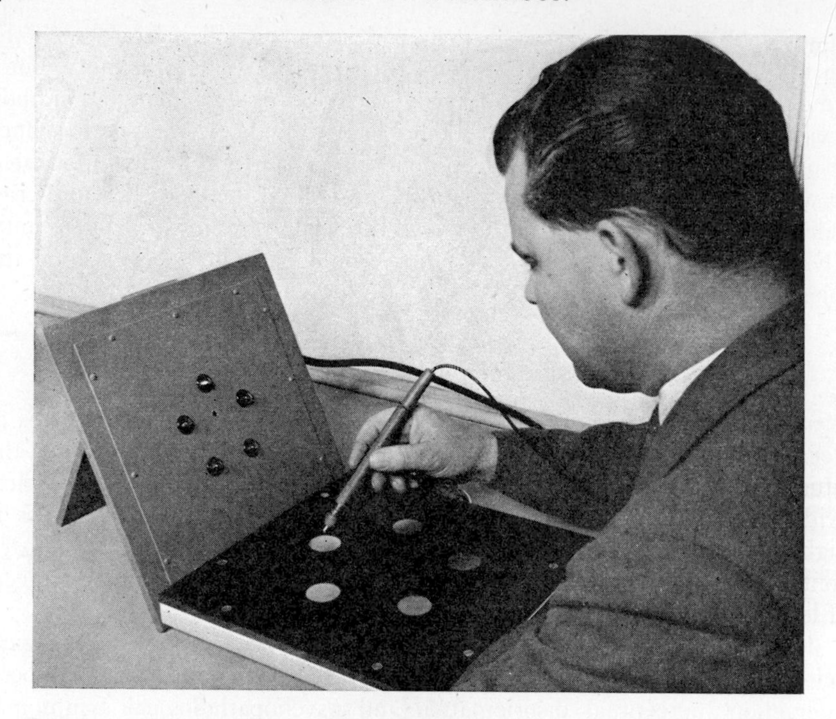

FIGURE 3. The five-choice test of serial reaction.

without sleep the number of correct taps is reduced. But the clearest effect is on what are called "gaps" or long pauses of $1\frac{1}{2}$ sec or more between responses. This is shown in Fig. 4. These again are the periodic lapses in the efficient functioning of the sleep-deprived man. Errors are scored when the wrong disc is tapped, the light still moving on. It was thought that these were little affected by sleep deprivation, but when analysed as a function of the total number of responses made, errors appeared almost as sensitive a measure as gaps. This was especially noticeable when subjects were well trained in the task. Presumably such subjects were responding more automatically and made errors when lapses occurred, instead of pausing as less practised subjects might do. A similar thing, of course, must happen if the task is a paced one, when errors of omission or commission will be the pattern of inefficiency under sleep deprivation.

2. *Effect of time*

Impairment of performance under sleep deprivation correlates with time spent awake (Williams, Lubin and Goodnow, 1959), but the relationship is no simple monotonic one. The force of the stress seems to wax and wane from hour to hour; in particular the impact is greater at night than during the day, so that performance on the first night may be worse than on the second day of wakefulness, that of the second night worse than on the third day, and so on. These fluctuations, then, are superimposed upon a steadily rising level of impairment up to a stage of about 100 hr without sleep. Beyond this point little in the way of sustained performance seems possible for most people.

In any study of performance under sleep deprivation the duration of the task is of great importance. Figure 4 shows clearly how the effect of loss of sleep

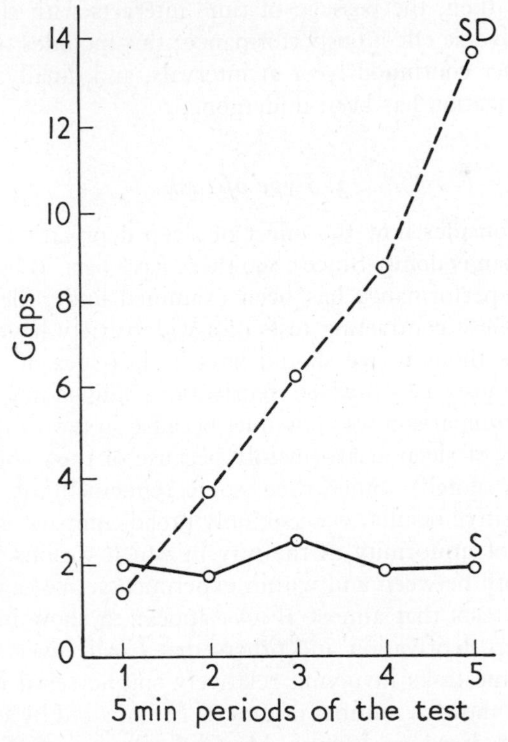

FIGURE 4. Effect of 30 hr sleep deprivation upon the score of gaps in the five-choice test of serial reaction. SD = sleep deprived; S = normal sleep.

increases sharply with time spent continuously at work and that even with tests as sensitive as those of serial reaction there appears to be no impairment during the first 5 min of work. This, we may note in passing, is probably the main reason why hardly any of the early workers in the field (they date back to Patrick and Gilbert in 1896) were able to show a clear effect of sleep deprivation on any task. Their tests were almost all of short duration. This implies that tasks requiring only intermittent activity will be little affected by the stress, although it should be noted here that Wilkinson (1959) found that rest pauses as short as 30 sec every 5 min in the test of serial reaction failed to circumvent the impairment seen in Fig 4.

Time is important in two other respects also. We know that, if the procedure of keeping subjects awake for one night and giving them the test of serial reaction the next day is repeated once a week for six weeks, the effects on performance increase progressively over the six weeks (Wilkinson, 1961b). This suggests that, with repeated administration of the test and of the experimental sleep deprivation, both lose novelty, and the effect of both is to increase the impact of sleeplessness on the test.

In summary, then, the passage of time interacts with sleep deprivation to increase its adverse effect on performance; this includes time awake, time on the task, either continuously or at intervals, and, finally, the number of times sleep deprivation has been undergone.

3. *Type of task*

We must now consider how the effect of sleep deprivation varies with the type of work a man is doing. Since 1896 there have been at least forty experiments in which performance has been examined under sleep deprivation, and in most of these contrasting tests of a wide variety have been used. By now, it might be thought, we should have a clear idea of the comparative vulnerability of most tasks to sleep deprivation. This is not so. Before 1940 little by way of comparison was possible, because so few of the tests showed any effect of loss of sleep at all (mainly because of their short duration, as we have already noted). Since then, some refinement of techniques has yielded more positive results; even so, only broad comparisons can be made, because of lack of uniformity in the way in which various tests have been administered, both between and within experiments. We can, however, distinguish certain tests that almost always appear to show impaired performance from sleep deprivation and others that hardly ever show it. In the first category come tasks involving relatively simple serial choice reaction. These have been shown a number of time to be impaired by as little as 30 hr without sleep (Ax, Fordyce, Loovas, Meredith, Pirojnikoff, Shmavonian and Wendahl, 1954; Wilkinson, 1958; 1959; 1961b; 1963a; 1964; Pepler, 1959; Corcoran, 1962). The five-choice test outlined above is a typical laboratory example of this type of task; others are crossing out all e's in a passage of prose, naming rows of different coloured squares, sorting cards and so on. Practical examples might include simple sorting operations of various kinds, the recording of coded messages and even copy typing.

Vigilance tasks are probably as sensitive as those of serial reaction, although there have been fewer demonstrations of their vulnerability (Ax *et al.*, 1954; Wilkinson, 1960; 1964). Tests of this sort typically require the subject to watch for an occasional small change in a relatively unchanging display; practical examples include inspection work in industry and look-out duties of various types. A third test which shows impairment after only 30 hr without sleep is pursuit-meter tracking (Ax *et al.*, 1954; Pepler, 1959). Here typically the subject has to follow the irregular movement of a target or pointer with a similar marker under his own control. Some tests of intermediate vulnerability may be noted that, though producing generally no effect after 30 hr sleep deprivation, begin to show impairment at about the 60 hr stage. These are adding or other simple calculations (Wilkinson, 1962; Loveland and Williams, 1963) and some tests of learning and relatively simple communication (Schein, 1957; Kornetsky, Mirsky, Kessler, and Dorff, 1959; Williams *et al.*, 1959). At the other end of the scale of vulnerability there are three general categories of test that appear rarely to show any significant effects of sleep deprivation even when this is prolonged for 60 to 100 hr: they involve problem solving, concept formation and complex decision processes of various kinds (Ax *et al.*, 1954; Williams *et al.*, 1959; Corcoran, 1964; Wilkinson, 1964).

We have already seen how the duration of the task can influence greatly the overall impairment of its performance. The main reason why it has been difficult even in recent years to make more precise comparisons of the vulnerability of various kinds of tasks is that, even within experiments, this factor has not been maintained constant. In a recent experiment, however, involving a 60 hr sleep deprivation task, duration has been controlled more adequately (Wilkinson, 1964). The design was such that performance in all the tests could be compared over a period of 20 to 30 min work, and the battery of tests included those of known vulnerability, such as serial reaction, to serve as a baseline for assessing the vulnerability of the others. Table I shows the tests used, arranged in the approximate order in which they were impaired on the second night or third day of wakefulness. These results illustrate one of the most important features of sleep deprivation: its effect depends crucially upon the kind of task being carried out. There is a wide range of the effect of loss of sleep, from almost complete impairment of function towards the end of 20 min vigilance or choice reaction to no effect at all in hectic complex decision taking, rote learning or a competitive game.

4. *The vulnerable task*

What, then, are the features of a task that make its performance sensitive to loss of sleep?

a. Low interest or incentive. When looking down the line of tests in Table I the first point to note is that the tests impaired most were the uninteresting ones. Confirming this impression, the subjects of this study were asked to rate the tasks in terms of how interesting they found them and

Table I. *Effect of up to 60 hr Sleep Deprivation on Various Tests and Scored Games, All of 20 to 30 Min Duration (Level of performance without sleep is expressed as a percentage of the control level of performance with normal sleep)*

Test	*Level of performance without sleep %* (*Control level* = 100%)
Serial reaction (gaps)	7
Vigilance (signals seen) whole test	34
Vigilance (signals seen) last half of test	4
Coded decision-taking (errors)	45
Serial reaction (errors)	47
Chess	51
Card sorting (errors)	60
Card sorting (speed)	76
Serial reaction (correct responses)	77
Darts	97
Rote learning (errors)	100
Table tennis	100
Tactical decision-taking (errors)	100

the order correlated well with that of vulnerability—the more interesting the less vulnerable. Although interest is a vague term, it implies that the carrying out of the test is rewarding in some way for the subject, that the test provides the subject with incentive to work. Fortunately we have independent and more precise evidence of the importance of this factor of incentive (Wilkinson, 1961b; 1964). Two versions of the serial reaction test described earlier were used, one with high feedback of knowledge of results and one with no such feedback (which is the normal form of the test). Figure 5 shows how with no knowledge of results the usual large effect of loss of sleep occurred. In the version of the test with knowledge of results this effect of sleep deprivation was reduced almost to zero. A similar effect has been shown to occur in a test of vigilance (Wilkinson, 1964). Apparently the ability of the task to evoke interest is of prime importance. We might put it this way: if a task provides clear and short-term goals the man has little difficulty in resisting the effects of at least moderate loss of sleep and performs with normal efficiency. If the goals are vague and distant, however, loss of sleep makes it increasingly difficult to maintain normal standards of performance. In other words, sleep deprivation appears to attack one of the basic abilities of the highest organisms—the capacity to neglect proximal goals in favour of distant but ultimately more rewarding ones.

b. Complexity. Again, on looking down the line of tests in Table I it appears that the effect of loss of sleep was least in the more complex tests. Certainly the tactical taking of decisions and the test of rote learning appear to involve more complexity than watching a screen for an occasional faint light or making a series of five simple responses. But we must look more closely at this matter of complexity. In the card-sorting test of this experiment complexity was varied on its own, much as was incentive in the experiment previously described. There were two categories of sorting, a four-choice version (where sorting was into suits) and a ten-choice version (where the cards were sorted into the numbers 1 to 10, court cards being

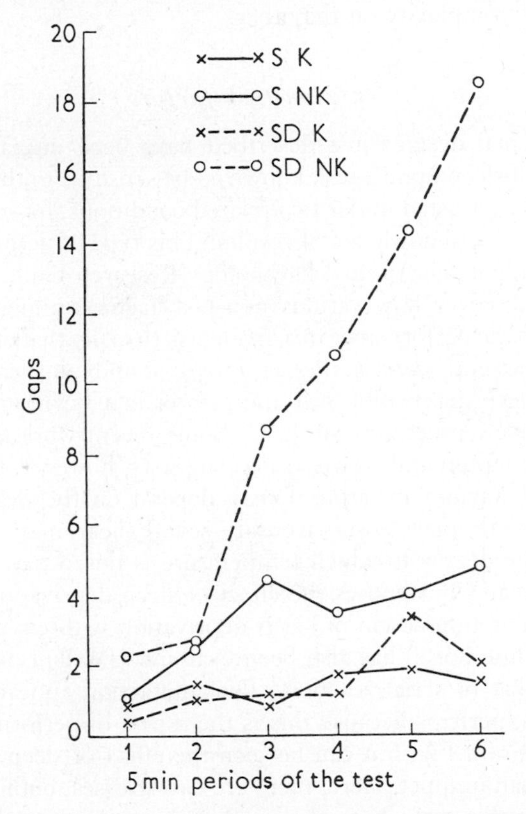

FIGURE 5. Effect of 30 hr sleep deprivation on the five-choice test of serial reaction, with and without knowledge of results. SD = sleep deprived; S = normal sleep; K = knowledge of results; NK = no knowledge of results. (From Wilkinson, 1961b.)

omitted). The effect of loss of sleep was greater in the ten-choice version. In other words, contrary to the impression gained from examining the battery of tests in Table I, the effect of sleep deprivation was greater in the more complex test. Why? The clue may lie in the interest reports of the subjects; they reported no preference on this count for either version of the test. The implication therefore is that, if interest remains constant, it will be the more complex form of test that will be the more vulnerable to sleep deprivation.

If, however, a task is complex in a way that makes it more interesting than a simple one (tactical decision as compared with, say, serial reaction), the added incentive may more than compensate for the deleterious influence of the complexity itself, so that the task may be one of those least affected by sleep deprivation.

It is suggested, then, that there are two independent features that make a task vulnerable to sleep deprivation: lack of incentive and interest, on the one hand, and complexity on the other.

5. *Non-task factors*

The experimental designs just described have been interaction ones. The effect of lack of sleep upon a task known to be sensitive to the stress has been examined and contrasted under two related conditions, for example with and without feedback of knowledge of results. This type of design has been used extensively both at the Applied Psychology Research Unit, Cambridge, and elsewhere to discover how various non-task factors influence the effect of loss of sleep on the performance of a given task (usually that of serial reaction).

a. Environmental warmth. Pepler (1959) found no clear evidence that the effect of sleep deprivation was any greater in a warm environment, that is at an effective temperature of 30·5°. Some recent work (Wilkinson, Fox, Goldsmith, Hampton and Lewis, 1964) suggests, however, that the effect of environmental warmth on arousal may depend on the degree of warmth. Moderate warmth may lower arousal; severe heat may elevate it. This implies that the degree to which temperature is raised may decide whether warmth potentiates or opposes the effect of sleep deprivation.

b. Noise. The interaction of sleep deprivation with environmental noise (continuous white noise) has also been examined (Wilkinson, 1963a). Again the test was that of serial reaction. The interaction appeared most clearly in error scores, perhaps because this is the aspect of performance that noise attacks most. From Fig. 6 it can be seen the effect of sleep deprivation was less in noise than in quiet. Here, then, are two stresses, both of which impair performance in this task, but whose effects cancel when they are applied together.

c. Physical exercise. A third influence, in a sense environmental, is that of physical exercise either accompanying or preceding performance of a task. Hasselman, Schaff and Metz (1960) showed that the effect of loss of sleep reaction was reduced if the subjects were simultaneously exercising on a bicycle ergometer. Lybrand, Andrews and Ross (1954) found similarly that physical exercise before the task reduced the effect of sleep deprivation on it. In studies of sleep deprivation it has often been observed incidentally that the best way to revive a flagging subject is to take him for a walk or otherwise exercise him physically. The two findings mentioned above confirm the importance of maintaining a reasonable level of muscular activity to counteract the influence of loss of sleep.

d. Drugs. Under the head of the internal environment we may, perhaps, consider how various drugs may modify the effect of sleeplessness. Colquhoun and Wilkinson (unpublished results) have found, in effect, that low doses

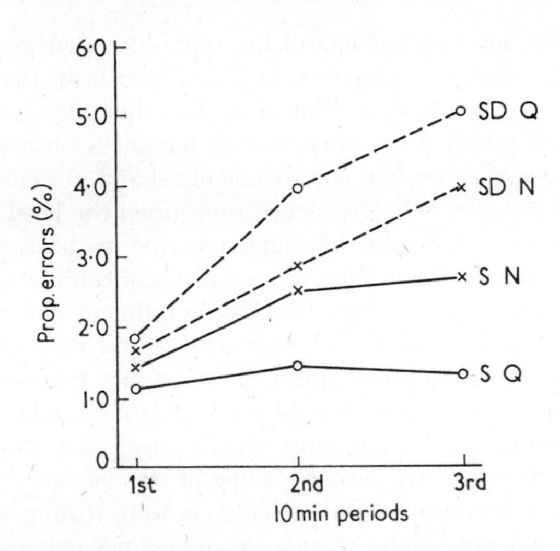

FIGURE 6. Effect of 30 hr sleep deprivation on the five-choice test, in noise (100 decibels "white" noise, open field) and in quiet conditions. Scores are in errors, expressed as a proportion of total responses made. SD = sleep deprived; S = normal sleep; N = noise; Q = quiet. (From Wilkinson, 1963a.)

of alcohol reduce the effect of the stress on the serial reaction task. High doses, on the other hand, potentiate it. Another drug examined was hyoscine (Wilkinson and Colquhoun, unpublished results), which, being an anti-emetic, may often in practice be administered to fatigued men. Although this drug impairs serial reaction as much as sleep deprivation does, there is no interaction; the effects of hyoscine and loss of sleep together are merely additive.

So far, however, the most effective drug for combating loss of sleep appears to be amphetamine. There have been several demonstrations of this (Tyler, 1947; Seashore and Ivy, 1953; Laties, 1961); the clearest, perhaps, was that of Kornetsky, Mirsky, Kessler, and Dorff (1959). The tests used were mainly of the simple serial reaction type. On most of them performance after 68 hr without sleep was returned to normal by 15 mg D-amphetamine. What was not made clear in this study, however, was the extent to which the same dose could improve performance on these tasks under conditions of normal sleep; this would have to be less than the improvement under sleep deprivation for an interaction to be demonstrated. Nevertheless it is clear that

D-amphetamine can oppose the influence of the stress, although it should be noted that this ability may decline with repeated administration of the drug during a vigil (Tyler, 1947).

6. *Level of arousal and "effort"*

Results of interaction experiments of the type discussed give us a modest over-view of the kind of influences, task and non-task, that potentiate or oppose the effect of loss of sleep. But more than this, they can provide some evidence derived purely from performance measures on a question raised in the previous sections dealing with biochemical and physiological changes, namely, that of the effect of sleep deprivation upon the level of arousal. We see from the interaction results that influences one might expect to raise the level of arousal, namely, knowledge of results, amphetamine and noise, seem to oppose the effect of loss of sleep. Depressing influences, on the other hand, such as large doses of alcohol, lack of incentive and time spent on an uninteresting test, appear to potentiate it. This implies that sleep deprivation itself has a depressing influence on the level of arousal, which was the conclusion that seemed to follow most clearly from the physiological and biochemical findings. To recapitulate, some of this evidence showed, it will be recalled that when sleep-deprived subjects were resting or carrying out some task that did not inspire them to maintain normal standards of performance, physiological measures such as heart and respiration rate, muscle tension, skin conductance and so on were lower than normal. This suggested a lowered state of arousal. When, however, performance was maintained, as we have seen it can be in stimulating situations, abnormally high levels of these measures were recorded. They were thought to reflect not heightened arousal but an increased expenditure of energy or effort necessary for this accomplishment in face of sleep deprivation. This implies that the stress may have two adverse effects—lowered performance or, to the extent that this is avoided, increased expenditure of effort and strain upon the energy-transfer systems. If this hypothesis is correct, measures both of performance and of physiological and biochemical changes must be made if we are to assess the impact of loss of sleep in any absolute sense. This point is particularly relevant to the next matter considered, namely, individual differences in susceptibility to the effects of loss of sleep.

7. *Individual differences*

Individuals differ considerably in the degree to which their performance is impaired by loss of sleep. This may not, however, be a sufficient criterion for assessing their intrinsic ability to withstand the effects of sleep deprivation, if some aspect of physiological cost as well as behavioural output has to be taken into consideration. A form of combined behavioural and physiological assessment of the individual response would seem to be needed, and this does not seem possible at our present state of knowledge. The best substitute may be a setting in which subjects are highly motivated to carry out some task continuously "until they drop". A study by Cappon and Banks (1960) may fall into this category. Their subjects were competitors in a staged "talkathon", in which wakefulness and a continuous flow of conversation had to be maintained for as long as possible, which turned out to be about 80 hr. The measure of ability to withstand the effects of sleep deprivation was simply the number of hours they kept going. This was correlated with various independent measures. The following "tentative conclusions" were reached: "The more neurotic the person, in terms of the Maudsley Medical Questionnaire, and the higher the nervous level of the subject, as measured by the Cornell Medical Index—Nervous, the higher the manifest anxiety level of the subject, as measured by the Taylor Manifest Anxiety Scale, and the lower the intelligence level, as measured by the Raven Progressive Matrices, the lower the subject's ability to withstand prolonged sleep deprivation."

For the rest we have to consider performance measures alone and accept that these may be reflecting not only the subject's capacity to function efficiently but also his willingness to make the effort to do so. Changes in performance due to loss of sleep do indeed vary widely from person to person, but attempts to correlate these differences with individual measures of personality or intelligence have, in general, met with little success. Loveland and Singer (1959) found that after 80 hr without sleep their subjects showed no characteristic change in their responses to Rorschach cards; by using the Rorschach scores alone it was impossible to predict the extent to which any man deprived of sleep would maintain performance on a wide variety of tasks. In a series of experiments over the last few years at the Applied Psychology Research Unit, Cambridge (many of which are referred to elsewhere in this chapter), we have recorded the effects of one night's loss of sleep on the performance of a large number of men in the five-choice serial reaction test. It is a matter of routine in this laboratory to take measurements on all subjects of their introversion-extroversion, neuroticism and intelligence. There appears to be no relationship between any of these scores and the individual tendency to show impaired performance with loss of sleep in five-choice test (Wilkinson, unpublished findings). A possible exception to this generally negative picture is a finding by Corcoran (1965) that some 60 hr sleep deprivation impaired the performance of extroverts more than that of introverts in a test of pursuit-tracking with blasts of loud noise as a punishment for errors. It is of interest here that the test concerned was, in contrast to the five-choice test, a rather tense and anxiety-provoking one.

8. *After-effects*

At the end of a period of prolonged wakefulness, subjects typically sleep for about 9 to 12 hr. Quantitatively, of course, this fails to make up for the total amount of sleep lost. The question arises, therefore, whether performance is likely to be still impaired after this restorative sleep. In a number of studies of the direct effects of sleep deprivation, performance measures have been continued into these days of recovery, but the results have been equivocal, some authors reporting an after-effect (Edwards, 1941; Smith, 1916; Williams, Granda, Jones, Lubin and Armington, 1962) and some not (Robinson and Hermann, 1922; Tyler, 1955). This ambiguity is probably due to the difficulties encountered in measuring the after-effect as a part of any experiment concerned chiefly with the assessment of performance during the period of sleeplessness itself. In a study designed exclusively to measure the after-effect, however, Wilkinson (1963b) found performance of both the five-choice serial reaction test and a test of vigilance to be significantly impaired on the day after the restorative sleep that terminated a 30 hr period of sleep deprivation. Two peculiarities of the results, however, raised doubts whether this result was due to a carry-over of the effect of sleep deprivation itself. For one thing, in both tests the effect appeared on the morning of the recovery day and had gone by the afternoon; secondly, the way in which performance was impaired was not typical of that occurring during the course of sleep deprivation. It was suggested, therefore, that, although there is an after-effect of sleep deprivation, this is due not to a failure of restorative sleep to make good the sleep that has been lost, but possibly to the disturbance of various bodily rhythms that the loss of sleep may have caused. A swift recovery from the effects of losing sleep may therefore turn not so much on the taking of additional sleep as on rapid resumption of normal diurnal patterns of sleep and activity.

9. *Partial deprivation*

Almost all that has been said so far has concerned the effects of acute sleep deprivation, that is, periods during which sleep is withheld completely. It might with justice be objected that this condition rarely occurs in practice; opportunities usually exist for snatching an hour or two of sleep from time to time even during the most prolonged and arduous of exercises. Why, then, is partial sleep deprivation not more fully represented in the literature? To my own knowledge there have only been four such studies. That of Susuki (1961) included no measures of performance and has been mentioned earlier under biochemical changes. Three others, by Smith (1916), Husband (1935) and Freeman (1932), could find no significant changes in performance due to curtailing sleep by more than 50% for periods of a week or more. Two of these experiments, however, employed only one subject and the other, Freeman's, only two. It can thus be seen that the topic has not been examined on the scale necessary to give reliable results. This is presumably because of the considerable resources required to keep a continuous watch over, as well as to test, a reasonably large sample of subjects for periods of a week or more.

10. *General behavioural changes*

A circumstance that must have irritated some of the early workers in the field of sleep deprivation was that, though singularly unsuccessful in registering any significant changes in performance on the part of their sleep-deprived subjects, they had only to observe them incidentally during the periods when no formal tests were being taken to see abundant evidence for a disorganisation of behaviour. They might have been excused the aggrieved feeling that only when nothing was being measured was there anything to measure. We now know that this was no coincidence. When formal tests are of short duration (as most of the earlier ones were), the mere placing of a sleepless man in such a situation appears to be enough to restore him to normal for the duration of the test. Before this was realised, however, it was thought that the more incidentally observed behavioural changes constituted the most important effects of sleep deprivation. Kleitman (1939), for example, says, "The increased sensitivity to pain, impairment of disposition, tendency to hallucinations, and other signs of this character are, in my opinion, the most outstanding and significant findings in all the studies of lack of sleep." Although we now know that performance, too, may be greatly modified by sleep deprivation, what Kleitman says could still be true. What is unfortunate is that most of the evidence for the occurrence of the phenomena he mentions is still based on incidental and subjective reports. Recently, however, there have been attempts to make more formal measurements of these phenomena, and they will be our main concern here. Their discussion will be organised under the three heads, social behaviour, perceptual anomalies and psychotic tendencies.

a. Social behaviour. The effect of sleep deprivation upon people's ability to work harmoniously together in a group is a matter of considerable practical importance, and yet there is comparatively little evidence on the question. Murray, Schein, Erikson, Hill and Cohen (1959) noted a tendency to change restlessly from one activity to another with lack of perserverance in group activities, and Laties (1961) found a social attitude of negative affect and behaviour. The most important consideration for matters of social interaction, however, is probably whether lack of sleep causes an increase in irritability and hostility towards others of a group. Incidental observations of sleepless subjects have led to contradictory reports on this question (Kleitman, 1939; Edwards, 1941; Redfearn, Eagles and Halliday, 1952; Williams *et al.*, 1959).

Recent studies incorporating rather more objective techniques are also not unanimous in their conclusions. In two experiments the method has been to administer a thematic apperception test, in which subjects are shown a series of illustrations and asked to write a dramatic story around them. Ax *et al.* (1954) found clear evidence of higher aggression content in the stories of sleep-deprived subjects, but Murray (1959) did not. Much, it is felt, may depend upon the attitude of experimenters towards subjects during a prolonged vigil, and with Murray's subjects it is known that gentle persuasion was the keynote. Finally an experiment of Laties (1961) is of interest in suggesting that, although covert irritability may be increased by loss of sleep, overt signs of hostility are not. The subjects carried out group-performance tests under normal conditions and during a 37 hr period of sleep deprivation and were assessed for signs of hostility on each occasion; no differences were found. When, however, the subjects were asked privately to check from a list of adjectives those that best described their current feeling, it was found that more adjectives of a hostile or aggressive nature were selected under conditions of sleep deprivation than after normal sleep.

b. Perceptual anomalies. Many of the studies already referred to in different contexts contain reports of perceptual anomalies described by sleepless men. Although differing in detail, the phenomena seem to be broadly the same in all reports. One particularly thorough analysis can be taken as typical. This, by Morris, Williams and Lubin (1960), is valuable because the authors made some attempt to quantify the data and correlate the incidence and nature of the various phenomena with the degree of sleep deprivation. Three classes of behavioural anomaly were distinguished: visual misperceptions, temporal disorientation and cognitive disorganisation. The authors tabulate instances of each of these on a five-point scale of increasing abnormality. The three tables are reproduced here (Tables II to IV). Other abnormalities noted by Morris *et al.* (1960) included tactile illusions, feelings of depersonalisation and spatial disorganisation. These phenomena were reported at interviews conducted at the 40 and 65 hr points of sleep deprivation. As with lapses in performance, the episodes were transitory and interspersed with periods of apparently normal functioning of perception.

c. Psychotic tendencies. The relevance of behavioural abnormalities to possible covert neurotic and psychotic tendencies in the individual is clearly a matter of some importance. Morris and Singer (1961) interviewed the subjects of the study just mentioned and concluded that "the kinds of alterations" (with sleep deprivation) "represent extensions or revelations of tendencies native to the individual personality but often covert, while the form and dynamic meaning of these disturbances and to some extent their degree are influenced by the setting and by interpersonal transactions." In line with this, it may be noted that several instances of overt psychotic symptomatology have occurred in an experiment involving prolonged sleep

Table II. *Visual Misperception Scale* (*From Morris* et al., 1960)

1. Eyes itching, burning, or tired; difficulty in seeing, blurred vision or diplopia.
2. Visual illusions: changes in or loss of shape, size, movement, colour or texture constancies; disturbed depth perception.
   Examples: "The floor seems wavy."
   "The light seems to flicker."
   "The size and colour of the chairs seems to change."
3. Labelling of illusions, but without doubt about their illusory character.
   Examples: "Looks like fog around the light."
   "That black mark looked like it was changing into different rock formations."
4. Labelling of illusions with some doubt about their reality.
   Examples: "I thought there was a fuzz around the bottle."
   "I thought steam was rising from the floor; so I tested my eyes to check whether it was real."
5. Labelling of illusions (hallucinations) with, for a time at least, belief in their reality.
   Examples: "I saw hair in my milk. The others said there wasn't any, but I still felt there was and would not drink it."
   "That (Rorschach card) looked like an envelope; I turned it over to check, and it had my name and address on it."

Table III. *Temporal Disorientation Scale* (*From Morris* et al., 1960)

1. Time seems to pass slowly or to be "different" in duration.
2. Occasional mistakes are made in thinking about time, with spontaneous correction.
3. Occasional mistakes as above, but subject does not recognise error until questioned.
4. More frequent mistakes, which subject believes correct, uncertain when confronted.
5. Gross disorientation in time, or unshaken belief in mistaken concept of time.

Table IV. *Cognitive Disorganisation Scale* (*From Morris* et al., 1960)

1. Slowing of mental processes; some difficulty in thinking of words (no undue interference with normal communication).
2. Occasional mistakes or failures in thinking and speech, which can be corrected easily.
3. Loses train of thought, forgetting what he was thinking or talking about; leaving statements incomplete, etc; sudden, unexplained shifts in trend of thought or speech; can correct with effort if challenged.
4. Some thoughts or statements completely incoherent; clarification not altogether possible; some confusion with reality of fantasies, dreams or intrusive thoughts.
5. Rambling, incoherent speech for brief periods, with failure to recognise errors; inability to straighten out jumble of incoherent thoughts when challenged.

deprivation of over 120 hr associated with a fatiguing programme of military manoeuvres (Tyler, 1955) and also with severe sleep deprivation of over 200 hr (Katz and Landis, 1935; Williams and Lubin, 1959; Luby *et al.*, 1960). Further evidence comes from Koranyi and Lehman (1960), who found that five out of six schizoid subjects, when kept awake for 100 hr, again manifested their acute psychotic picture "as it had been observed at the time of their admission to the hospital." An excellent description of what is called "the psychosis of sleep deprivation" has been given by West, Janszen, Lester and Cornelisoon (1962). These authors distinguish two phases: a prodrome in the first four to five days of sleeplessness and then a psychosis becoming more obvious from the fifth night on. The psychotic phase is said to resemble the clinical picture of the reactive type paranoid schizophrenia by day and that of toxic delirium at night. A final point of considerable significance may be that Luby *et al.* (1960) found the incidence of such psychotic behaviour patterns to coincide broadly with biochemical signs (see above) of exhaustion of the emergency energy transfer systems.

E. Conclusions

Probably the most important fact about sleep deprivation as a stress is the reversibility of its effect, certainly upon performance and behaviour and probably upon physiological and biochemical function also. For this reason, measurements taken in these areas may vary widely in the degree and the way in which they differ from normal as a function of the situation of the sleep-deprived subject when they are taken; in particular it is important to know what he is doing at the time of measurement. Is he passive or active, and, if active, what type of activity and how efficiently is he engaged in it? In general, biochemical studies of sleep deprivation and, to a lesser extent, physiological ones have not taken these factors into account; as a result the findings have been equivocal or contradictory. The same was true for performance studies until about the turn of the century, but since then progress has been made not only in defining the types of tasks that are most and least affected, but also in distinguishing how many independent variables can influence the effect of lack of sleep upon performance. The position can be summarised as below.

The impact of sleep deprivation on performance varies widely with the nature of the work being carried out, being in general more pronounced as the task (*a*) ceases to evoke interest and incentive and (*b*) becomes more complex. Of these features, the former is much the more influential, so that the most vulnerable tasks appear all to be simple uninteresting ones, whereas among the least affected tasks may be found highly complex activities that can resist the effects of severe sleep deprivation because of the interest and incentive they evoke, often in spite of their complexity. The pattern of impaired performance is one of periodic lapses in efficiency rather than any continuous depression of performance. The influence of time is important in various ways: impairment appears to be roughly a monotonic function of the number of hours spent awake, but subject to a superimposed variation due to the influence of the diurnal cycle of bodily activity and efficiency. The effect of loss of sleep becomes greater also with time spent on the task, both during a single test and over a series of spaced ones; familiarity with the task and with experimental sleep deprivation itself appears to be an important causal factor here. Individuals differ greatly in the degree to which their performance is impaired by sleep deprivation; with occasional exceptions, these differences have not been found to correlate in any obvious way with standard measures of personality, intelligence or neuroticism. Other factors that influence the effect of loss of sleep performance are noise, physical exercise, amphetamine and small doses of alcohol which reduce the effect; moderate warmth and large doses of alcohol may potentiate it; hyoscine has no effect on it. An effect of sleep deprivation even after restorative sleep has been demonstrated, appears to be due to a disturbance of diurnal bodily rhythms rather than to any carry-over of the effect of sleep deprivation itself. Little is known of the effect of partial loss of sleep, that is, a reduction in the hours of sleep over a period of weeks rather than complete absence of sleep for a shorter period.

Sleep deprivation itself reduces the level of arousal or activation of the body. If the incentive is there, however, this can be overcome, so that arousal and therefore performance are maintained at normal levels. This may require additional effort, which may be reflected in heightened levels of the so-called physiological arousal indices and also, perhaps, in increased activity of the biochemical energy-transfer systems. There are limits to which the body can compensate in these ways for the inefficiency that prolonged wakefulness imposes, and these may be reached at about the 100 hr point of deprivation.

From about the 40 hr point of wakefulness various disorders of behaviour and perception are to be observed. Socially there is an increase in covert irritability, lack of perserverance in activities pursued and in general a negative kind of response. Anomalies of perception may occur, particularly during periods of relative inactivity. Coinciding, perhaps, with the exhaustion of emergency energy-transfer systems at around the 100 hr point of sleep deprivation, there supervenes a picture of disordered behaviour, which resembles that of paranoid psychosis, but also appears reversible when a sufficient occasion demands.

F. Addendum

This addendum is concerned with a new development, selective sleep deprivation. There are now thought to be two importantly different kinds of sleep.

One has been called forebrain sleep, because it appears to require the mediation of the cerebral cortex. It is best distinguished by high amplitude, low frequency waves in the electroencephalogram, as recorded from the scalp. The other, hindbrain sleep, appears to be independent of the cortex and can be identified by an electroencephalogram similar to that of wakefulness but accompanied by drastic loss of muscle tone and by eye movements suggestive of dreaming. Dement (1960) selectively deprived his subjects of the latter sleep, believing as he did so that he was preventing their dreaming. As a result, more of this kind of sleep appeared on subsequent nights; this immediately suggests that there is a need for such hindbrain sleep and that forebrain sleep alone cannot satisfy. It follows, as Oswald (1964) has pointed out, that we may now have two kinds of sleep deprivation to consider, forebrain and hindbrain. At present the implications of these developments are more of theoretical than practical relevance. But should it become possible, perhaps by pharmacological means, to influence the kind of sleep taken, it may become important operationally to distinguish the effects of losing one or the other.

REFERENCES

Armington, J. C., and Mitnick, L. L. (1959). *J. appl. Physiol.* 14, 247.
Ax, A., Fordyce, W., Loovas, I., Meredith, W., Pirojnikoff, L., Shmavonian, B., and Wendahl, R. (1954). Unpublished manuscript. (Abstract in *Amer. Psychologist* (1954), 9, 324.)
Ax, A., and Luby, E. D. (1961). *Arch. gen. Psychiat.* 4, 55.
Bjerner, B. (1949). *Acta physiol. scand.* 19, Suppl. 65.
Blake, H., and Gerard, R. W. (1937). *Amer. J. Physiol.* 119, 692.
Bliss, E. L., Clark, L. D., and West, C. D. (1959). *A.M.A. Arch. Neurol. Psychiat.* 81, 348.
Burch, N. R., and Greiner, T. H. (1958). *J. Psychol.* 45, 3.
Cappon, D., and Banks, R. (1960). *Arch. gen. Psychiat.* 2, 346.
Corcoran, D. W. J. (1962). *Quart. J. exp. Psychol.* 14, 178.
Corcoran, D. W. J. (1964). *Brit. J. Psychol.* 55, 307.
Dement, W. (1960) *Science* 131, 1705.
Edwards, A. S. (1941). *Amer. J. Psychol.* 54, 80.
Freeman, G. L. (1932). *J. exp. Psychol.* 15, 267.
Hasselman, M., Schaff, G., and Metz, B. (1960). *C.R. Soc. Biol. Paris* 154, 197.
Hebb, D. O. (1949). "The Organization of Behavior." Wiley, New York.
Husband, R. W. (1935). *J. exp. Psychol.* 18, 792.
Katz, S. E., and Landis, C. (1935). *Arch. Neurol. Psychiat.* 34, 307.
Kleitman, N. (1923). *Amer. J. Physiol.* 66, 67.
Kleitman, N., Cooperman, N. R., and Mullin, F. J. (1934). *Amer. J. Physiol.* 107, 589.
Kleitman, N. (1939). "Sleep and Wakefulness", p. 320. University of Chicago Press, Chicago. (See also 2nd Ed., 1963.)
Koranyi, E. K., and Lehman, H. E. (1960). *A.M.A. Arch. gen. Psychiat.* 2, 534.
Kornetsky, C., Mirsky, A. F., Kessler, Edith K., and Dorff, J. E. (1959). *J. Pharmacol. exp. Ther.* 127, 46.

Laties, V. A. (1961). *J. psychiat. Res.* **1**, 12.
Leonard, J. A. (1959). *M.R.C.*, Applied Psychology Research Unit, Report No. 326/59.
Loveland, N. T., and Singer, M. T. (1959). *J. proj. Tech.* **23**, 323.
Loveland, N. T., and Williams, H. L. (1963). *Percep. mot. Skills.* **16**, 923.
Luby, E. D., Frohman, C. E., Grisell, J. L., Lenzo, J. E., and Gottlieb, J. S. (1960). *Psychosomat. M.* **22**, 182.
Luby, E. D., Grisell, J. L., Frohman, C. E., Lees, H., Cohen, B. D., and Gottlieb, J. S. (1962). *Ann. N.Y. Acad. Sci.* **96**, 71.
Lybrand, W. A., Andrews, T. G., and Ross, S. (1954). *Amer. J. Psychol.* **67**, 704.
Malmo, R. B. (1959). *Psychol Rev.* 66, 367.
Malmo, R. B., and Surwillo, W. W. (1960). *Psychol. Monogr.* 74 (Whole No. 502).
Mangold, R., Sokoloff, L., Connor, E., Kleinermann, J., Therman, P., and Kety, S. S. (1955). *J. clin. Invest.* **34**, 1092.
Morris, G. O., and Singer, M. T. (1961). *Arch. gen. Psychiat.* **5**, 453.
Morris, G. O., Williams, H. L., and Lubin, A. (1960). *A.M.A. Arch. gen. Psychiat.* **2**, 247.
Murawski, B. J., and Crabbe, J. (1960). *J. appl. Physiol.* **15**, 280.
Murray, E. J., Williams, H. L., and Lubin, A. (1958). *J. exp. Psychol.* **56**, 271.
Murray, E. J. (1959). *J. abnorm. Psychol.* **59**, 95.
Murray, E. J., Schein, E. H., Erikson, K. T., Hill, W. F., and Cohen, M. (1959). *J. social Psychol.* **49**, 229.
Oswald, I. (1964). "The Scientific Basis of Medicine Annual Reviews", p. 102.
Patrick, G. T. W., and Gilbert, J. A. (1896). *Psychol. Rev.* **3**, 469.
Pepler, R. D. (1959). *J. comp. physiol. Psychol.* **52**, 446.
Pincus, G., and Hoagland, H. (1943). *J. Aviat. Med.* **14**, 173.
Rakestraw, N. W., and Whittier, F. O. (1923). *Proc. Soc. exp. Biol.* **21**, 5.
Redfearn, J. W. T., Eagles, J. B., and Halliday, A. M. (1952). *Army Operational Research Group (U.K.)*, Report No. 18/52.
Robinson, E. S., and Hermann, S. O. (1922). *J. exp. Psychol.* **5**, 19.
Rodin, E. A., Luby, E. D., and Gottlieb, J. S. (1962). *Electroenceph. clin. Neurophysiol.* **14**, 544.
Schein, E. H. (1957). *J. appl. Psychol.* **41**, 247.
Scholander, T. (1961). *Acta physiol. scand.* **51**, 325.
Schumacher, G. A., Goodell, H., Hardy, J. D., and Wolff, H. G. (1940). *Science* **92**, 110.
Seashore, R. H., and Ivy, A. C. (1953). *Psychol. Monogr.* **67**, (Whole No. 365).
Smith, M. (1916). *Brit. J. Psychol.* 8, 327.
Susuki, I. (1961). *J. Sci. and Labour* (Tokyo) 37, 166.
Tyler, D. B., Marx, W., and Goodman, J. (1946). *Proc. Soc. exp. Biol. Med.* **62**, 38.
Tyler, D. B. (1947). *Amer. J. Physiol.* **150**, 253.
Tyler, D. B., Goodman, J., and Rothman, T. (1947). *Amer. J. Physiol.* **149**, 185.
Tyler, D. B. (1955). *Dis. nerv. Syst.* 16, 2.
Von Frey. (1926). *Hand. normal pathol. Physiol.* **11**, 94.
West, L. J., Janszen, H. H., Lester, B. K., and Cornelisoon, F. S. (1962). *Ann. N.Y. Acad. Sci.* 96, 66.
Wilkinson, R. T. (1958). *M.R.C.*, Applied Psychology Research Unit, Report No. 323/58.
Wilkinson, R. T. (1959). *Ergonomics* **2**, 373.
Wilkinson, R. T. (1960). *Quart. J. exp. Psychol.* **12**, 36.
Wilkinson, R. T. (1961a). *In* "The Nature of Sleep" (G. E. W. Wolstenholme and Maeve O'Connor, eds.), pp. 329–42. Churchill, London.
Wilkinson, R. T. (1961b). *J. exp. Psychol.* **62**, 263.
Wilkinson, R. T. (1962). *J. exp. Psychol.* 64, 565.
Wilkinson, R. T. (1963a). *J. exp. Psychol.* 66, 332.

Wilkinson, R. T. (1963b). *J. exp. Psychol.* 66, 439.
Wilkinson, R. T. (1964). *Ergonomics* 7.
Wilkinson, R. T., Fox, R. H., Goldsmith, R., Hampton, I. R. G., and Lewis, H. E. (1964). *J. appl. Physiol.* 19, 287.
Williams, H. L., Lubin, A., and Goodnow, J. J. (1959). *Psychol. Monogr.* 73, (Whole No. 484).
Williams, H. L. and Lubin, A. (1959). *Walter Reed Army Institute of Research Report.*
Williams, H. L., Granda, A. L., Jones, R. C., Lubin, A., and Armington, J. C. (1962). *Electroenceph. clin. Neurophysiol.* 14, 64.

CHAPTER 15

Fatigue and Monotony

A. T. Welford

A. The Problem of Definition

The study of fatigue stands at one of the traditional meeting-points between physiology and psychology. Until the early years of the present century research centred mainly round the decrements of performance that occur in the course of prolonged muscular work, and the division of labour between the two disciplines could be simply conceived: physiology studied the neuromuscular mechanism itself, psychology the accompanying subjective feelings of discomfort and exhaustion. During the past fifty years or so interest in the psychological field has shifted from a predominant concern with conscious processes to an almost exclusive concentration on observable behaviour and measurable performance. In its early stages this shift served to strengthen rather than diminish psychological interest in fatigue; here, it seemed, there were striking changes of performance that could be related closely to definable conditions. Soon, however, it became apparent that neither in its physiological nor in its psychological aspect is fatigue a simple phenomenon. The neuromuscular mechanism involved is complex, the changes that might be regarded as fatigue effects are manifold. Behavioural changes, though clear in outline, vary in important detail from one set of circumstances to another. The situation became even more confused when attempts were made to see analogies to neuromuscular fatigue in purely mental operations carried on for long periods.

These difficulties have led some to wish to abandon the term "fatigue". Yet there is need for a term to cover those changes of performance that take place over a period of time during which some part of the "mechanism", whether sensory, central or muscular, becomes chronically overloaded. If fatigue is defined in this way, the task of the investigator ceases to be one of finding a single "entity" that can be labelled "fatigue": instead he has, first, to study the detailed changes of performance brought about by such overloading and, secondly, to seek to explain them in terms of changes of function in any of the many bodily mechanisms concerned.

This chapter will first re-examine in these terms the classical type of experiment on neuromuscular fatigue and will then cover the much more complex problem of so-called mental fatigue.

B. Neuromuscular Fatigue

Work on neuromuscular fatigue has been summarised by Bartley and Chute (1947), Darcus (1953), Hemingway (1953) and others, and it will only be briefly discussed here.

Many classical experiments on fatigue have used an apparatus known as an ergograph. Typically an arm and hand are strapped in position in such a way that only one finger is free to move. With this finger the subject is required to depress a lever that can be loaded to different extents by means of a spring or weight. The lever is depressed and released in time with a metronome or other timing device giving regularly repeated signals.

If the loading is light or the rate slow or both, the task can be continued indefinitely, but if the load is substantial and the rate more frequent the depressions of the lever will, after a time, begin to diminish in amplitude and eventually fall to zero. A typical ergographic record is shown in Fig. 1. The decline of performance begins sooner and proceeds more rapidly as the load or the frequency increase. If, when performance has ceased, the subject is allowed to rest for a period and then tries again, recovery will be found to have occurred: after a short rest it may be partial, in the sense that amplitude may

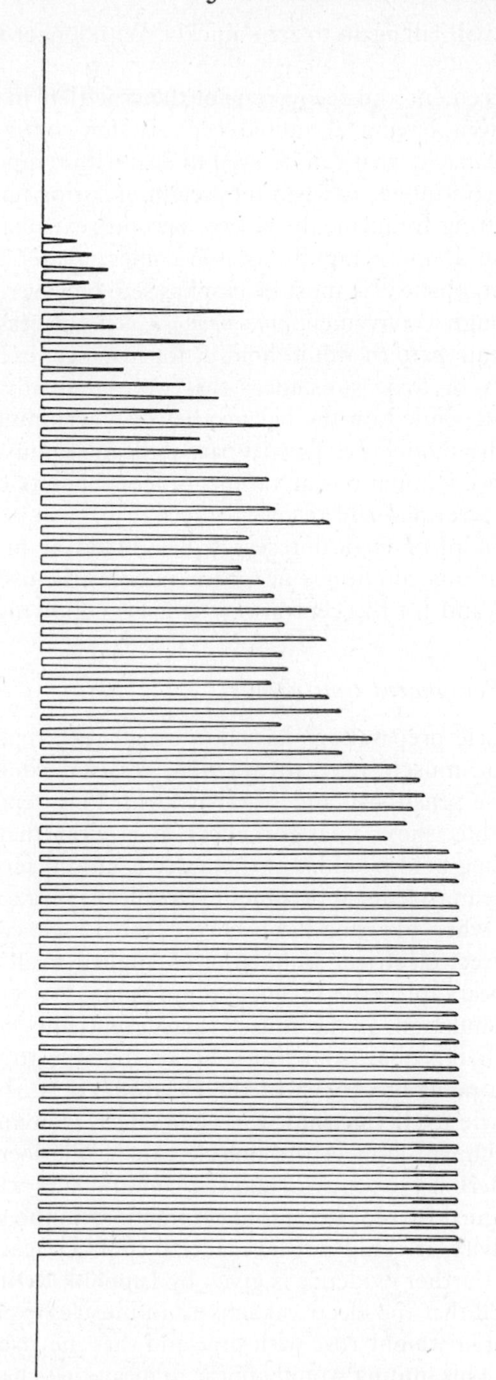

FIGURE 1. Section of ergograph record. Each vertical stroke was made by a flexion of the right forefinger against a lever weighted in such a way that the force required to move the point of contact with the finger was 14 lb. The time marker below shows 1 sec intervals.

not be reattained in full and will fall again to zero quickly. With longer rests, recovery will be complete.

The whole pattern of decrement and recovery is of the type that may be envisaged as that of a system having a limited capacity for continuous operation and some reserve capacity that can be used to deal with temporary overloads. If the overload is continued, as when the weight or spring tension is relatively heavy and the lifting frequent, the reserve becomes exhausted—slowly if the overload is slight, more rapidly as it becomes greater. Rest allows the reserve to be re-established. It must be emphasised, however, that such a conceptual model, though a convenient *aide memoire*, is not necessarily true in a literal sense: the same pattern would hold if, for instance, exercise gradually built up inhibitory or toxic substances that were gradually dissipated and if work output depended on the balance between accumulation and dispersion. Nor does such a model specify what part of the neuromuscular system is limiting performance: limitations at various different points could produce the same pattern of overload and recovery effects.

A considerable volume of physiological research has aimed at finding the locus of fatigue under various conditions in (*a*) the muscles themselves, (*b*) the myoneural junctions and (*c*) the central mechanisms supplying the muscles.

1. *Peripheral limitations*

Classical work on nerve-muscle preparations has shown that, with repeated stimulation of the nerve, the muscle may after a time cease to contract. Although nerve cells can lose sensitivity and reactivity with long-repeated rapid stimulation and can thus show a fatigue effect, it is clear that the contraction ceases to occur long before failure of the nerve itself can account for it. It also occurs before the muscle itself becomes incapable of contraction, as is shown by the fact that, when blocking has occurred, the muscle can be made to contract again if direct electrical stimulation is applied to it. The site of the blocking thus appears to be the myoneural junction.

Even direct electrical stimulation of the muscle may eventually fail to secure contraction. Although this may sometimes be attributable to local conditions developing at the point of contact of the electrode, it is usually taken to imply that the muscle itself can show a fatigue effect. Convincing evidence that in the intact human subject the muscle can be the locus of fatigue has been given by Merton (1956), who showed conditions in which, when a muscle had been fatigued by repeated rapid contractions to the point of complete voluntary inactivity, it could not be made to contract again by direct electrical stimulation. Further evidence is given by Lippold, Redfearn and Vuco (1960), who showed that the electrical activity in muscles exerting a static force against a spring or weight rose with time and that the rise was steeper the heavier the load. This finding would appear to mean that changes in the muscles rendered contraction less powerful for any given level of efferent stimulation, so that neural activity had to be increased to maintain the required output of muscular power. Both Merton (1956) and Lippold *et al.* (1960) found that cutting off the blood supply to the fatigued muscle or muscles prevented recovery during a period of relaxation. It therefore seems clear that the lowered performance of the muscles was either due to lack of oxygen or other substances or to accumulation of "fatigue products", which would have been corrected by blood flow. If so, it is understandable that fatigue effects are found to be greater with static contraction than with rhythmic, since blood flow is likely to be greater during the latter than the former. Although direct evidence about the effects of sustained contraction on blood flow in muscles is equivocal (see Hemingway, 1953, p. 74), we can perhaps expect that the force of the contraction would be critical in determining the extent to which blood flow is restricted, so that results would depend on the precise conditions of different experiments.

The fact that maintenance of blood-sugar level is important for athletic endurance (Edwards, Margaria and Dill, 1934; Douglas and Koch, 1951) is usually taken as further evidence that failing fuel supply to muscles is a direct cause of fatigue. The argument is not conclusive, however, since lowered blood-sugar can affect a wide range of bodily mechanisms, neural and central as well as muscular.

2. *Central limitations*

In his classical studies of the scratch reflex in dogs, Sherrington (1906) found that after a period of repeated elicitation the reflex ceased, though the same muscles would still respond to a different reflex. Clearly there could be no question of myoneural junctions or other effector parts of the mechanism constituting the limiting factor; the limitation was presumably sensory or central. The cessation came later when the reflex was elicited by stimulating various slightly different points on the skin rather than by repeated stimulation of the same point, but it came eventually nevertheless, indicating central rather than sensory origin. Sherrington's results have since been directly confirmed by Lloyd (1942), who found that action potentials in the efferent nerve of the reflex diminished with repeated stimulation of the afferent nerve. We may note in passing that Sherrington found the fatigue effects to be highly specific: if a reflex had ceased after repeated elicitation by stimuli applied to one point on the skin, it could be restored immediately by shifting to a different point. There is no question here of regarding fatigue, as has sometimes been done, as due to the general circulation in the blood stream of "fatigue products" resulting from exercise and of these affecting the central nervous system.

The general importance of central factors in neuromuscular fatigue has been stressed by Reid (1928). Using an ergograph with human subjects he showed that, although voluntary contraction had ceased, the muscle could still be made to contract either by direct electrical stimulation or by electrical stimulation of the efferent nerve trunk. It seems clear that the essential locus of fatigue here was central. Reid (1928) further showed, however, that recovery was much less after a period during which voluntary effort ceased, with the muscle concerned directly stimulated electrically and thus kept contracted, than after a period in which the muscle was rested. This result appears to imply that, although the fatigue was central, local conditions in the muscle nevertheless influenced it, presumably by means of afferent impulses. If this conclusion is correct, the questions reopened in a new way are of how physical or chemical changes in muscle resulting from exercise influence fatigue and of why continuous contraction causes fatigue effects more quickly than does rhythmic: although conditions in the muscles may not limit performance directly, they cause signals to the spinal cord and brain that effect conditions there. What are the subjective counterparts (if any) of these afferent impulses is not certain, though it seems reasonable to link them with the feelings of discomfort and pain that mount rapidly as severe muscular contraction continues. If so, it means that fatigue decrements in muscular performance will depend to some extent on a subject's sensitivity to and toleration of pain.

It must be emphasised that Reid's work did not imply that the limitation of performance in fatigue was invariably central. His results showed that when a series of exceptionally rapid voluntary contractions had fallen to zero direct stimulation of the muscle or the nerve trunk did not restore contraction or did so only partly. It seems clear that with rapid or intense contractions a truly peripheral muscular fatigue can be produced, but that with conditions nearer to those of everyday life central limitations are more likely.

Two further points about neuromuscular fatigue indicate the importance of central factors.

a. Motivational factors. Figure 1 shows a number of minor variations typical of ergographic records; the rate of decline is not smooth, but is interrupted by temporary partial recoveries. To some extent such variation is to be expected from the random fluctuations of function to be found in almost any complex biological mechanism, but subjectively some, at least, correspond to periods of special effort, and it is easy to show that they can be produced by urging the subject to "try harder".

The effect of incentives upon fatigue effects has been vividly illustrated in an experiment by Schwab (1953), who required subjects to hang on a horizontal bar as long as they could. He found that with instructions to hold on "as long as possible" the average length of time before letting go was less than 1 min. With strong urging and suggestion the time was raised to rather over 1 min. With the reward promised of a $5 bill for bettering their previous records, subjects managed to hang on for an average of nearly 2 min. A further motivational effect was shown by Ash (1914), who found that, after performance on an ergograph had ceased, it could be made to begin again not only by lightening the weight, but also merely by leading the subject to believe that the weight had been lightened. These results are important, as showing that fatigue effects are to some extent under voluntary control in the sense, perhaps, that the subject sets levels of effort he is willing to make and of discomfort he is prepared to bear. Such motivational variation is not strictly attributable to the system that fatigues, but must be taken into account when measuring fatigue effects.

b. Recruitment. It is easy to observe in ergograph studies that when a load is placed upon a small group of muscles other muscles spontaneously become active and that, as fatigue proceeds, the activity in other muscles spreads until almost the whole body is involved: the subject may tense his legs and grit his teeth in the effort to depress one middle finger so strapped in position that these activities cannot possibly help. The classical demonstration of this phenomenon is by Ash (1914). A more recent demonstration of the same kind of effect has been given by Lundervold (1958), using electromyographic recordings in a typewriting task. In normal circumstances of everyday life this recruitment would be adaptive in the sense that other muscles could take the load off the group that was becoming fatigued, but this is not so in ergograph experiments; the fact that the phenomenon nevertheless occurs implies that it is largely involuntary. The extra muscles become active in a specific order (Seyffarth, 1940; Denny-Brown, 1949), presumably along lines of functional or neural proximity. One may assume the relatively simple neurological model that focal activity in a central area concerned with one muscle or group of muscles tends to spread to surrounding areas, the amount of spread becoming greater the longer the focal activity continues or the more intense it becomes. Lippold *et al.* (1960) showed that as other muscles become active, electrical activity in the fatigued muscle may diminish, presumably implying that activity in areas surrounding a focal central area may continue after it has ceased in the focal area itself. It may be noted in this context that excessively exercised muscles tend to go into contraction more readily than rested ones, sometimes showing spasm, as in writer's or telegraphist's cramps. This appears to imply that a focal area, when it has recovered from acute fatigue, may be left hypersensitive, a condition found in some studies (other than those concerned with fatigue) to occur in nerve tissue subjected to prolonged stimulation.

To sum up, the position apparently reached in the study of neuromuscular fatigue is that, in the intact organism, changes in the muscles brought about by prolonged or repeated contraction can, according to circumstances, have one of two limiting effects. Either the muscles themselves become temporarily incapable of further contraction or the condition of the muscles produces afferent stimuli and these in turn affect the central mechanisms and lead to a cessation of efferent impulses. Which effect occurs first depends on factors at present not entirely clear. It may, however, tentatively be suggested that peripheral limitations are likely to succeed intense activity and central limitations to result from activity less intense but more prolonged. Simsonson (1965), using a different but parallel set of results, has arrived independently at similar conclusions about the mental nature of neuromuscular fatigue.

C. Mental Fatigue

If the term "mental fatigue" is to have a meaning in line with that of neuromuscular fatigue, it must denote the impairment of some brain mechanism as a result of long-continued use. The impairment must be reversible in the sense that it disappears with rest and may take the form of lowered sensitivity or responsiveness or capacity. The last of these may show as a reduction in either the amount of information that can be handled at any one instant, and thus in reduced "mental power", or in the amount that can be handled in a given period and so in slowness of perception, choice and so on.

Such a definition enables a distinction to be made between mental fatigue and several other central changes, such as adaptation, habituation, satiation, inhibition and monotony or boredom; all of these lead to decrement of performance with time. Adaptation implies a loss of sensitivity or discriminating power over one part of a range of possible stimulus values, but simultaneous gains over another: there is not so much a lowering of sensitivity as a shift in the point of maximum sensitivity. Habituation denotes a learnt ignoring of stimuli. Satiation is a state in which action ceases because the need or appetite that gave rise to it has been satisfied. Inhibition, although often loosely used, strictly means the reduction of one process by the activity of some opposing process. Boredom or monotony refers to a state in which the organism is underloaded, not overloaded: it seems as if a certain throughput of information is necessary to maintain full efficiency, and typically boring situations seem to be those in which attention is required but little information is conveyed; the classical bore compels his hearer to listen to conversation that is insignificant in content.

The distinction between fatigue and these states is easy to make formally, but is often difficult to draw in practice. For example, even neuromuscular fatigue can in a sense be viewed as a central inhibitory state brought about by afferent impulses from the muscles, although it should be noted that Sherrington was able to distinguish between patterns of decrement in a reflex due to what he termed fatigue and due to what he identified as inhibition. The greatest difficulty arises in distinguishing fatigue from monotony or boredom: many tasks used in studies of fatigue are repetitious and thus liable to become monotonous, whereas some used for studying monotony require actions or decisions to be repeated frequently enough for them to be a possible cause of fatigue. It is indeed reasonable to suppose that some tasks are both fatiguing and boring: some parts of the subject's central mechanisms may be overloaded even though the overall throughput of information is low. Usually, however, it seems fairly clear which of the two effects will limit performance sooner.

We shall here survey work that has been regarded, probably correctly, as studying fatigue and then, to set this work in perspective, turn more briefly to work on so-called "vigilance" tasks, in which conditions are clearly such as to produce monotony or boredom.

1. *Phenomena of mental fatigue*

Bartlett's Ferrier Lecture to the Royal Society (1943) was a landmark in the development of ideas about fatigue and indeed about human performance generally. In particular, it showed that the phenomena of fatigue are more varied and more complex than is often supposed. On the basis of this lecture and of work done since, we may recognise four main types of change that can come about in mental fatigue.

a. Sensory or perceptual changes. The classical studies of visual fatigue have been summarised and discussed by Bartley and Chute (1947) and by Weston (1953). They have, for the most part, been concerned with the possible fatigue of eye muscles and its relation to feelings of eye strain under conditions of low illumination, glare or close attention to detail. It is doubtful if the effects are to be wholly, or even mainly, attributed to the eye muscles: some probably result from frowning and general muscular tension built up as a result of concentrated efforts to see under difficult conditions. Be this as it may, it seems clear that there are some more strictly sensory or perceptual effects. For instance, Berger and Mahneke (1954) found that visual acuity (cancelling Landolt rings) and critical flicker frequency (CFF) both fell when tests were made continuously for 55 min, but rose again after 5 min rest. Examples of their results are shown in Fig. 2. Saldanha (1955, 1957) found that repeated settings made on the vernier scale of a calliper gauge for $\frac{1}{2}$ hr or more became less regular, and thus less accurate, with time, but that accuracy returned after $\frac{1}{2}$ hr of rest. The motor components in all these tasks were trivial, and thus it is clear that there was loss of fine differentiation either spatially (visual acuity or vernier settings) or temporally (CFF). Why this occurred is not clear. CFF is often used as an indicator of central impairment, falling with brain injury or malfunction, but these results could have been due both to some change in criteria of judgement and to true loss of discriminatory power. Recent studies of sensory thresholds have shown that these factors can have closely similar effects.

b. Slowing of performance. One of the most frequently observed fatigue effects is the slowing of sensory-motor performance. It is often suggested

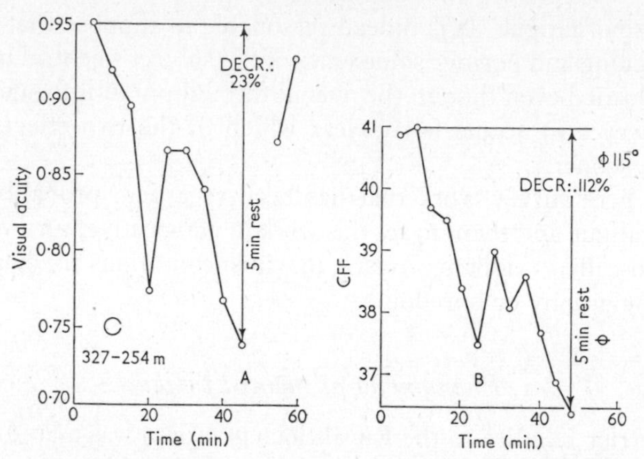

FIGURE 2. Examples of fatigue effects obtained by Berger and Mahneke (1954) with repeated measurements of visual functions.

A. *Visual acuity*. For each determination a Landolt ring was brought progressively closer to the subject until the direction of the gap could be recognised: ninety determinations were made continuously over a period of 55 min and twenty more after an interval of 5 min. Retinal illumination was held constant at 3·5 lux. Each point is the mean of ten determinations by one subject.

B. *CFF*. Flicker was produced by a rotating disc illuminating a diffusing glass surface: 120 determinations were made continuously over a period of 55 min and further ones after an interval of 10 min. The flickering target subtended 1·15° at the subject's eye. Each point is the mean of ten determinations by one subject.

that this may be due to muscular fatigue, but there is no doubt that central factors are often, and probably mainly, involved. An indication of the central locus of such a fatigue effect is contained in the results of an experiment by Singleton (1953) who used a serial choice-reaction task. The apparatus for this is shown in Fig. 3. Subjects sat in a chair and pushed a joystick from a central position along slots in four directions in response to lights at the ends of the cross on the display (shown top left of Fig. 3). Upon the subject's reaching the end of the correct slot, the light went out; on his return to the centre another came on, until sixty-four responses had been completed. He was told to work at this task as fast as possible. Three variations of the task were presented; they were, in ascending order of difficulty, (i) "Direct" with the joystick having to be pushed away when the top light came on, to the left for the left light, and so on; (ii) "180°" with the joystick pushed away in response to the bottom light, to the left for the right light, and so on; and (iii) "270°" with the joystick pushed away in response to the left light, to the left for the bottom light, and so on. The times per response gradually lengthened during each run, but, as can be seen from Fig. 4, the lengthening was much more in time spent at the centre, that is, in deciding which way to move, than in the actual execution of movements. Moreover, this lengthening increased with the difficulty of the condition, implying that the fatigue effect became greater as the demands of the central task rose.

Slowness may cause several complications when the subject cannot, as he could in Singleton's experiment, work at his own speed, but is paced by the task. The classical example of such a task used in the study of fatigue is the Cambridge Cockpit, which was the basis of experiments by Craik, Drew and Davis (Davis 1948). Subjects were tested for 2 hr spells in a simulated aircraft cockpit under blind flying conditions and had to deal with a series of manoeuvres. Although in a general sense subjects could control the timing of these, the complications of the task were such that, once begun, many of

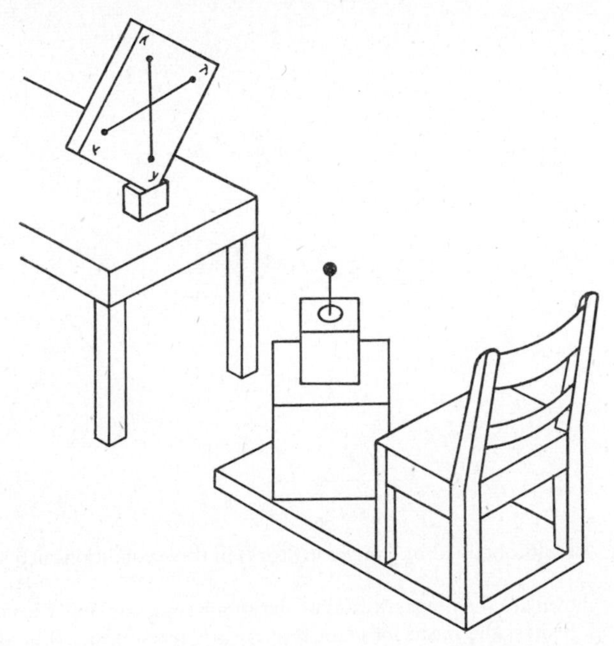

FIGURE 3. Four-choice serial reaction apparatus used by Singleton (1953).

the actions required in the manoeuvre were in effect externally paced. Under such conditions subjects can react to slowing in one of two ways.

(i) Some of the actions required can be omitted. Davis found this tendency to be characteristic of some of his subjects during the latter part of their 2 hr spells. He also noted that most subjects tended to pay less and less attention to the more "peripheral" parts of the task as the spell continued, giving their main attention increasingly to the controls in constant use. For example, the fuel indicator had to be reset every 10 min, but while the test progressed it was more and more often neglected, as shown in Fig. 5. We may note that Bursill (1958) found a similar tendency for peripheral items to be neglected under conditions of high temperature in a task in which subjects had to track a moving target and respond at the same time to signal lights at various distances from the target centre. Such omissions may perhaps be regarded as spontaneous attempts to simplify the task and a parallel to tendencies that seem to come with advancing age (Welford, 1962a).

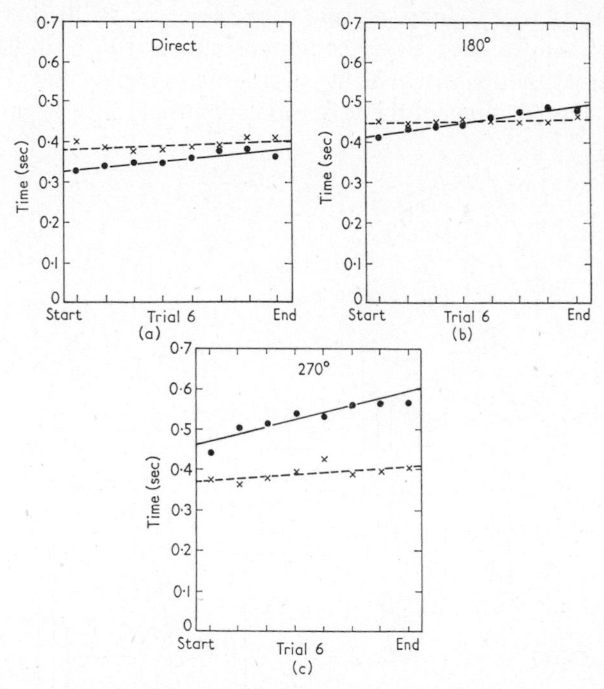

FIGURE 4. Results obtained by Singleton (1953) in three conditions of a serial reaction task.

The results shown are for the sixth trial of sixty-four reactions under each of the three conditions. Each point is the mean for eight reactions by ten subjects. The dots and solid lines are for times between the end of one movement and the beginning of the next, with the joystick in a central position, and are essentially reaction times. The crosses and broken lines are for the movement times from centre to end of slot and back.

(ii) If, alternatively, the subject tries to complete all the actions required in the time available, he will have to hurry and may not have time to make his decisions and judgements accurately. In tasks like that of the Cambridge Cockpit, any slowing that results in a longer time being needed to observe various instruments accurately may be felt as a "stickiness of attention". Performance under these conditions will tend to suffer a disruption that builds up in a vicious circle: the longer time taken to observe an instrument means that the resulting error will tend to be larger before correcting action can be taken; when action is eventually taken it may, in order to make up time, be poorly controlled and require subsidiary corrections; these take further time and mean that subsequent correcting actions have to be larger again. The result, as Davis showed in many of his subjects, is that the onset of fatigue may lead to marked overactivity, often coupled with signs of anxiety. These in turn may direct attention from the task to worrying about whether it is being performed adequately and thus lead to still further slowing and disruption. An example of this kind of disruption is shown in Fig. 6, taken from the records of another experiment by Davis, in which subjects had to bring a pointer from one position to another by means of a velocity control.

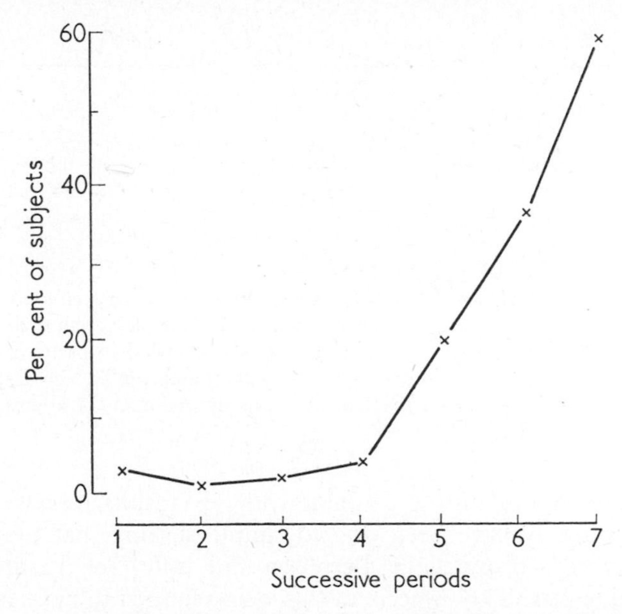

FIGURE 5. Percentage of subjects, in Davis's (1948) experiments with the Cambridge Cockpit, who omitted to reset the fuel indicator during successive periods of a 2 hr test.

c. Irregularity of timing. Long-continued performance tends to become not only slower but also less regular. To some extent irregularity may be more apparent than real: the distributions of times for individual cycles of repetitive tasks tend to be skew, with a tail of long times and with a variance proportional to the mean; any overall slowing will therefore increase the variance and the number of what seem to be unusually long times. It has, however, been suggested by Bills (1931) that irregularity is due rather to occasional "blocking". That is to say, every now and then a short gap appears in an otherwise rapid performance, and the frequency of such gaps increases when the task is continued for a long time. On this view the greater part of the distribution might be only a little affected by slowing, but there should be a marked increase in the tail.

One source of such blocking in paced tasks is easy to understand. For example, Vince (1949) showed that subjects required to respond by pressing a key to signals at regular intervals kept pace up to a certain rate, but at higher rates gradually fell behind until eventually they stopped "to make a fresh start". Bills was, however, primarily considering unpaced tasks, such as alternate addition and subtraction of 3 from a list of digits, colour naming, substituting letters for digits according to a code or giving opposites of words. He found that the frequency of times exceeding twice the subject's average

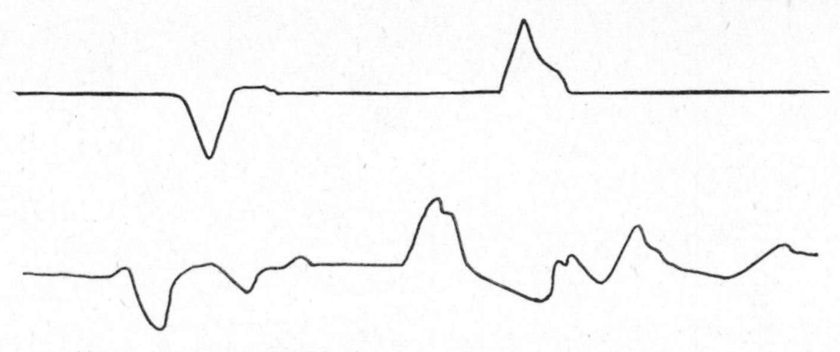

FIGURE 6. Records of two different types of response in attempting to move a pointer on to a target by means of a velocity control. The records are plots against time (from left to right) of the movements of the control. The upper record shows a normal skilful response with a movement of the control to get the pointer going and a second movement to stop it. The lower record shows a disorganised response made in the attempt to achieve the same result.

time tended to increase during 7 min of work. His results are not convincing and could probably have been due to simple slowing, but much clearer evidence has been obtained by Bertelson and Joffe (1963) using a serial choice-reaction task. The subjects in this were required to press one of four keys in response to the figures 1 to 4 shown in random order on a "Nixie" tube. Each response brought on the next figure, so that the task was continuous but unpaced. Samples of reaction times were scored at the second minute of work and at the end of 30 min. No change was found among a group of thirty-five subjects in the averages of the shortest or the median reaction times, but there was a marked increase in the average of the longest. The percentage of "blocks" (defined as reaction times longer than twice the mean, excluding responses in which errors were made) rose rapidly during the first 5 min of work and slowly thereafter. More important, there was a clear tendency for reaction time and errors to rise during the responses immediately before a block and to fall immediately after, as shown in Fig. 7. The results are consistent with the view that some kind of fatigue effect builds up gradually over a series of responses and is dissipated by the block.

It should be noted that on this view there are two fatigue effects involved: a short-term effect dissipated at each block and a longer-term effect causing a rise in the frequency of blocks. The longer-term effect could perhaps be regarded as due to recovery during a block being not quite complete, so that the time taken to build up to the next block is shorter than it would otherwise have been.

The cause of such blocking is not clear. Perhaps the most obvious suggestion is that some part of the sensory-motor mechanism becomes momentarily inoperative, although to say this is to do little more than restate

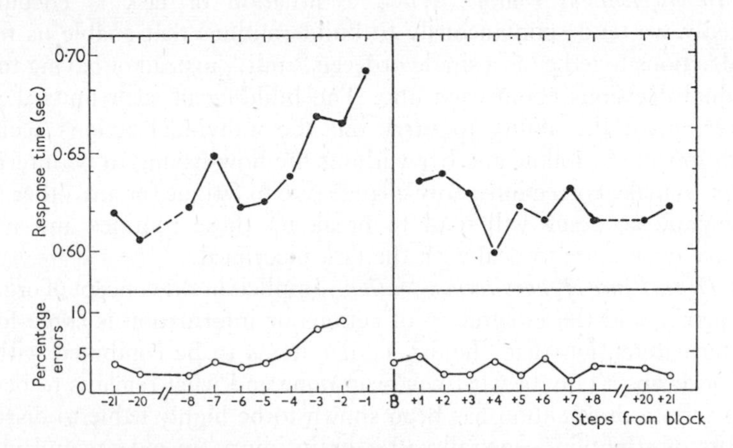

FIGURE 7. Response times and errors made before and after a "block", in a serial reaction experiment by Bertelson and Joffe (1963). The graph was constructed by taking for each of twenty-eight subjects the last eleven blocks observed (and not immediately associated with an error) during 30 min work and plotting the mean reaction times and errors for the first to eighth reactions before and after the block. The twentieth and twenty-first reactions before and after are also shown as an indication of reactions well clear of a block.

the phenomenon. Broadbent (1958) has suggested that the mechanism that selects information relevant to the task in hand from the whole mass of information impinging on the organism becomes temporarily ineffective and allows irrelevant signals to gain attention. This would neatly account for the increase of distractability often observed in fatigued subjects, although there is an alternative explanation available in the likelihood that stimuli arising from hard seats, awkward postures or tensed muscles would become more insistent with time. In other words, distraction could be due not to any failure of central mechanisms directly involved in performance, but to increased competition from ancillary stimuli.

d. Disorganisation of performance. Bartlett (1943), with the preliminary results of the Cambridge Cockpit experiments in mind, suggested that fatigued subjects may sometimes perform correct actions, but in the wrong order. In other words, the co-ordination of their performance, the ordering of individual actions into "larger units", has broken down. This line of thought has not been followed up to the extent it deserves, probably because it is far from easy to study the kinds of complex performance in which such breakdowns might show. It does, however, tally well with the mild confusion, inability for sustained thought and impairment of judgement often observed in states of fatigue. Two possible explanations seem likely to repay further research.

(i) *Impairment of routine.* When a situation or task is encountered repeatedly we tend spontaneously to build routines that enable us to treat several actions together as a single ordered "unit", instead of having to make individual decisions about each one. The building of such units depends, however, upon the ability to carry out the individual actions accurately enough for one to follow another without the flow having to be interrupted in order to make corrections. Any change due to fatigue (or any other factor) that impairs accuracy will tend to break up these routines and make it necessary once more to deal with the task piecemeal.

(ii) *Disturbance of short-term retention.* Implicit in the concept of organised performance and the integration of actions or information is some form of short-term retention that "holds" earlier items to be combined with later ones and keeps a tally of what has been done and what remains to be done. Such short-term retention has been shown to be highly liable to disruption by shifts of attention, especially after brain injury, in old age and in other conditions in which some organic impairment can be presumed. There is some evidence of a similar breakdown with fatigue from tests on civilian aircrews. In one experiment (Welford, Brown and Gabb, 1950) radio officers were tested with a type of electrical problem before and after flights from the United Kingdom to Africa, India, Australia or the Far East. For each problem subjects were given a box with six terminals on the top, a circuit diagram and a resistance meter and had to find out which terminal on the box corresponded to each on the diagram by taking readings on the meter. Subjects took many more readings if they were meeting the task for the first time after a flight than before, and it was clear that while taking one reading they were tending to forget others already obtained and so were having to take some readings several times. The results are shown on the left of Fig. 8 (the results on the right of this figure are discussed later). This experiment was on crews of unpressurised aircraft, and the results might have been partly due to chronic anoxia. This question does not arise, however, with an experiment by Kay (1953) on radio officers, stewards and stewardesses of pressurised aircraft on the Atlantic and Australian routes from the United Kingdom. Kay's task involved use of a row of twelve light bulbs, each with a corresponding morse key. The lights came on in random order at 1·5 sec intervals, and the subject had, in different trials, to respond by pressing the keys corresponding to the lights that had been on 1, 2, 3 or 4 earlier in the series. Each subject was tested on only one occasion, either just before going out on a flight or immediately on return. The results are shown in Table I. The younger subjects were, on average, consistently poorer after flights. Surprisingly, the older were not so, but scored consistently less than the younger both before and after flights. It looks as if the effects of fatigue and age did not summate, perhaps because the lower achievement of the older subjects meant that they were exhausting themselves less. A rather similar pattern of results, with older

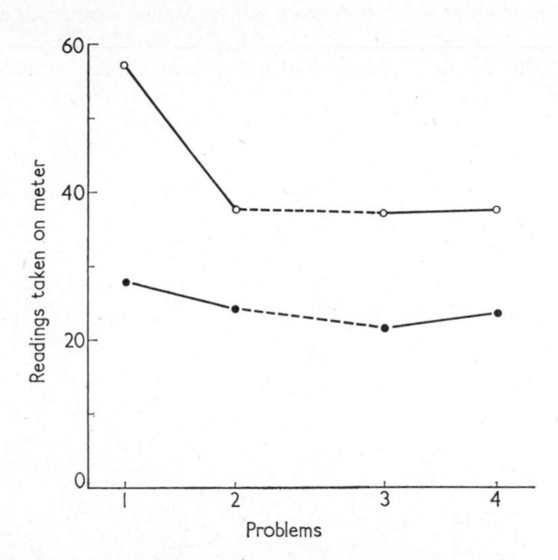

FIGURE 8. Numbers of readings taken on a meter by radio officers of civilian aircraft in solving electrical problems (Welford, Brown and Gabb, 1950).

The upper curve shows results for twelve subjects who solved their first two problems immediately on return from a flight and their last two after at least eight days' stand-down. The lower curve shows results for twelve subjects whose first two problems were after stand-down and whose last two were after flights.

subjects achieving less but showing less fatigue effect, was obtained by Botwinick and Shock (1952) with an adding task.

Four further points from experiments on fatigue may be mentioned more briefly.

e. Temporary improvement of performance. Somewhat surprisingly, the first sign of oncoming fatigue is sometimes an improvement of performance, in the sense that it becomes more active and achievement rises: deterioration does not set in until later. For example, in another experiment by Welford, Brown and Gabb (1950) on the same radio officers as took part in the experiment that formed the basis of Fig. 8, performance at a plotting task was better after a trip with a relatively easy schedule than it was before going out. Performance was, however, poorer after a trip with a more arduous schedule than it was before flight, and the rise after an easy trip occurred only with the relatively easy plotting task: it did not occur with the more difficult electrical problems. Nor did it appear among stewards tested with the plotting task. The temporary improvement of performance seems to be confined to narrow limits, which depend on the degree of fatigue, the difficulty of the task and the grade of the subjects.

Table I. *Results Obtained by Kay (1953) in a Running Short-term Memory Task given to Civilian Aircrew, either before or immediately after flight. (Each figure in the table is the percentage of correct responses out of 36 for each subject. Signals were paced at 1·5 sec intervals.)*

Subjects under 30 years of age	*No. of subjects*	*Number of lights back in series to which response had to correspond* 1	2	3	4
Radio officers					
Before flight	5	100	69	69	41
After flight	8	97	63	53	38
Stewards					
Before flight	13	97	74	50	44
After flight	15	87	70	42	32
Stewardesses					
Before flight	14	98	77	46	41
After flight	18	85	55	40	30
Subjects over 30 years of age					
Radio officers					
Before flight	15	99	58	41	33
After flight	13	90	64	46	36
Stewards					
Before flight	11	89	64	34	34
After flight	13	87	62	34	31

f. "Transfer" of fatigue effects. The results in Fig. 8 show not only that the radio officers who met the test first after a flight did worse than those who met it first before going out, but that they continued to do worse when retested after a week of "stand-down" than did the other group when retested after a flight. This means that the performance on the first test, whether affected by fatigue or not, had somehow "transferred" to the second. The probable explanation is that subjects learnt on the first occasion a technique that was affected by whether or not they had just been on a flight and that they continued to use the same technique on the second occasion. We have, on this view, an example of fatigue affecting performance long after the fatigue state itself has passed. Similar results were obtained for the radio officers and the stewards with the plotting task already mentioned.

g. Reduction of fatigue effects by familiarity. Fig. 8 also shows that the differences between performances of those tested before and after flight were much greater for the first than for the second problem. Similar results appeared in several other experiments and imply that fatigue effects are greater for novel than for familiar tasks: some capacities called into play to deal with new material seem especially susceptible to fatigue. What these capacities are we cannot at present say.

h. Specificity of fatigue effects. Fatigue effects usually seem to be highly specific to the performances that produced them, leaving other performances unaffected. As a consequence, tests in which a subject is taken off his main task and put for a brief period on to another to assess his state of fatigue are seldom successful: little change of performance at the tests occurs until such extreme states of exhaustion are reached that no test is needed to supplement common observation. However, examples do exist of what appears to be fatigue produced by one task affecting performance at another. The experiments on aircrews mentioned above are a case in point. There is also the experiment by Rey and Rey (1963), who found that CFF became lower, reaction time longer and rate of tapping less regular after 45 min of work at a cancellation task. They further found that cancellation was less accurate and fatigue effects were greater when the task was done under fluorescent illumination fed from a 50 cycle source than under similar illumination fed from a 10,000 cycle source. Complementary to these results is that of Saldanha (1957), who found that accuracy at his vernier setting task returned more after half an hour of rest than after a similar period spent cancelling Landolt rings. Such examples as these deserve to be noted and followed up. They suggested that fatigue effects do spread to some extent, perhaps first to related functions and gradually to more remote ones. The nature and extent of such spread is, however, as yet almost entirely unknown, and as measures of fatigue, tests on different tasks seem inevitably less effective than do direct measures of the performance concerned.

2. *Explanatory models of mental fatigue*

The attempt to postulate more fundamental explanations of mental fatigue than those so far considered is made difficult by the fact that many other influences, such as local muscular fatigue or, as with the aircrews, anoxia or loss of sleep, may affect performance as well as any true mental fatigue. Taking the evidence as a whole, however, two main hypotheses about the nature of mental fatigue seem reasonable.

a. Local neural impairment. The traditional assumption is that some group of nerve cells concerned with the performance that fatigues, or with some essential link in it, becomes insensitive or unresponsive through continued activity. Such a view explains well the similarity of some fatigue effects, such as fall of CFF, to those of brain injury. It can account for slowing of performance by assuming that some stage in the sensory-motor chain requires a stronger stimulus to operate it and that a given level of stimulation can be integrated over time. Blocking is accounted for by assuming that the breakdown may be of only short duration. Loss of short-term retention would result if the self-maintaining neuronal circuits, on which it must almost certainly depend, became insensitive, so allowing the memory traces to decay.

Rest, on this view, would have its effect by allowing for recovery of the nerve cells involved (including the replenishment of their supplies of oxygen). Overactivity and improved performance in the early stages of fatigue could be accounted for either by analogy with the recruitment effect found in neuromuscular fatigue or by assuming that, faced with incipient failure, the subject makes compensatory increases of effort that may, for a time, more than offset losses due to fatigue. Such compensatory efforts are perhaps seen in the results of Bills and Shapin (1936), who found that deterioration of performance could be postponed by pacing the task a little faster than the comfortable rate instead of allowing the subject to work at his own speed.

b. Increase of neural "noise". An alternative view can be based on one put forward by Crawford (1961) as a possible explanation of fatigue in car drivers. He suggested that such fatigue might result from the accumulated after-effects of stresses and annoyances from other road users during prolonged spells of driving. If so, it could be regarded as due not to underfunctioning of particular brain mechanisms but to overactivity in the brain or parts of it. It has been recognised for some years now that stress, effort and emotion tend to produce increased "arousal", accompanied by a rise of muscular tonus and of ambient, non-specific activity in the brain—what one might reasonably term "neural tonus" (Lindsley, 1951; Hebb, 1955; Duffy, 1957; Malmo, 1959). Moderate rises in such ambient activity appear to make the organism more responsive than normal (Freeman, 1933), probably because the cells in the brain are brought to a state in which they are more readily fired by stimuli from sense organs or other parts of the brain. Severe rises have, however, been found to impair performance, especially high-grade mental activity. Several suggestions of why this should be so have been made, but the simplest is that high levels of ambient neural activity increase the sensitivity of cells in the brain to such an extent that they tend to fire spontaneously (Welford, 1962b). If this happened, the brain would become "noisy" with the random spontaneous activity, and signals would be blurred. With severe rises, firing might occur to such an extent that an appreciable number of the cells would be active and thus not available for carrying messages, so that the capacity of the system would be reduced. If this happened, the effects would be most pronounced for the highest-grade activities, because it is they that would require the greatest capacity.

Such rises in sensitivity could result directly from the neural activity produced by immediately present stresses or sometimes perhaps from the facilitative after effects, often lasting several minutes, of prolonged stimulation of nerve cells. The cells would be made hypersensitive; though in moderate degrees this would raise the sensitivity of the system, in more severe degrees it would make cells liable to be fired by any random activity in the brain and especially by after-discharges from previous activity. Discussions in this area have tended to stress general arousal or activation, but the same principles could clearly apply to restricted brain mechanisms.

On this view, fatigue results when ambient activity or hypersensitivity, either general or local, rises beyond some optimum level. Such a theory would account for sensory fatigue directly in terms of the blurring effects of neural noise. It would account for slowing of performance by arguing that the effective strength of stimuli from one part of the sensory-motor mechanism to another should be measured in terms of signal-to-noise ratio and that a rise in the noise level thus implies a need for more powerful signals in order to maintain the ratio required. Any randomness from moment to moment in the level of ambient activity would mean brief periods of specially intense noise, which might account for blocking or, at a lower level, brief periods of facilitation which might allow unwanted signals to cause distraction. Loss of short-term retention would be assumed to result from disturbance rather than decay of the memory traces. Rest would allow excess activity or sensitivity to die away. Overactivity and improved performance during the early stages of fatigue would be accounted for directly by the facilitatory effects of ambient activity or of continued stimulation. This view could also account well for the irritability and difficulty of relaxing or sleeping often observed after a long period of taxing mental work.

Present data on fatigue seem to be equivocal in the sense that they are consistent, given plausible supplementary hypotheses, with either view. Distinction between the two "models" appears to need a substantial amount of new research carried out by a combination of physiological and psychological methods. The two models may well both operate in different circumstances, and it seems reasonable to suggest that new research should aim at establishing the conditions in which each applies rather than justifying one theory to the exclusion of the other.

D. Vigilance

We must now digress from studies of fatigue to consider research on so-called "viligance tasks", which are in a sense complementary to those of fatigue and throw light on the problem of distinguishing fatigue from monotony or boredom.

1. *Experimental findings with vigilance tasks*

The extensive experimental work on the maintenance of vigilance that has been done since World War II arose from the fact that when radar and other watchkeepers look for infrequent signals they tend to miss them when they come if they have been on watch for an hour or so. Research in this area has gained attention not only because of its interest to the armed services, but for its obvious implications for inspection and monitoring work in industry and because it raises important theoretical issues.

The initial experiments on synthetic radar displays and other laboratory tasks made it clear that the proportion of signals detected fell sharply over a period of $\frac{1}{2}$ hr or so when the signals were faint, brief and infrequent. With synthetic radar displays showing signals at unpredictable intervals, it could be argued that the subject, in order to be sure of seeing them when they occurred, would have to be making a rapid and continuous series of checks. These could be fatiguing, and the fall in performance might therefore be a fatigue effect. Other laboratory tasks have, however presented the subject with clearly observable signals at regular intervals and required him to detect occasional signals that are slightly different from most of the others. The intervals have then been too long (1 to 5 sec) to make a fatigue theory tenable: some other factor is clearly involved.

Of the "pure" laboratory tasks probably the best known is the original Clock Test used by Mackworth (1950). The subject sat facing a circular dial on which a 6-in pointer moved in steps of 12′ of arc once per sec. Occasionally the pointer made a jump of 24′, and the subject's task was to report these by pressing a key. The double movements occurred twelve times in 20 min (1 in 200 movements on average) at irregular intervals. Other tasks have involved the use of a wide variety of signals, such as faint spots of light (e.g. Wilkinson, 1961) or small changes of sound (e.g. Mackworth, 1950), or occasional features of a regularly presented series of signals, such as the detection of "odd, even, odd" in a series of digits (e.g. Bakan, 1952). The decrements seem to occur with both visual and auditory signals and may be shown not only by missing signals altogether, but by slower responses to those detected (e.g Wallis and Samuel, 1961). The fall in performance is not due simply to lack of activity on the subject's part; thus, Whittenburg, Ross and Andrews (1956), when they repeated the Clock Test requiring the subject to respond to every jump, making different responses to small and large, found decrement still to occur in the detection of the occasional large jumps. Nor is it due to lack of readiness for the signals, as Wilkinson (1961) showed that decrements still occurred when warnings of signals were given or even when the subject himself determined the times at which they appeared: the subject, needless to say, did not know whether or not any particular signal would require a response, but he had every opportunity to be ready if it did.

Mackworth showed that performance could be temporarily restored by a telephone message during the "watch". Decrements could be reduced by raising the signal strength in synthetic radar displays and prevented altogether by telling the subject, each time after a signal had appeared, whether or not he had detected it correctly—in other words by giving him what is known as "knowledge of results" of his performance. Subsequent work has shown that detection is improved even if knowledge of results is incomplete (Wilkinson, 1964) or false (Loeb and Schmidt, 1963). It is also improved by brief rest pauses (Bergum and Lehr, 1962) or when, in a series of signals some of which are "wanted" and some not, the proportion of wanted signals is increased (Colquhoun, 1961). Several experiments have explored the possibility of raising detection rate by combining the vigilance task with another—in other words raising the wanted signal rate by giving signals from two different tasks. The results, however, are equivocal (Wallis and Samuel, 1961), probably because any improvement due to the extra signals from the second task may be offset by the subject having to divide his attention between the two tasks. The detection rate in the primary task does, however, rise if knowledge of results is given about the secondary—the effects of knowledge of results seem to spread to both tasks (Baker, 1961).

2. *Theories of vigilance*

Theories to account for decrement of vigilance have been critically discussed by Broadbent (1958) and by Broadbent and Gregory (1963), and no attempt will be made to go over them again here. The main present theoretical question is whether or not the experimental evidence can be adequately accounted for by assuming that the detection of signals is impaired because of a lowering of the "arousal" discussed under our second theory of fatigue. Such a lowering could reduce the sensitivity of perceptual mechanisms to the extent that signals might sometimes fail to "get through" at all—in short, it might lead to another variety of "blocking". At first sight this theory is inadequate, since it is commonly held that arousal level depends on the amount of sensory stimulation impinging on the organism. If this were true, we should expect decrements to be less if detection of visual signals were attempted against a background of loud auditory noise, which might be expected to raise the general level of stimulation without masking the signals. Detection is improved a little under these conditions, but the characteristic decline still occurs (Dardano, 1962). Further, the activity involved in those experiments already noted in which subjects had to respond to all signals, both wanted and unwanted, would, as Broadbent (1958) has pointed out, have provided a considerable stream of kinaesthetic and tactile stimulation, yet did not prevent the infrequent wanted signals being often responded to as if they were the more frequent unwanted signals.

Such negative evidence can, however, be regarded not as rendering impossible the theory of vigilance as a state of lowered arousal, but as throwing doubt on the assumption that arousal is a function of the level of stimulation impinging on the organism. It is, indeed, clear from other studies of arousal and from common observation that such an assumption is untenable. If it were true, we should be unable to sleep in a room overlooking a noisy street, and many people who are used to such noise sleep well. The stimulation that causes arousal may reasonably be identified as that to which the organism has not habituated, that is to say, stimuli conveying information of significance to the organism. This is not quite the same thing as saying that arousal is a function of presented information in the "information theory" sense, because information thus defined deals with objective signals without taking the state of the organism into account. Admittedly such a view, if it is not to be circular, requires us to specify what is "of significance" to the subject using criteria other than arousal effects, but it seems to be possible in principle to do this by taking account of the relationships between present events and the subject's previous experience and future aims. In most experiments these will be sufficiently defined by the instructions.

In these terms, the stimuli which will be ineffective in maintaining arousal are likely to be, first, the ones that are repeated or continuous, because it is to these that habituation most easily occurs. Occasional changes of activity or interruptions, just because they are occasional and therefore of potential significance, restore alertness, but would probably cease to do so if repeated frequently, since one can habituate to repeated change. Rest pauses permit a shift of attention to new stimuli, that is, to ones more likely to be alerting. The beneficial effects of an increase in the frequency of wanted signals could be attributed directly to an increase of significant stimulation, and the same could be true of knowledge of results in the sense that the information given is both a guide to future performance and an indicator of one's present level of efficiency.

Perhaps the most significant direct evidence in favour of this theory is Mackworth's (1950) finding that no appreciable decrement in performance occurred during a 2 hr watch in subjects who had been given 10 mg of benzedrine shortly before the experimental session began: the drug, which has a known stimulating effect on the arousal mechanism, seems to have prevented any fall in vigilance. Further evidence is contained in the improved performance at a vigilance task found when body temperature was raised, a condition in which subjects showed signs of hyperactivity, such as restlessness and irritability, and in which performance at an adding task was impaired (Fox, Goldsmith, Hampton and Wilkinson, 1963). Indirectly the theory has the advantage that the alternative theories put forward can all be conceived as specifying particular conditions under which arousal diminishes or ways in which diminished arousal affects performance. Thus the blocking proposed by Broadbent (1958) as the reason for signals being missed has already been mentioned as one possible result of lowered arousal. Expectation effects suggested by Deese (1955) and others as affecting vigilance could well be one factor lowering arousal, as Deese himself recognised. Expectation has been found in many studies of other tasks (e.g. Freeman, 1933) to influence the level of "activation", which can be roughly equated with arousal. Motivation, suggested by Mackworth (1950) as a factor in vigilance, is now generally assumed to be intimately bound up with arousal (Hebb, 1955). Lowered arousal has also been regarded by Broadbent and Gregory (1963) as the most likely explanation of their finding that subjects engaged in vigilance tasks not only detected fewer signals as time went on, but also showed less tendency to report signals when in fact none were present. They argue that, for detection to occur with any given degree of confidence, the signal and any neural noise in the system must together exceed a critical level, and that if the average neural noise level falls, as it is likely to do in states of lowered arousal, the critical value will less often be reached.

3. *Vigilance and fatigue*

We are now in a position to consider the view that some at least of the decrements observed in "mental fatigue" are due to the monotony or boredom of the tasks rather than to any true fatigue effect. If we can equate the effects of monotony or boredom with the decrements observed in the performance of vigilance tasks, such a view of fatigue would mean that the general arousal of the subject, or the activation level in some essential part of the sensory-motor mechanism, gradually diminished owing to the lack of significance of a repetitious task and that this fall more than offset any effects of effort and incentives. Such a view would often be the same as saying that the subject became habituated to the task or that his initial positive interest in it became satiated, so that it no longer "stimulated" him. Such a theory, it is clear, is diametrically opposite to the second of those we have already outlined as a possible explanation of fatigue.

An indication that not all mental fatigue can be accounted for in terms of diminished arousal is found if we compare the effects of benzedrine in Mackworth's experiments and in Davis's (1948) Cambridge Cockpit experiments. We have already noted that benzedrine prevented any appreciable deterioration of performance in Mackworth's subjects over a period of 2 hr. Davis, on the other hand, found that benzedrine had no consistent effect on the deterioration of performance in his task over the same period of time. The lead given by these two sets of experiments could well be extended by trying the effects of other factors preventing loss of vigilance on performance at tasks normally regarded as causing mental fatigue: in particular the effects of knowledge of results would seem worth studying. As in trying to distinguish between theories of fatigue, there seems to be a need for co-operative work between physiologists and psychologists aimed not at establishing one view to the exclusion of the other but at delimiting the areas in which each holds true.

E. Fatigue at Work

Surprisingly little is known about fatigue outside the laboratory in the wider contexts of industry and everyday life. In spite of the fact that industry spends many millions a year on "fatigue allowances", these cannot be calculated on any rational basis and are often little more than political stratagems to mitigate an otherwise unacceptable piece-rate. What little is known may be outlined under three heads.

1. *Output measures at different times during a shift*

Presumptive evidence of some kind of fatigue effect during an industrial shift is contained in the classical reports of the Industrial Fatigue Research Board (later renamed Industrial Health Research Board). These reports showed not only that shorter working shifts led to higher hourly output (Osborne, 1919; Vernon, 1919), but also that a net reduction in working hours could sometimes lead to a net rise in total output (Vernon, 1920a, b). It seems as if industrial workers tend to anticipate fatigue and to distribute their efforts over a working shift in much the same way as do long-distance runners and cyclists (Ward, 1950) and subjects in some experimental tasks (Bills and Brown, 1929): they work faster throughout if the shift is short.

It was also found in many operations that the rate of production rose a little at the beginning of the shift, presumably owing to some "warm-up" effect, fell towards the end of the morning, recovered somewhat during the lunch break and fell again during the afternoon. The falls could be largely prevented by introducing brief rest-pauses of 2 to 15 min shortly before the fall would otherwise have begun (Vernon and Bedford, 1924; Wyatt and Fraser, 1925; Wyatt, 1927). The early work has been summarised by Chambers (1961) and other writers on industrial psychology.

The interpretation of these findings is far from easy. When work involves heavy muscular effort, or is carried out in hot and humid conditions, heat stress may slow work down or make pauses necessary. This, however cannot be the complete explanation of the changes found, because many of the jobs were light and done under reasonably easy environmental conditions. The fact that "spurts" were sometimes found to occur during a shift, or even right at the end, need not weigh against a fatigue hypothesis in view of what has been found about motivation effects in laboratory studies of fatigue. Some effects may arise from physiological rhythms over the 24 hrs and from physiological changes related to time from last meal; in this connection we may note that Wyatt and Fraser (1925) found breaks more effective if refreshments were taken. Again, however, such factors are not a sufficient explanation of the changes in production observed.

Apart from the suggestion of a true fatigue effect, two main alternative explanations have been offered.

a. Monotony. Studies by Vernon (1924) and by Wyatt and Langdon (1932) showed that output per hour could be raised by switching from one job to another, and Wyatt and Fraser (1928; also Wyatt, Fraser and Stock, 1929) showed that switching also reduced irregularity in the speed of individual cycles in a repetitive operation. These findings have been taken to support the hypothesis that the falls of production towards the end of a shift are due to monotony, although in view of what is now known of the specificity of fatigue effects, they could still reasonably be attributed to fatigue. There is, indeed, some support for this view in the finding by Wyatt and Fraser (1928) that the benefits of switching were reduced if the jobs concerned were closely similar.

More telling evidence in favour of a monotony theory is that obtained by Wyatt and Langdon (1937), who compared output figures with answers to questionnaire items about the incidence of boredom and showed that output could be raised by "music while you work". Wyatt and his co-workers have also shown that decrements of performance tend to be greater among workers of higher intelligence who, it might reasonably be presumed, are less "absorbed" by their work.

It seems fair to say that present evidence does not enable us to distinguish clearly between fatigue and monotony as explanations of these findings; it will be necessary to do much more detailed studies based on the factors found to be significant in laboratory experiments.

b. Ancillary activities. Dudley (1958) has suggested that the variations in output over a shift may not be due to fatigue or monotony but to the way in which the job is organised. In his studies of repetition work he found the usual output curves showing a rise at the beginning of the shift and a fall towards the end. However, when he made detailed studies of the work done, he found that the times for individual work cycles did not vary. In short, the initial rise and later fall were due to production work being interspersed at the beginning and end of the shift with other activities, such as getting out and putting away tools. These studies emphasise the importance of looking at jobs in detail, but they do not necessarily preclude an explanation in terms of fatigue. We have seen in the experiment by Singleton (1953) that there was little, if any, change in the speed of actual movements made during a working spell: fatigue effects showed in lengthening of the times between one movement and the next. Although the time scale in Singleton's study was much shorter than in Dudley's, it seems possible that fatigue effects did not affect cycle times in the operations Dudley studied, but tended to make the men turn briefly to other activities as a means of taking a rest. Such rests might not be taken consciously: the whole process was more likely to be unwitting. Certainly Dudley's explanation seems hardly adequate to account for the substantial effects of brief rest pauses found by Vernon, Wyatt and their colleagues.

2. *Performance of long-distance lorry drivers*

Several researches on long-distance lorry drivers in the United States have been summarised by McFarland and Moseley (1954) and discussed together with some other results by Crawford (1961). Lorry drivers tested with various psychomotor tasks after long periods of driving have shown changes of performance similar to those found in laboratory experiments on fatigue. Similar changes have also been found in studies of driving performance. Lauer and Suhr (1958) have shown that ill effects can be greatly reduced by frequent rests. On the other hand, accidents and near accidents have been found to be more frequent at the beginning of a trip, or after only a few hours of driving, than at the end of a long haul: in one study, McFarland and Moseley observed twenty-two out of forty-eight near accidents to occur during the first 2 hr of driving and only four in the last 2 hr of a 9 hr haul.

It is difficult to say how far the effects of prolonged driving are to be attributed to true fatigue effects and how far to monotony. We have already mentioned that Crawford (1961) has suggested that fatigue may result from stress and annoyance at other road users. On the other hand, certain "hypnotic" effects of long-distance driving have been observed and indicate that monotony may be important. It is perhaps reasonable to suggest that driving is a task both fatiguing and sometimes, especially on long American roads, monotonous, and that the observed effects are a balance between or combination of the two. This view would provide a plausible explanation of the hallucinations often reported by long-distance drivers: the fatigue effects of stress would tend to make the brain overactive, but the impairment of sensitivity due to monotony would mean that spontaneous activity would be less firmly controlled by sensory input than it would be in normal circumstances.

Crawford has provided a number of suggestions for further research in this field; if followed up, they would give us valuable information not only about the changes that come with long-distance driving but also about several general aspects of fatigue.

3. *Long-term industrial fatigue*

About the long-term tiredness, lassitude and lack of enthusiasm often termed "chronic fatigue" there is little systematic knowledge. Subjectively, such states seem to arise from long-continued overload, leading to mild chronic overtenseness, which in severe forms can lead to inability to concentrate or make decisions, irritability and feelings of futility—in other words they seem to be examples of the second type of mental fatigue effect outlined earlier. It may be that, just as in discussing the work of Bills (1931) and of Bertelson and Joffe (1963) it was necessary to postulate short-term and long-term effects operating over seconds and minutes, respectively, so there may be still longer-term effects extending over, say, a week or a year and dissipated by rests at weekends and annual holidays. Such a hierarchical organisation almost certainly exists for performance and for behaviour generally (Welford, 1962c), and perhaps fatigue effects can occur at any of the same hierarchical levels.

On the other hand, some of these effects appear to be coupled with low morale and thus with various factors of social and industrial organisation, such as type of leadership, seeing "results" for one's work and avoidance of holdups in the flow of production. Factors of this kind could reasonably link long-term morale to the same kind of mechanisms that we have considered in relation to monotony and boredom. Holdups in production, leading to waiting and idleness, are an obvious case in point. Seeing results of work has a clear affinity to the "knowledge of results", which has been shown to maintain viligance. The same is perhaps true of leadership: it is often claimed that morale is better with a "democratic" type of leadership, meaning that the worker, though taking orders from management, is also readily able to exert an effect on management. Looking at such a leadership relationship in cybernetic terms, we can say that the servo loop from management to worker and back is completed, just as it is by knowledge of the results of performance.

F. The Present Research Position

In 1951, at the end of a symposium on fatigue held by the Ergonomics Research Society, I suggested that the position then could be described as "the end of the beginning" (Welford 1953). Many classical experimental studies existed, and it was becoming possible to step back from them to see how they looked in the perspectives of the new theoretical concepts emerging in psychology at the time. Now, some fourteen years later, it seems that considerable progress has been made, partly by the gathering of new knowledge and partly by sorting out facts and ideas put forward previously. At the present time four points seem to stand out.

First, fatigue is now clearly recognised as a blanket term covering a variety of processes in many different bodily mechanisms. These are, to a much greater extent than had previously been realised, neural mechanisms in the brain. Much of the confusion that has arisen hitherto in discussions on fatigue has been due to the fact that closely similar phenomena in the areas of slowing, blocking, disorganisation of performance and phases of hyperactivity can have widely different causes. We cannot distinguish between these by studying the crude phenomena alone, but need to gather ancillary evidence and to consider the precise nature of the tasks being performed. Secondly, progress has been made in drawing a distinction between fatigue and monotony or boredom in terms of overloading as opposed to underloading, but methods of separating the two in experimental studies are still not fully worked out and the interpretation of much previous work must be in some doubt. Thirdly, the main principles of how to prevent the adverse effects of monotony have been made fairly clear by the important researches done in the study of "vigilance". Fatigue effects still need considerable further investigation, but there are at least two theoretical "models", one postulating underactivity and one overactivity in the brain, that have enough support from other physiological and psychological studies and are sufficiently different from one another to provide a powerful stimulus to research. Lastly, much of the advance that has been made has resulted from attention to details of performance and of experimental conditions; the necessity of this as a means of working out the "mechanism" behind any observed effect is becoming increasingly clear.

There is still much work to be done. Many further studies are needed in the experimental laboratory to distinguish between theoretically different positions. We still know little about industrial fatigue. We know practically nothing about chronic fatigue as it impinges on everyday experience.

Apart from direct studies of fatigue itself, questions arise about how fatigue is affected by other activities and agents. What, for example, is the effect of training upon susceptibility to fatigue? We know that physical training can increase blood supply to muscles, ventilation rate and other physiological processes in such a way that the capacity of the organism for muscular work is raised and its fatigability thus reduced. Are there analogous changes in mental operations? There is abundant evidence that practice at many sensory-motor and mental tasks leads to improved techniques and "strategies" that make performance better organised, and Wilkinson (1964) has suggested that the same processes can account for the limited improvements with practice found in some vigilance tasks. Whether, however, practice leads to any true increase in capacity, as opposed to improved use of capacity, remains doubtful. Further studies of mental fatigue might well be an appropriate means of obtaining information on this point and on the influence of many other factors such as stress, incentives and drugs.

Our ignorance of fatigue and of its relations to other factors has, however, been much more sharply defined in recent years. Developments of knowledge and theory in many fields have combined to prepare the ground for a new combined physiological and psychological attack that in another decade could lead to a much more definitive understanding not only of fatigue itself but thereby also of many other aspects of human performance and behaviour.

REFERENCES

Ash, I. E. (1914). *Arch. Psychol.* No. 31.

Bakan, P. (1952). *U.S.A.F. Human Resources Center Research Note* 52–7, Lackland Air Force Base, Cited by Broadbent (1958).

Baker, C. H. (1961). *Ergonomics*, 4, 311.

Bartlett, F. C. (1943). *Proc. Roy. Soc.* B, 131, 247.

Bartley, S. H., and Chute, E. (1947). "Fatigue and Impairment in Man." McGraw Hill, New York and London.

Berger, C., and Mahneke, A. (1954). *Amer. J. Psychol.* 67, 509.

Bergum, B. O., and Lehr, D. J. (1962). *J. appl. Psychol.* 46, 425.

Bertelson, P., and Joffe, Rachel (1963). *Ergonomics* 6, 109.

Bills, A. G. (1931). *Amer. J. Psychol.* 43, 230.

Bills, A. G., and Brown, C. (1929). *J. exp. Psychol.* 12, 301.

Bills, A. G., and Shapin, M. J. (1936). *J. gen. Psychol.* 15, 335.

Botwinick, J., and Shock, N. W. (1952). *J. Geront.* 7, 41.

Broadbent, D. E. (1958). "Perception and Communication." Pergamon Press, London, Paris, New York, Los Angeles.

Broadbent, D. E., and Gregory, M. (1963). *Brit. J. Psychol.* 54, 309.

Bursill, A. E. (1958). *Quart. J. exp. Psychol.* 10, 113.

Chambers, E. G. (1961). *Occup. Psychol.* 35, 44.

Colquhoun, W. P. (1961). *Ergonomics* 4, 41.

Crawford, A. (1961). *Ergonomics* 4, 143.

Darcus, H. D. (1953). *In* "Symposium on Fatigue" (W. F. Floyd and A. T. Welford, eds.), p. 59. H. K. Lewis & Co., London.

Dardano, J. F. (1962). *J. appl. Psychol.* 46, 106.

Davis, D. R. (1948). "Pilot Error." Air Ministry Publication A.P. 3139A. H.M.S.O., London.

Davis, D. R. (1949). *Quart. J. exp. Psychol.* 1, 136.

Deese, J. (1955). *Psychol. Rev.* 62, 359.

Denny-Brown, D. (1949). *Arch. Neurol. Psychiat.* 61, 99.

Douglas, C. G., and Koch, A. C. E. (1951). *J. Physiol.* 114, 208.

Dudley, N. A. (1958). *Inst. Prod. Engrs. J.* 37, 303.

Duffy, E. (1957). *Psychol. Rev.* 64, 265.

Edwards, H. T., Margaria, R., and Dill, D. B. (1934). *Amer. J. Physiol.* 108, 203.

Fox, R. H., Goldsmith, R., Hampton, I. F. G., and Wilkinson, R.T. (1963). *J. Physiol.* 167, 22P.

Freeman, G. L. (1933). *Amer. J. Psychol.* 45, 17.

Hebb, D. O. (1955). *Psychol. Rev.* 62, 243.

Hemingway, A. (1953). *In* "Symposium on Fatigue" (W. F. Floyd and A. T. Welford, eds.), p. 69. H. K. Lewis & Co., London.

Kay, H. (1953). "Experimental Studies of Adult Learning." Unpublished thesis, Cambridge University Library.

Lauer, A. R., and Suhr, V. W. (1958). Cited by Crawford (1961).

Lindsley, D. B. (1951). *In* "Handbook of Experimental Psychology" (S. S. Stevens, ed.), p. 473. John Wiley & Sons, New York; Chapman and Hall, London.

Lippold, O. C. J., Redfearn, J. W. T., and Vuco, J. (1960). *Ergonomics* 3, 121.

Lloyd, D. P. C. (1942). *J. Neurophysiol.* 5, 153.

Loeb, M., and Schmidt, E. A. (1963). *Ergonomics* 6, 75.

Lundervold, A. (1958). *Ergonomics* 1, 226.
McFarland, R. A., and Moseley, A. L. (1954). "Human Factors in Highway Transport Safety." Harvard School of Public Health, Boston.
Mackworth, N. H. (1950). "Researches on the Measurement of Human Performance", *M.R.C. Memor.* Special Report Series No. 268. H.M.S.O., London.
Malmo, R. B. (1959). *Psychol. Rev.* 66, 367.
Merton, P. A. (1956). *Brit. med. Bull.* 12, 219.
Osborne, E. E. (1919). *Industrial Fatigue Research Board Report* No. 2. H.M.S.O., London.
Reid, C. (1928). *Quart. J. exp. Physiol.* 19, 17.
Rey, P., and Rey, J-P. (1963). *Ergonomics* 6, 393.
Saldanha, E. L. (1955). *M.R.C. Applied Psychology Research Unit Report* No. 243.
Saldanha, E. L. (1957). *M.R.C. Applied Psychology Research Unit Report* No. 289.
Schwab, R. S. (1953). *In* "Symposium on Fatigue" (W. F. Floyd and A. T. Welford, eds.), p. 143. H. K. Lewis & Co., London.
Seyffarth, H. (1940). Cited by Darcus (1953).
Sherrington, C. (1906). "The Integrative Action of the Nervous System." Reset edition 1947. Cambridge University Press.
Simonson, E. (1965). *In* "Behavior, Aging and the Nervous System" (A. T. Welford and J. E. Birren, eds.), ch. 21. Charles C. Thomas, Springfield, Illinois.
Singleton, W. T. (1953). *In* "Symposium on Fatigue" (W. F. Floyd and A. T. Welford, eds.), p. 163. H. K. Lewis & Co., London.
Vernon, H. M. (1919). *Industrial Fatigue Research Board Report* No. 1. H.M.S.O., London.
Vernon, H. M. (1920a). *Industrial Fatigue Research Board Report* No. 5. H.M.S.O., London.
Vernon, H. M. (1920b). *Industrial Fatigue Research Board Report* No. 6. H.M.S.O., London.
Vernon, H. M. (1924). *Industrial Fatigue Research Board Report* No. 26. H.M.S.O., London.
Vernon, H. M., and Bedford, T. (1924). *Industrial Fatigue Research Board Report* No. 25. H.M.S.O., London.
Vince, Margaret A. (1949). *Brit. J. Psychol.* 40, 23.
Wallis, D., and Samuel, J. A. (1961). *Ergonomics* 4, 155.
Ward, N. (1950). *Brit. J. Psychol.* 40, 212.
Welford, A. T. (1953). *In* "Symposium on Fatigue" (W. F. Floyd and A. T. Welford, eds.), p. 183. H. K. Lewis & Co., London.
Welford, A. T. (1962a). *Lancet* i, 335.
Welford, A. T. (1962b). *Nature, Lond.* 194, 365.
Welford, A. T. (1962c). *In* "Society: Problems and Methods of Study," (A. T. Welford, M. Argyle, D. V. Glass and J. N. Morris, eds.), p. 153. Routledge & Kegan Paul, London.
Welford, A. T., Brown, Ruth A., and Gabb, J. E. (1950). *Brit. J. Psychol.* 40, 195.
Weston,, H. C. (1953). *In* "Symposium on Fatigue" (W. F. Floyd and A. T. Welford, eds.), p. 117. H. K. Lewis & Co., London.
Whittenburg, J. A., Ross, S., and Andrews, T. G. (1956). *Percept. Mot. Skills* 6, 109.
Wilkinson, R. T. (1961). *Ergonomics* 4, 259.
Wilkinson, R. T. (1964). *Ergonomics* 7, 63.
Wyatt, S. (1927). *Industrial Fatigue Research Board Report* No. 42. H.M.S.O., London.
Wyatt, S., and Fraser, J. A. (1925). *Industrial Fatigue Research Board Report* No. 32, H.M.S.O., London.
Wyatt, S., and Fraser, J. A. (1928). *Industrial Fatigue Research Board Report* No. 52. H.M.S.O., London.

Wyatt, S., and Langdon, J. N. (1932). *M.R.C. Industrial Health Research Board Report* No. 63. H.M.S.O., London.

Wyatt, S. and Langdon, J. N. (1937). *M.R.C. Industrial Health Research Board Report* No. 77. H.M.S.O., London.

Wyatt, S., Fraser, J. A., and Stock, F. G. L. (1929). *Industrial Fatigue Research Board Report* No. 56, H.M.S.O., London.

CHAPTER 16

Emotion

S. M. HILTON

A. Introductory

It may seem hardly proper for a British physiologist to admit the existence of emotion and even more outrageous to include it as a fit subject for study and discussion. Yet worthy investigators have made it their life's work, and even

Sherrington performed experiments to test a theory of emotion. It is, naturally enough, in the study of the functional significance of emotional expression, as well as of the central nervous mechanisms underlying this form of behaviour, that the biologist, and especially the physiologist, has had a major role to play. In this context "emotional expression" does not refer simply to the patterns of activation of the facial muscles common to so many mammalian species, as documented by Darwin (1872), nor even to the widespread activation of the skeletal musculature that characterises such overt reactions as flight or attack; it extends to the various visceral and hormonal components of these complex responses and the elucidation of their biological role. Everyone who writes on this subject has to express his debt to Cannon (1929) for his book "Bodily Changes in Pain, Hunger, Fear and Rage", which is a landmark in the field and still one of the most stimulating accounts of the subject. It no longer needs emphasising that the reactions are adaptive, in so far as they lead to the preservation of the organism. Cannon put this idea on a firm foundation by showing how all the bodily changes then known to occur as components of these reactions could be viewed within this framework. The experimental basis for his conclusions was provided by the detailed studies on cats carried out mainly by Cannon himself and his many co-workers, but their relevance to the human subject also, in health and disease, was comprehensively documented. Likewise the present account will have to rest largely on the results of animal experiments, in the confidence that the emergent patterns of reaction will have traces or even stronger reflections in the similar conditions of man. This assumption has been justified at whatever point it has been tested.

In this account we deal not with the alimentary and sexual reactions, whose biological implications are self-evident, but with the emergency reactions—as they have been known after Cannon—whose functional significance is most easily appreciated by reference to animal behaviour. Thus, there is no difficulty in understanding the value of the reactions of flight or attack for preservation of the individual animal, and the fascination for the physiologist then comes with the attempt to unravel the complicated stories, first of the direction and extent of the multitudinous visceral and hormonal changes occurring at different stages of these reactions and secondly of the interconnections of these changes and the various ways in which they contribute to the maximum efficiency of the organism during the emergency.

B. Anatomical Substratum

The regions of the central nervous system essential to the integration and performance of these emergency reactions have been established in several mammalian species, and the work so far carried out on man, though necessarily more meagre, points to the same general anatomical pattern.

1. *Nervous structures in hypothalamus and midbrain subserving defence reactions*

It was early recognised, after the pioneer work of Goltz (1892) with decorticate dogs, that the cerebral hemispheres are not essential for this form of behaviour. Cannon and Britton (1925) emphasised the significance of diencephalic structures for these responses, for which they coined the term "sham rage". The most extensive and detailed use of cerebral ablation, by Bard and his collaborators (Bard, 1928; Bard and Rioch, 1937; Bard and Macht, 1958), led finally to the conclusion that a large part of the brain stem plays a role in the mediation of most features of these responses, but that the hypothalamus must be intact for the behaviour to be well organised and of anything like normal intensity. The responses of these chronic decorticate or decerebrate preparations are elicited by mild cutaneous stimulation. Even when the level of decerebration is as caudal as at the beginning of the pons, pulling the skin of the back is sufficient to elicit most features of the "rage" reaction (Keller, 1932), and Bard and Macht (1958) obtained some features of the reaction, in response to electrical stimulation of the tail and to a loud noise, when the line of section was so far caudal as to remove the rostral part of the pons.

More precise demarcation of the brain-stem regions integrating these reactions came with the utilisation of localised electrical stimulation by means of implanted electrodes. Hess (1949) and his collaborators, in work carried out over many years, charted the regions from which various characteristic patterns of behaviour could be elicited in cats. Responses that began as alerting and culminated, if stimulation was sufficiently intense, in flight or attack were obtained most readily from that part of the hypothalamus just medial, ventral and lateral to the fornix (Hess and Brügger, 1943). The fully fledged responses were indistinguishable from those that would be called "fear" or "rage" in an animal responding to a natural environmental stimulus, but Hess and Brügger (1943) preferred to use the collective term "defence reaction" (*Abwehrreaktion*) for them. The excitable region for this reaction in the hypothalamus, as located by them, is illustrated in Fig. 1. Their results indicated that the caudal part of the hypothalamic region connected with the central grey matter at the midbrain level. Abrahams, Hilton and Zbrożyna (1960b) found the region to be even larger, extending laterally at the midbrain level to occupy an area of the tegmentum ventral to the superior colliculi. Their investigation had begun from an interest in the sympathetic vasodilator nerve supply to the skeletal muscles, which was known not to be involved in any of the homoeostatic circulatory reflexes, but could be activated by electrical stimulation within the hypothalamus and dorsal midbrain (Uvnäs, 1954; Lindgren, 1955). There proved to be such a remarkable identity of the regions, in the hypothalamus, central grey matter and midbrain tegmentum, from which defence reactions were elicited in the conscious

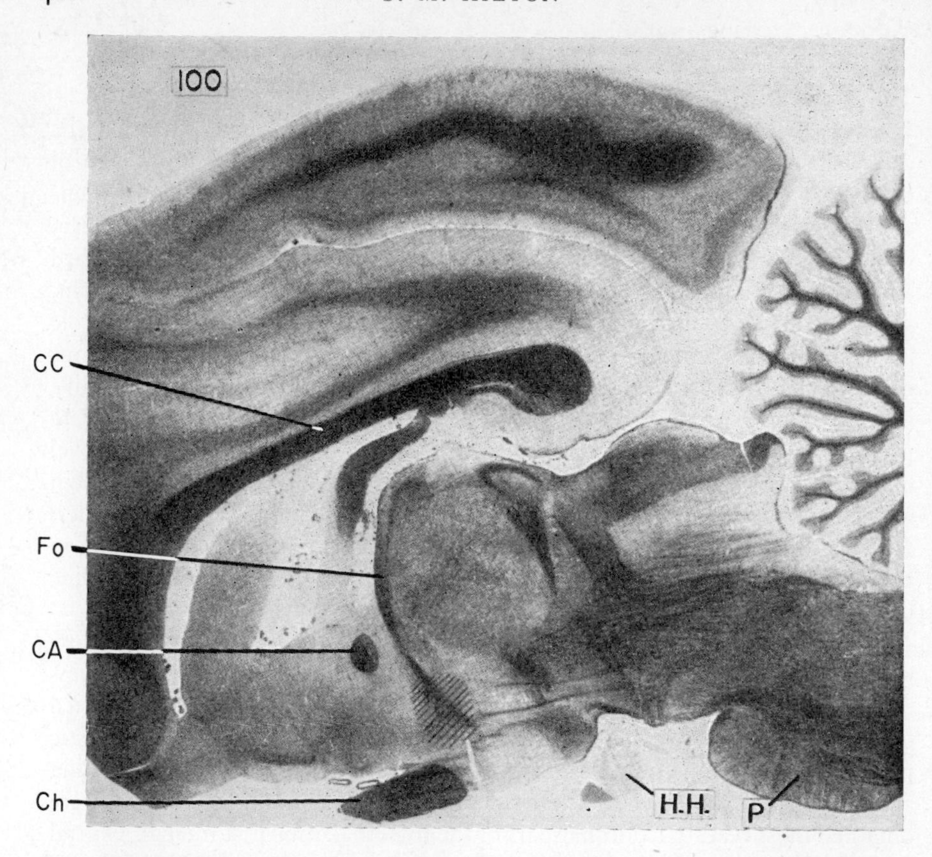

FIGURE 1. Paramedian sagittal section of cat's brain. Hatched area indicates perifornical region in hypothalamus from which defence reactions were most readily elicited. CA, anterior commissure; CC, corpus callosum; Ch, optic chiasma; Fo, descending column of fornix; HH, posterior hypothalamus; P, pons. (From Hess and Brügger, 1943.)

animal, and the active vasodilatation provoked in skeletal muscle of the same animal under anaesthetic, that this vasodilatation was virtually the best objective index of the reaction itself. It thus proved an invaluable aid to mapping the exact borders of the excitable regions for the whole defence reaction, because of the large amount of information that could be gained from exploratory experiments on anaesthetised animals with stereotactically oriented microelectrodes. The regions thus mapped are shown diagrammatically in Fig. 2.

Similar experiments on monkeys, though rather less detailed so far, have provided evidence of the same general topographical organisation (Masserman, 1943; Lilly, 1958, 1960). Recently, some results of electrical stimulation in the medial hypothalamus of human subjects have been reported (Heath and Mickle, 1960; Sem-Jacobsen and Torkildsen, 1960). Mild stimulation produced feelings variously recorded as restlessness, anxiety, depression fright and horror; stronger stimulation in the posterior hypothalamus produced "rage" reactions. Almost everyone who has carried out such experiments on animals has been impelled to conclude that electrical stimulation of these regions of the brain stem, though clearly the stimulus is abnormal, must produce changes within the brain (and thence throughout the organism) that are hardly distinguishable from those due to natural stimulation. It is beyond the scope of this chapter to discuss the interesting neurophysiological implications of this conclusion, but it seems proper at least to emphasise that, by use of some of the simplest techniques known to physiology, we can begin to explore the mechanisms responsible for phenomena that have often been thought the preserve of the psychologist or even the philosopher.

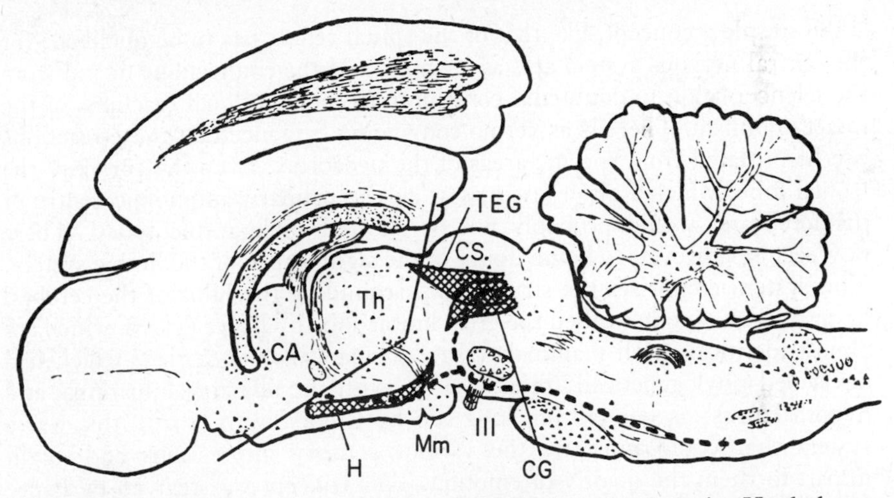

FIGURE 2. Diagrammatic paramedian sagittal section of cat's brain. Hatched areas represent regions in hypothalamus (H), central grey matter (CG) and mesencephalic tegmentum (TEG) apparently functioning as reflex centres for defence reactions. Heavy interrupted line represents vasodilator pathway from centre to periphery. CA, anterior commissure; CS, superior colliculus; Mm, mammillary body; Th, thalamus; III, oculmotor nerve. (Modified from Abrahams *et al.* (1960b), by permission of *Journal of Physiology*.)

It is already clear that the function of these regions of the brain stem is not solely motor. Indeed, the ease with which stereotyped responses are elicited by natural stimuli in the various decerebrate preparations led Abrahams *et al.* (1960b) to conclude that these regions are reflex centres in the usually accepted sense. Cannon (1929) had already discussed the emergency reactions in these terms. It seemed curious therefore that interest in this concept should have waned at a time when so much evidence had accumulated for the existence of the sensory connections appropriate to this reflex function. As was pointed out by Abrahams, Hilton and Macolm (1962), the discovery of sensory pathways impinging on the brain stem—the afferent collateral system of Starzl, Taylor and Magoun (1951)—had only been discussed since that time in relation to the concept of the reticular activating system. Abrahams *et al* (1962) confirmed the findings of earlier workers and extended them a little, to show that potentials could be evoked in all parts of the integrative centre for the defence reaction, in the hypothalamus and midbrain, as a response to cutaneous, auditory and visual stimuli. As they concluded, connections undoubtedly exist that would enable these regions to function as a reflex centre for the defence reaction.

2. *Role of the limbic system*

So simple a concept, like that of the spinal reflex, has to be qualified; for the central nervous system acts as a whole, and the connections to and from the telencephalon undoubtedly contribute to the emotional reactions of the intact organism. There is as yet no convincing evidence that these reactions can be initiated from specific areas of the neocortex. Even the cortex of the frontal lobe, which Fulton (1949) saw as the primary autonomic centre of the forebrain, would probably not be allotted such a role today. There now seems, however, to be no doubt of the significance of the limbic system. This system comprises the structures surrounding the hilus of the cerebral hemispheres, together called the great limbic lobe by Broca (1878), which are almost identical in all mammals (Fig. 3). They include cortex, which had developed phylogenetically in association with the olfactory apparatus, and hypothalamus; as pointed out long ago by Elliot Smith (1919), this was a system that could integrate the various sensory impressions and might impart to them the quality of emotion. This concept was revived by Papez (1937), who gave it a much more definitive form. On the basis of clinical and experimental evidence, he proposed the view that the hypothalamus, anterior thalamic nuclei, cingulate gyrus and hippocampus, through their interconnections, form a circuit that can be excited from the neocortex, through the hippocampus, as well as through the hypothalamus, which "is accessible to both visceral and somatic impressions". Subsequent work has given strong support to this ingenious hypothesis, the region of the limbic system most directly connected with the hypothalamus and midbrain proving to be in the amygdala, a group of nuclei subjacent to the hippocampus, within the temporal lobe. Gastaut, Vigouroux, Corriol and Badier (1951) first showed that electrical stimulation of the amygdala of the cat produced defence reactions, and this has since been confirmed many times. According to Hilton and Zbrożyna (1963), the full reaction, including active vasodilatation in skeletal muscle, is produced by stimulation in the nucleus basalis amygdalae, pars medialis, and the efferent pathway from here to the hypothalamus takes a direct route along a narrow ventral band, as had been suggested by Ursin and Kaada (1960). The stria terminalis, which connects the anterior hypothalamus with the amygdala and had already been reported to contain fibres initiating defence reactions in the cat (Gastaut *et al.*, 1951; Fernandez de Molina and Hunsperger, 1959), was found to be afferent to the amygdala, thus perhaps completing a circuit with positive feedback (Hilton and

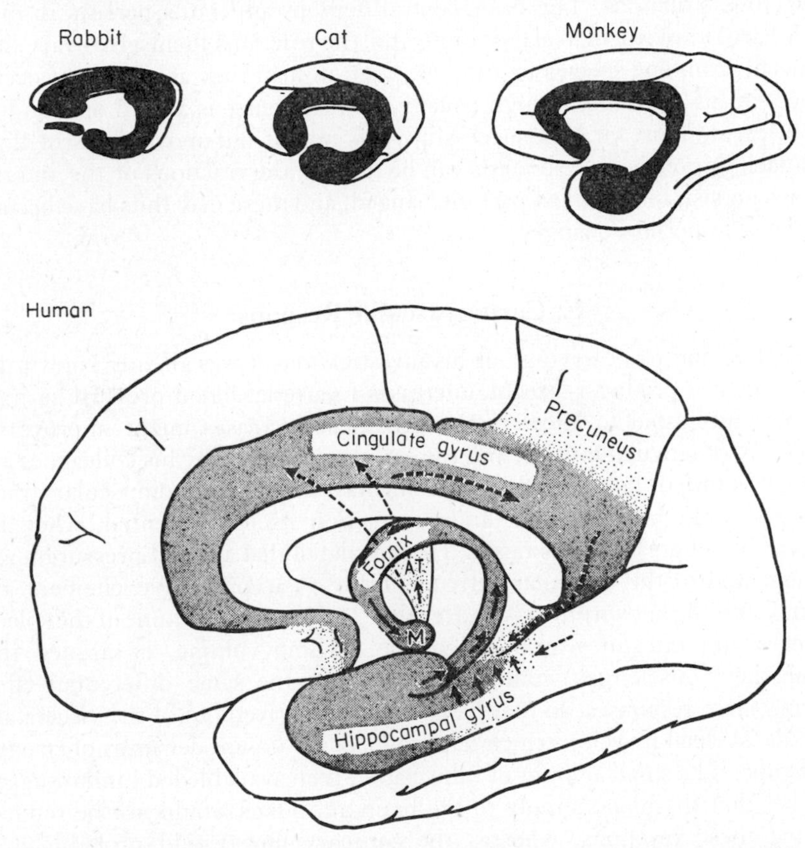

FIGURE 3. Above: medial aspect of brains of rabbit, cat and monkey, roughly proportional to size. Limbic lobe, surrounding hilus, shown in black. Below: medial aspect of human brain. Shaded area indicates some of the circuits in the limbic system, after Papez (1937). Mm, mammillary body; AT, anterior thalamic nuclei. (Modified from MacLean, 1954.)

Zbrożyna, 1963). Again, there is similar evidence implicating these regions of the limbic system in higher mammals. Indeed, some of the earliest experimental material was obtained by Klüver and Bucy (1939), who showed complete absence of "fear" and "rage" reactions in macaques after bilateral excision of the temporal lobes. The few studies incorporating electrical stimulation in the human amygdala have yielded similar reactions, including increases in blood pressure and heart rate (Heath, Monroe and Mickle, 1955; Chapman, 1960; Sem-Jacobsen, as quoted by Bovard, 1962). In these ways a firm edifice is being built. The evidence shows how certain specific regions of the brain became specialised early in vertebrate phylogeny for the mobilisation of all the resources of the organism in response to the signals of dangerous situations. The basic central nervous apparatus persists in man, with hardly any further elaboration, and the reactions themselves vary little in detail from one species to another. Even in man they are frequent occurrences in his daily life, though usually the occurrence is a mild one, and we have learnt, more or less successfully, to inhibit our movements of flight or attack. Nevertheless, so far as can be judged, the reactions of the internal organs persist, for the most part unchanged, and these may thus have become the most important changes.

C. Cardiovascular Response

When Cannon was carrying out his investigations it was already known that "excitement" produces sizeable increases in arterial blood pressure in man. He was interested to know how much these increases might improve the efficiency of fatigued muscle, and the early experiments by his colleagues and himself showed how, in the experimental animal, neuromuscular transmission could be facilitated, and contraction itself augmented, when the arterial blood pressure was raised. He speculated that a raised pressure might be essential to the muscular activities likely to accompany excitement and pain. From the known effects of adrenaline, a decrease in volume of the spleen, kidneys and intestines and an increase in limb volume, he argued that sympathetic discharges would probably exert the same differential effect during these responses, so that blood would be driven out of the viscera and into the skeletal muscles, which have to meet the urgent demands of struggle or escape. The small amount of information then available led him to suggest further that the blood supply to the brain and lungs would not be reduced during these reactions, whereas the coronary flow would probably even increase. He concluded: "This shifting of the blood so that there is an assured adequate supply to structures essential for the preservation of the individual may reasonably be interpreted as a fact of prime biological significance."

1. *Uniqueness of the muscle vasodilatation*

The results of subsequent investigations have gradually revealed the remarkable accuracy of Cannon's insight. Vasodilator nerve fibres in the sympathetic innervation of skeletal muscle were, in fact, described some ten years later

(Burn, 1938). These special fibres were then shown not to be activated in baroreceptor (that is, depressor) reflexes (Folkow and Uvnäs, 1948), but it was only with the observations of Eliasson, Folkow, Lindgren and Uvnäs (1951) that electrical stimulation in the hypothalamus became known to produce a large muscle vasodilatation through this innervation. They remarked that this muscle vasodilatation was accompanied by vasoconstriction in the skin and intestine, tachycardia, contraction of the spleen, pupillary dilatation and retraction of the nictitating membranes. They suggested that the vasodilator nerve fibres were being activated as part of the patterned response characteristic of the emergency reactions of "fear" and "rage". When Abrahams *et al.* (1960b) demarcated those parts of the hypothalamus and midbrain from which the muscle vasodilatation is evoked, they showed that these were indeed the regions responsible for the reactions of alerting, flight and attack in the conscious animal. As might be expected, the vasodilatation was also obtained as part of the so-called "pseudaffective" (or sham rage) reflex in decerebrate cats in which the hypothalamus had been spared. It seems clear that the vasodilatation is an integral part of the defence reaction; this is worth emphasising because, so far as we know at present, it is unique to this reaction and hence the best single index of its occurrence. By means of a technique in conscious animals for recording the changes of femoral venous blood temperature occurring during such vasodilator responses, it has been shown that the responses occur during defence reactions elicited either by hypothalamic stimulation or by natural environmental stimuli. It seems particularly significant that the vascular response is fully developed during the early alerting stage of the behavioural reaction, when the only outward signs are pupillary dilatation, pricking of the ears and an increase in rate of respiration (Abrahams, Hilton and Zbrożyna, 1960; 1964).

2. *Response considered as a preparatory reflex*

As already indicated, when the muscle vasodilatation is elicited in anaesthetised animals by electrical stimulation within the brain-stem region integrating the defence reaction, it is but one component of a complex pattern of cardiovascular adjustment. In addition to the features noted above, the contractile force of the heart and the cardiac output increase (Rosén, 1961a), and there is also evidence of widespread venoconstriction (Folkow, Mellander and Öberg, 1961; Hilton, 1963). Clearly, the cardiovascular reserves are being mobilised. A greatly increased cardiac output is being directed chiefly to the skeletal musculature, and the experiments on conscious animals suggest that this reaction occurs sufficiently early for the organism to be fully prepared within a few seconds of receiving a sudden stimulus to meet the immediate demands of widespread muscular activity, as in flight or attack. Abrahams *et al.* (1964) therefore suggested that this pattern of cardiovascular response should be classed as a preparatory reflex. The significance of such a reflex for survival of the individual is easily understood, yet it must be emphasised that, in order to mobilise the resources of the organism, such a preparatory reflex establishes for the time being a new state of internal equilibrium that represents a radical departure from the status quo. Mechanisms ordinarily operating to maintain the constancy of the internal milieu will therefore have to suffer a temporary abrogation.

This point is underlined by considering one further feature of the cardiovascular response. While it is developing, the pulse pressure and heart rate are increasing simultaneously. From such observations as these, which have been made many times, Bard (1960) was led to conclude that baroreceptor reflexes must be inhibited. This was tested recently (Hilton, 1963) by observing the well-known depressor effect of raising the pressure in a blind sac preparation of the carotid sinus on one side and showing that this effect is strongly inhibited and may be unobtainable during stimulation in the hypothalamic region for the defence reaction. The reflex bradycardia is similarly inhibited. Thus, the hypothalamus imposes its own pattern of reaction and overrides the homoeostatic reflex organised at a lower (medullary) level of the neuraxis. This suggests the potency of such a basic reaction not only for promoting the continued existence of the individual by helping it deal with an emergency, but also, if the reaction should be prolonged, for embarrassing the organism by weakening some of its defences and depleting its reserves. Gantt (1960) and his co-workers have carried out many experiments showing that the tachycardia and hypertension produced by painful stimuli in dogs can be readily conditioned and that these conditioned responses are remarkably stable. The few experiments carried out on conditioned muscle vasodilator responses by Abrahams *et al.* (1964) point in the same direction. It would be beyond the scope of this chapter to consider questions of so-called psychosomatic disease; but it can be seen that we are approaching the time when our understanding of the normal physiological reaction will be sufficiently detailed for us to test whether we can make the extrapolation from physiology to pathogenesis.

It would be presumptuous to make such a claim if we did not already know enough to feel sure that the pattern of response occurs in other species than the cat, including man. It is known, for instance, that a similar cardiovascular response, including the active muscle vasodilatation, is provoked in anaesthetised dogs on stimulation of the same regions of the brain stem as those from which the response is elicted in the cat (Eliasson, Lindgren and Uvnäs, 1952; Lindgren and Uvnäs, 1953; Lindgren, 1955). The cardiac responses have been obtained in conscious dogs on stimulation of points in the posterior hypothalamus (Smith, Jabbur, Rushmer and Lasher, 1960) and when they are subjected to the stimulus of a sudden shower of water (Charlier, Guz, Keatinge and Wilcken, 1962). The last-named investigators also reported that the arterial blood pressure rose, though total peripheral resistance was unchanged or fell. It is not surprising, therefore, that the appropriate cardiac responses in dogs trained to run on a treadmill are recorded when the experimenter is observed to move his hand to the treadmill switch (Rushmer, Smith and Lasher, 1960).

Brod, Fencl, Hejl and Jirka (1959) were the first to point out that a similar overall pattern of response occurs also in man. It is not necessary to document the statement that "anxiety" in man causes a rise in arterial blood pressure and heart rate. Besides this, the muscle blood flow increases (Wilkins and Eichna, 1941; Golenhofen and Hildebrandt, 1957), owing mainly to vasodilatation (Brod *et al.*, 1959). This vasodilatation has since been shown to be produced by atropine-sensitive nerve fibres of the sympathetic outflow (Blair, Glover, Greenfield and Roddie, 1959), although circulating adrenaline may make a significant contribution in some subjects (Barcroft, Brod, Hejl, Hirsjärvi and Kitchin, 1960). The cardiac output is increased and the total peripheral resistance reduced (Stead, Warren, Merrill and Brannon, 1945; Hickham, Cargill and Golden, 1948; Brod *et al.*, 1959).

This pattern of cardiovascular response in man, as in animals, is elicited by a variety of stimuli, which may become numerous during an individual lifetime because of the ease with which conditioning occurs. The response must be such a frequent occurrence as to invite speculation about the role it plays in health and disease. We need to learn as much as possible about it in all its details, both in further studies on man, so as to discover how the pattern may differ from that found in animals, and by simultaneously extending the studies on animals to complete the picture of the whole cardiovascular response. For instance, hardly anything is known as yet of the changes occurring in the profoundly important triad of organs—brain, heart and lungs. These are all regions in which the possibilities for central nervous control of the circulation have not appeared very impressive. Nevertheless, electrical stimulation of those parts of the hypothalamus and midbrain that integrate the defence reaction leads, amongst all the other changes, to widespread activation of the brain, including the cerebral cortex (Starzl, Taylor and Magoun, 1951; French, von Amerongen and Magoun, 1952). Since activity in central nervous structures leads to a local functional vasodilatation (Penfield and Jasper, 1955), we would expect widespread cerebral vasodilatation to be a feature of the whole response. In fact, Geiger and Sigg (1955) have already reported that stimulation for 5 sec in the hypothalamus, just anterior and lateral to the mamillary bodies, causes a 25 to 100% increase in cerebral blood flow lasting for 5 to 6 min. Some contribution to this increase will be made by the rise in arterial blood pressure, and this will also promote some increase in coronary blood flow. The increased heart rate and contractile force of the heart as well will add to the increase in coronary flow, quite apart from the vexed question of whether there is in addition a sympathetic vasodilator nerve supply to the coronary vessels that could be activated during such responses as the defence reactions. There is no information about whether changes occur in the pulmonary vascular bed during these reactions, but the rate of flow must greatly increase, and it would be important to investigate whether pulmonary vasomotor innervation makes a contribution to this end or, perhaps more pertinently, to provide the most favourable conditions for gaseous exchange.

3. *Chemoceptor inflow and response during exercise*

The whole pattern of response has been discussed so far as a preparatory one, hence anticipatory of the actions that immediately follow. This is emphasised by the ease with which it may be conditioned. Nevertheless, we may reasonably ask whether any means exists for maintaining the response during muscular exertion if this should indeed be provoked. Needless to say, Cannon also took this consideration into account. He pointed to the evidence for release of adrenaline after mild asphyxiation in experimental animals as indicating a possible mechanism for producing effects during vigorous exertions that are similar to those "produced in pain and excitement". The general consensus today would not support the suggestion that asphyxia was a likely accompaniment of even the most severe physical exertion, but we may follow up Cannon's main line of reasoning by taking into account a reflex whose existence was not known to him, the chemoceptor reflex.

Hilton and Joels (1965) have recently tried to test whether there is any interaction between the preparatory defence reflex and the chemoceptor reflex. In their experiments the integrative region for the defence reaction in the hypothalamus was stimulated electrically, as before, and the chemoceptor response was elicited by close arterial injection of cyanide into the region of the carotid bifurcation. The effects on respiration were observed as well as those on arterial blood pressure. It is well known that cyanide, by stimulating the chemoceptor endings, produces a reflex hypernoea and rise in arterial blood pressure. When the injection was preceded by hypothalamic stimulation the reflex response was frequently facilitated. It was never inhibited, as the baroreceptor reflex had been found to be. This facilitation suggests that chemoceptor afferent inflow may, on its own, activate the very regions of autonomic control within the brain stem that are brought into play by hypothalamic stimulation. Indeed, the possibility that the chemoceptor inflow has functional connections with the regions integrating the defence reactions has already been indicated by the experiments of Bizzi, Libretti, Malliani and Zanchetti (1961). They showed that, in cats decerebrated at a high level with the diencephalon left intact, pseudaffective responses are readily provoked on excitation of the carotid chemoceptors by lobeline injected close arterially or on letting the animal breathe a mixture low in oxygen (5 to 12% O_2 in N_2.) It has been known since Bard's detailed description of the reactions of decerebrate cats that pseudaffective responses are abnormally easily elicited, even, for instance, by gentle stroking (Bard, 1928), but these results do suggest that in the normal animal the chemoceptor inflow can cause some degree of activation of the defence reaction, just as can any mild form of cutaneous stimulation. This conclusion has several implications. The one most relevant to the present discussion is the possibility that the chemoceptor drive may maintain the visceral components of the defence reaction, if the "anticipated" motor behaviour does occur. If this were so, it might be expected that oxygen lack would be a more effective respiratory stimulus during exercise than at rest, a fact indeed already well established in man.

A further point might be of great importance to physiology. Abrahams *et al.* (1960b) found that the defence reaction, in common with other reactions integrated at the diencephalic level, could not be obtained as a reflex response in anaesthetised animals. Of the autonomic responses they were observing, only the rise in arterial blood pressure and pupillary dilation remained, even on stimulation of the whole sciatic nerve. The muscle vasodilatation so characteristic of the reaction was never obtained. This vasodilatation was elicted, however, together with the various other manifestations of the whole reaction, as part of the pseudaffective reflex in the high decerebrate cat, when the volatile anaesthetic had been blown off. Thus, in so far as the chemoceptor afferent input is excitor to this high-level reflex, its reflex effects will also be modified by anaesthetics, so that experiments will have to be carried out on this reflex response in unanaesthetised preparations before we can be sure that we know all its component parts. We may also find the threshold levels of stimulation to be much smaller than have been thought up to the present. The results of experiments carried out so far on anaesthetised animals might turn out to be more serious artefacts than has hitherto been realised.

D. Splenic Contraction and Polycythaemia

In cats acclimatised to their surroundings the stimulus presented by the sight of a dog for 1 min causes a polycythaemia (Izquierdo and Cannon, 1928), the red cell count often taking more than 10 min to return to normal. This effect is abolished by section of the splenic nerves and was attributed by those authors to the splenic contraction that other workers, most notably Barcroft and his colleagues, had shown to occur under various conditions, including anoxia, muscular exercise and "anxiety" (Barcroft and Barcroft, 1923; Barcroft, Harris, Orahovats and Weiss, 1925; Barcroft and Stevens, 1927).

The utility of the increased oxygen-carrying capacity of the blood seems obvious enough, but there is, in fact, no evidence as yet that splenic contraction is a mechanism of any significance in man. Perhaps the storage capacity of venous reservoirs is adequate to meet the demands of severe exertion.

E. Respiratory Changes

Rapid and deep respiration is an obvious feature of the response to pain and emotional excitement. This is also the response obtained on electrical stimulation of the midbrain and posterior hypothalamic regions integrating the defence reaction in the cat (Hess, 1949). Tachypnoea, rather than hyperpnoea, is the usual response to stimulation in the perifornical area of the hypothalamic region, and this particular pattern of respiration has been related to the hissing that is a prominent feature of the reaction in the cat. But the combination of tachypnoea and hyperpnoea, being characteristic of muscular exertion, in which condition its utility seems self-evident, led Cannon to two interesting speculations. One is that the important effect of these changes, as part of a defence reaction, is to produce an anticipatory reduction of blood P_{CO_2}. Some findings of Douglas and Haldane (1909), quoted in this connection, showed that forced breathing for a few minutes would relieve the subsequent respiratory distress of severe muscular exertion. All later work has gone to show that P_{CO_2} during effort remains remarkably constant, except in heavy work, when it falls; it is thus difficult to understand why this first speculation has received so little attention, particularly since it would point to a possible site of the higher centre in the brain involved in the respiratory adjustments of muscular exercise.

Having accepted the desirability of minimising the respiratory distress of muscular exertion, Cannon proceeded to consider the contribution that might be made by dilatation of the bronchioles. The air flow could certainly be relatively impeded by bronchiolar constriction, as in asthma, and adrenaline, which is discharged during strong defence reactions, was known to dilate bronchioles; so, he suggested, might not such dilatation be one of the mechanisms that help to prevent a healthy man becoming "winded" during severe exertion?

F. Gastrointestinal Secretion and Contraction

Changes in activity of the gastrointestinal tract have been discussed less in relation to short-term effects seen during defence reactions than in connection with the abnormalities of function that may develop in chronic emotional states, especially in man. Though there must be a relationship between these sets of changes, this has provided little incentive for a study of the short term effects. Perhaps they have in any event seemed rather uninteresting, since most have appeared to be inhibitory. Indeed, there is little to add to the conclusions documented by Cannon (1929) that pain, like fear and anxiety, will lead in animals and men to drying up of salivary, gastric and possibly even pancreatic secretion and to inhibition of peristaltic movement throughout the gastrointestinal tract.

From more recent studies of their patient, Tom, with a gastric fistula, Wolf and Wolff (1943) thought they could distinguish between the effect of anger and anxiety, which increased gastric motility, acid secretion and vascularity, and fear, which diminished them. It has likewise been reported that the former states lead to hyperactivity of the colon and the latter to quiescence (Grace, 1950). But such distinctions between emotional states are difficult to make with precision, and opposite effects have been reported, for instance by Crider and Walker (1948). If we knew the effects on patterns of activity of the gastrointestinal tract that follow electrical stimulation of those parts of the hypothalamus of established biological function, a firmer basis might be provided for criticism of Cannon's views. It has been shown that increases in tone and motility of the stomach, duodenum, jejunum, colon and rectum, but not of the ileum, can be elicited by stimulation in the tuberal region of the hypothalamus (Ström and Uvnäs, 1950), but the precise location of the regions was not established. Moreover, in such investigations the whole pattern of visceral response needs to be studied. Eliasson (1954), for instance, has shown that gastric motility is inhibited on stimulation of some parts of the hypothalamus and midbrain and excited from others, but it is uncertain what conclusions may consequently be drawn about the regulation of gastrointestinal function. Increased gastric activity occurs during vomiting and can hence be related to a general inhibition of digestive activity. Not surprisingly, Cannon (1929) discusses the evidence that vomiting may occur as a feature of a strong defence reaction, and this could provide an interpretation of various findings by Wolf and Wolff (1943).

G. Release of Adrenaline

1. *Effects on cardiovascular system*

In 1925 Cannon and Britton showed that, during pseudaffective reactions in a decorticate cat, the adrenal medullary hormones are released in sufficient amounts to speed the rate of the denervated heart. Not only have these observations been confirmed, but also they were soon matched by others showing release of the hormones on hypothalamic stimulation of anaesthetised cats (Houssay and Molinelli, 1925; Magoun, Ranson and Hetherington, 1937). Grant, Lindgren, Rosén and Uvnäs (1958) measured the amounts of adrenaline and noradrenaline released on stimulation of the hypothalamic region from which the vasodilator fibres to skeletal muscle were activated. The mean values for the output of adrenaline from one adrenal gland, at rest and during stimulation, were 0·06 and 0·32 μg/kg per min, respectively, and for noradrenaline were 0·19 nd 0·35 μg/kg per min. The figures for the release during stimulation correspond closely to those given by von Euler and Folkow (1953) and Celander (1954) for the release in prolonged asphyxia, which is probably the most powerful natural stimulus. Much the same amounts were released on stimulation in the region of the vasodilator pathway in the medulla oblongata, and finally it was shown that this release would make no contribution to the increase in muscle blood flow unless the electrical stimulation of the brain stem were intense (Lindgren, Rosén and Uvnäs, 1959a and b). They appear also to have no significant effect on the rate of the innervated heart, though they will make a contribution to the increase in contractile force (Rosén, 1961b). Some results of Folkow, Johansson and Mellander (1961) obtained in experiments on the capacity vessels of the hindquarters of the cat would indicate that they are hardly affected by the amounts of noradrenaline released during defence reactions, but there is no corresponding information about adrenaline. Thus, from what is known so far of the cardiovascular effects of these catecholamines in the cat, the amounts released during defence reactions would only be expected to contribute significantly to the increase in contractile force of the heart. In man, the relative contributions of hormone and nervous system may well be different. Certainly, the muscle vasodilatation is sometimes produced largely by circulating adrenaline (Barcroft, Brod, Hejl, Hirsjärvi and Kitchin, 1960), and it would be most interesting to know if the amounts released could also have an important effect on venous tone. That this is possible is shown by recent observations of venoconstriction in man produced by small intravenous doses of adrenaline (Sharpey-Schafer and Ginsberg, 1962).

2. *Effects on fatigue of muscles and nerves*

In animal experiments adrenaline in adequate doses has been found an effective antidote to muscular fatigue. It has long been known to overcome fatigue of the neuromuscular junction and to antagonise the blocking action of curare, besides restoring the contractions of fatigued muscle (Oliver and Schaefer, 1895; Gruber, 1922a, b). Some augmentation, both of transmission and contraction, are also seen in unfatigued muscles, though naturally less markedly so (Bowman and Zammis, 1958). These are direct effects of adrenaline and are independent of changes in blood flow. The former effect, which is due to a reversal of the failure of presynaptic conduction (Corkill and Tiegs, 1933; Krnjević and Miledi, 1958), requires such large amounts of adrenaline for its demonstration that it could hardly be exhibited during defence reactions. Bowman and Zaimis (1958) showed that the amounts necessary to increase the tension developed by unfatigued fast muscles, such as the tibialis anterior, were also rather higher than would ever be expected physiologically. The soleus, a slow muscle in the cat, was much more sensitive, being affected by 0·06 to 0·5 μg/kg intravenous adrenaline. Curiously, however, the contraction time was much reduced, and less tension was developed. In man most muscles are mixtures of slow and fast-contracting fibres; there could hardly be sufficient circulating adrenaline to increase the tone or contractile force of unfatigued skeletal muscles. Whether such an effect will be produced in fatigued muscles is still an open question.

These actions of adrenaline on the contractile force of the heart and skeletal muscle may be compared with its stimulating effect on the activity of the cerebral cortex, first described by Bonvallet, Dell and Hiebel (1954), who believed that this effect was primarily due to a direct stimulating action on the mesencephalic reticular formation. Some workers have shared this view, others have not. In particular, Baust, Niemczyk and Vieth (1963) hold that the rise in arterial blood pressure caused by adrenaline is the stimulus acting on the cells of the recticular formation. But all agree that circulating adrenaline can cause alerting and affect awareness. Until now, however, the doses of adrenaline used (usually 5 μg/kg in the cat) have been much higher than those to be expected even during a strong defence reaction, so that the physiological significance of the phenomenon is still uncertain. The mere fact, however, that adrenaline can lead in excitable tissues to such ubiquitous effects, which will ameliorate and may ward off fatigue, remains strongly suggestive of a supporting role for this hormone during emotional excitement.

3. *Effects on movement of potassium*

Behind this idea, which was one that Cannon advanced most forcefully, lay the knowledge that adrenaline exerts a powerful effect on carbohydrate metabolism: liver glycogen is mobilised (Vosburgh and Richards, 1903), and this so readily that adrenaline leads to hyperglycaemia in the cat, for instance, when infused at a rate of 0·05 μg/kg per min (Cori, Fisher and Cori, 1935) and similarly in man when 0·025 μg/kg per min are infused (Cori and Buchwald, 1930). Anaerobic glycolysis is stimulated in skeletal musclc, with the output of large amounts of lactate (Cori, 1925). These basic findings have stood the test of time, as has the conclusion that fat forms the major fuel for the increased metabolic rate induced by adrenaline (Cori and Cori, 1928). They are supported by a wealth of biochemical information, which nevertheless does not need to be discussed here, since it does not lead to any larger generalisations about the effects of emotional excitement in man or other animals. One other effect of adrenaline, however, is worth mentioning, that is, its ability to induce hyperkalaemia. It was established by D'Silva (1936) and Marenzi and Gerschman (1937) that most of the potassium comes from the liver and that, during the resulting hyperkalaemia, the uptake of potassium by skeletal muscle increases (Marenzi and Gerschman, 1936). The evidence for this movement of potassium was reviewed by Fenn (1940), who also pointed out that the potassium released from muscle during contraction is probably taken up rapidly by liver. He suggested that a physiological role of the potassium movement induced by adrenaline might be to replenish the muscle loss during a burst of activity and to maintain relatively normal potassium distribution during prolonged muscular exercise.

A further point of interest comes from the fact that moderate rises of blood potassium cause a vasodilatation in skeletal muscle, possibly sufficient to make a major contribution to functional vasodilatation in this tissue (Dawes, 1941; Kjellmer, 1961). It has been known for some time that the generalised muscle vasodilatation that occurs in man in response to infusions of adrenaline is due, in fact, not to the catecholamine itself, but to some as yet unidentified vasodilator substance, or substances, produced by the infused material (Barcroft and Swan, 1953). It has been suggested elsewhere that plasma kinin formation might be responsible for this effect (Hilton, 1960), but perhaps potassium plays a role here, too, in which event the muscle vasodilatation produced during defence reactions in man, in so far as it results from circulating adrenaline, may truly anticipate the vasodilatation of exercise itself.

4. *Effects on blood-clotting*

Since the early experiments of Vosburgh and Richards (1903) it has been known that adrenaline injected into the animal shortens the clotting time of blood. Cannon (1929) was already able to draw on evidence that this is a phenomenon occuring under natural conditions, such as strenuous exercise, when adrenaline is liberated (Hartman, 1927; Mills, Necheles and Chu, 1928). In his own experiments secretion of adrenaline elicited by splanchnic nerve stimulation, or reflexly by sciatic nerve stimulation, was as effective as injected adrenaline in reducing the clotting time. As little as 1 μg/kg halved the clotting time in 20 min: it took 30 to 40 min to return to normal. If an intact animal was enraged, larger and more prolonged effects were obtained. Even today the mechanism of this action of adrenaline is not understood, but a provocative footnote has been added by observations that after a subcutaneous injection of adrenaline there is a great increase in the fibrinolytic activity of plasma, which is maintained for an hour or more (Biggs, MacFarlane and Pilling, 1947). Similar changes occur after severe muscular exercise. It would probably be wiser to postpone speculation about the functional significance of these changes until more is known as to how they are produced and the relationship, if any, between them.

H. Pituitary Hormones

1. *Adrenocorticotrophic hormone (ACTH)*

The association between several biological stimuli and release of ACTH has been documented many times in recent years, and it should now suffice to state that ACTH release would be expected as part of the defence reaction. The conclusion has not previously been put in this simple form, despite clear statements such as that of Sayers (1937) to the effect that neural mechanisms involving the brain-stem reticular formation play a major role. Harris (1960) has pointed out that the hypothalamus appears to act as an integrating mechanism where effects of afferent impulses, emotional states and humoral agents interact, and he suggested that there may be a close link between the patterns of endocrine activity and emotional behaviour integrated in this region of the brain. But no reference was made to the body of information accumulated so far about defence reactions and their central nervous control.

To select a few of the most relevant findings, it may be noted, first, that in monkeys and men the pituitary-adrenocortical system, judged by plasma 17-hydroxycorticoid levels, is remarkably sensitive to stimuli leading to distress or arousal. Mason (1959), for instance, observed striking elevations when monkeys were simply moved into a new room or cage, as well as in conditioning experiments in which the stimulus had become a "threat" that an electric shock would follow. Similar increases were reported for man, arising acutely in an aircrew on an 11,000 mile non-stop flight and as a more persistent change in patients during the first week after thoracotomy. These increases, which represented a doubling of the normal level, were of the same order as those reported in previous investigations of patients in anxiety states (Hamburg, Sabshin, Board, Grinker, Korchin, Basowitz, Heath and Persky, 1958). Hodges, Jones and Stockham (1962) report similar increases in the blood of students taken just after sitting for the second M.B. (London) oral examination. They also found the blood levels of ACTH to have increased to 6 to 9 mu/ml and pointed out that these levels are as high as those found in patients with Cushing's disease.

It has been known since the pioneer experiments of de Groot and Harris (1950) on unanaesthetised rabbits that electrical stimulation of the hypothalamus activates the pituitary-adrenocortical system and that bilaterial lesions in the hypothalamus would abolish the lymphopenic response to emotional excitation. Most workers have put the site, both of effective stimulation and of effective lesions, in the posterior tuberal region of the hypothalamus (Harris, 1960). Mason (1958), working on monkeys, elicited an increase in plasma 17-hydroxysteroids on stimulation of the periventricular region of the anterior hypothalamus. It may not seem surprising that the results of lesion experiments do not give a clear localisation; stimulation experiments could give more accurate indications, but no one has yet performed the heroic labour of using precise stimulation techniques in an attempt to map the whole brain-stem region from which ACTH secretion is elicited. The present information, most notably that provided by Porter (1954) on the basis of lymphopenic responses, would be consistent with a considerable overlap of the ACTH-releasing area with that integrating the defence reaction. Relevant to this is the observation that plasma 17-hydroxycorticoid levels are raised in the conscious monkey as much by electrical stimulation within the amygdaloid complex as within the hypothalamus (Mason, 1959). Results of lesions and stimulation experiments implicating the median eminence (Harris, 1960) may be interpreted as an indication that the final common path lies in this region.

One basic question remains, even after these many investigations: What is the physiological role of the adrenocortical hormone so readily released? A recent answer (Beck and McGarry, 1962) was that its role in carbohydrate metabolism is a minor one and that it is chiefly concerned with the distribution of body water and electrolytes, maintenance of blood pressure and glomerular filtration rate and the renal regulation of water excretion. Nevertheless, as the authors also state, the influence of cortisol in accelerating gluconecogenesis is well established, and there is also evidence that the hyperglycaemia it causes is partly due to inhibition of pyruvate catabolism. This could mean that the products of anaerobic glycolysis in muscle would be available for resynthesis of glucose. Thus, release of ACTH will cooperate in some of its final effects with adrenaline to ensure the supply of adequate fuel for intense muscular activity. It may be recalled that there is much evidence to suggest that adrenaline facilitates the release of ACTH.

2. *Vasopressin (Antidiuretic hormone, ADH)*

It has long been known that operative procedures or muscular exercise will reduce the rate of urinary secretion in man. Rydin and Verney (1938) showed the antidiuretic effect in dogs of forced running on a treadmill, an unpleasant noise or weak electrical stimulation of the skin. This effect was shown to be independent of adrenaline release and the sympathetic innervation of the kidneys. It is due to a reflex excitation of the neurohypophysis (O'Connor and Verney, 1942; O'Connor, 1946). Harris (1955) emphasised that the stimuli used by these and other workers will all have elicited some degree of what is loosely called "emotional stress". Even the effect of sudden exercise cannot be due to the running itself, for the antidiuretic effect was found to wear off as the exercise was repeated (Rydin and Verney, 1938). Circulatory collapse in man, produced by sudden movement on a tilt-table from the horizontal to the vertical position, evokes an antidiuretic effect, probably by release of vasopressin (Brun, Knudsen and Raaschon, 1946), and the hormone has been demonstrated in the blood of human subjects during a faint due to venesection (Noble and Taylor, 1953). Since the faint itself seems to initiate the release of ADH, this has been taken as additional evidence that the hormone is liberated during so-called "stress". But what exactly is the "stress" in this experimental situation? More specifically, what is the stimulus and does it act by initiating the defence reaction? A fall of blood pressure on its own, for example, can lead to ADH release in anaesthetised animals (Ginsburg and Brown, 1957). Recently, an attempt has been made to test whether ADH is released during defence reactions in the cat (Bisset, Hilton and Poisner, 1963). The suggestion had already been made that pressor reactions obtained in *encéphale isolé* preparations, on electrical stimulation in the midbrain reticular formation, were due to ADH (Sharpless and Rothballer, 1961). Bisset *et al.* (1963) assayed the amounts of ADH released into the external jugular venous blood in anaesthetised cats on discrete electrical stimulation within the hypothalamus. They found restricted regions from which hormone release was elicited, but did not obtain any when stimulating the zone from which defence reactions are most readily elicited. It is possible that the anaesthetic had blocked the pathway from the integrative centre for the defence reaction to the supraoptic nucleus, but until this is established we have no firm evidence that ADH release is a necessary component of the defence reaction itself.

I. Conclusion

There is little left to add to what has been written in the preceding sections about the functional significance in emergency of the defence reaction and its various components. The diverse visceral and hormonal features seem best regarded as the preparatory stage of a reflex, which, in civilised man, may be expressed in this way alone. Curiously enough, Cannon regarded these reactions none the less as homoeostatic, for he was emphasising the long-term view; in the short run, though, they are the very antithesis, since they establish a new equilibrium only made possible by an interruption of short-term homoeostatic mechanisms. This was illustrated in dealing with the pattern of cardiovascular response by reference to the inhibition of the baroreceptor reflex, which would otherwise interfere. Some of the consequences of release of ACTH and adrenaline also are incompatible with short-term homoeostasis. The powerful effects of such hormones can greatly disturb the internal milieu, particularly if no muscular exertion occurs. Their potentially harmful effect in prolonged or chronic reactions should not need to be reiterated. Thus, though responses of the organism leading to changes of the internal milieu may still be physiological, they entail unavoidable risks.

This does not detract from the imaginative hypotheses that Cannon deployed and to which reference has frequently been made here. As to the bodily responses themselves, it need hardly be said that there are still uncertainties about matters of fact, as well as of significance, and that these are multiplied with the increasing range of bodily responses to be discussed. The responses included under the heading of the defence reaction are no doubt much more complicated than can be imagined even today, but this is not to deny the growing certainty about the validity of the theoretical approach, which carries with it the forceful reminder that physiology has deep roots in biology.

Most of the bodily responses fall into place regardless of subtle psychological distinctions, and from the physiologist's point of view there has seemed little need to distinguish between "fear" and "rage" or their frequent counterparts in experimental psychology, "flight" and "attack". Nor would such distinctions help to decide the equivocal significance of such components as changes in blood-clotting or the role of ADH; on the contrary, in this branch of biology, physiology may at present be of greater assistance than psychology. It seems fitting to conclude a contemporary account of this subject with the words that Cannon used to introduce the first edition of his book, almost fifty years ago. "Fear, rage, and pain, and the pangs of hunger are all primitive experiences which human beings share with the lower animals. These experiences are properly classed as among the most powerful that determine the action of men and beasts. A knowledge of the conditions which attend these experiences, therefore, is of general and fundamental importance in the interpretation of behaviour."

REFERENCES

Abrahams, V. C., Hilton, S. M., and Malcolm, J. L. (1962). *J. Physiol. Lond.* **164**, 1.
Abrahams, V. C., Hilton, S. M., and Zbrożyna, A. W. (1960a). *J. Physiol. Lond.* **152**, 54P.
Abrahams, V. C., Hilton, S. M., and Zbrożyna, A. W. (1960b). *J. Physiol. Lond.* **154**, 491.
Abrahams, V. C., Hilton, S. M., and Zbrożyna, A. W. (1964). *J. Physiol. Lond.* (In press.)
Barcroft, J., and Barcroft, H. (1923). *J. Physiol. Lond.* **58**, 138.
Barcroft, H., Brod, J., Hejl, Z., Hirsjärvi, E. A., and Kitchin, A. H. (1960). *Clin. Sci.* **19**, 577.
Barcroft, J., Harris, H. A., Orahovats, D., and Weiss, R. (1925). *J. Physiol. Lond.* **60**, 443.
Barcroft, J., and Stevens, J. G. (1927). *J. Physiol. Lond.* **64**, 1.
Barcroft, H., and Swan, H. J. C. (1953). "Sympathetic Control of Human Blood Vessels." Edward Arnold, London.
Bard, P. (1928). *Amer. J. Physiol.* **84**, 490.
Bard, P. (1960). *Physiol. Rev.* **40**, Suppl. 4, 3.
Bard, P., and Macht, M. B. (1958). *In* "Neurological Basis of Behaviour", p. 55. Churchill, London.
Bard, P., and Rioch, D. McK. (1937). *Johns Hopk. Hosp. Bull.* **60**, 65.
Baust, W., Niemczyk, H., and Vieth, J. (1963). *Electroenceph. clin. Neurophysiol.* **15**, 63.
Beck, J. C., and McGarry, E. E. (1962). *Brit. med. Bull.* **18**, 134.
Biggs, R., MacFarlane, R. G., and Pilling, J. (1947). *Lancet.* i, 402.
Bisset, G. W., Hilton, S. M., and Poisner, A. J. (1963). *J. Physiol. Lond.* **169**, 40P.
Bizzi, E., Libretti, A., Malliani, A., and Zanchetti, A. (1961). *Amer. J. Physiol.* **200**, 923.
Blair, D. A., Glover, W. E., Greenfield, A. D. M., and Roddie, I. C. (1959). *J. Physiol. Lond.* **148**, 633.
Bonvallet, M., Dell, P., and Hiebel, G. (1954). *Electroenceph. clin. Neurophysiol.* **6**, 119.
Bovard, E. W. (1962). *Perspect. Biol. Med.* **6**, 116.
Bowman, W. C., and Zaimis, E. (1958). *J. Physiol. Lond.* **144**, 92.

Broca, P. (1878). *Rev. anthrop.* 3 ser. 1, 385.
Brod, J., Fencl, V., Hejl, Z., and Jirka, J. (1959). *Clin. Sci.* 18, 269.
Brun, C., Knudsen, E. O. E., and Raaschon, F. (1946). *J. clin. Invest.* 25, 568.
Burn, J. H. (1938). *Physiol. Rev.* 18, 137.
Cannon, W. B. (1929). "Bodily Changes in Pain, Hunger, Fear and Rage." 2nd Edition. Appleton, New York.
Cannon, W. B., and Britton, S. W. (1925). *Amer. J. Physiol.* 72, 283.
Celander, O. (1954). *Acta physiol. scand.* 32, Suppl. 116.
Chapman, W. P. (1960). *In* "Electrical Studies on the Unanaesthetized Brain" (E. R. Ramey and D. S. O'Doherty, eds.), p. 334. Paul B. Hoeber, New York.
Charlier, A., Guz, A., Keatinge, W. R., and Wilcken, D. (1962). *J. Physiol. Lond.* 164, 17P.
Cori, C. F. (1925). *J. biol. Chem.* 63, 253.
Cori, C. F., and Buchwald, K. W. (1930). *Amer. J. Physiol.* 95, 71.
Cori, C. F., and Cori, G. T. (1928). *J. biol. Chem.* 79, 309.
Cori, C. F., Fisher, R. E., and Cori, G. T. (1935). *Amer. J. Physiol.* 114, 53.
Corkill, A. B., and Tiegs, O. W. (1933). *J. Physiol. Lond.* 78, 161.
Crider, R. M., and Walker, S. M. (1948). *Arch. Surg. Chicago.* 57, 1.
Darwin, C. (1872). "The Expression of the Emotions in Man and Animals." John Murray, London.
Dawes, G. S. (1941). *J. Physiol. Lond.* 99, 224.
Douglas, C. G., and Haldane, J. S. (1909). *J. Physiol.* 39, 1.
D'Silva, J. L. (1936). *J. Physiol. Lond.* 86, 219.
Eliasson, S. (1954). *Acta physiol. scand.* 30, 199.
Eliasson, S., Folkow, B., Lindgren, P., and Uvnäs, B. (1951). *Acta physiol. scand.* 23, 333.
Eliasson, S., Lindgren, P., and Uvnäs, B. (1952). *Acta physiol. scand.* 27, 18.
Fenn, W. O. (1940). *Physiol. Rev.* 20, 377.
Fernandez de Molina, A., and Hunsperger, R. W. (1959). *J. Physiol. Lond.* 145, 251.
Folkow, B., Johansson, B., and Mellander, S. (1961). *Acta physiol. scand.* 53, 99.
Folkow, B., Mellander, S., and Öberg, B. (1961). *Acta physiol. scand.* 53, 7.
Folkow, B., and Uvnäs, B. (1948). *Acta physiol. scand.* 15, 389.
French, J. D., von Amerongen, F. K., and Magoun, H. W. (1952). *Arch. Neurol. Psychiat., Chicago* 68, 577.
Fulton, J. F. (1949). "Physiology of the Nervous System." Oxford University Press, New York.
Gantt, W. H. (1960). *Physiol. Rev.* 40, Suppl. 4, 266.
Gastaut, H., Vigouroux, R., Corriol, J., and Badier, M. (1951). *J. Physiol. Path. gén.* 43, 470.
Geiger, A., and Sigg, E. (1955). *Trans. Amer. neurol. Ass.* 80, 127.
Ginsburg, M., and Brown, L. M. (1957). *In* "The Neurohypophysis" (H. Heller, ed.), p. 109. Butterworth, London.
Golenhofen, K., and Hildebrandt, G. (1957). *Pflüg. Arch. ges. Physiol.* 263, 637.
Goltz, F. M. (1892). *Pflüg. Arch. ges. Physiol.* 51, 570.
Grace, W. A. (1950). *Proc. Ass. Res. nerv. Dis.* 29, 679.
Grant, R., Lindgren, P., Rosén, A., and Uvnäs, B. (1958). *Acta physiol. scand.* 43, 135.
Groot, J. de, and Harris, G. W. (1950). *J. Physiol. Lond.* 111, 335.
Gruber, C. M. (1922a). *Amer. J. Physiol.* 61, 475.
Gruber, C. M. (1922b). *Amer. J. Physiol.* 62, 438.
Hamburg, D. A., Sabshin, M. A., Board, F. A., Grinker, R. R., Korchin, S. J., Basowitz, H., Heath, H., and Persky, H. (1958). *Arch. Neurol. Psychiat., Chicago.* 79, 415.
Harris, G. W. (1955). "Neural Control of the Pituitary gland." Edward Arnold, London.
Harris, G. W. (1960). *In* "Handbook of Physiology", Sec. 1, vol. II (J. Field, ed.), p. 1007. Williams and Wilkins, Baltimore.

Hartman, F. A. (1927). *Amer. J. Physiol.* **80**, 716.
Heath, R. G., and Mickle, W. A. (1960). *In* "Electrical Studies on the Unanaesthetized Brain" (E. R. Ramey and D. S. O'Doherty, eds.), p. 214. Paul B. Hoeber, New York.
Heath, R. G., Monroe, R. R., and Mickle, W. A. (1955). *Amer. J. Psychiat.* **111**, 862.
Hess, W. R. (1949). "Das Zwischenhirn." Schwabe, Basel.
Hess, W. R., and Brügger, M. (1943). *Helv. physiol. acta.* **1**, 33.
Hickam, J. B., Cargill, W. H., and Golden, A. (1948). *J. clin. Invest.* **27**, 290.
Hilton, S. M. (1960). *In* "Polypeptides which Affect Smooth Muscles and Blood Vessels" (M. Shachter, ed.), p. 258. Pergamon Press, London.
Hilton, S. M. (1963). *J. Physiol. Lond.* **165**, 56P.
Hilton, S. M., and Joels, N. (1965). *J. Physiol. Lond.* **176**, 20P.
Hilton, S. M., and Zbrożyna, A. W. (1963). *J. Physiol. Lond.* **165**, 160.
Hodges, J. R., Jones, M. T., and Stockham, M. A. (1962). *Nature, Lond.* **193**, 1187.
Houssay, B. A., and Molinelli, E. A. (1925). *C.R. Soc. Biol., Paris.* **93**, 1454.
Izquierdo, J. J., and Cannon, W. B. (1928). *Amer. J. Physiol.* **84**, 545.
Keller, A. D. (1932). *Amer. J. Physiol.* **100**, 576.
Kjellmer, I. (1961). *Med. exp.* **5**, 56.
Klüver, H., and Bucy, P. C. (1939). *Arch. Neurol. Psychiat., Chicago*, **42**, 979.
Krnjević, K., and Miledi, R. (1958). *J. Physiol. Lond.* **141**, 291.
Lilly, J. C. (1958). *In* "Reticular Formation of the Brain" (H. H. Jasper, L. D. Proctor, R. S. Knighton, W. C. Noshay, and R. T. Costello, eds.), p. 705. Churchill, London.
Lilly, J. C. (1960). *In* "Electrical Studies on the Unanaesthetized Brain" (E. R. Ramey and D. S. O'Doherty, eds.), p. 70. Paul B. Hoeber, New York.
Lindgren, P. (1955). *Acta physiol. scand.* **35**, Suppl. 121.
Lindgren, P., Rosén, A., and Uvnäs, B. (1959a). *Acta physiol. scand.* **47**, 233.
Lindgren, P., Rosén, A., and Uvnäs, B. (1959b). *Acta physiol. scand.* **47**, 243.
Lindgren, P., and Uvnäs, B. (1953). *Circulation Res.* **1**, 479.
MacLean, P. D. (1954). *In* "Recent Developments in Psychosomatic Medicine (E. D. Wittkown and R. A. Cleghorn, eds.), p. 101. Pitman, London.
Magoun, H. W., Ranson, S. W., and Hetherington, A. (1937). *Amer. J. Physiol.* **119**, 615.
Marenzi, A. D., and Gerschman, R. (1936). *Rev. Soc. argent. Biol.* **12**, 424.
Marenzi, A. D., and Gerschman, R. (1937). *C.R. Soc. Biol., Paris.* **124**, 382.
Mason, J. W. (1958). *J. appl. Physiol.* **12**, 130.
Mason, J. W. (1959). *Recent Progr. Hormone Res.* **15**, 345.
Masserman, J. H. (1943). "Behavior and Neurosis". University of Chicago Press, Chicago.
Mills, C. A., Necheles, H., and Chu, M. K. (1928). *Chin. J. Physiol.* **2**, 219.
Noble, R. L., and Taylor, N. B. G. (1953). *J. Physiol. Lond.* **122**, 220.
O'Connor, W. J. (1946). *Quart. J. exp. Physiol.* **33**, 149.
O'Connor, W. J., and Verney, E. B. (1942). *Quart. J. exp. Physiol.* **31**, 393.
Oliver, G., and Schaefer, E. A. (1895). *J. Physiol. Lond.* **18**, 230.
Papez, J. W. (1937). *Arch. Neurol. Psychiat., Chicago* **38**, 725.
Penfield, W., and Jasper, H. (1955). "Epilepsy and the Functional Anatomy of the Brain." Churchill, London.
Porter, R. W. (1954). *Recent Progr. Hormone Res.* **10**, 1.
Rosén, A. (1961a). *Acta physiol. scand.* **52**, 291.
Rosén, A. (1961b). *Acta physiol. scand.* **53**, 270.
Rushmer, R. F., Smith, O. A., and Lasher, E. P. (1960). *Physiol. Rev.* **40**, Suppl. 4, 27.
Rydin, H., and Verney, E. B. (1938). *Quart. J. exp. Physiol.* **27**, 343.
Sayers, G. (1947). *In* "Hormones in Blood" (G. E. W. Wolstenholme and E. C. P. Miller, eds.), p. 138. Ciba Foundation Colloquia on Endocrinology, II. Churchill, London.

Sem-Jacobsen, C. W., and Torkildsen, A. (1960). *In* "Electrical Studies on the Unanaesthetized Brain" (E. R. Ramey and D. S. O'Doherty, eds.), p. 275. Paul Hoeber, New York.
Sharpey-Schafer, E. P., and Ginsberg, J. (1962). *Lancet* ii, 1337.
Sharpless, S. K., and Rothballer, A. B. (1961). *Amer. J. Physiol.* **200**, 909.
Smith, G. Elliot. (1919). *Brit. med. J.* i, 758.
Smith, O. A., Jabbur, S. J., Rushmer, R. F., and Lasher, E. P. (1960). *Physiol. Rev.* **40**, Suppl. 4, 136.
Starzl, T. E., Taylor, C. W., and Magoun, H. W. (1951). *J. Neurophysiol.* **14**, 479.
Stead, E. A., Warren, J. V., Merrill, A. J., and Brannon, E. S. (1945). *J. clin. Invest.* **24**, 326.
Ström, G., and Uvnäs, B. (1950). *Acta physiol. scand.* **21**, 90.
Ursin, H., and Kaada, B. R. (1960). *Exp. Neurol.* **2**, 109.
Uvnäs, B. (1954). *Physiol. Rev.* 34, 608.
von Euler, U. S., and Folkow, B. (1953). *Arch. exp. Path. Pharmak.* **219**, 242.
Vosburgh, C. H., and Richards, A. N. (1903). *Amer. J. Physiol.* **9**, 35.
Wilkins, R. W., and Eichna, L. W. (1941). *Johns Hopk. Hosp. Bull.* 68, 425.
Wolf, S., and Wolff, H. (1943). "Human Gastric Function." Oxford University Press, New York.

CHAPTER 17

Irreversible Changes†

R. W. Brauer

† The opinions or assertions contained here are those of the author and are not to be construed as official or reflecting the views of the Navy Department or the Naval Service at large.

A. Introductory

1. *General concepts*

The ability to adapt to varying environments is a fundamental property of the human species, bred into it over long times of evolutionary change. It must not be assumed that efforts at "adaptation" pass necessarily without leaving any trace in the biology of the individual subject.† Mankind is approaching a time when the physical environments in which individuals spend their lives become increasingly amenable to control, and hence potentially to choice, but the price that may have to be paid for adaptation emerges as one of the pieces of information that come under any consideration of what may constitute optimal environments in time and space. Similar considerations are raised by man's increasing penetration into exotic environments. Thus, study of this subject should take its place alongside those older and more developed aspects of environmental physiology that deal with the here and now of survival.

Multicellular biological systems respond to environmental changes in ways that tend to minimise the effects of any stress upon the internal environments in which their constituent cells function (Prosser, 1958). It is all too easy to jump from this concept to the assumption that, when the external conditions return to their original values, the biological system, too, will reverse its regulatory efforts and return to its point of departure as though nothing had happened; it may seem tempting to relegate any residual changes observed under such conditions to the realm of the pathological and to attribute them to injury sustained by the system as a result of excessive environmental stress. Such a philosophy has had the effect of discouraging efforts to study the frequently not very conspicuous, though potentially important, residual changes that may be left behind after a change in environmental experiences. Only in recent years, perhaps largely as a result of concern with the effects of ionising radiation, has this subject begun to receive the attention it deserves (Handler, 1961).

Although it is true that biological systems are frequently excellent regulatory machines, it is also true that they are inherently time-binding machines. All processes associated with growth, development and ageing are intrinsically linked with the lapse of time. The biological organism thus becomes a

† To avoid semantic problems, it may be pointed out that in these pages the term "adaptation" will designate the process of adjustment of an individual to a new environment. When the concept of "price of adaptation" is used, on the other hand, one implicitly hypothesises this as a neutral—or a more nearly neutral—environment, departure from which imposes a load on the individual. When used in this sense the term approaches in meaning the concept as used by H. Selye. In the present context, however, it is intended even under these conditions to undertake experimental tests in a fashion somewhat different from that adopted by him, since, as will be seen, the basic hypotheses of a "neutral" environment seem to me to be of doubtful validity.

calendar reflecting the passage of time. It is not altogether clear yet what it is that turns the leaves of this kind of calendar. Evidence from the study of various genetic systems tends to show that in some important ways these mechanisms are determined by the genetic make-up of the system. From studies of early development in mammals and in insects, however, it is becoming increasingly clear that the old debate about "nature versus nurture" is likely not to be resolved in favour of either alternative, but rather in terms of "nature and nurture", the observed biological events being the result of an interplay of genetic make-up with the successive sets of conditions to which the individual is exposed.

At the other end of life, when the processes of ageing culminate in death, a similar realisation has not been so generally accepted; there are still many who like to speak about "intrinsic ageing" and to refer to any deviations from a presumed ideal pattern as evidence of pathological ageing. It would seem eminently plausible however, to question an attitude that postulates so sharp a discontinuity in the very nature of the individual, a sharp break between the early part of life, when environmental factors are important in moulding, and in the latter part of life when supposedly the individual runs its course unperturbed by ("non-pathological") experiences, that it is bound to accumulate as it passes through life. The logical weakness of this position is most easily seen when one attempts to determine at what point in time ageing begins. Depending upon one's choice of definition, this point will fall at almost any part of the life span, beginning from the time when organogenesis draws to a close in the early embryo (Medawar, 1955). There is much evidence to indicate that a complex biological organism like man does not age all in one piece, but rather that different organ systems age at different rates (for documentation see below). In this sense, one can properly say that such an organism always represents an age chimera, parts of which have already reached a stage of advanced senility while others are still in the prime of life (Medawar, 1955). As experimental results in this field accumulate, it is becoming clear that the concept of ageing as a unitary process is erroneous; ageing represents the sum of concurrent processes that, each by itself and all together, represent the time-binding machinery of such an organism.

It is in such a system, then, and against such a theoretical background, that one must look for the degree to which one or another environmental change imposes irreversible alterations upon this or that organ system and for the processes that contribute their separate shares to changing the pace at which the biological clock (or calendar) marks time. The very fact that biological time is not easily or uniquely translatable into terms of physical time thus turns out to be an expression of the degree to which the environment is capable of leaving its imprint upon the development of the individual.

Experimental approaches

There would seem to be two main ways in which the effects of environmental stress can alter irreversibly the biology of an individual: the most obvious one is by producing a truly non-reversible change in the tissues, as by deletion of cells or by the deposition of materials that henceforth cannot or will not be removed. The second way entails changes that may in effect be reversible, though perhaps only slowly, but may, while they persist, so alter the biology of the individual that its responses to subsequent living experiences differ recognisably from those of the unexposed subject, that its path of development in time is altered and that, in fact, it will never return to the state it would have reached had the experience not taken place. Immunological mechanisms and certain behavioural system mechanisms are among the more obvious examples of this aspect of biological memory.

Questions that must thus be answered (and must eventually be resolved by enquiring into the complex problem posed by the irreversible effects of the environment on man) include definition of the types of such change sustained, the effects of such changes on the biology of the individual, the degree to which normal or tolerable environments produce such changes and perhaps the mechanisms underlying the fixation of such changes in the biological make-up of the individual. The important question of genetic selection and its effects on succeeding generations hardly falls within the scope of this chapter.

Experimental approaches to the study of non-reversible effects of environmental factors, as thus conceived, appear to group themselves under four major headings:

a. exploration of processes of recovery from environmental stress (or from adaptation thereto), with a view to measuring rates of reversal and singling out slowly reversed or non-reversed changes;

b. exploration of residual lesions left after ostensible recovery from strain;

c. population studies aimed at measuring and describing changes in ageing patterns within population groups existing under various environments;

d. exploration of phenomena imprinting a change upon the young individual and hence altering the course of its development.

B. The Concept of Reversibility

1. *Muscular contraction and exercise conditions*

To illustrate the gradations between reversible and irreversible change, and to emphasise the importance of rate constants as well as the problem of what precisely might constitute intrinsically non-reversible change, let us consider the behaviour of skeletal muscle stimulated to contract. A single muscle twitch leaves behind a series of changes in constituents of the muscle membrane, of its contractile substance and of its energy-producing system, which are reversed in a matter of a few milliseconds or at most within a few seconds. Repeating this process on a suitable schedule will lead to changes in the performance characteristics of the muscle, referred to as "exercise conditioning" (Denny-Brown, 1960). These are associated with more lasting changes in the chemical make-up of the muscle. Significant differences in the results produced are found to depend upon whether the exercise conditioning involved numerous contractions against relatively light loads or far fewer contractions, perhaps, but against loads that approach or even exceed the maximum against which the muscle can contract (Cornelius, Burnham and Hill, 1963). Endurance conditioning of the former type involves biochemical changes that are primarily lodged in the energy-producing enzyme systems. Interruption of the exercise schedule in this situation leads to reversal of this change, that is, to loss of the exercise conditioning, in a matter of a few days at most. By contrast, the isometric or maximum load conditioning entails an increase in the amount of contractile substance and in the number of muscle fibres (Cornelius *et al.*, 1963). Exercise conditioning of this type is reversed far more slowly than the former kind when the exercise schedule is interrupted: it is a matter of months before reversal is completed. Moreover, no more than one or two maximum load exercises per week suffice to maintain indefinitely this type of conditioning (weight lifters' conditioning, for instance).

Thus, in a single tissue, a single type of stimulus produces phenomena that illustrate several points fundamental to this discussion. In the first place, the wide range of time constants of recovery illustrates the fact that the concept of reversibility is inherently fluid, as applied to the effects of external conditions upon living systems. Clearly, when the recovery times extend to the point at which they approach or exceed the life expectancy of the individual, even a theoretically reversible change becomes, in fact, nonreversible. Further, the change, once imposed, can alter responses to subsequent exposure (or stimulation) in a manner that in effect renders the single first change the point of origin of a chain of events, as a result of which the organism, as it exists in a real environment, has become permanently, which is to say biologically irreversibly, impaired.

A second important aspect of the muscle experiments described is the fact that by far the greater part of the biochemical changes accompanying a single twitch, or even a series of twitches, is rapidly reversed. It is only a minute, and so far a not at all well comprehended, residue of these changes that remains behind to become manifest in the more lasting changes of muscle performance described above. Such a relationship will be found to be an almost invariable characteristic of the reactions to environmental changes considered in this chapter. It is their subtlety, when compared to the gross changes associated with the more prompt and transient adaptive responses, that has tended to discourage much systematic study of these residual changes.

Severe muscular exercise results in changes in the blood enzyme profile strongly suggestive of some degree of cell destruction in the horse (Cardinet, Fowler and Tyler, 1963), the rat (Atland and Highman, 1961) and man (Schlang, 1961; Haloren and Kontinen, 1962). Indeed, in the rat at least these indications are borne out by histological examination (Atland and Highman, 1961), which shows evidence of scattered foci of tissue injury in severely exercised muscle. At present it would seem difficult to establish the precise minimal degree of muscular exercise at which such tissue reaction still occurs. One may wonder legitimately whether a minimal degree of tissue destruction may not be inescapably associated with normal life, as well as with exercise conditioning. Here again, the muscle example helps to pose, if not necessarily to answer, a question basic to our subject. Is tissue injury a necessary accompaniment of irreversible tissue change by environmental factors? Note should be taken of how tenuous is the line in this instance between injury as a normal event associated with life in a real world and "injury" as an event of pathological significance that removes the phenomenon from proper consideration in an essentially physiological context. In the aged the total number of muscle fibres has been shown to be reduced (Rubinstein, 1960); so far, however, there is no information to indicate whether such reduction is related to the exercise history of the subject. The basic fact alone, however, serves to focus attention on the cell population aspect of irreversible changes imposed by the environment. The coupling of cell destruction with cell restoration, as also the degree of slack possible in such coupling, represents another basic aspect of the problem of injury fixation in multicellular systems. In muscle this aspect is underscored by recent evidence that compensatory hypertrophy of muscle in rabbits after unilateral denervation of a limb is much more reduced in aged than in young adult animals (Drahota and Gutmann, 1962).

2. *Cold adaptation*

Environments to which land-dwelling animals become adapted in the course of their evolution are characterised by rather wide fluctuations in a number of important variables. One might therefore expect that adaptation to these "natural" variables should tend to be reversible to a particularly high degree. For inhabitants of the temperate zones of the earth, exposures to low temperatures exemplify such a condition. This theoretical expectation is confirmed to some extent by the results of studies on rodents. Mice maintained in a 10° environment survived far longer when subsequently stressed by exposure to − 17° than mice that had been maintained at 30° (Hart, 1953). Transferring adapted mice to a low temperature to the 30° environment results in a rapid loss of their ability to survive a test temperature of $-17°$; at the end of less than four weeks at 30° the survival times at $-17°$ for the formerly cold-adapted animals have returned to values indistinguishable from those of animals maintained throughout at 30°. Similar results have been obtained with rats (Sellers, Reichman and Thomas, 1951). In interpreting such results one may well raise the question whether here the reversal of cold acclimatisation really is merely a passive loss of cold adaptation, which is reversal in the ordinary sense, or whether it represents, in fact, re-adaptation of the animals to a warm environment. A second question, more pertinent to the present enquiry, is about the criterion of reversibility. The mere fact that the ability to survive a $-17°$ temperature is lost does not prove that all of the induced changes associated with cold exposure and adaptation thereto have been fully reversed. One would have liked to know whether the entire cycle might not have changed the life expectancy of the animals, whether there might not have been induced changes in various organ systems that, although not dramatically obvious in the animals shortly after exposure, might yet have altered permanently the animals' ability to resist various types of stress or affected the development of some of the physiological changes associated with ageing.

In studying local adaptation to cold in the hands of Gaspé fishermen, the degree of this adaptation was found to vary little between the time of maximal cold exposure during the fishing season and the period at home, when one would expect such cold exposure to be minimal (LeBlanc, 1962). This observation suggests that cold acclimatisation of this type in man may persist at least over a four months' span. However, the authors themselves suggest that since these four months without fishing operations fall in the winter time, the possibility cannot be altogether discounted that these men in their relatively primitive housing might still experience sufficient generalised chilling to account for some persistence of the local adaptation.

Seasonal variations in cold adaptation have had some study in at least two other situations. In Australian aborigines of the central plateau, the time courses of the changes in temperatures of various parts of the body during sleep in a cold environment have been reported to be the same toward the end of the summer and during the winter. Since closely related tribes from the northern tropical zone of Australia fail to show similar patterns, more or less resembling Caucasian subjects in this respect, one may infer that one is confronted here with a cold adaptation that is not measurably reversed over several months of exposure to a relatively warmer environment (Scholander, Hammel, Hart, LeMessurier and Steen, 1959). By contrast, the diving women of Korea show a remarkable seasonal variation in basal metabolic rate and in maximal tissue insulation (Hong, 1963). Although during the winter the values of these variables were considerably higher in the diving women than in the non-diving control subjects, during the summer months differences between the two groups disappeared, the divers dropping to the levels characteristic of the non-divers. However, the diving women, who continue their occupation summer and winter always maintained higher shivering thresholds than the controls. Here, therefore, cold exposure varying in degree with the season was able to elicit not only some major changes that proved apparently reversible, but also changes in at least one variable that showed continued maintenance of some degree of cold adaptation throughout the year.

Rats reared under conditions known to induce cold acclimatisation have a perceptibly lower mean life-span than controls reared in a more nearly thermoneutral environment (Kibler, Silsby and Johnson, 1963; Johnson, Kintner and Kibler, 1963). Except for some local lesions attributable to cold exposure of the feet and possibly of the ears, these animals failed to show any specific pathological signs. The difference in life span is associated merely with a change in the time of onset of diseases normally occurring in rats ageing in a warm environment, the several conditions appearing somewhat sooner in the animals reared in the cold. Despite an increase in food intake throughout their lives, these rats failed to reach the body size of controls reared under more nearly thermoneutral conditions. These animals, therefore, although developing a marked increase in cold tolerance, paid a rather considerable price for their adaptation. There is a question whether these animals "failed to ever acclimatise adequately to their exposure conditions" or whether the changes observed merely represent a secondary effect of the adaptive changes acquired, as might be inferred from the fact that they did, in fact, survive nearly a year and a half in the cold environment. More important, however, this experiment illustrates the fact that certain non-reversible processes can be profoundly influenced by exposure to this particular cold environment. What remains to be resolved is the question whether the effects observed represent a secondary modification of irreversible processes going on in any event, as a concomitant of ageing, or whether the need to make the metabolic adaptations required for survival in the cold environment imposed irreversible or poorly reversible changes upon the animals.

3. *Adaptation to mountain altitudes*

A third example of the extent to which studies of the rates of reversal of adaptative change can contribute to the subject is the body of information about adaptation to high altitudes. Here the emphasis was placed almost from the outset upon the use of stress tolerance tests. The existence of adequate adaptation, especially in natives at the high altitudes, was thenceforth being taken for granted. The changes that take place when man or other mammals are subjected to low oxygen tensions are, in the aggregate, improved tissue oxygenation under the prevailing atmospheric pressures. A most dramatic feature among these changes is the haematological response.

This results eventually in a perceptibly higher oxygen-carrying capacity per unit volume of blood and in a somewhat expanded blood volume. All studies indicate that, when altitude-adapted subjects are brought to sea-level, these haematological changes are reversed about as promptly as is compatible with the life span of the erythrocytes: in man, within eight weeks of transfer to sea-level, reticulocytes, plasma-iron turnover rates, blood volume, erythrocyte counts and haemoglobin content have all reached values not significantly different from those of permanent dwellers at sea-level (Hurtado, Menno and Delgado, 1945). Similar conclusions are suggested by results on rats (Atland, 1949), rabbits (Rotta, 1958) and guinea-pigs (Gordon and Kleinberg, 1937). Longer persistence of some related changes is implied by the observations that erythroprotoporphyrin levels are still perceptibly elevated in altitude-reared men after seven months at sea-level (Reynafarje, Lozano and Valdevieso, 1959) and that after six months to one year such subjects show somewhat lower haematocrit and haemoglobin values than those found in sea-level controls (Hurtado *et al.*, 1945). These values too, however, are said to return to normal by the end of two years' residence at sea-level.

Another characteristic change in man, and one that may conceivably be related in part to the polycythaemia, is an increased pulmonary arterial pressure (Rotta, Carep, Velasquez, Hurtado, Salazar and Chavez, 1952, 1956) and a (perhaps causally linked) hypertrophy of the right ventricle (Rotta, 1947). These changes give rise to a characteristic shift in the electrical axis of the heart, as well as to recognisable X-ray changes of the cardiac contour. In previously well-adapted altitude subjects all of these changes are completely reversed within three months of transfer to sea-level (Alzamora and Monge, 1955). In Monge's disease, when these changes are far more exaggerated, return to sea-level is said to result in normalisation within at most three years.

Hypoxia leads to prompt changes in the amounts of such enzymes as succinoxidase, whose concentration is increased in many tissues, as well as in metabolite patterns suggesting an increased capacity of the citric acid cycle (Tappan, Reynafarje, Potter and Hurtado, 1957b; Reynafarje, 1962). These changes, in rats and guinea-pigs at least, are reversed within a few weeks after return to sea-level oxygen partial pressures. In man, blood glucose, lactate and pyruvate levels likewise recover at about the same rate, so that they have become normal by the end of eight weeks (Monge, 1949). In altitude-adapted rabbits the oxygen tension of the urine is elevated for several weeks after return to sea-level, but approaches control values by the end of four weeks (Verzar, 1945).

By and large, then, the altitude-related changes considered so far have been shown to be reversed promptly, and it seems fair to conclude that a great part of the recovery, or reversal, is effected within the period of roughly three months required for return of the haematological values to those of sea-level. Such a conclusion, however, is not, or not yet, permissible for certain other changes. Thus, though there appears to be prompt and complete reversal of the enhanced CO_2 sensitivity of the respiratory centre established during the first few days of altitude exposure (and well stabilised after ten to seventeen days' sojourn at an altitude of 12,400 ft [Kellogg, Pace, Archibald and Vaughan, 1957]), long time residents in the Andean high valleys have been reported to show respiratory acidosis even after four months at sea-level (Monge, Contreras, Velasquez, Reynafarje and Chavez, 1948). Relative bradycardia has likewise been reported to persist for at least this length of time, though the evidence here may prove not quite so convincing (Monge *et al.*, 1948). The relative increase in number and diameter of capillaries in many tissues of altitude-acclimatised rats is said to persist "for months" after return to sea-level (Monge *et al.*, 1948). Retinoscopic observations in rabbits have brought to light similar evidence and seem to indicate that, if one compensates for the effects of the high haematocrit in the altitude-adapted animals, the capillary changes persist for at least twelve weeks at sea-level with little evidence of reversal (Monge *et al.*, 1948; Huerkamp and Opitz, 1950; Opitz, 1951). A high content of muscle myoglobin, especially in the heart, has been found to persist in altitude-acclimatised guinea-pigs for at least ten weeks at sea-level (Tappan and Reynafarje, 1957a).

Taken altogether, it appears that these findings surely cannot be said to prove the complete reversibility of all altitude-induced physiological changes in man. Such an inference has been suggested, but should perhaps be viewed as an exaggeration to offset the earlier hyperbole that "Andean man" should be viewed as a distinct variant of the race (or even the species), differentiated under the influence of "exceptionally severe climatic aggression". The facts would seem to justify the more modest conclusion that most of the more conspicuous changes brought about in the physiological make-up of altitude-exposed man are indeed reversible, but that even the limited variety of descriptive material so far available carries the suggestion that certain changes are far more slowly resolved than the rest, if indeed they are ever fully resolved. Some of these changes, in particular the high capillary density, may be far from deleterious and may play a significant role in the characteristically low incidence of embolic disease among the highland population of Peru (Hultgren and Sprickard, 1960), though assuredly other factors may also contribute to this, and thus play a role in modifying the pattern of ageing in this group.

In this context consideration must also be given to the complex of conditions collectively referred to as chronic mountain sickness. This includes the vague assembly of deteriorative phenomena experienced by some individuals who apparently have made a satisfactory initial adaptation to life at high altitudes. These symptoms progress and may increase in severity until they seriously disable the subject (Monge, 1928). Though relieved by return to sea-level, many such individuals can never again safely ascend to the altitudes where previously they may have lived for years. It is not at all clear whether the onset of this condition represents a constitutional inability to adapt fully, which becomes obvious with advancing age, or whether one is faced with an altitude-induced change in the physiological make-up of these individuals, resulting in a functional defect that remains occult at sea-level, but is unmasked by renewed exposure to a high altitude. Perusal of the clinical material would seem to lend weight to the latter view, especially because of the frequently rather extended latent period when the individual appears well adapted before the first symptoms appear. The same considerations apply in the more extreme condition commonly referred to as Monge's disease (Monge, 1937). This appears clearly to represent the result of "de-adaptation", often after long periods of comfortable sojourn at altitudes and is characterised by low arterial oxygen saturation and by increases, frequently dramatic, in erythrocyte mass and concentration. The aetiology appears sometimes to be of pulmonary origin (Hurtado, 1942), but other organs, notably the central nervous system (Chiodi, 1960), can also be the primary site of change. These subjects in general show remission of all overt symptoms on return to sea-level, but most of them experience relapses after relatively short periods upon return to a high altitude. Here again, it seems a plausible inference that the disease represents an alteration induced by exposure of susceptible individuals to high altitudes, which is not readily reversed and masked at sea-level, but is unmasked by subsequent altitude stress.

It may be permissible to point out the failure of this relatively large body of experimental and clinical studies, represented by the work on altitude adaptation, to reveal more conclusively the existence of residual effects except when gross clinical disturbances are present. Perusal of the material suggests several reasons for this. In part, the emphasis of most of the work has been on the most obvious manifestations of altitude acclimatisation, and these, it would appear, are generally, in fact, fairly promptly reversible. More subtle changes, such as might be expected to persist, are none too readily singled out from the maze of other changes including those due to cold, poor nutrition, filth-borne diseases, drug effects and social stress. The fact that none of these factors, all of them known to be operative and to differ widely between sea-level and mountain populations in Peru, were found to be expressed in this body of observations would seem to justify the thought that the methodology employed, however excellent in its technical details, is not suitable for demonstrating the type of changes being considered here. Subsequent sections will include discussions of instances when other methods more specifically adapted to revealing irreversible change were successfully applied to other stress situations; it may be possible that this will suggest fresh approaches to the altitude problem.

C. Residual Lesions

1. *Decompression sickness and aseptic bone necrosis*

One approach, perhaps the most direct one, involves study of the residuum of change left behind by an environmental experience after subsidence of the transient disturbances associated with transfer from one environment to another. A straightforward example of this type of observation is presented by the bone lesions observed in caisson workers and others working relatively long hours at elevated pressures. The lesion was clearly described as early as 1930 (Kahlstrom, Burton and Pheminster, 1939). It consists of areas of aseptic bone necrosis, largely confined to the shafts and heads of the long bones of the extremities, often surrounded by areas of recalcification, which decrease the size of the original lesion. The end effects of these changes depend upon their location. They tend to be silent lesions when located in the diaphyses. When they involve weight-bearing surfaces, as epiphyseal lesions frequently do, severe arthritic changes can ensue, culminating in disabling arthritis deformans (MacCallum, Stanger, Walder and Paton, 1954). On longitudinal section the epiphyseal lesions often have a characteristic fan-shaped appearance (Kahlstrom *et al.*, 1939; MacCallum *et al.*, 1954). Diaphyseal lesions usually consist of scattered foci, which may assume considerable extent (Kahlstrom *et al.*, 1939). It has been noted that these lesions show some tendency to occur symmetrically, that is, to involve both shoulders and so on (Allan, 1943).

All observers agree that these bone necroses are causally linked to the formation of gas bubbles during decompression after relatively long periods of sojourn under pressures greater than 2 atm. However, the question remains whether these particular lesions are the results of intravascular bubble formation or of extravascular bubble formation in the medullary tissues of the bones. An early experiment designed to clarify this question (Kahlstrom *et al.*, 1939) is in retrospect open to many objections. Evidence for extravascular bubble formation has been presented for other tissues, but primarily for rather extreme rates of pressure fluctuation (Gersh, 1945). Intravascular bubbles have been demonstrated more frequently and appear to be formed readily under conditions such as might be expected to have prevailed in those circumstances (Harvey, 1945). The correspondence of the shape of the epiphyseal lesions with the vascular pattern of this bone region has been noted (Kahlstrom *et al.*, 1939; MacCallum *et al.*, 1954), supporting the view that intravascular miliary air embolisation is the immediate cause of the lesions seen there.

It is important here to note that these lesions are not confined to those who have suffered attacks of acute decompression sickness, but have been found also in numerous individuals who have effected many uneventful transitions between low and high pressures (Bell, Edson and Hornick, 1942); moreover in many symptomless subjects with a history of caisson work silent diaphyseal bone lesions have been detected by radiography. That these lesions are to all intents and purposes permanent is shown by their discovery in subjects whose caisson work ended many years previously.

In all, then, the situation described represents an instance of a well-defined lesion that can be produced by one (James, 1945) or by many transitions from high to low pressure, a lesion not necessarily associated with any other subjective disturbances during the decompression, non-reversible and apparently without further consequences unless brought to light by mechanical stress, when lesions occur in the vicinity of weight-bearing surfaces.

Certain other aspects of decompression sickness appear to warrant inclusion at this point. One of these is the tendency of "bends" pains to recur in specific sites in any particular individual. This appears to occur in divers (Behnke, 1942; Rivera, 1963), but has not been substantiated in personnel exposed in altitude chambers (Henry, 1946). A pertinent phenomenon is the predilection of certain individuals in a given group at risk to suffer attacks of the bends. Thus, in an apparently fairly uniform population involved in tunnelling operations, statistical analysis showed that 2% of the group accounted for 43% of the attacks, and that of these about one-quarter sustained three or more attacks (Golding-Campbell, Griffiths, Hempleman, Paton and Walder, 1960). These statistics could be utilised equally well to support either of two opposite views of this matter: one might conceive of these relations as an expression of constitutional bends-proneness in the affected individuals, or one might infer that a single bends episode entails local changes that henceforth render the subject more susceptible to subsequent attacks of decompression sickness. It seems plausible to suggest that recurrences in identical anatomical sites and the occasional association of these effects with previous local trauma favour the view that one is here confronted with possibly irreversible subtle changes that give rise to a latent lesion unmasked by subsequent decompression.

A most interesting further phenomenon uncovered by examination of the data of Golding-Campbell *et al.* (1960) is the diminished frequency of bends in a previously unexposed population subjected to daily caisson work succeeded by decompression. This "adaptation to decompression" appeared to be complete in a matter of two weeks. The effect, however, is rapidly reversible, so that a period of no more than ten days without caisson work suffices to restore the original bends frequency.

The two sets of observations reviewed in the last two paragraphs may at first sight appear to be in mutual conflict. In fact, however, this is not so; despite a decreasing frequency of bends in the group as a whole, on a day-to-day basis the individuals who have once sustained a bends episode henceforth show up as more susceptible than the average of the remainder of the group. It would appear from this that "adaptation to decompression" and "bends proneness" reflect two different sets of processes, both possibly triggered by exposures to a high-pressure environment and then to decompression.

2. *Auditory acuity*

This may be considered an example of a somewhat traumatising severe stress, with residual changes that are clearly pathological; an instance of change associated with the normal daily environment of man in a modern civilisation is afforded by studies on auditory acuity. Such work, incidentally, also represents an example of one type of population-oriented approach to the study of residual injury.

In all human populations careful examination has shown that the relation between auditory threshold and test-sound frequency changes continuously with age. At all frequencies there is a progressive decrease, but this is greater the higher the frequency (Glorig and Nixon, 1960). Dividing a population according to whether they habitually work in noisy or in relatively quiet environments reveals that the decrease in auditory acuity progresses considerably more rapidly in the "noisy work" group (Nixon and Glorig, 1961). Such changes were perceptible mainly in workers employed in environments that resulted in temporary hearing changes at the end of a single working period (Ward, Glorig and Sklar, 1959), marked immediately after the noise exposure, but in general almost completely reversed within 30 min. Thus, here again one encounters the phenomenon of a somewhat marked change imposed by an environmental strain, with subsequent ostensible complete recovery that yet leaves a minute residue, which becomes perceptible with the lapse of time. There is an indication that even short periods of repeated noise exposure can permanently alter the time course of the changes in the audiogram. Thus, separating a population into those who had undergone basic military training a number of years previously and those who had not had such an experience revealed considerably more advanced auditory deterioration in the former group (Glorig, Summerfeld and Nixon, 1960).

These several sets of results are still incomplete: information is needed about the time course of changes in the audiogram when noise exposure was kept to an absolute minimum, that is, about what might be called the intrinsic course of ageing of auditory perception. A situation possibly approaching this condition has been reported recently. The population in question is a forest-dwelling tribe in the central Sudan, the Mabaan, living under conditions in which noise approaches only on rare festive occasions the level of 10 decibels, a level frequently encountered and considered normal by Americans living and working in "non-noisy" occupations. Tests of auditory acuity in this tribal group showed but slight differences from that of their American counterparts when the 20-year-old members were compared; however, the time course of auditory deterioration from this point on is remarkably retarded in the Mabaan. Thus, 60-year-old Mabaan men showed audiograms that compared favourably with those of 40-year-old American men not exposed to noise (Rosen, Bergman, Plester, El Mofty and Hamad-Satti, 1962). Some caution is needed in the interpretation of these results in the present context, not only because of the possibility of racial differences (no racially similar group living in a modern industrial setting has been examined by the same methods), but also because of evidence that certain age-linked changes other than those of auditory acuity (e.g. blood pressure) likewise progress more slowly in the Mabaan than in American population groups. Thus, causal relations other than those immediately pertinent to the present discussion might still be found to be at the base of the well-preserved sense of hearing in the aged Mabaan. Yet as they stand at present, the findings suggest that the so-called normal noise level surrounding American rural populations is not inert, but may produce a measurable increased rate of deterioration in auditory acuity.

It is not clear at present whether the auditory changes described above represent the accumulation of anatomical changes in the peripheral sense organ, most notably perhaps of the hair cells of the organ of Corti rather than of the sound conduction system (Jorgensen, 1961), or whether they represent changes in the central nervous system affecting the perception of sound. Thus, in this particular instance the fairly well-described sensory phenomena cannot yet be matched conclusively with morphological or physiological changes that might contribute to an analysis of the basic mechanisms by which the organism fixes environmentally imposed changes.

3. *Ionising radiation*

The above analysis, however, applies far more closely to the residual effects of exposure to ionising radiation, the stress that more than any other perhaps has served to focus attention on the existence and the potential importance of subtle delayed effects persisting after apparent recovery from an acute stress episode.

Though the tumorigenic effects of exposure to ionising radiation have been recognised since the turn of the century, the more diffuse late effects of this agent have received close investigation during the past two decades. The basic phenomenon can be illustrated by two types of measurement. If, after a relatively brief period of exposure to X-rays at a sublethal level, test animals (generally mice) are allowed to recover for a sufficient interval to dissipate all of the acute effects of the irradiation and are then tested for radiation resistance, the animals give distinctly lower LD_{50} values than non-pretreated controls (Krebs, Brauer and Kalbach, 1959). If the radiation is delivered in relatively large separate doses at dose rates such that each exposure is complete in a matter of an hour or two, the ratio of initial dose to residual injury as expressed by the shift in LD_{50} is nearly constant and remains independent of the interval between pre-irradiation and test dose, once this interval exceeds the acute recovery period (Krebs *et al.*, 1959). Thus, the "nonrecuperable fraction" of radiation injury, for X-rays and on LAF_1 mice, for instance, consistently equals 8 to 12% of the priming dose.

A second quantifiable change wrought by exposure to ionising radiation is a reduction in life span of the exposed population. Here again the findings suggest a roughly linear relation between dose and the decrease in mean life span. The relations here, however, are somewhat more complicated: there are suggestions that in caged rats, at least, this effect is somewhat dependent upon the age of the animals at the time of exposure (Jones and Kimeldorf, 1963); further, one or another specific mode of death may assume disproportionate importance in irradiated animals of various strains (Kaplan, 1948; Lesher, Grahn and Sallese, 1957; Cohn and Gutman, 1963). However, even if specific tumour susceptibilities are ruled out, the general effect of shortening life span persists (Upton, Kastenbaum, and Conklin, 1963) and the age dependence appears to be no more than an expression of the time interval required for expressing a given impairment under the peculiar pattern of stresses to which caged laboratory animals are subjected. Thus, for instance, renal impairment may be an important element in these late deaths, but can be unmasked at a much earlier stage by suitable test procedures (Krebs, 1960).

Phenomena of these two types have been singled out here merely because they are most readily quantifiable. In fact, close observation has revealed a whole galaxy of changes in such pre-irradiated animals, including not only tumours of diverse types but also vascular (Casarett, 1958), renal (Lawson, Billings and Bennett, 1959; Krebs, 1963), hair (Chase, 1963), bone marrow (Patt and Quastler, 1963), reproductive organ (Kohn, Kallman, Berdjis and DeOme, 1957; Rugh and Wolff, 1956), endocrine organ (Gains, Brown, Hansen and Howland, 1963) and other effects. Most of these are not unique to irradiated animals, but rather seem to occur at an earlier age in irradiated than in control animals, giving rise to a complex question of a gerontological nature, discussed in more detail below (Handler, 1961).

Further light on the processes involved in the production of these residual effects is cast on studying the kinetics of recovery from radiation injury by the split-dose method (Paterson, Gilbert and Mathews, 1952; Krebs *et al.*, 1959). These show that recovery from a single large dose can be described in terms of a larger fraction that is dissipated roughly as an exponential of the time plus a non-reversible fraction corresponding to that already described (Blair, 1952; Kaplan and Brown, 1952). If, however, the dose is broken up into smaller fractions delivered at shorter intervals for a given total dose, recovery rates may be slower, and the non-recuperable fraction may decrease, until with sufficiently small individual doses recovery is represented by a constant daily amount, unrelated to dose, of a magnitude in the neighbourhood of 50 r/day (Krebs and Brauer, 1963). On comparison with cell-clone culture findings, Krebs and Brauer (1963) suggest that this mechanism represents the ability of individual cells to recover from radiation damage (or to tolerate radiation damage), as evidenced by the characteristic shoulder (Elkind, 1961) found in dose-mortality studies of such cultures. A more refined analysis of such events can be attempted and involves dose-rate phenomena and the question of the role of multiple versus single hits; such treatment, however, goes beyond the scope of discussion here (Krebs and Brauer, 1963). The faster initial recovery rates observed in animals subjected to relatively large X-ray doses at higher dose rates, it has been suggested, represent recovery not in terms of individual cells, but rather in terms of the ability of the cell populations constituting various tissues to compensate for cell loss, either functionally (as perhaps in the central nervous system) or by cell regeneration, as most prominently occurs in the bone marrow, for instance (Lajtha, 1961). The effects of radiations such as fast neutrons, protons and the like, which show higher linear rates of energy transfer (LET), fit into this picture in so far as the tissue-regeneration type of recovery after such exposure, though possibly slower than with X- and γ-rays (Melville, Conte, Slater and Upton, 1957), leaves a non-recuperable fraction of the same magnitude as the photon radiations (Krebs and Brauer, 1962). Cumulation of injury at low dose rates of high LET radiations is considerably faster than with X- or γ-rays, corresponding to nearly shoulderless dose-survival curves in cell cultures (Andrews and Berry, 1962).

All of these effects are viewed at present as primarily the expression of damage to the nuclear chromatin, either as point mutations—a most important aspect of the inter-generation genetic effects of radiation exposure (Failla and McClement, 1957)—or as chromatin loss resulting from one of several types of chromosome damage (Muller, 1963). The possible role of these mechanisms in relation to the delayed effects of radiation exposure has been vigorously, if not conclusively, debated by geneticists. The alternatives are induced mosaicism as against multifocal diffuse cell loss (Gowen, 1961). An eloquent case has been made in favour of the paramount importance of chromatin loss in the residual radiation lesion (Muller, 1963). Experimentally, chromosome aberrations have been found months after radiation exposure in rat liver stimulated to regenerate (Leong, Pessotti and Krebs, 1961) as well as in circulating lymphocytes (Nowell and Cole, 1963). The persistence of such cells is compatible with cell-culture findings, which show that cells with this type of defect can undergo at least two or three mitoses before the clone dies out (Puck, Marcus and Cieciura, 1956).

4. *Physiological changes*

To understand the transition from such cytogenetic mechanisms to physiological effects in the multicellular individual is by no means easy or clear. Virtually nothing is known about the possible effects of multiple diverse point mutations in somatic tissues, although such inferences as might be permissible from findings on insects tend to discount any prime role for such effects, except possibly in certain types of tumorigenesis (Muller, 1963; Gowen, 1961). In connection with discussions of cell loss resulting from chromatin ablation or possibly from specific point mutations the argument revolves around cell population dynamics. For the regenerating liver there appears to be a clear threshold defining the fraction of the total liver substance that must be resected before an adequate stimulus for compensatory liver cell multiplication is established (Bucher, 1963). In keeping with this, liver regeneration has been found to cease when liver mass is somewhat below that for control rats of the same age (Leong *et al.*, 1961). Residual tissue loss then persists because of the slightly defective "setting" of the feedback mechanism that adjusts liver weight to body weight. Similar mechanisms might well be called into play in other tissues subjected to diffuse cell deletion by ionising radiation. Such a defect will be most perceptible where the cell population is derived from a group of stem cells whose replenishment is imperfect or even impossible. Examples of such a mechanism are the testicular atrophy characteristic of irradiated male rats and mice (Kohn *et al.*, 1957) and the loss of melanocytes giving rise to the characteristic greying of the hair in irradiated animals (Chase, 1949). Another more complex example is the loss of functional reserve of the bone marrow, as revealed by studies of haematological recovery in dogs in the course of a series of widely spaced radiation exposures (Baum and Alpen, 1959; Baum, 1961).

It seems virtually certain that these are not the only mechanisms by which radiation injury fixed in the tissues is translated into end effects that give rise to the impairment characterising the syndrome of delayed radiation effects. Among other possible mediating mechanisms mention may be made of the capillary changes reported to occur widely throughout irradiated tissues (Casarett, 1959); it has been suggested (but remains to be proved) that such vascular changes could give rise to focal micronecroses, resulting in widespread diffuse secondary loss of a few cells here and there, simulating some of the changes claimed to be associated with senescence (Casarett, 1959). Another possible causal chain involves primary disturbances in one or another of the endocrine glands resulting in secondary changes elsewhere. Examples of this may be some of the changes in renal function (Krebs, 1960, 1963) and possibly the impaired secondary growth in realimented hypocalorically reared rats (Carroll and Brauer, 1961), as well as some of the late changes in irradiated mice that may be related to ovarian (Sobel and Furth, 1948) or testicular injury.

D. Life Tables and Related Approaches

1. *Calendar functions and ionising radiation*

Before leaving consideration of the residual effects of ionising radiation it seems worth while returning briefly to the subject of the life-span shortening effects of exposure to radiation of this type. Though the discussion above was couched in terms of "reduction of mean life span", more effective statistics can be derived that use all the data contained in a life (or a mortality) table. A convenient function for this purpose is the logarithm of the death rate *per capita* of survivors of any given age, a function known as the Gompertz function, (Gompertz, 1825). The function derives its importance from the observation that for many different species the increase in Gompertz functions is approximately linear with age after puberty (Pearl, 1928). Such a relation invites extension of the concepts of "stages of development" and of "physiological age", properly used during the period of individual development, to the mature individual: the Gompertz function, because of its linear relation to time in a "normal" population, is thus conceived as a measure of physiological age. The effect of radiation exposure upon the Gompertz function is approximately that of increasing it at the time of exposure, without altering the subsequent slope of the regression of this on the age of the individuals composing the test population (Sacher, 1956). Interpreted in terms of the concept presented above, such a transposition could be described by saying that the physiological age of the members of the test population had been abruptly increased by an amount related to the dose of radiation received and that thereafter the population continues to age at the same rate as they would have if they had achieved this "physiological age" undisturbed in the course of time. It is this type of concept, supported by certain superficial resemblances (greying of the hair, for instance), that has invited much speculation about the relation of radiation to normal or to premature ageing (Handler, 1961).

However, once one introduces the concept of a calendar function, and so it is that the Gompertz function is employed in this kind of reasoning, it becomes immediately evident that other variables varying monotonously with chronological age in normal populations must also be considered. Physiological age could in theory be defined with reference to any of these, and "normal" ageing in this sense would imply an orderly progression of all such variables, each at its own "normal" rate as time advances. Accelerated ageing, or discontinuous change in physiological age, implies that shifts in all these variables due to any cause whatever are so related that the final position represents identical shifts in physiological age of all variables. If this condition is not fulfilled, the concept loses much of its meaning, and what is changed is no longer a "physiological age" but rather a whole pattern of ageing.

With radiation effects, there seems little doubt that precisely this second alternative applies. To cite two examples: the radiation resistance of mice is a calendar function in the sense defined above,† decreasing monotonously with age after thirty weeks (Lindop and Rotblat, 1962). The decrease in radiation resistance represented by the non-recuperable fraction of a given X-ray dose pattern is such that, using this variable as the basis for calculating physiological age, one would arrive at an ageing effect of 100 days at most caused by two X-ray doses of 400 r each. Calculations of the same kind based on life span shortening would yield values exceeding 200 days. Viewed against a maximal life span of about 800 days, it is clear that the discrepancy is extremely large. A similar discrepancy (though in the opposite sense) would be derived from calculations based on the change in renal concentrating ability (Lawson, Billings, and Bennett, 1959) and from present data on chromosome aberrations in the livers of rats (Leong *et al.*, 1961). In mice exorbitantly high chromosome aberration frequencies have been reported for two strains; the discrepancies, though still real, are not as large as in the preceding examples (Crowley and Curtis, 1963; Curtis and Crowley, 1963; Curtis, 1963).

These instances suggest that the late effects of radiation cannot be equated with a speeding up of "normal" ageing, but must be viewed as producing a specific shift in the exposed animal's pattern of ageing. This conclusion is supported by the results of post-mortem examinations; though showing few changes unique to irradiated animals, these suggest that the relative importance of various lesions at death, and the time of onset of various deleterious conditions or diseases, differ to widely different degrees in control animals and in animals of the same strain subjected to various types or patterns of irradiation. Examples of such discrepancies are amyloid disease (Lesher, Grahn, and Sallese, 1957), skin changes, arteriosclerotic changes in major blood vessels (Gold, 1961), testicular atrophy (Kohn, Kallman, Berdjis and DeOme, 1957), incidence of non-killing tumours (Upton, Kastenbaum and Conklin, 1963) and response of connective tissue to thermal or chemical induction of contracture (Darden and Upton, 1964).

From what has been said so far it should be clear that a function such as

† These relations are, indeed, even more complex. Whereas life span shortening and non-recuperable fraction of injury appear to be linear functions of dose of radiation (all other conditions being constant), the relation between LD_{50} and age is not linear throughout the adult life span. Thus, even if a fit could be construed at one time, such a relation could not be expected to hold throughout the life span of the animal. Parenthetically, it may be mentioned that non-linear relation between age and values of a possible calendar function appears to be not uncommon. Thus, in particular, the curve representing changes in connective tissue properties, as first proposed as a calendar function by F. Verzar, has since been found to flatten out to such an extent during the latter half of life as to prove itself nearly useless as a calendar function for gerontological experimentation.

that of Gompertz can conceal a good deal of information if it is interpreted too literally. At the same time, it should be recognised as a potent integrating tool, capable of bringing together large bodies of facts in a form that makes possible some judgement about whether further enquiry of the type described above is warranted or not. This character shows up clearly in the two reasonably promising theoretical interpretations of the Gompertz function so far attempted. According to one of these, this function is the sum of mortalities due to a number of diseases whose onset is linked to age, and perhaps to environment (Simms, 1946)—an extension of the evolutionist view that susceptibility to various diseases will tend to be pushed by selection toward or beyond the end of the reproductive period. The other view explains the Gompertz function as the end result of decreasing ability to compensate for environmentally imposed fluctuations in the metabolic state of the animal (Sacher, 1956)—a stochastic approach that merely renders the irreversible progress explicit and relates it to the here and now of the environment.

2. *Prisoner of war findings*

It may be pertinent here to illustrate the application of this function as a problem-revealing device in two instances of human populations. Consider the mortality increase of American internees in Java during the years 1944 and 1945 (Bergman, 1948). When the pertinent results are represented on a Gompertz diagram, the shift in the force of mortality is seen to affect all age groups to roughly the same extent, giving the impression that the entire Gompertz curve has been shifted upward without changing its slope, in a manner most closely represented by suggesting that each of these prisoners of war, by the very act of having been interned, had his age abruptly increased by almost twelve years during the first year of internment and by another six or so during the second year. When these prisoners were released from captivity many of the outward signs of deterioration during imprisonment were found to be reversible. From our present point of view, however, it is important to note that some residue of the experience was left behind. This was expressed outwardly by the fact that these men appeared "more grey and wrinkled than fits their real age." One might expect to find quantitative expression of the existence of such a residue by examining the force of mortality during subsequent years. At the present time only severely limited results of this kind are available. In a six year follow-up of United States Armed Forces Personnel, most prisoners of war failed to show any detectable residual injury of this type. Only the group interned by the Japanese, representing largely prisoners taken in the Philippines, were materially affected and showed a persistent change of the same type as that found during the captivity in the Australian group previously described, though of far smaller magnitude (Pearson, 1946; Jones, 1957).

3. *Secular trends*

Another indication that residual effects of environmental experiences can be shown by analysing statistics of this type may be found in indications from the study of groups of different birth ages in several of the Western countries. The secular trend to decreased mortality during the first quartile of life, due presumably to improved sanitation and improved nutrition and perhaps to improved medical care, is associated with a perceptible downward shift of the curves representing the force of mortality as a function of age, the shift being by roughly equal amounts for all age groups, suggesting once again that in a sense the physiological age of these populations has been shifted against the chronological age (Jones, 1960). This shift is at least correlated with, if not causally linked to, effects, imposed at much earlier age that are not reversed by many years of subsequent survival.

So far as I know, findings from experiments testing this hypothesis for specific environmental factors imposed upon properly controlled, and sufficiently large, experimental populations are available at the present time only to a limited extent. Some interesting results have recently been published about various wild and domesticated animal populations; they indicate little beyond the fact that for the larger and better cared for species, both among mammals and birds, the relations between force of mortality and age approach those described so far (Comfort, 1956, 1959). Small mammals and small birds in the wild frequently show curves indicating that the pressure of predation upon them is so heavy that the force of mortality is virtually constant, and high, throughout their lifetime, a lifetime found by comparison with those found in some instances for the same animals under conditions of captivity to be quite small (Bourlière, 1959). Even these findings, it would seem, may undergo extensive modification when we have available the results of ethological studies involving prolonged observation of mammalian or avian populations consisting largely of individually marked animals.

Together with the scarcity of information involving adequate human populations divisible into subgroups existing in predictably different environments, and the hitherto virtually complete lack of laboratory studies of life tables and pathology at death of genetically uniform populations living under controlled sets of environmental condiditions, these few findings suggest that here indeed is a rich field for future investigation.

E. Studies in Nutrition

In one area only have studies been undertaken that, partly at least, satisfy these requirements, the field of experimental nutrition. The general situation here is similar in character to others discussed in the preceding pages: many of the immediate effects of nutritional strain are fairly readily and dramatically reversible upon removal of the dietary strain by supplementation with those nutrients in short supply in the experimental diet.

1. *Japanese school children after 1945*

Numerous examples could be presented from both animal and human nutrition experiments. One of the most dramatic ones, and one pertinent to the immediate discussion at this point, is the observation that the somewhat severe malnutrition of Japanese school children during the World War II entailed a marked reduction in their growth (Sams, 1945–51). This was shown by retardation both in stature and in weight below pre-war standards, to an extent corresponding roughly to a retardation of development by six months to a year in children ranging from 7 to 12 years of age. Resumption of a more adequate diet, aided perhaps by supplementation with school lunches, was associated with a rapid reversal of this trend during the post-war years. By 1949 pre-war levels of the main variables had been restored and were, in fact, in the process of being progressively exceeded by youngsters of this age group. Children in the 12- to 18-year age group were not given dietary supplements in school; reliance was placed instead upon improved food availability in the home. These children conspicuously failed to catch up with pre-war standards. There is unfortunately no evidence about whether this group would have responded to supplementation.

2. *Vitamin E deficiency*

An extensive literature dealing with deficiencies less complex than the one exemplified by these children suggests the generalisation that if the deficiency of a given essential factor is not carried to the point at which gross pathological changes arise, relief of the deficiency will lead to reversal of the clinical picture. The dramatic nature of this kind of response, and its sociological and medical importance, have tended to minimise the attention given to less dramatic residual effects that might persist as a result of any kind of a bout of nutritional deficiency. In monkeys, for instance, protracted vitamin E deficiency results in changes in brain morphology similar to those commonly described as senile, consisting in particular of deposition of certain strongly basophilic lipoprotein particles resembling those designated as lipofuscin (Einarson, 1962). The deposition of this pigment is apparently not reversed by vitamin E supplementation. In the monkey this process appears to remain for some time at the level of a simple quantitative pigment increase without causing irreparable impairment of neuronal function. If the condition is allowed to continue without relief by supplementation until after marked muscular dystrophy has set in, the morphological changes in muscle and central nervous system, with the exception of the lipofuscin deposits, still appear to be reversed fairly rapidly, but then a certain degree of functional impairment will persist as a residue of the nutritional injury sustained. In the rat the whole process follows a more dramatic course, in which irreparable functional disturbance is practically uninfluenced by the subsequent administration of vitamin E except at the very beginning of the process (Einarson and Ringsted, 1938).

3. *Calorie shortage*

Even more instructive from the point of view of the present discussion is a series of studies dealing with the effects of calorie deficiency substantially uncomplicated by a deficiency of specific nutrients. This preparation was first utilised by McCay and co-workers to demonstrate that rats reared under conditions of such low calorie intake show remarkably extended life spans when compared with control animals fed *ad libitum* (McCay, Crowell and Maynard, 1935). It has since been shown that the effectiveness of this procedure in extending the life span of rats is not wholly dependent on stunting of growth or on prevention of sexual maturation (Berg and Simms, 1960). A most important finding, from the point of view of the discussion here, is that, when careful stock was taken of the pathology of these animals, it was found that the shift in the survival curves was found to be less than the shift in the time of onset of what had been considered major lesions of senescent rats (periarteritis, glomerulonephritis, myocardial degeneration), and tumour onset was delayed to an extent that may actually prove to be less than the life span extension (Berg and Simms, 1961). These results, therefore, furnish another illustration of modification of the pattern of ageing by an environmental factor, here nutrition. It is interesting to note that closely similar conclusions had been reached previously by McCay working with a far less healthy rat colony, in which the "diseases of old age" included pulmonary and middle ear infections (McCay, Sperling and Barnes, 1943).

4. *Patterns of ageing*

Another related series of investigations, in which the quality as well as the quantity of diet supplied to rats was made a variable, furnished indications that the normal stock diet used for rat-colony maintenance was perceptibly less favourable for maintenance of long life than certain synthetic diets, giving rise to some interesting considerations in line with those raised above in this chapter to the effect that normal environments are not by any means necessarily neutral (Ross, 1961).

Further indications that extension of life span is not an immediate consequence of severe stunting of growth in nutritionally restricted animals are furnished by observations on genetically obese mice. In such subjects, restriction of the diet merely to the point of preventing the development of obesity, though allowing growth to the full stature attained by siblings either free of or heterozygous for the obese trait, restored the life span of the obese animals to values that compared favourably with those of non-restricted non-obese animals (Lane and Dickie, 1958).

These manipulations all involve modification by nutritional factors of processes that result in irreversible impairment of the animal and are certainly major factors contributing to the physiology of ageing in the rat. The nutritionally restricted animal, however, affords an even more direct contribution to the discussion of potentially irreversible effects of nutritional stress. In his original studies, McCay showed that animals reared on a low calorie regime respond to an increase in their food supply by rapid and marked growth, even if the period of nutritional restriction has been somewhat protracted (McCay *et al.*, 1943). As measured in terms of total body weight, animals restricted in this fashion for a period of a year or more and then placed on an unrestricted regime quickly increase their weight and reach final equilibrium weights close to those reached by animals that have never been restricted. Such a finding is in line with other results already alluded to, indicating that the effects even of rather severe nutritional stress are superficially reversible. With the nutritional restriction sustained for 800 days or more, the growth response to re-alimentation rapidly decreases in magnitude. Such animals fail to approach the body weight of normally reared siblings of similar age (age reference here must be in terms of fractions of the revelant mean life span rather than of chronologic age, since 800-day-old rats on a normal regime are rare and in any event show severe senile changes).

One might argue that this late failure of the ability to respond to re-alimentation is an expression of old age in these nutritionally retarded rats, rather than proof of the fixation of irreversible changes as a result of a nutritional stress; there is other evidence to suggest that even at a much earlier stage such nutritional restriction leaves traces that are not fully reversed. This type of effect is exemplified by changes in long-bone length. Thus, for example, the tibias of re-alimented animals that have attained normal body weight are considerably shorter than those of normally reared siblings (Barnes, Sperling and McCay, 1947). Similarly, such re-alimented animals show disproportionately short tails, apparently associated with resorption of one or two of the tail vertebrae (Carroll, unpublished findings). These findings are all the more interesting because as it has been shown that in the rat the epiphyses of the long bones do not close even late in life, but rather enter an inactive stage, in which real new bone formation ceases in the region, though apparently the epiphyses remain capable of reactivation.

However limited may be the body of information about the irreversible effects of nutritional stress, such facts as are available, and as have been exemplified in the above discussion, suggest two concepts. In the first place, the effects of nutritional stress of many different kinds can apparently be reversed to a remarkable extent by correction of the nutritional deficiency, unless the process has been allowed to go on to the point where gross pathological lesions have developed that might be expected to leave behind scars after healing. In every instance, however, when such information has been sought with suitable methods, evidence has been found of earlier and more subtle residual changes that persist even if the nutritional deficiency is not sustained to the point of gross organic lesions. The significance of such effects must be sought not so much in any acute changes in the properties of the organism as in the accumulation of more subtle changes with time, in a modification of ageing patterns. A second point raised by the nutritional findings is the observation, especially striking in the calorie restriction experiments, that manipulation of the nutrition of the individual is capable of modifying the time course of processes going on even in the absence of nutritional stress. To the extent that processes of this type are precisely those irreversible changes associated with ageing of the individual, nutritional manipulation is seen to affect the lapse of biological time; since the various processes of this kind are not all affected to the same extent, such manipulation affects patterns of ageing in ways that are both practically and theoretically of importance.

F. Modifications of Individual Development

1. *Early handling effects*

The recurrent question about the role of injury in producing irreversible changes by environmental strain would appear to be almost completely sidestepped when the interaction takes place not with a mature individual but with a growing one. The embryologist is familiar with the occurrence of critical stages in development, when changes in the environment of the embryo, or the imposition on it of any one among many kinds of stress, will radically alter the course of subsequent development of one or another organ system. Such responsiveness, however, is not restricted to these earliest stages in an individual's career, as may be illustrated by the effect of early handling of the young upon subsequent behavioural processes. If young rats during the first week or two of their lives are subjected to even gentle handling, it is found that they grow into young adults that perform better in certain types of learning situations and that, when observed in various types of experimental environments, show a distinctly greater tendency to explore than unhandled control animals (Hunt and Otis, 1955; Ruegamer, Bernstein, and Benjamin, 1954). This situation has been described as representing "decreased emotionality" of these handled animals and, when first observed, was attributed to enrichment of the animal's store of experience. Subsequent studies have failed to bear out this perhaps somewhat anthropomorphic interpretation. A similar effect can be brought about by mild cooling of the animals, allowing a slight drop in rectal temperature; this effect can be prevented, all other handling procedures being the same, by maintaining the animal's body temperature constant throughout the handling periods (Schaefer, Weingarten and Town, 1962). Interestingly enough, more severe cooling at this stage has the converse effect (Hutchings, 1963). If the handling is postponed until about a week after the animals' birth, this particular series of changes can no longer be produced. One might thus argue, in view of the notorious immaturity of the neonatal rat, that this effect is little more than an extension into extrauterine life of one already well established for the embryo. It is indeed possible that these effects represent not so much an accretion of experience on the animal's part as a change of the course of development of parts of the rat's brain, imposed by a stress during a period when it is well known that considerable development is still going on in the phylogenetically younger parts of the central nervous system. Other experiments, however, indicate that this situation may represent merely one example of a more complex body of phenomena.

2. *Socialisation and structuring of the environment*

It has, for instance, been shown that rats considerably older than those discussed so far, when reared in the ordinary non-structured cage environment, develop learning behaviour, and social behaviour, perceptibly inferior to what is developed by the same animals if they are reared in richer more highly structured environments, such as a larger cage with diversified play, feeding and housing areas (Denenberg, 1962, 1963). Something like critical periods can be vaguely discerned in this situation and seem to correlate with the level of behavioural development of these animals. It is clear, however, that this type of manipulation of the environment, without producing gross organic changes, is capable of producing far-reaching and persistent changes in the functioning of the central nervous system of these young animals. Carrying this type of procedure yet one step farther, it has been shown that in adult rats profound organic changes can be brought about by manipulation of the social environment. Thus, in ageing multiple-caged rats of a certain strain the incidence of mammary tumours was far lower than is found in the same strain caged singly (Muhlbock, 1957). Analysis of this situation has revealed that the effect is related to the increased incidence of pseudopregnancy in the multiple caged animals and that the link between multiple caging and low incidence of this particular tumour involves the olfactory sense of the animals.

3. *Genetic considerations*

This type of study serves to provide a convenient framework in which the question of the interplay between genetic determinism and environmental imprinting can be considered. For instance, in two strains of deermice identical early handling procedures elicit opposite final effects on learning performance in the adult (King and Eleftherion, 1959). The inherently wilder and more emotional strain is set back by this procedure, whereas the more stable strain is favourably affected in the same manner as the rats discussed above. Genetic differences similarly have been shown to be associated with somewhat different responses to standardised environmental variations in the early environment of dogs of different breeds (Scott, 1945). An extreme instance of this type of situation, socialisation in sheep, is based on a behaviour pattern normally developed with such uniformity that it has entered into the language to designate one type of human behaviour. Yet the development of this pattern can be interrupted altogether by removing a lamb from the flock at birth and rearing it on a bottle (milk) for seven to ten days (Scott, 1945; Colias, 1956; Moore, 1960). The animal then does not follow the flock and shows a remarkable degree of independence. This drastic change in behaviour pattern can, in fact, be initiated by even a few hours of separation of the mother sheep from the lamb. If in this type of situation the mother, too, is isolated from the herd, the effect of such brief separation can be virtually nullified. In this particular instance a genetically determined behaviour pattern can find expression only provided the environment in which it is to mature is kept constant and can be prevented from being expressed by some minor manipulation of the environment, dramatised in this particular situation by the extreme dependence of socialisation upon the reactions of the mother. Though there are numerous other examples suggestive of the same type of mechanism as one ascends the evolutionary tree to primates and to man, the examples become more complex and less easily manipulated, so that at the present time they are not nearly so well described.

These several illustrations demonstrate that in the young obvious non-reversible changes can be brought about by relatively minor and transient manipulation of the environment of the animal, and that the interplay between genetic background and environment in this situation resolves itself into a matter of genetically predetermined behaviour patterns whose expression depends upon the sequence of environments to which the animal is exposed. Further, the limited amount of information available contains suggestions that effects of this type are by no means confined to the earliest phases in the development of an individual.

The examples chosen to illustrate this last subject came from the field of behavioural sciences because here questions of reversibility, and of cumulative effects flowing from a change once introduced, have received the most direct study. There are suggestions, however, of similar effects in numerous other areas. Examples that come to mind include phenocopies resulting from environmental stress at suitable times (Goldschmidt, 1949), temperature-modified patterns of hair pigmentation, which in some cases seem to persist for considerable periods of time, as in the Himalayan rabbit (Daneel, 1941), and the complex patterns of interaction between host and infectious agents when studied in a suitable genetic framework (Gowen, 1950). Other more pertinent and important examples could readily be drawn from the field of immunology, but would require treatment beyond the scope of the present chapter. Burnett (1959) gives an incisive treatment that raises points peculiarly pertinent to the subject of this discussion.

G. Conclusions

Having presented a few selected examples to illustrate the type and magnitude of irreversible effects impressed upon the individual by various environmental stresses, I must now examine briefly the extent to which the findings illuminate and support, or eliminate, the concepts enunciated at the beginning of this chapter and consider to what extent these considerations are pertinent to future work.

Two general patterns of response have been found relevant to this discussion. The first, and most common, involves a major environmental stress resulting in the impression of fairly drastic changes upon the individual. When the stress is then removed, most, but not all, of these changes are found to be reversible. The residue left behind, and constituting the non-reversible effects, is usually neither large nor immediately conspicuous when compared to the reversible portion. Further, characteristically such changes as would be picked up eventually were often different in kind from those conspicuous during imposition or removal of the strain. Changes of this type are therefore likely to be missed unless a specific effort is made to isolate and observe them.

The second pattern found to result in non-reversible changes involved the modification of a path of development in time, either by somewhat extended exposure to a specific "abnormal" environment or by a stress applied at a critical period in the individual's career. The close relationship between this last situation and the first one described is evident, the link being constituted by acceptance of the concept that the development of an individual does not stop at or shortly after puberty, but that "ageing" is merely a special name given to the later phases of this process. The concept of calendar functions, measurable indices that vary more or less continuously with time, is a direct consequence of such a view. The illustrations presented show that ageing is not a unitary process, but a composite of various calendar functions, whose time courses can be varied relative to each other, thus giving rise to diverse patterns of ageing. This constitutes the biological documentation for a continuity of development and for its dependence on the environment.

A number of mechanisms by which environmentally impressed changes can become fixed may now be discerned. Some of these involve systems specifically adapted to retention of information. Examples of this are the central nervous system and the immune apparatus. The chains of developmental processes so closely linked to the way in which the genetic make-up of the individual finds expression (and here again known late-acting genes serve as a reminder that this, though predominantly, is not exclusively a feature of early life) provide another mechanism by which an event may condition future development so as to alter irreversibly the biology of the individual. Effects of a second group involve the deposition of products that henceforth persist in the organism and may merely cause neutral pigmentation changes or may have far-reaching morphological or pharmacological consequences (work in toxicology furnishes lead and radium deposition as perhaps the classical examples of this kind of retention). Changes of a third group entail cell death and cell injury with incomplete regeneration. This may also be at the base of certain vascular and endocrine changes that, persisting, entail more widespread secondary changes and serve to remind us of the importance of the hitherto but imperfectly understood field of amplifying and mediating mechanisms that translate minute primary lesions into eventually much more important effects. Changes of this third type have also raised the important question of the role of injury in the production of non-reversible changes and pointed to the conclusion that it is not by any means permissible to presume that the normal environment is necessarily a "neutral" one.

The specific examples presented have been chosen to illustrate a number of experimental approaches to the study of non-reversible residual effects. Among these methods the study of the kinetics of reversal has emerged as not a very powerful tool relatively, because of the need to concentrate on the quantitatively most important reversible changes, which often are qualitatively different from those that persist. The methods aiming specifically at revealing non-reversible lesions or changes proved far more sensitive, provided the most scrupulous attention was given to proper control populations, and were shown to be even further improved by the application of specific "unmasking" techniques. Such techniques aim to reveal latent functional defects by imposing suitably selected, preferably fairly specific, strains on prestressed and control populations; if applied in a suitable time framework, they have been shown to anticipate changes that become evident in unchallenged subjects at a later date.

The close relation of the phenomena with which this chapter is concerned to experimental gerontology has been underscored by a series of observations on calendar functions and their modification by environmental stress. The use of life tables illustrates this concept and serves to demonstrate both the problem-revealing power of this approach and its limitations as a device for exposing the mechanism or the true significance of the observed events. At the same time it would appear that, when supplemented by determination of the course of other calendar functions, and with proper provision and respect for adequate controls and appreciation for potentially effective factors in the environment, this general approach can become extraordinarily powerful and of the first importance in the design of meaningful experiments. The approach blends continuously into the approach from the side of early development and the study of imprinting phenomena.

The significance for human biology of the field thus sketched out seems not inconsiderable. It clearly must be a part of the consideration of both tolerable and optimal environments once significant proportions of a population have to live their lives in controlled or in artificial environments. The methodology that has emerged as pertinent should, further, furnish a basis for the future development of studies of chronic drug effects, of atmospheric or industrial toxic contaminants and of experimental nutrition. Though considerations of this kind have clearly been at the base of much nutritional work, a review of the literature dealing with chronic toxicity studies reveals a situation in which such concepts have been conspicuously slighted, leaving the field in a far from satisfactory state (Barnes and Day, 1954). The implications of these same concepts for experimental and observational gerontology have already been touched upon and there is reason to hope that work on testing the kind of reasoning advanced promises both fundamental and practical major progress in this field. One might note that multi-environmental experimental designs in gerontology are so extraordinarily costly in terms of investigator time, space and money as to place a high premium on the evaluation of natural experiments in human populations, thus closely tying the subject in with the general objectives of this book.

REFERENCES

Allan, J. H. (1943). *J. Aviat. Med.* 14, 105.
Alzamora, R., and Monge, M. C. (1955). *Anu. Fac. Lima*, 38, 65.
Andrews, J. R., and Berry, R. J. (1962). *Radiation Res.* 16, 76.
Atland, P. D. (1949). *J. Aviat. Med.* 20, 187.
Atland, P. D., and Highman, B. (1961). *Amer. J. Physiol.* 201, 393.
Barnes, J. M., and Day, F. P. (1954). *Pharmacol. Rev.* 6, 19.
Barnes, L. L., Sperling, G., and McCay, C. M. (1947). *J. Gerontol.* 2, 240.
Baum, S. (1961). *Amer. J. Physiol.* 200, 155.
Baum, S., and Alpen, E. L. (1959). *Radiation Res.* 11, 844.
Behnke, A. R. (1942). *Milit. Surg.* 90, 9.
Bell, A. L. L., Edson, G. N., and Hornick, N. (1942). *Radiology* 38, 698.
Berg, B. N., and Simms, H. S. (1960). *J. Nutr.* 71, 255.
Berg, B. N., and Simms, H. S. (1961). *J. Nutr.* 74, 23.
Bergman, R. A. M. (1948). *J. Gerontol.* 3, 14.
Blair, H. A. (1952). AEC Project, University of Rochester Report UR-206.

Bourlieré, F. (1959). *In* "The Life Span of Animals" (G. E. W. Wolstenholme and M. O'Connor, eds.), p. 90. Little Brown & Company, Boston.
Bucher, N. L. R. (1963). *Int. Rev. Cytol.* **15**, 245.
Burnet, M. F. (1959). "The Clonal Selection Theory of Acquired Immunity." Cambridge University Press.
Cardinet, S. H., Fowler, M. E., and Tyler, W. S. (1963). *Amer. J. vet. Res.* **24**, 980.
Carroll, H. W., and Brauer, R. W. (1961). *Amer. J. Physiol.* **201**, 1078.
Casarett, G. W. (1959). *In* "Radiobiology at the Cellular Level" (T. G. Hennessy, ed.). Pergamon Press, London.
Casarett, G. W. (1958). AEC Project University of Rochester, UR-521.
Chase, H. B. (1949). *J. Morph.* **84**, 57.
Chase, H. B. (1963). *In* "Cellular Basis and Aetiology of Somatic Late Effects of Ionizing Radiation" (R. J. C. Harris, ed.) p. 309. Academic Press, New York and London.
Chiodi, H. (1960). *Anu. Fac. Med. Lima*, **43**, 437.
Cohn, H. I., and Gutman, P. H. (1963). *Radiation Res.* **18**, 348.
Colias, N. E. (1956). *Ecology*, **37**, 228.
Comfort, A. (1956). "The Biology of Senescence." Routledge, London.
Comfort, A. (1959). *In* "The Life Span of Animals" (G. E. W. Wolstenholme and M. O'Connor, eds.), p. 35. Ciba Foundation Colloquium. Churchill, London.
Cornelius, C. E., Burnham, L. G., and Hill, H. E. (1963). *J. Am. vet. Med. Ass.*, **142**, 639.
Crowley, C., and Curtis, H. J. (1963). *Proc. nat. Acad. Sci. Wash.* **49**, 626.
Curtis, H. J. (1963). *Proc. Am. phil. Soc.* **107**, 5.
Curtis, H. J., and Crowley, C. (1963). *Radiation Res.* **19**, 337.
Daneel, R. (1941). *Ergebn. Biol.* **18**, 55.
Darden, E. B., and Upton, A. C. (1964). *J. Gerontol.* **19**, 62.
Denenberg, V. H. (1962). *J. comp. physiol. Psychol.* **55**, 813.
Denenberg, V. H. (1963). *Sci. Amer.* **208**, 138.
Denny-Brown, D. (1960). *In* "Neuromuscular Disorders – The Motor Unit and its Disorders" (R. D. Adams, L. M. Eaton and G. M. Shy, eds.), p. 147. Williams & Willkins, Baltimore.
Drahota, Zh., and Gutmann, G. (1962). *Gerontologia*, **6**, 81.
Einarson, L. (1962). *In* "Biological Aspects of Aging" (N. Shock, ed.), p. 131. Columbia University Press, New York.
Einarson, L., and Ringsted, A. (1938). Thesis. Ejnar Munksgard, Copenhagen.
Elkind, M. M. (1961). *In* "Fundamental Aspects of Radiosensitivity", p. 220, Brookhaven Symposium Biol., 14. Brookhaven Nat. Laboratory, Upton, New York.
Failla, G., and McClement, P. (1957). *Amer. J. Roentgenol.* **78**, 146.
Gains, F. M., Brown, T. D., Hansen, C. L., and Howland, J. W. (1963). AEC Project University of Rochester, UR-610.
Gersh, I. (1945). *J. cell. comp. Physiol.* **26**, 101.
Glorig, A., and Nixon, J. (1960). *Ann. Otol., etc., St. Louis*, **69**, 497.
Glorig, A., Summerfeld, A., and Nixon, J. (1960). *In* "Distribution of Hearing Levels in Non-Noise-Exposed Populations" (L. Cramer, ed.), p. 150. Proceedings of the 3rd International Congress on Acoustics, Elsevier, Amsterdam.
Gold, H. (1961). *A.M.A. Arch. Pathol.* **71**, 268.
Golding-Campbell, F., Griffiths, P., Hempleman, H. V., Paton, W. D. M., and Walder, D. W. (1960). *Brit. J. industr. Med.* **17**, 167.
Goldschmidt, R. B. (1949). *Sci. Amer.* **181**, 46.
Gompertz, B. (1825). *Phil. Trans.* 513.
Gordon, A. S., and Kleinberg, W. (1937). *Proc. Soc. exp. Biol. N.Y.* **37**, 507.
Gowen, J. W. (1950). *Scientia* **85**, 145.

Gowen, J. W. (1961). *Fed. Proc.* 20, *Suppl.* 8, 35.
Haloren, P. I., and Kontinen, A. (1962). *Nature, Lond.* **193**, 942.
Handler, P. (1961). *Fed. Proc.* **20**, *Suppl.* 8.
Hart, J. S. (1953). *Canad. J. Res.* 31, 112.
Harvey, E. N. (1945). *Harvey Lect.* **40**, 41.
Henry, F. M. (1946). *J. Aviat. Med.* 17, 28.
Hong, Suk Ki. (1963). *Fed. Proc.* **22**, 831.
Huerkamp, B., and Optiz, E. (1950). *Pflüg. Arch. ges. Physiol.* **252**, 129.
Hultgren, H., and Sprickhard, W. (1960). *Stanf. med. Bull.* 18, 76.
Hunt, H. F., and Otis, L. S. (1955). *Amer. Psychol.* **10**, 432.
Hurtado, A. (1942). *J. Amer. Med. Ass.* **120**, 12.
Hurtado, A., Menno, C., and Delgado, E. (1945). *Arch. int. Med.* 75, 284.
Hutchings, D. E. (1963). *Trans. N.Y. Acad. Sci.* **25**, 890.
James, C. C. M. (1945). *Lancet* ii, 6.
Johnson, H. D., Kintner, L. D., and Kibler, H. H. (1963). *J. Gerontol.* 18, 25.
Jones, D. C., and Kimeldorf, D. J. (1963). USNRDL-TR-673.
Jones, H. B. (1957). *Adv. Med. Phy.* 4, 281.
Jones, H. B. (1960). *In* "Proceedings of the 4th Berkeley Symposium", p. 267. 4th Edition. University of California Press, Berkeley.
Jorgensen, M. B. (1961). *Arch. Otolaryng.* 74, 164.
Kahlstrom, S. F., Burton, C. C., and Pheminster, D. B. (1939). *Surg. Gynec. Obstet.* **68**, 129.
Kaplan, H. S. (1948). *J. nat. Cancer Inst.* 8, 191.
Kaplan, H. S., and Brown, M. B. (1952). *J. nat. Cancer Inst.* **12**, 765.
Kellogg, R. H., Pace, N., Archibald, E. R., and Vaughan, B. E. (1957). *J. appl. Physiol.* **11**, 65.
Kibler, H. H., Silsby, H. D., and Johnson, H. D. (1963). *J. Gerontol.* 18, 235.
King, J. A., and Eleftherion, B. E. (1959). *J. comp. physiol. Psychol.* 52, 82.
Kohn, H. I., Kallman, R. F., Berdjis, C. C., and DeOme, K. B. (1957). *Radiation Res.* 7, 407.
Krebs, J. S. (1960). *Radiation Res.* **12**, 450.
Krebs, J. S. (1963). *Radiation Res.* **19**, 246.
Krebs, J. S., and Brauer, R. W. (1962). *Radiation Res.* 17, 855.
Krebs, J. S., and Brauer, R. W. (1963). *In* "Biological Effects of Neutron Irradiations", Intern. Atomic Energy Comm. Symposium. Intern. Atomic Energy Agency, New York.
Krebs, J. S., Brauer, R. W., and Kalbach, H. (1959). *Radiation Res.* **10**, 80.
Lajtha, L. G. (1961). *Prog. Biophys.* **11**, 79.
Lane, P. W., and Dickie, M. M. (1958). *J. Nutr.* 64, 549.
Lawson, B. G., Billings, M. S., and Bennett, L. R. (1959). *J. nat. Cancer Inst.* **22**, 1059.
LeBlanc, J. (1962). *J. appl. Physiol.* 17, 950.
Leong, G. F., Pessotti, R. L., and Krebs, J. S. (1961). *J. nat. Cancer Inst.* **27**, 131.
Lesher, S., Grahn, D., and Sallese, A. (1957). *J. nat. Cancer Inst.* **19**, 1119.
Lindop, P. J., and Rotblat, J. (1962). *Brit. J. Radiol.* 35, 23.
MacCallum, R. I., Stanger, J. K., Walder, D. N., and Paton, W. D. M. (1954). *J. Bone Jt. Surg.* 36B, 606.
McCay, C. M., Crowell, M. F., and Maynard, L. A. (1935). *J. Nutr.* **10**, 63.
McCay, C. M., Sperling, G., and Barnes, L. L. (1943). *Arch. Biochem.* **2**, 469.
Medawar, P. B. (1955). *In* "Aging – General Aspects" (G. E. W. Wolstenholme and M. P. Cameron, eds.), p. 5. Ciba Foundation Colloquium. Churchill, London.
Melville, G. S., Conte, F. P., Slater, M., and Upton, A. C. (1957). *Brit. J. Radiol.* **30**, 196.

Monge, M. C. (1928). *Anu. Fac. Med. Lima*, **11**, 1.
Monge, M. C. (1937). *Arch. Int. Med.* **59**, 32.
Monge, M. C. (1949). *Anu. Fac. Med. Lima*, **32**, 1.
Monge, M. C., Contreras, G., Valesquez, T., Reynafarje, C., and Chavez, R. (1948). *Anu. Fac. Med. Lima*, **31**, 431.
Moore, A. (1960). *Amer. Psychol.* **15**, 413.
Mühlbock, O. (1957). *In* "The Methodology of the Study of Aging" (G. E. W. Wolstenholme and C. M. O'Conner, eds.), p. 115. Ciba Foundation Colloquium. Churchill, London.
Muller, H. J. (1963). *In* "Cellular Basis and Aetiology of Late Somatic Effects of Ionizing Radiation" (R. J. C. Harris, ed.), p. 237. Academic Press, New York and London.
Nixon, J. C., and Glorig, A. (1961). *J. acoust. Soc. Amer.* **33**, 904.
Nowell, P. C., and Cole, L. J. (1963). *Science*, **141**, 524.
Opitz, E. (1951). *Exp. Med. Surg.*, **9**, 389.
Paterson, E., Gilbert, E. W., and Mathews, J. (1952). *Brit. J. Radiol.* **25**, 427.
Patt, H. M., and Quastler, H. (1963). *Physiol. Rev.* **43**, 357.
Pearl, R. (1928). *In* "The Rate of Living." A. A. Knopf, New York.
Pearson, E. F. (1946). *Ann. int. Med.* **24**, 988.
Prosser, C. L. (1958). *In* "Physiological Adaptation" (C. L. Prosser, ed.), p. 166. Am. Physiol. Soc., Washington.
Puck, T. T., Marcus, P. I., and Cieciura, S. J. (1956). *J. exp. Med.* **103**, 653.
Reynafarje, B. (1962). *Sch. av. Med.* SAM-TDR-62–89.
Reynafarje, C., Lozano, R., and Valdevieso, J. (1959). *Blood* **14**, 433.
Rivera, J. C. (1963). U.S. Navy Experimental Diving Unit, Research Report, 1–63.
Rosen, S., Bergman, M., Plester, D., El Mofty, A., and Hamad-Satti, M. (1962). *Trans. Amer. otol. Soc.* **50**, 135.
Ross, M. H. (1961). *J. Nutr.* **75**, 197.
Rotta, A. (1947). *Amer. Heart J.* **33**, 669.
Rotta, A. (1958). Com. al 1er Congreso Peruano de Cardiologia, Lima.
Rotta, A., Carep, A., Velasquez, T., Hurtado, A., Salazar, A., and Chavez, R. (1952). *Rev. argent. Cardiol.* **19**, 374.
Rotta, A., Carep, A., Velasquez, T., Hurtado, A., Salazar, A., and Chavez, R. (1956). *J. appl. Physiol.* **9**, 328.
Rubinstein, L. J. (1960). *In* "Structure and Function of Muscle, III" (G. H. Bourne, ed.), p. 203. Academic Press, New York and London.
Ruegamer, W. R., Bernstein, L., and Benjamin, J. (1954). *Science* **120**, 84.
Rugh, R., and Wolff, J. (1956). *Fert. Steril.* **7**, 546.
Sacher, G. A. (1956). *Radiology* **67**, 250.
Sams, C. S. (1945–51). Reports of Public Health and Welfare Section, GHA, SCAP.
Schaefer, T., Weingarten, F. S., and Town, J. C. (1962). *Science* **135**, 41.
Schlang, H. A. (1961). *Amer. J. med. Sci.* **242**, 338.
Scholander, P. F., Hammel, H. T., Hart, J. S., LeMessurier, D. H., and Steen, J. (1959). *J. appl. Physiol.* **13**, 211.
Scott, J. P. (1945). *Comp. Psychol. Monogr.* 18(4), 1.
Scott, J. P. (1958). *Psychosom. Med.* **20**, 42.
Sellers, E. A., Reichman, S., and Thomas, N. (1951). *Amer. J. Physiol.* **167**, 644.
Simms, H. S. (1946). *J. Gerontol.* **1**, 13.
Sobel, H. J., and Furth, J. (1948). *Endocrinology* **42**, 436.
Tappan, D., and Reynafarje, B. (1957a). *Amer. J. Physiol.* **190**, 99.
Tappan, D., Reynafarje, B., Potter, V. R., and Hurtado, A. (1957b). *Amer. J. Physiol.* **190**, 93.

Upton, A. C., Kastenbaum, M. A., and Conklin, J. W. (1963). *In* "Cellular Basis and Aetiology of the Late Somatic Effects of Ionizing Radiation" (R. J. C. Harris, ed.), p. 285. Academic Press, New York and London.

Verzar, F. (1945). *In* "Hohenklima – Forschungen" (F. Verzar, ed.), p. 23. Benno Schwabe, Basel.

Ward, W. D., Glorig, A., and Sklar, D. L. (1959). *J. acoust. Soc. Amer.* **31**, 600.

Author Index

(Numbers in italics refer to pages on which references are listed in bibliographies at the end of each chapter.)

G

I

J

L

M

N

O

P

Q

R

T

Subject Index

A

B

C

D

F

G

H

I

K

R

S

T

U

V

W